2 **He** $1s^2$
			5 [illegible] 2.0 **B** $2s^2\,2p^1$	4.7 2.5 **C** $2s^2\,2p^2$	7 3.0 **N** $2s^2\,2p^3$	8 3.5 **O** $2s^2\,2p^4$	9 3.95 **F** $2s^2\,2p^5$	10 **Ne** $2s^2\,2p^6$
			13 4.25 1.5 **Al** $3s^2\,3p^1$	14 4.8 1.8 **Si** $3s^2\,3p^2$	15 2.1 **P** $3s^2\,3p^3$	16 2.5 **S** $3s^2\,3p^4$	17 3.0 **Cl** $3s^2\,3p^5$	18 **Ar** $3s^2\,3p^6$
4.5 **Ni** $4s^2$	29 4.4 1.9 **Cu** $3d^{10}\,4s^1$	30 4.2 1.5 **Zn** $3d^{10}\,4s^2$	31 4.0 1.5 **Ga** $3d^{10}\,4s^2\,4p^1$	32 4.8 1.8 **Ge** $3d^{10}\,4s^2\,4p^2$	33 5.1 2.0 **As** $3d^{10}\,4s^2\,4p^3$	34 4.7 2.4 **Se** $3d^{10}\,4s^2\,4p^4$	35 2.8 **Br** $3d^{10}\,4s^2\,4p^5$	36 **Kr** $3d^{10}\,4s^2\,4p^6$
4.8 **Pd** $5s^0$	47 4.3 1.8 **Ag** $4d^{10}\,5s^1$	48 4.1 1.5 **Cd** $4d^{10}\,5s^2$	49 3.8 1.5 **In** $4d^{10}\,5s^2\,5p^1$	50 4.4 1.8 **Sn** $4d^{10}\,5s^2\,5p^2$	51 4.1 1.8 **Sb** $4d^{10}\,5s^2\,5p^3$	52 4.7 2.1 **Te** $4d^{10}\,5s^2\,5p^4$	53 2.5 **I** $4d^{10}\,5s^2\,5p^5$	54 **Xe** $4d^{10}\,5s^2\,5p^6$
5.3 **Pt** $6s^1$	79 4.3 2.3 **Au** $5d^{10}\,6s^1$	80 4.5 1.8 **Hg** $5d^{10}\,6s^2$	81 3.7 1.7 **Tl** $5d^{10}\,6s^2\,6p^1$	82 4.0 1.7 **Pb** $5d^{10}\,6s^2\,6p^2$	83 4.4 1.8 **Bi** $5d^{10}\,6s^2\,6p^3$	84 2.0 **Po** $5d^{10}\,6s^2\,6p^4$	85 2.2 **At** $5d^{10}\,6s^2\,6p^5$	86 **Rn** $5d^{10}\,6s^2\,6p^6$

Eu $5d^0\,6s^2$	64 1.2 **Gd** $4f^7\,5d^1\,6s^2$	65 1.2 **Tb** $4f^9\,5d^0\,6s^2$	66 1.2 **Dy** $4f^{10}\,5d^0\,6s^2$	67 1.2 **Ho** $4f^{11}\,5d^0\,6s^2$	68 1.2 **Er** $4f^{12}\,5d^0\,6s^2$	69 1.2 **Tm** $4f^{13}\,5d^0\,6s^2$	70 1.1 **Yb** $4f^{14}\,5d^0\,6s^2$	71 1.2 **Lu** $4f^{14}\,5d^1\,6s^2$

* Recommended by V. S. Fomenko: *Handbook of Thermionic Properties*, ed. by G. V. Samsonov (Plenum Press, New York 1966). For more values see Tables 1.2, 1.3, and 1.5 of [22]

** Recommended by W. Gordy, W. J. O. Thomas: J. Chem-Phys. **24**, 439 (1955)

Topics in Applied Physics Volume 27

Topics in Applied Physics Founded by Helmut K. V. Lotsch

1 **Dye Lasers** 2nd Edition
Editor: F. P. Schäfer

2 **Laser Spectroscopy** of Atoms and Molecules. Editor: H. Walther

3 **Numerical and Asymptotic Techniques in Electromagnetics** Editor: R. Mittra

4 **Interactions on Metal Surfaces**
Editor: R. Gomer

5 **Mössbauer Spectroscopy**
Editor: U. Gonser

6 **Picture Processing and Digital Filtering**
2nd Edition. Editor: T. S. Huang

7 **Integrated Optics** Editor: T. Tamir

8 **Light Scattering in Solids**
Editor: M. Cardona

9 **Laser Speckle** and Related Phenomena
Editor: J. C. Dainty

10 **Transient Electromagnetic Fields**
Editor: L. B. Felsen

11 **Digital Picture Analysis**
Editor: A. Rosenfeld

12 **Turbulence** 2nd Edition
Editor: P. Bradshaw

13 **High-Resolution Laser Spectroscopy**
Editor: K. Shimoda

14 **Laser Monitoring of the Atmosphere**
Editor: E. D. Hinkley

15 **Radiatonless Processes** in Molecules and Condensed Phases. Editor: F. K. Fong

16 **Nonlinear Infrared Generation**
Editor: Y.-R. Shen

17 **Electroluminescence** Editor: J. I. Pankove

18 **Ultrashort Light Pulses**
Picosecond Techniques and Applications
Editor: S. L. Shapiro

19 **Optical and Infrared Detectors**
Editor: R. J. Keyes

20 **Holographic Recording Materials**
Editor: H. M. Smith

21 **Solid Electrolytes** Editor: S. Geller

22 **X-Ray Optics.** Applications to Solids
Editor: H.-J. Queisser

23 **Optical Data Processing.** Applications
Editor: D. Casasent

24 **Acoustic Surface Waves**
Editor: A. A. Oliner

25 **Laser Beam Propagation in the Atmosphere**
Editor: J. W. Strohbehn

26 **Photoemission in Solids I**
General Principles
Editors: M. Cardona and L. Ley

27 **Photoemission in Solids II.** Case Studies
Editors: L. Ley and M. Cardona

28 **Hydrogen in Metals I.** Basic Properties
Editors: G. Alefeld and J. Völkl

29 **Hydrogen in Metals II**
Application-Oriented Properties
Editors: G. Alefeld and J. Völkl

30 **Excimer Lasers** Editor: Ch. K. Rhodes

31 **Solar Energy Conversion.** Solid-State Physics Aspects. Editor: B. O. Seraphin

32 **Image Reconstruction from Projections**
Implementation and Applications
Editor: G. T. Herman

33 **Electrets** Editor: G. M. Sessler

34 **Nonlinear Methods of Spectral Analysis**
Editor: S. Haykin

35 **Uranium Enrichment**
Editor: S. Villani

36 **Amorphous Semiconductors**
Editor: M. H. Brodsky

37 **Thermally Stimulated Relaxation in Solids**
Editor: P. Bräunlich

38 **Charge-Coupled Devices**
Editor: D. F. Barbe

Photoemission in Solids II

Case Studies

Edited by L. Ley and M. Cardona

With Contributions by

Y. Baer M. Campagna M. Cardona W. D. Grobman
H. Höchst S. Hüfner E. E. Koch C. Kunz
L. Ley R. A. Pollak P. Steiner G. K. Wertheim

With 214 Figures

Springer-Verlag Berlin Heidelberg New York 1979

Dr. *Lothar Ley*
Professor Dr. *Manuel Cardona*

Max-Planck-Institut für Festkörperforschung, Heisenbergstraße 1
D-7000 Stuttgart 80, Fed. Rep. of Germany

ISBN 3-540-09202-1 Springer-Verlag Berlin Heidelberg New York
ISBN 0-387-09202-1 Springer-Verlag New York Heidelberg Berlin

Library of Congress Cataloging in Publication Data. Main entry under title: Photoemission in solids. (Topics in applied physics; v. 26–27). Includes bibliographies and index. Contents: 1. General principles. – 2. Case studies. 1. Photoelectron spectroscopy. 2. Solids-Spectra. 3. Photoemission. I. Ley, Lothar, 1943–, II. Cardona, Manuel, 1934–. QC454.P48P49 530.4′1 78-2503

Monophoto typesetting, offset printing and bookbinding: Brühlsche Universitätsdruckerei, Lahn-Gießen
2153/3130-543210

Preface

This book constitutes the continuation of Volume 26 of the series Topics in Applied Physics (*Photoemission in Solids I*). In the first volume we discussed the general principles underlying the phenomena of photoemission and photoelectron spectroscopy, including a brief review of the experimental techniques. Such topics as the general formal theory of photoemission, the three-step model, the theory of photoionization cross sections, one-electron excitations and phenomena beyond the one-electron approximation were treated by some of the leading specialists in the field. The emphasis of the present volume lies on the discussion of photoelectron spectra of specific families of materials and the information that can be obtained from such spectra about their electronic structure. The largest contribution, Chap. 2, refers to semiconductors. It contains extensive background discussion on the band structures of the most common types of semiconductors. The vast amount of knowledge accummulated for these materials, due in part to their practical applications, makes them ideal to exemplify the methodology and the scope of photoelectron spectroscopy. Successive chapters cover transition metals and their compounds, rare earths, organic molecular crystals of the type which show characteristic solid-state effects and, last but not least, simple metals. In addition, Chap. 6 discusses photoemission experiments for which the use of synchrotron radiation is of the essence. For convenience of the users we have reproduced in this volume the periodic table with work functions and the table of binding energies which already appeared in TAP 26.

The range of information obtained with photoelectron spectroscopy is so wide that this book should be of interest to both students and practitioners of solid-state physics interested in the electronic structure of solids. While it would be impossible to compile an exhaustive materials bibliography within the space limitations of the volume (such compilation would anyway make the volume rather dull), we believe enough references are included to help the research worker muddle his way through the literature of specific types of solids.

We have found the task of editing these volumes an extremely rewarding experience. The exchange of ideas and information with the various authors has been rather intensive. We thank them all once more for their cooperation and patience. We would also like to thank again the colleagues of the institutions to whom we owe our expertise in the field, the Max-Planck-Institut für Festkörperforschung, the Deutsches Elektronen-Synchrotron (DESY), and the University of California, Berkeley. We should also thank the staffs of various companies involved in the manufacturing of photoelectron spectrometers, especially those whose equipment we use. Without them the enormous development which has taken place in the field within the past ten years would not have been possible.

Stuttgart, December 1978

Lothar Ley
Manuel Cardona

Contents

1. Introduction. By L. Ley and M. Cardona 1
1.1 Survey of Previous Volume 4
1.2 Contents of Present Volume 8
References 9

2. Photoemission in Semiconductors
By L. Ley, M. Cardona, and R. A. Pollak (With 97 Figures) 11
2.1 Background 11
2.1.1 Historical Survey 13
2.2 Band Structure of Semiconductors 15
2.2.1 Tetrahedral Semiconductors 15
2.2.2 Semiconductors with an Average of Five Valence Electrons per Atom 28
2.2.3 Selenium, Tellurium, and the V_2VI_3 Compounds 30
2.2.4 Transition Metal Dichalcogenides 32
2.3 Methods Complementary to Photoelectron Spectroscopy 40
2.3.1 Optical Absorption, Reflection, and Modulation Spectroscopy 40
2.3.2 Characteristic Electron Energy Losses 43
2.3.3 X-Ray Emission Spectroscopy 45
2.4 Volume Photoemission: Angular Integrated EDC's from Valence Bands 47
2.4.1 Band-Structure Regime: Germanium 51
2.4.2 XPS Regime: Tetrahedral Semiconductors 55
2.4.3 XPS Regime: IV–VI Compounds 62
2.4.4 Partial Density of Valence States: Copper and Silver Halides; Chalcopyrites; Transition Metal, Rare Earth, and Actinide Compounds 67
2.4.5 Layer Structures: Transition Metal Dichalcogenides . . . 72
2.4.6 Layer Structures: SnS_2, $SnSe_2$, PbI_2, GaS, GaSe 75
2.5 Photoemission and Density of Conduction States 78
2.5.1 Secondary Electron Tails 79
2.5.2 Partial Yield Spectroscopy 79
2.6 Angular Resolved Photoemission from the Lead Salts 80

2.7 Amorphous Semiconductors 85
2.7.1 Tetrahedrally Coordinated Amorphous Semiconductors . . 87
a) Amorphous Si and Ge 87
b) Amorphous III–V Compounds 100
2.7.2 Amorphous Semiconductors with an Average of Five Valence Electrons per Atom 104
2.7.3 Amorphous Group VI Semiconductors 111
2.7.4 Gap States in Amorphous Semiconductors 114
2.8 Ionicity 118
2.8.1 An Ionicity Scale Based on Valence Band Spectra 121
2.8.2 Binding Energy Shift and Charge Transfer 126
2.9 Photoemission Spectroscopy of Semiconductor Surfaces 130
2.9.1 Semiconductor Surface States 131
2.9.2 Silicon Surface States 133
a) Photoemission from Si(111) 2 × 1 and 7 × 7 Surfaces . . . 135
b) Electronic Structure Theory of Si(111) Surfaces 141
2.9.3 Surface States of Group III–V Semiconductors 148
2.9.4 Surface Chemistry of Semiconductors — Si(111): H and Si(111): SiH_3 151
2.9.5 Interface States: Metal-Semiconductor Electrical Barriers . 154
References 158

3. Unfilled Inner Shells: Transition Metals and Compounds
By S. Hüfner (With 25 Figures) 173
3.1 Overview 173
3.2 Transition Metal Compounds 176
3.2.1 The Hubbard Model 176
3.2.2 Final State Effects in Photoemission Spectra 177
a) Satellites 177
b) Multiplet and Crystal-Field Splitting 179
3.2.3 Transition Metal Oxides 183
a) MnO, CoO, NiO: Mott Insulators 183
b) VO_2: A Nonmetal-Metal Transition 188
c) ReO_3: A Typical Metal 189
3.2.4 Miscellaneous Compounds 191
3.2.5 The Correlation Energy U 191
3.3 *d*-Band Metals: Introduction 192
3.3.1 The Noble Metals: Cu, Ag, Au 194
3.3.2 The Ferromagnets: Fe, Co, Ni 200
3.3.3 Nonmagnetic d-Band Metals 205
3.4 Alloys 206
3.4.1 Dilute Alloys: The Friedel-Anderson Model 206
3.4.2 Concentrated Alloys: The Coherent Potential Approximation 210
3.5 Intermetallic Compounds 212
3.6 Summary, Outlook 212
References 213

4. Unfilled Inner Shells: Rare Earths and Their Compounds
By M. Campagna, G.K. Wertheim, and Y. Baer (With 35 Figures) . . . 217
4.1 Background 217
4.1.1 Where Are the $4f$ Levels Located? 217
4.1.2 Multiplet Intensities Versus Total Photoelectric Cross Sections at 1.5 keV 218
4.1.3 Renormalized Atom Scheme and Thermodynamics 221
4.1.4 Multiplet and Satellite Structure in Photoemission from Core Levels Other than $4f$ 226
4.2 Techniques 227
4.2.1 The Need of High Resolution in Rare-Earth Studies 227
4.2.2 Sample Preparation 228
a) Pure Metals 228
b) Chalcogenides, Borides, and Alloys 229
4.3 Results 229
4.3.1 Metals 229
a) Identification of the Outermost Levels 229
b) The Light Rare Earths 230
c) The Heavy Rare Earths 233
d) Cerium 235
e) The $4f$ Promotion Energy 237
4.3.2 Compounds and Alloys: Stable $4f^n$ Configurations 237
a) Rare-Earth Halides 237
b) Chalcogenides and Pnictides 238
c) Phonon Broadening in EuO 243
d) Interatomic Auger Transitions in Rare-Earth Borides . . 245
e) Rare-Earth Intermetallics 249
f) $4s$ and $5s$ Multiplet Splittings 250
g) Spectra of $3d$ and $4d$ Electrons of Rare-Earth Solids . . 251
h) $4f$ and $4d$ Binding Energy: Atom Versus Solid 253
4.3.3 Intermediate Valence (IV) Compounds 254
a) The Intra-Atomic Coulomb Correlation Energy U_{eff} . . 257
4.4 Conclusions and Outlook 257
References 258

5. Photoemission from Organic Molecular Crystals
By W.D. Grobman and E.E. Koch (With 14 Figures) 261
5.1 Some Experimental Aspects of Photoemission from Organic Molecular Crystals 262
5.1.1 Charging Effects 262
5.1.2 Secondary Electron Background 264
5.1.3 Electron Attenuation Length (Escape Depth) $\lambda_e(E)$ 264
5.1.4 Vacuum Requirements 265
5.1.5 Effects of the Transmission Function of the Electron Energy Analyzer 265
5.2 Band Formation in Linear Alkanes 266

5.3 Aromatic Hydrocarbons 267
5.3.1 Acene 268
5.3.2 Organometallic Phenyl Compounds 270
5.3.3 Anthracene 272
5.4 Photoemission Induced by Exciton Annihilation 275
5.5 Photoemission from Biological Materials 278
5.5.1 Phthalocyanines 278
5.5.2 Nucleic Acid Bases 280
5.6 Valence Orbital Spectroscopy of Molecular Organic Conductors 280
5.6.1 Valence Bands of TTF-TCNQ and Related Compounds . . 280
5.6.2 Valence Bands of $(SN)_x$ 285
5.6.3 The Absence of a Fermi Edge in Photoemission Spectra of Organic "Metals" 287
5.7 Core Orbital Spectroscopy of Organic Molecular Crystals 288
5.7.1 Solid-State Effects on Core Levels in Charge Transfer Salts . 288
5.7.2 Core Level Spectroscopy and Charge Transfer in TTF-TCNQ 292
5.7.3 Conclusions 293
References 294

6. Synchrotron Radiation: Overview. By C. Kunz (With 33 Figures) . . 299
6.1 Overview 300
6.2 Properties of Synchrotron Radiation 301
6.2.1 Basic Equations 301
6.2.2 Comparison with Other Sources 305
6.2.3 Evolution of Synchrotron Sources 306
6.3 Arrangement of Experiments 310
6.3.1 Layout of Laboratories 310
6.3.2 Monochromators 311
6.4 Spectroscopic Techniques 313
6.4.1 Spectroscopy of Directly Excited Electrons 313
6.4.2 Energy Distribution Curves (EDC) 314
6.4.3 Constant Final-State Spectroscopy (CFS) 316
6.4.4 Constant Initial-State Spectroscopy (CIS) 317
6.4.5 Angular Resolved Photoemission (ARP, ARPES) 319
6.4.6 Secondary Processes 319
6.4.7 Photoelectron Yield Spectroscopy (PEYS) 322
6.4.8 Yield Spectroscopy at Oblique Incidence 323
6.5 Applications of Yield Spectroscopy 326
6.5.1 Anisotropy in the Absorption Coefficient of Se 326
6.5.2 Investigation of Alloys 328
6.5.3 Investigation of Liquid Metals 329
6.6 Experiments Investigating Occupied and Empty States 330
6.6.1 Valence Bands in Rare-Gas Solids 330
6.6.2 Conduction Band State from Angular Dependent Photoemission 333

6.7 Experiments on Relaxation Processes and Excitons 335
6.7.1 Phonon Broadening of Core Lines 335
6.7.2 Exciton Effects with Core Excitations 337
6.7.3 Energy Transfer Processes 339
6.8 Surface States and Adsorbates 341
6.8.1 Surface Core Excitons on NaCl 341
6.8.2 Adsorbates and Oxidation 343
References 344

7. Simple Metals
By P. Steiner, H. Höchst, and S. Hüfner (With 10 Figures) 349
7.1 Historical Background 349
7.2 Theory of the Photoelectron Spectrum 351
7.3 Core Level Spectra 357
7.4 Valence Band Spectra 364
7.5 Summary 369
References 370

Appendix: Table of Core-Level Binding Energies 373

Additional References with Titles 385

Subject Index 389

Contents of **Photoemission in Solids I**

General Principles (Topics in Applied Physics, Vol. 26)

1. Introduction. By M. Cardona and L. Ley (With 26 Figures)

1.1 Historical Remarks

1.1.1 The Photoelectric Effect in the Visible and Near uv: The Early Days

1.1.2 Photoemissive Materials: Photocathodes

1.1.3 Photoemission and the Electronic Structure of Solids

1.1.4 X-Ray Photoelectron Spectroscopy (ESCA, XPS)

1.2 The Work Function

1.2.1 Methods to Determine the Work Function

1.2.2 Thermionic Emission

1.2.3 Contact Potential: The Kelvin Method

The Break Point of the Retarding Potential Curve

The Electron Beam Method

1.2.4 Photoyield Near Threshold

1.2.5 Quantum Yield as a Function of Temperature

1.2.6 Total Photoelectric Yield

1.2.7 Threshold of Energy Distribution Curves (EDC)

1.2.8 Field Emission

1.2.9 Calorimetric Method

1.2.10 Effusion Method

1.3 Theory of the Work Function

1.3.1 Simple Metals

1.3.2 Simple Metals: Surface Dipole Contribution

1.3.3 Volume and Temperature Dependence of the Work Function

1.3.4 Effect of Adsorbed Alkali Metal Layers

1.3.5 Transition Metals

1.3.6 Semiconductors

1.3.7 Numerological and Phenomenological Theories

1.4 Techniques of Photoemission

1.4.1 The Photon Source

1.4.2 Energy Analyzers

1.4.3 Sample Preparation

Cleaning Procedures

1.5 Core Levels

1.5.1 Elemental Analysis

1.5.2 Chemical Shifts
Theoretical Models for the Calculation of Binding Energy Shifts
Core Level Shifts of Rare Gas Atoms Implanted in Noble Metals
Binding Energies in Ionic Solids
Chemical Shifts in Alloys
1.5.3 The Width of Core Levels
1.5.4 The Core Level Cross Sections
1.6 The Interpretation of Valence Band Spectra
1.6.1 The Three-Step Model of Photoemission
1.6.2 Beyond the Isotropic Three-Step Model
References

2. Theory of Photoemission: Independent Particle Model
By W. L. Schaich (With 2 Figures)
2.1 Formal Approaches
2.1.1 Quadratic Response
2.1.2 Many-Body Features
2.2 Independent Particle Reduction
2.2.1 Golden Rule Form
2.2.2 Comparison With Scattering Theory
2.2.3 Theoretical Ingredients
2.3 Model Calculations
2.3.1 Simplification of Transverse Periodicity
2.3.2 Volume Effect Limit
2.3.3 Surface Effects
2.4 Summary
References

3. The Calculation of Photoionization Cross Sections: An Atomic View
By S. T. Manson (With 16 Figures)
3.1 Theory of Atomic Photoabsorption
3.1.1 General Theory
3.1.2 Reduction of the Matrix Element to the Dipole Approximation
3.1.3 Alternate Forms of the Dipole Matrix Element
3.1.4 Relationship to Density of States
3.2 Central Field Calculations
3.3 Accurate Calculations of Photoionization Cross Sections
3.3.1 Hartree-Fock Calculations

3.3.2 Beyond the Hartree-Fock Calculation: The Effects of Correlation
3.4 Concluding Remarks
References

4. Many-Electron and Final-State Effects: Beyond the One-Electron Picture. By D. A. Shirley (With 10 Figures)
4.1 Multiplet Splitting
4.1.1 Theory
4.1.2 Transition Metals
4.1.3 Rare Earths
4.2 Relaxation
4.2.1 The Energy Sum Rule
4.2.2 Relaxation Energies
Atomic Relaxation
Extra-Atomic Relaxation
4.3 Electron Correlation Effects
4.3.1 The Configuration Interaction Formalism
Final-State Configuration Interaction (FSCI)
Continuum-State Configuration Interaction (CSCI)
Initial-State Configuration Interaction (ISCI)
4.3.2 Case Studies
Final-State Configuration Interactions:
The $4p$ Shell of Xe-Like Ions
Continuum-State Configuration Interaction: The $5p^6\,6s^2$ Shell
Initial-State Configuration: Two Closed-Shell Cases
4.4 Inelastic Process
4.4.1 Intrinsic and Extrinsic Structure
4.4.2 Surface Sensitivity
References

5. Fermi Surface Excitations in X-Ray Photoemission Line Shapes from Metals. By G. K. Wertheim and P. H. Citrin (With 22 Figures)
5.1 Overview
5.2 Historical Background
5.2.1 The X-Ray Edge Problem
5.2.2 X-Ray Emission and Photoemission Spectra
5.3 The X-Ray Photoemission Line Shape
5.3.1 Behavior Near the Singularity
5.3.2 Extrinsic Effects in XPS
5.3.3 Data Analysis

5.4 Discussion of Experimental Results
5.4.1 The Simple Metals Li, Na, Mg, and Al
5.4.2 The Noble Metals
5.4.3 The *s–p* Metals Cd, In, Sn, and Pb
5.4.4 The Transition Metals and Alloys
5.5 Summary
References

6. Angular Dependent Photoemission. By N. V. Smith (With 14 Figures)
6.1 Preliminary Discussion
6.1.1 Energetics
6.1.2 Theoretical Perspective
6.2 Experimental Systems
6.2.1 General Considerations
6.2.2 Movable Analyzer
6.2.3 Modified Analyzer
6.2.4 Multidetecting Systems
6.3 Theoretical Approaches
6.3.1 Pseudopotential Model
6.3.2 Orbital Information
6.3.3 One-Step Theories
6.4 Selected Results
6.4.1 Layer Compounds
6.4.2 Three-Dimensional Band Structures
6.4.3 Normal Emission
6.4.4 Nonnormal CFS
References

Appendix: Table of Core-Level Binding Energies

Contens of **Photoemission in Solids II**

Additional References with Titles

Subject Index

Contributors

Baer, Yves
Institut für Festkörperphysik, Eidgenössische Technische Hochschule, CH-Zürich, Switzerland

Campagna, Maurice
Institut für Festkörperforschung, Kernforschungsanlage Jülich, D-5170 Jülich 1, Fed. Rep. of Germany

Cardona, Manuel
Max-Planck-Institut für Festkörperforschung, Heisenbergstraße 1, D-7000 Stuttgart 80, Fed. Rep. of Germany

Grobman, Warren David
IBM Thomas J. Watson Research Center, Yorktown Heights, NY 10598, USA

Höchst, Hartmut
Fachbereich 11, Physik, Universität des Saarlandes, D-6600 Saarbrücken, Fed. Rep. of Germany

Hüfner, Stefan
Fachbereich 11, Physik, Universität des Saarlandes, D-6600 Saarbrücken, Fed. Rep. of Germany

Koch, Ernst-Eckhard
Synchrotronstrahlungsgruppe F41, Deutsches Elektronen Synchrotron DESY D-2000 Hamburg 52, Fed. Rep. of Germany

Kunz, Christof
II. Institut für Experimentalphysik, Universität Hamburg, Luruper Chaussee 149, D-2000 Hamburg 50, Fed. Rep. of Germany

Ley, Lothar
Max-Planck-Institut für Festkörperforschung, Heisenbergstraße 1, D-7000 Stuttgart 80, Fed. of Germany

Pollak, Roger A.
Thomas J. Watson IBM Research and Development Center, Yorktown Heights, NY 10598, USA

Steiner, Paul
Fachbereich 11, Physik, Universität des Saarlandes, D-6600 Saarbrücken, Fed. Rep. of Germany

Wertheim, Gunther K.
Bell Laboratories, 600 Mountain Avenue, Murray Hill, NJ 07974, USA

1. Introduction

L. Ley and M. Cardona

Caminante, no hay camino
sino estelas en la mar

Antonio Machado

This is the second of a series of two volumes devoted to photoemission in solids with particular emphasis on photoelectron spectroscopy. Photoelectron spectroscopy is one of a number of spectroscopic techniques involving photons and electrons (see [Ref. 1.1, Fig. 1.1]): monochromatic, possibly polarized photons impinge on a sample and, as a result, electrons are ejected. Their energy is then measured with a suitable analyzer. The photoelectron spectrum yields information about the electronic levels of the solid in its photoexcited states. Thus photoelectron spectroscopy is mainly a technique for the investigation of the electronic structure. Recently, however, it has also become possible to observe effects of the vibronic structure (phonons) in the photoelectron spectra of solids, (see [Ref. 1.1, Sect. 1.4.3]).

Photoelectron spectroscopy has undergone an unprecedented development within the past ten years and has become one of the most popular areas of research in solid state physics. The reason for this development is to be found mainly in the increasing improvement of experimental techniques and the commercial availability of photoelectron spectrometers. Measurements are nowadays performed in a routine way under ultrahigh vacuum, thus eliminating one of the main sources of unreliability of photoelectron spectra: the surface contamination. The escape depth of the photoelectrons is very small (5–50 Å) and it becomes imperative to work with ultraclean surfaces.

It is customary to subdivide the field of photoelectron spectroscopy into two categories depending on the type of photon source used. When lamps are used for excitation with uv photons one speaks of ultraviolet photoelectron spectroscopy or UPS. If X-ray tubes are used one calls it XPS (X-ray photoelectron spectroscopy). A table of the gas discharge and X-ray lines conventionally used can be found in [Ref. 1.1, Table 1.7]. Because of the high work functions of most materials ($\phi \gtrsim 4$ eV) little data can be obtained in the wavelength region where air is transparent. Hence UPS is usually performed in the vacuum ultraviolet. The upper limit of 11.8 eV on the photon energy is imposed in some spectrometers by the use of a LiF window between the lamp and the sample. The modern trend, however, is to disregard the region below 11.8 eV and to operate without a window, mostly by using the He I (21.2 eV) and the He II (40.8 eV) lines. XPS is presently done mainly with Al K_α (1486.6 eV) radiation. A particularly annoying problem, the natural linewidth of the Al K_α line and its satellite structure, can be solved with an X-ray monochromator.

The increasing availability of electron synchrotrons and storage rings for spectroscopic work using synchrotron radiation seems to be changing the trends of work in photoelectron spectroscopy (see Chap. 6). Synchrotron radiation as produced in storage rings is ideal for photoelectron spectroscopy work. It is intense, linearly (or circularly) polarized, produced in ultrahigh vacuum and covers the photon energy range from photoelectric thresholds to the hard X-rays for big machines (electron energies 2 GeV). Smaller machines, which can be built as "dedicated" sources of synchrotron radiation at a moderate cost, cover the region up to $h\nu \sim 1000$ eV. Thus synchrotron radiation, after suitable monochromatization, can be used advantageously in photoelectron spectroscopy, especially in experiments which require the continuous variation of the exciting photon energy. Synchrotron radiation bridges the traditional gap between XPS and UPS making the distinction, based on the source used, basically meaningless.

The number of parameters at our disposal for variation in a photoelectron spectroscopy experiment is very large. The photon energy, polarization and angle of incidence of the exciting source can be varied while for the emitted electrons the energy and the spin polarization can be analyzed as a function of polar and azimuthal angles of emergence. The sample orientation, or actually that of the photoemitting surface, can be changed, provided we know how to prepare such a surface with the required cleanliness and perfection. The number of parameters and the volume of data obtained by varying them is actually prohibitingly large so that an intelligent choice must be made in order to make the analysis possible. Thus arise the various techniques discussed in our two volumes. The conventional EDC's (energy distribution curves) are run for photons of a given energy at fixed polarization, and angle of incidence, as a function of the electron energy with the electrons being collected either over a wide solid angle (angular integrated) or angular resolved. In the latter case the EDC's are taken using the azimuthal and polar take-off angles as parameters, (see [Ref. 1.1, Chap. 6]). Measurements can also be performed for a fixed electron energy and varying the energy of the exciting photon (constant final-state or CFS spectroscopy, see Chap. 6) if synchrotron radiation is available. Another alternative is to vary both, photon and electron energies keeping their difference constant. Thus one obtains the technique of constant initial-state spectroscopy (CIS) which also requires the use of synchrotron radiation. The technique of spin-polarized photoemission, in which the spin of the photoelectrons is analyzed, has been left out of our two volumes as it is covered elsewhere [1.2].

The traditional emphasis of photoelectron spectroscopy lies in the investigation of the electronic structure of valence electrons and core levels. It is generally believed that for sufficiently big photon energies ($\gtrsim 25$ eV) the angular integrated EDC's represent the density of occupied (valence) states somewhat modulated by the appropriate matrix elements (see Chap. 7). At lower photon energies the spectra are also affected by the properties of the final state: the

EDC's corresponding to the energy distribution of the joint density of states (EDJDOS) are obtained (see [Ref. 1.1, Chaps. 1 and 6]). The matrix elements mentioned above depend on the exciting photon energy. This fact can be advantageously used to extract the partial densities of valence states, i.e., to split the density of states into components of a given atomic parentage (see Chap. 2). The angular resolved EDC's yield, in principle, the complete energy versus wavevector curves in the case of two-dimensional solids (layer structures or surface states, see [Ref. 1.1, Chap. 6]).

The investigation of core levels lies usually in the domain of XPS although a few of the outermost *d* core levels can be also excited with uv radiation (in particular He II). Core level XPS can be used for qualitative or quantitative chemical analysis (electron spectroscopy for chemical analysis "ESCA", see [Ref. 1.1, Sect. 1.5.1]). The core levels, in particular their chemical shifts, contain also information about the electronic structure and bonding (see [Ref. 1.1, Sect. 1.2.2]). The line shape of the core levels contains additionally a number of interesting effects, especially the asymmetry due to low-energy electron excitations near the Fermi surface, the so-called Mahan, Nozières, De Dominicis effect (see [Ref. 1.1, Chap. 5]). The linewidths of the core lines reflect the lifetime of the core holes plus a contribution due to phonon broadening, see [Ref. 1.1, Sect. 1.5.3]. Core levels can be accompanied by a number of satellites due to simultaneous excitations of other electronic levels (shake up, shake off [Ref. 1.1, Sect. 4.3]) and single or multiple excitation of plasmons, see Chap. 7. Also, central to the photoelectron spectroscopy of core levels is the matter of electronic *relaxation:* the binding energies being measured are not those of the one-electron core states corresponding to the ejected electron, but those of the atom left behind with a core hole. The difference in these energies is produced by the remaining electrons "relaxing" upon the creation of the core hole (see [Ref. 1.1, Sects. 1.5, 4.2]). The main part of this relaxation ($\sim$20 eV typically) is of atomic origin and thus the same in the solid as in an isolated atom (intra-atomic relaxation). Part of the relaxation, however, is produced by the valence electrons and thus affected by solid-state bonding (extra-atomic relaxation). Extra-atomic relaxation can amount to a few eV.

As already mentioned, the phenomenon of photoemission takes place within a depth which varies between $\sim$5 Å (UPS) and $\sim$50 Å (XPS) (see [Ref. 1.1, Fig. 4.10]). Hence it is expected to be strongly sensitive to the condition and properties of the surface. In the XPS case 50 Å suffice to observe bulk properties provided the surface is clean. For UPS with escape depths of 5 Å, typically two monolayers, the electronic properties of these two monolayers, which can differ significantly from those of the bulk, are usually measured. Hence UPS is particularly appropriate for the investigation of the electronic properties of surfaces, in particular surface states. With the exception of Chap. 2 (surfaces of semiconductors) we have not dealt with the question of photoemission from surface states since it is adequately covered by another recent monograph [1.3]. We should point out, however, that besides specific surfaces the bulk electronic states also appear in UPS. This is particularly true if the measurements are angular integrated or on polycrystalline material; the corresponding angular integration tends to wash out preferentially features due to surface states.

Articles related to the general principles of photoemission and photoelectron spectroscopy are contained in [1.1], while the present volume concentrates on case studies of specific families of materials (semiconductors, transition metals and their compounds, rare-earth metals and compounds, organic materials, simple metals). Experiments which are specific to the use of synchrotron radiation as a source are also covered. In the following an overview of *Photoemission in Solids* is given, covering both the previous and the present volume.

1.1 Survey of Previous Volume [1.1]

This volume is concerned with the general principles of photoemission. Chapter 1 contains an extensive historical survey by M. Cardona and L. Ley of the work on electron photoemission from the discovery of the phenomenon late last century to the present day. This survey includes the discovery and development of photocathodes and the evolution of the various photoemission spectroscopy techniques, leading up to present day UPS and XPS. The remainder of Chap. 1 is mainly reserved for aspects of photoemission or photoelectron spectroscopy which, in the judgement of the authors, are not sufficiently covered by other contributions to the two volumes on *Photoemission in Solids.* Among those topics is the question of the photoelectric threshold and the work function. This parameter, rather central to the phenomenon of photoemission and important to several aspects of photoelectron spectroscopy, is usually left out of modern reviews and textbooks. The various methods used for determining the work function, some based on photoemission, others on a number of related phenomena, are discussed in [Ref. 1.1, Sect. 1.2.1]. A tabulation of the work functions recommended by *Fomenko* [1.4] is added. This tabulation, arranged in the form of a periodic table, is reproduced for convenience in the present volume (Table in the inside cover).

The theory of the work function, with emphasis on its microscopic aspects, is discussed in [Ref. 1.1, Sect. 1.3]. It is shown in [Ref. 1.1, Sect. 1.3.1] that, at least for simple metals, the work function arises mainly from exchange and correlation effects. These effects place a so-called exchange and correlation hole around each electron which makes it feel more strongly the nuclear attraction than the compensating repulsion by the other electrons. The calculation of work functions thus requires the use of many-body techniques which can get very involved if the *actual* crystal potentials are taken into account. Of particular interest is the contribution of surface dipole layers and surface states which lead to a dependence of the work function on the crystallographic surface under consideration. In view of the difficulties involved in evaluating the work function from first principles, a number of phenomenological approaches to derive it and to relate it to other parameters of the solid and its constituent atoms have been attempted, they are briefly reviewed in [Ref. 1.1, Sect. 1.3.7].

Experimental details are absent from most of the articles in this Series. They have been therefore briefly reviewed in [Ref. 1.1, Sect. 1.4], including the delicate question of sample preparation [Ref. 1.1, Sect. 1.4.3].

The binding energy and the width of core levels and their cross sections for photoelectron production are discussed in [Ref. 1.1, Sects. 1.5.1–4]. The binding energy is determined by the one-electron core energies and by relaxation effects. The shifts in binding energies with chemical binding (core shifts) are discussed in [Ref. 1.1, Sect. 1.5.2] for rare gases implanted into noble metals, ionic solids, and metallic alloys. The widths of the core levels are believed to have two contributions, an extra-atomic, temperature-dependent one due to phonons, and an essentially intra-atomic one due to the lifetime of the photoexcited core hole. The latter has in its turn two contributions: Auger decay and radiative recombination. The width of core levels is discussed in [Ref. 1.1, Sect. 1.5.3]. The cross section for photoexcitation of electrons from core levels in the XPS case (Al K_α and Mg K_α radiation) is treated in [Ref. 1.1, Sect. 1.5.4]. While a general discussion of photoionization cross sections is given in [Ref. 1.1, Chap. 3], we felt that the XPS case mentioned above is so important, in particular for quantitative chemical analysis, so as to warrant a separate review.

An introduction to the analysis of valence-band spectra is given in [Ref. 1.1, Sect. 1.6], in particular the old controversy of direct vs nondirect transitions, which has recently flared up again as work on angular resolved photoemission has started.

W. Scheich explores the microscopic, formal theory of photoemission in [Ref. 1.1, Chap. 2]. The photoemission process is treated as a quadratic response phenomenon; the photocurrent outside the material is proportional to the square of the electromagnetic field generating it, i.e., to the exciting light intensity.

Using time-dependent second-order perturbation theory applied to the exact unperturbed many-body states, one obtains an expression for the photocurrent as a function of a *three-particle* correlation function of the current operator (ensemble average of the product of three current operators at different points and times over the ground state). This formula, in all its generality, is useless to interpret photoemission experiments, but it provides a starting point for further approximations and simplifications. It can also be transformed into an expression involving time-ordered products of current operators (Green's function) to which the standard techniques of many-body theory can, in principle, be applied.

In order to make further progress, drastic assumptions must be made concerning both the nature of the spatial dependence of the electromagnetic field and the electronic states. In the independent particle approximation an expression of the golden rule type is obtained: the photocurrent is written as a sum of terms involving a transition rate between initial and final states, a density of states, and an energy conserving δ function. Further approximations, including the assumption of a one-electron potential with the translational symmetry of the crystal parallel to the surface, and the use of the corresponding two-dimensional Bloch's theorem, lead to an expression involving matrix elements of the momentum operator over initial and final-state one-electron wave functions. The final-state wave functions are time-reversed LEED (low energy electron diffraction) wave functions. The next approximation, the assumption of a large escape depth, leads to an expression which can be directly related to the three-step model of photoemission, involving an electron excitation *in the solid*, transport to the surface, and transmission through it. The procedure, though fraught with dubious or invalid assumptions, provides at least a formal path to derive the highly useful three-step model from first principles.

The general formalism described treats in principle automatically surface effects, those due to surface states (wave functions), as much as those due to the crystal potential variation near the surface. However, in deriving the three-step model, valid strictly speaking only for the volume, but including the transmission coefficient of the surface, it appears that it is not possible to separate unambiguously surface and volume effects.

Within the context of the three-step model the concept of photoionization cross section, step number one, plays an important role. Here one has to distinguish between low-energy excitations, near the photoelectric threshold, and excitations at higher photon energies. Crystal potential effects are very important in determining the spectral dependence of the excitation probability near threshold. The various atomic levels taking part in these excitations are strongly mixed and it is difficult to assign *partial* cross sections corresponding to independent atomic states. The excitation probability, i.e., the absorption coefficient, is usually obtained by means of a pseudopotential band-structure calculation. A large body of experimental information for the absorption spectra of solids is available in this region ($\lesssim 6$ eV) and a number of monographs covering it have appeared [1.5, 6]. Well above threshold the "photoionization" cross sections have essentially atomic character, especially for transitions from core levels. Some anomalies can, however, appear near threshold for core level absorption in metals (see [Ref. 1.1, Chap. 5]). Also, band-structure effects appear near the threshold for transitions from core levels. These effects, however, do not extend more than ~ 10 eV above threshold.

In view of the relevance of atomic photoionization cross sections to the theory of photoemission Chap. 3 of [1.1] by S.T. Manson treats the theory of these cross sections. The discussion is confined to the nonrelativistic case which applies to photon energies < 2 keV. This region includes the photon energies under present use for UPS and XPS. The cross sections for higher energies have been reviewed elsewhere [1.7]. Calculations performed under the central field approximation and generalizations to include correlation effects are also discussed in [Ref. 1.1, Chap. 3]. The results of computations for a number of elements across the periodic table are compared with experimental absorption data and, in a few cases, with the results of EDC measurements as a function of photon energy performed with synchrotron radiation.

The theory of photoemission presented in [Ref. 1.1, Chap. 2] contains in its most general form all sorts of many-body effects. Its generality, however, prevents it from being of much use to treat specific effects of many-body interactions. These effects are usually superimposed onto the theoretical framework of Chap. 2 in a semiphenomenological manner which reverts to the three-step model. Specific many-body effects are discussed in [Ref. 1.1, Chap. 4] by D. A. Shirley, including the question of multiplet splittings. Such splittings are observed for core level transitions in atoms with unfilled outer shells. They are due to exchange coupling between the outer shell and the core hole. Multiplet splittings are of particular importance in transition metals, with unfilled *d*-shells (Chap. 3), and in rare earths (unfilled *f*-shells, Chap. 4). Another effect of at least partial many-body character discussed in [Ref. 1.1, Chap. 4] is relaxation, in its intra- and extra-atomic forms. Extra-atomic relaxation in metals is reduced, on the basis of the virtual excit on model, to a Slater integral between the core hole and the first unoccupied atomic orbital. Also discussed in this chapter is the subject of configuration interaction, leading in particular to shake-up and shake-off satellites (final-state configuration interaction). The intensities of the core lines and their satellites are strongly modified by configuration interaction effects within the initial state. A brief discussion of the problem of plasmon satellites follows. Such satellites can, in principle, be produced at the instant of photoemission by the sudden core potential (intrinsic plasmons) or by the excited electrons on their way to the surface in the spirit of the three-step model (extrinsic plasmons). This question is picked up again in Chap. 7. Chapter 4 of [1.1] closes with a discussion of the sensitivity to surfaces. In this connection a universal curve for the escape depth of photoemitted electrons as a function of energy is also given [Ref. 1.1, Fig. 4.10]. The shape of core level lines in metals, in particular the effect of Anderson's orthogonality theorem is discussed in [Ref. 1.1, Chap. 5] by G. K. Wertheim and P. H. Citrin. This theorem, which states that the *ground state* of the electrons in the presence of the core hole is completely orthogonal to the corresponding ground state without the core hole, excludes that ground state as a final state. Instead a continuum of states, involving low energy excitations of electrons around the Fermi sphere, is obtained. The phenomenon is ultimately related to the Mahan-Nozières-De Dominicis (MND) shape of the core absorption and emission edges. In fact, in [Ref. 1.1, Chap. 5] the resulting smearing of the core line is described by convoluting it with the MND shape $(\omega-\omega_0)^{\alpha-1}$. In this manner, the asymmetric Doniach and Šunjič core line shape is obtained. Chapter 7 uses instead of $(\omega-\omega_0)^{\alpha-1}$ the slightly modified expression $e^{-(\omega-\omega_0)/\zeta}(\omega-\omega_0)^{\alpha-1}$ which includes a "cutoff" energy ζ in exponential form to make the line shape integrable [Ref. 1.1, Chap. 5]. The lifetime of the core hole in a Lorentzian fashion and the *phonon broadening* as a Gaussian are also treated in [Ref. 1.1, Chap. 5]. Detailed fits to the observed core spectra of alkali metals, the noble metals, Al, Mg, Cd, Sn, and Pb are presented. A similar interpretation is attempted for a few transition metals (Pd, Pt, Ir). Finally core levels of dilute alloys show that the core level asymmetry can, at least in some cases, be due to excitations of the electron gas in the host material; for the core levels of Pt diluted into Ag the same asymmetry index as for Ag is found.

N. V. Smith discusses angular resolved photoemission in [Ref. 1.1, Chap. 6]. The question of the electron detector is central to this technique. The author discusses the experimental problems of the various types of analyzers used, from the movable type to angular-multichannel systems, which at the time of writing the article are just beginning to appear. Angular resolved photoemission became "respectable" when it was realized that it measured directly the complete energy dispersion relations of the occupied bands $E(\boldsymbol{k})$ in two-dimensional structures (layer structures, surface states). A number of case studies for layer type crystals (GaSe, InSe, MoSe) are presented. Also, the question of angular resolved photoemission and its interpretation in the three-dimensional case is briefly touched upon. This is, at the time of writing the present article, still subject to considerable controversy, especially to what extent is the component of the $\boldsymbol{k}$ vector perpendicular to the surface conserved. The chapter concludes with a discussion of angular resolved constant initial state spectroscopy which can be used to study the electronic structure of *empty* states.

Also included in [1.1] for reference is a table of binding energies of core levels up to the limit which can be excited with Al K_α radiation (1486 eV).

1.2 Contents of Present Volume

Chapter 2, by L. Ley, M. Cardona, and R. A. Pollak, discusses photoelectron spectroscopy as applied to semiconductors. These materials, especially those of the diamond-zincblende structure, have been extensively studied and their electronic bands are well known. Thus they are ideal for testing the capabilities of the photoemission techniques. The varying ionicities of the diamond-zincblende family offer the possibility of checking the relationship between ionicity, core shifts, and systematic variations in the valence bands. Also, recently our knowledge of electronic properties of the surfaces has improved enormously, both theoretically and experimentally. Hence semiconductors are also appropriate to elucidate the effects of surfaces and surface states on the photoelectron spectra. This chapter includes an extensive historical note about semiconductors and a detailed discussion of their band structures with a large number of figures and references. A comprehensive discussion of amorphous semiconductors is also given.

Chapter 3, by S. Hüfner, is devoted to transition metals and their compounds. Emphasis is on the subject of energy band structure and hybridization between the *d* and other levels. The subject of core levels and multiplet splittings is deemphasized, since it has been recently covered elsewhere [1.8]. The chapter also treats alloys and intermetallic compounds.

Chapter 4 treats the rare earth metals and their compounds, materials with partially filled $4f$ shells in which electron correlation effects become very important. Emphasis here is on the positions of the $4f$ multiplets and the valence of the rare earth atom, including the cases of mixed valence or valence fluctuations. The $4s$ and $5s$ multiplet splittings of rare earth atoms, due to exchange interaction with the corresponding core spectra of the $3d$ and $4d$ electrons, are also discussed.

Chapter 5, by W. D. Grobman and E. E. Koch, treats photoemission from organic solids. The emphasis here is not on the molecular effects, which have been exhaustively studied and are covered in other monographs [1.9, 10], but on typical solid-state effects, especially for materials which have been the object of great recent attention by solid-state physicists such as TTF-TCNQ and $(SN)_x$. Both valence bands and core level spectra are discussed, in particular the relationship between core shift and charge transfer in charge transfer compounds. Experimental details specific of photoelectron spectroscopy on organic solids are also touched upon.

Chapter 6, by C. Kunz, discusses synchrotron radiation as applied to photoelectron spectroscopy. After an extensive presentation of the properties of electromagnetic radiation emitted by synchrotrons and storage rings, its applications to synchrotron radiation, including laboratory layouts and specific technical details, are discussed. The remainder of the article concentrates on photoemission experiments which require the properties of synchrotron radiation, in particular its tunability and polarization, such as constant initial-state spectroscopy, constant final-state spectroscopy and yield spectroscopy. Examples of these various types of measurements are given.

Chapter 7, by P. Steiner, H. Höchst, and S. Hüfner, discusses photoelectron spectroscopy of simple metals (alkalies, Al, Be, Mg). A detailed treatment of photoelectron spectroscopy for these metals was almost left out of the Series. We were fortunate to include this last-minute contribution, which contains a considerable amount of unpublished material. Simple metals are ideal for the study of many-body effects such as the Mahan-Nozières-De Dominicis effect and the intrinsic and extrinsic production of plasmons. The chapter also covers the XPS spectra of the valence band, a subject till recently rather elusive in spite of its presumable simplicity. A decomposition of the densities of valence states into partial *s*, *p*, and *d* components, with different photoionization cross sections, provides a reasonable interpretation of the observed XPS spectra.

References

1.1 *Photoemission in Solids I. General Principles*, ed. by M. Cardona, L. Ley, Topics in Applied Physics, Vol. 26 (Springer, Berlin, Heidelberg, New York 1978)

1.2 M.Campagna, D.T.Pierce, F.Meier, K.Sattler, H.C.Siegmann: *Adv. in Electronics and Electron Physics*, Vol. 41 (Academic Press, New York 1976) p. 113
1.3 *Photoemission from Surfaces*, ed. by B.Feuerbacher, B.Fitton, R.F.Willis (Wiley, London 1977)
1.4 V.S.Fomenko: *Handbook of Thermionic Properties*, ed. by G.V.Samsonov (Plenum Press, New York 1970)
1.5 *Optical Properties of Solids*, ed. by F.Abelès (North-Holland, Amsterdam, London 1972)
1.6 *Atomic Structure and Properties of Solids*, ed. by E.Burstein (Academic Press, New York 1972)
1.7 J.W.Cooper: In *Atomic Inner Shell Processes*, Vol. I, ed. by B.Crassemann (Academic Press, New York 1975) p. 159
1.8 R.E.Watson, M.L.Pearlman: In *Structure and Bonding*, Vol. 24 (Springer, Berlin, Heidelberg, New York 1975) p. 83
1.9 K.Siegbahn, C.Nordling, G.Johansson, J.Hedman, P.F.Hedén, K.Hamrin, U.Gelius, T.Bergmark, L.O.Werme, R.Manne, Y.Bear: *ESCA Applied to Free Molecules* (North-Holland, Amsterdam 1969)
1.10 D.W.Turner, C.Baker, A.D.Barker, C.R.Brundle: *Molecular Photoelectron Spectroscopy* (Wiley-Interscience, London 1970)

2. Photoemission in Semiconductors

L. Ley, M. Cardona, and R. A. Pollak

With 97 Figures

2.1 Background

The semiconductors are no doubt the most widely used family of materials in electronics and optoelectronics. Their range of applications covers rectifiers, negative temperature coefficient resistors (thermistors), thermionic generators, thyristors, transistors, integrated circuits, photovoltaic generators, photoconductive cells, photocathodes, thermionic emitters, light emitting diodes, lasers, xerographic copying machines, and others. The structurally simpler members of the family (germanium, silicon...) also are among the most studied and best understood solids from a basic point of view. This fact is due to a happy conjunction of a number of reasons. On the one hand, their structural simplicity makes them amenable to microscopic or nearly microscopic theoretical treatment. The results of this treatment, however, are by no means simple. On the other hand, they exhibit a wide variety of possibilities and effects which makes the theoretical treatment interesting per se. The wide range of applications justifies support for basic research on semiconductors in an attempt to understand the fundamental principles underlying the semiconducting devices and thus to be able to improve them and to tailor-make them to specifications. Also, because of the structural simplicity and their applied interest, extremely perfect and pure crystals of some of the most commonly used semiconductors have become available as a result of a highly sophisticated and varied technology of crystal growth and subsequent treatment.

The definition of a semiconductor is not unambiguous. Semiconductors characterize themselves by an energy gap for electron excitations in their pure or intrinsic state. This gap must be sufficiently small for some carriers to be excited at room temperature, otherwise the material becomes an insulator. A gap of about 2 eV is a natural dividing line between semiconductors and insulators. We must point out, however, that doped insulators behave in the same manner as semiconductors and thus may be referred to as large band gap semiconductors. In this manner diamond or ZnO, insulating in their pure state (5.5 and 3.3 eV gaps, respectively) can become semiconductors if doped.

Similarly, it is not always possible to distinguish the properties of a highly doped semiconductor from those of a poorly conducting metal or a semimetal. Gray tin, for instance, believed for many years to be a semiconductor because of its isomorphism to germanium, is now believed to be a semimetal. Heavily doped germanium and silicon (carrier concentrations higher than $\simeq 10^{19}\,\mathrm{cm}^{-3}$)

are sometimes called metallic. What makes these materials semiconductors is the possibility, at least in theory according to their band structure, of making them intrinsic with an energy gap for electronic excitations $\lesssim 2$ eV.

Photoemission in semiconductors has found practical applications for cathodes of photocells since the discovery, in a purely empirical manner, of the Cs_3Sb photocathode [2.1a]. The amount of basic knowledge accumulated for semiconductors has led in recent years to the development, in a predictive way, of the negative electron affinity photocathodes [2.1b] (see [Ref. 2.2, Table 1.1 and Fig. 1.5]). Photoemission, in particular photoelectron spectroscopy, has been used since the early 1960's to obtain information about the band structure of semiconductors and to investigate the mechanism of photoemission itself [2.3, 4]. Advances in theoretical band-structure calculations [2.5] and a number of complementary experimental techniques, such as optical absorption [2.6, 7], modulation spectroscopy [2.8], electron energy losses [2.9], made these materials particularly suitable for such studies. The initial work was confined to the spectral distribution of the photoelectric yield and to energy distribution curves (EDC's) for exciting energies up to 6 eV. This work was soon extended to the vacuum ultraviolet with the upper limit of $\simeq 10$ eV imposed by a LiF window. Around 1970 this window fell as investigations began to be carried out with the He-resonance lamps (21.2 eV, UPS) AlK_α radiation (XPS), and synchrotron radiation (see Chap. 6).

Photoemission in semiconductors is discussed on a number of occasions throughout [2.2] and the present volume. In [Ref. 2.2, Sects. 1.3. 6,7] for instance, we discussed the work function of semiconductors and its relationship to the crystallographic surface structure; the shape of the photoyield curve is discussed in [Ref. 2.2, Sect. 1.2.4], while angular resolved photoemission is treated in [Ref. 2.2, Chap. 6]. This technique yields directly the band structure of two-dimensional materials and has been widely applied to layer-type semiconductors such as GaSe. Chapter 3 discusses a number of transition metal compound semiconductors, especially oxides (MnO, CoO, NiO). Semiconducting chalcogenides and pnictides of the rare earths are treated in Chap. 4, while Chap. 6 discusses some experiments performed on semiconductors with synchrotron radiation, in particular partial yield spectra (see Fig. 6.20).

In this chapter we discuss the aspects of photoelectron spectroscopy of semiconductors not treated in other parts of this volume and [2.2]. The main emphasis is on the relationship between photoelectron spectra and band structure, especially that of the valence bands. Along these lines densities of valence states can be obtained, with appropriate care, from angular integrated EDC's at sufficiently high photon energies. Variation of either the photon energy or the angle of incidence of the photons produces sometimes information on the atomic character of the wavefunctions involved, i.e., partial densities of states in the sense of the tight-binding approximation. If the photon energy is decreased below $\simeq 20\,\mathrm{eV}$, the EDC spectra obtained differ from the density of valence states as the details of the conduction bands (final states) become important.

Another aspect discussed is the possibility of obtaining structural information from the photoelectron spectra. Particularly appropriate in this respect is a comparison of EDC's for a given semiconductor both in the amorphous and the crystalline modification. Chemical shifts in the core levels can also be used, in principle, to obtain indirectly structural information, especially concerning the ionicity of the materials. The relationship between ionicity and core shifts, and the possibility of using the splitting of certain structures observed in the EDC's of valence electrons, to define an ionicity scale are also discussed in this chapter. Recent advances in the field of angular resolved photoemission of three-dimensional semiconductors, are illustrated with examples for the lead chalcogenides.

This chapter ends with a brief review of the work on photoemission related to surface states. This work, rather recent and most of it in a state of flux, is illustrated with examples related to silicon and GaAs. Surface states calculations, surface reconstruction, and metal-semiconductor interfaces are discussed in relationship to UPS spectra.

2.1.1 Historical Survey

It is not our purpose to give a detailed survey of the history of semiconductor science. We shall, however, give a few notes so as to add perspective to the present discussion of photoelectron spectroscopy. A survey of the work till 1955 can be found in [2.10]. An excellent chronological document of the basic work performed on semiconductors since 1950 can be found in the proceedings of the international conferences on the physics of semiconductors, held biannually since 1971 [2.11.I–XIII]. A summary of the highlights of the first eleven conferences is found in [2.12]. The proceedings of the international conferences on amorphous semiconductors [2.13.I–VIII] are also an interesting historical document.

The first observation of a typical semiconductor property is attributed to *Faraday* [2.14] who, in 1833 observed a negative temperature coefficient for the resistivity of Ag_2S. Rectification at contacts involving semiconductors was discovered in 1874 by *Braun* [2.15]. The discovery of radio waves by Hertz shortly thereafter created a demand for detectors for which semiconductor-metal point contacts were particularly suited. Thus interest in semiconductors research was spurred, for the first time, by a well defined applied goal. The interest, however, quickly vanished with the development of the vacuum tubes.

In the early twenties the Hall effect, which had been discovered in 1879, began to be investigated in semiconductors [2.16]. It was soon realized that these materials possessed a large Hall coefficient, varying over wide limits, of either positive or negative sign. These facts were correlated by *Wagner* [2.17] to small variations in the stoichiometry of the samples. An excess of oxygen in Cu_2O produced a positive Hall coefficient while for oxygen deficient samples this coefficient was negative. These facts provided the empirical foundation for our present understanding of intrinsic and extrinsic behavior in the conductivity of semiconductors. The quantum mechanical formulation required to understand these phenomena in terms of the energy band structure of semiconductors was presented by *Wilson* in 1931 [2.18].

The advent of World War II and the invention of radar renewed interest in detectors, this time for microwave frequencies at which vacuum tubes were not adequate. Fortunately, investigations in this direction soon focused on two very simple semiconductors, germanium and silicon. The investigations into the basic mechanisms of rectification continued after the war, leading in 1947 to the demonstration of the existence of surface states in semiconductors [2.19]. These studies culminated in the invention of the transistor in 1948 [2.20]. In view of the high commercial potential of this discovery, work on semiconductors proceeded now at an accelerating pace. Large single crystals of germanium were first grown in 1950 [2.21], while, shortly thereafter, the discovery of zone refining by *Pfann* [2.22] greatly improved the purity of these crystals.

The birth of modern semiconductors physics is to be found in the mid fifties with the advent of the first realistic band structure calculations for germanium [2.23]. Such calculations revealed the complicated nature of the band edges of these materials and the associated effective masses. These effects were substantiated experimentally by means of cyclotron resonance [2.24] and other transport measurements [2.25].

Subsequent work aimed at achieving a greater understanding of the microscopic properties of germanium and silicon, and at expanding these investigations to other semiconducting materials, many of which found application as electronic components. With the development of the photoresist microelectronics technology, however, considerable materials retrenchment into silicon has taken place in recent years.

The availability of band structure calculations opened the way for the interpretation of the optical absorption edge of these materials [2.26]. Optical studies concerned themselves mainly with the lowest indirect and direct absorption edges until the late fifties and early sixties. At this time absorption spectra began to be investigated by means of reflectance measurements and the structure found in them was related to the calculated band structures [2.27, 28]. Thus electronic states deep into the valence band, of primary concern in photoelectron spectroscopy, were first probed. A number of other developments in the field of optical properties of semiconductors followed. Among them we mention, as being of relevance to photoelectron spectroscopy, the advent of modulation spectroscopy [2.8, 29a] and the investigation of the optical properties of semiconductors due to transitions from core levels with synchrotron radiation [2.29b].

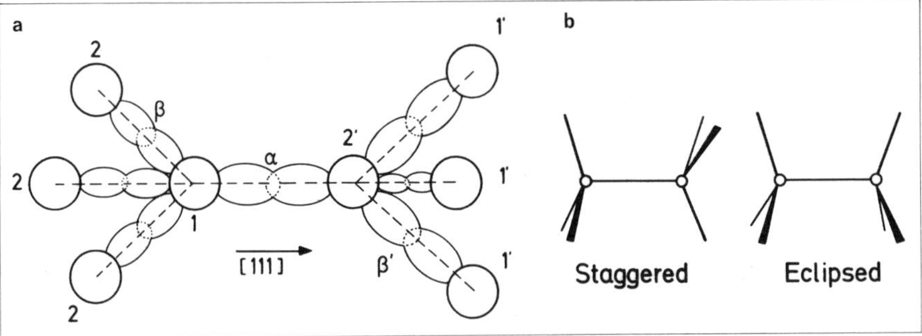

Fig. 2.1. a Schematic diagram of the tetrahedral structure of germanium including the sp^3 hybrid bonds α and β. Each 1–2 bond is parallel to a 2′–1′ bond. For zincblende structures (ZnS) the position 1 is occupied by a Zn and 2 by an S or equivalent atom. For wurtzite the 2,2,2 atoms are rotated by 60° around [111] so as to obtain the so-called eclipsed configuration. **b** Perspective view of staggered and eclipsed configurations mentioned above

2.2 Band Structure of Semiconductors

We discuss in the following sections the band structure of the main types of semiconductors of interest in photoelectron spectroscopy.

2.2.1 Tetrahedral Semiconductors

As already mentioned germanium and silicon because of their simplicity and technological importance, occupy a central role as prototypes in the physics of semiconductors. Their tetrahedral structure is schematically shown in Fig. 2.1 and can be described as two interpenetrating face centered cubic lattices (1 and 2 in Fig. 2.1), displaced from each other by a quarter of the diagonal of the cubic unit cell. The four valence electrons are shared between adjacent atoms in a covalent manner giving rise to "bonding" and "antibonding" states (this nomenclature is, as we shall see, not exact. There exists always a small degree of bonding-antibonding mixture except at the point of the Brillouin zone at which parity is an operation of the group of the $\boldsymbol{k}$ vector). The crystal potential produces a repulsion between the bonding and the antibonding states: the bonding states are lowered and form the valence bands while the antibonding states form the conduction bands. In germanium and silicon a small gap ($\lesssim 2$ eV) exists between bonding and antibonding states and the materials are semiconductors. Diamond, with the same crystal structure, has a gap of $\simeq 5.5$ eV and thus, if intrinsic, must be referred to as an insulator. For the remaining member of the group IV tetrahedral family, gray tin, bonding and antibonding states overlap: the material is a semimetal. We present in Fig. 2.2 the band structure of germanium calculated recently by *Chelikowsky* and *Cohen* [2.30a] with the nonlocal empirical pseudopotential method (EPM). Because of the amount of experimental information which went into adjusting the

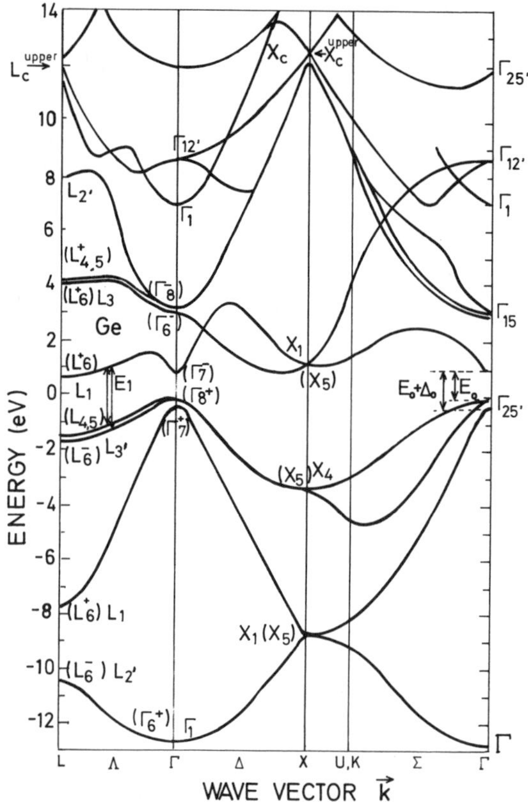

Fig. 2.2. Band structure of germanium calculated by *Chelikowsky* and *Cohen* [2.30a] with the nonlocal empirical pseudopotential method. The bands above 6 eV are from [2.133] (nonlocal EPM but without spin-orbit interaction). The letters in parentheses denote double group symmetries, those without parentheses single group symmetries. The location of the E_0, $E_0+\Delta_0$, and E_1 gaps are indicated by arrows

parameters of the pseudopotential, we believe this band structure to be highly reliable. Quite acceptable band structures can also be obtained from "first principles" without adjustable parameters (sometimes a few parameters are still slightly adjusted so as to fit a few experimental data). For germanium the most successful of the first principles techniques is the orthogonalized plane waves method (OPW) [2.23, 30b, c]. These calculations have the advantage, from the point of view of photoelectron spectroscopy, of yielding the position of the one-electron core states with respect to the valence bands. In Fig. 2.2 the top of the valence bands occurs at the center (Γ point) of the Brillouin zone (BZ) and has $\Gamma_{25'}$ orbital symmetry, split into Γ_8^+ and Γ_7^+ by spin-orbit interaction. This splitting is 0.3 eV for Ge. The lowest conduction band minimum has L_1 orbital symmetry. Thus germanium is a semiconductor with an indirect ($\Gamma \rightarrow L$) band gap of $\simeq 0.6$ eV.

The shape of the conduction bands of the germanium family varies rather drastically from one material to another. While germanium has the lowest conduction band minimum at L, silicon, for instance, has it near Δ and gray tin at Γ, degenerate with the top of the valence bands. The valence bands, however, are rather similar for all materials of the family: it has indeed even been suggested that the top valence bands from Γ to L and from Γ to X are nearly the same for all germanium and zincblende-type semiconductors [2.30d].

In the pseudopotential method the energy bands are generated by applying a small pseudopotential to the bands of the metallic-like free electron model. This description turns out to be very economical for a large number of purposes [2.30a, d]. Another possible description is based on the tight-binding or LCAO (linear combination of atomic orbitals) method. In this description the starting point is opposite to that of the pseudopotential method; one starts from atomic orbitals which interact with each other as they are brought together to form the solid. This interaction can be parametrized and adjusted to fit experimental data. The method, however, is not very practical to represent *both* valence and *conduction* bands, because a large number of parameters are required [2.31]. Nevertheless, the essential features of the *valence* bands of germanium can be represented with only a small number of ETBM (empirical tight-binding method) parameters, involving only the s- and p-valence states [2.32, 33]. The number of parameters can be reduced even further if hybrid orbitals directed along the bonds [2.33], or even better, only *bonding* linear combinations of these, are used as a basis [2.34, 35], (BOM or bond orbital model). Doing the latter, implies neglecting the *antibonding* combinations, i.e., the conduction bands completely. This method is closely related to that introduced by Weaire and Thorpe [2.36, 37] to investigate the electronic structures of tetrahedral amorphous semiconductors (see Sect. 2.7).

Since the basis functions are only the four sp^3 hybrid bonds, the model yields a rather simple 4×4 secular matrix which along high symmetry directions ([111], [100]) can even be diagonalized by hand. In its simplest version it requires only *one* adjustable parameter V_1, the matrix element between neighboring bonds on the same atom, in order to determine the shapes of the valence bands. The zero of energy which, of course, determines the photoemission threshold, is then unspecified but, that does not affect the shape of the valence bands. The valence bands so obtained, however, differ considerably from reality (see Fig. 2.3). In particular, the top valence bands show no dispersion whatsoever, a fact which indicates that they are unhybridized p-states (at Γ the valence bands dehybridize into p-like $\Gamma_{25'}$ and s-like Γ_1 states). The shortcomings of this model can be fixed by introducing an additional matrix element V' between parallel second neighbor bonds ($\beta - \beta'$ in Fig. 2.1). The effect of this extra parameter is shown in Fig. 2.3: the bands are now quite similar to the valence bands of Fig. 2.2. The top valence bands remain almost exclusively p-like (along [111] and [100] *exactly*), the lowest band is s-like with some p-component and the middle band is p-like with s-admixture.

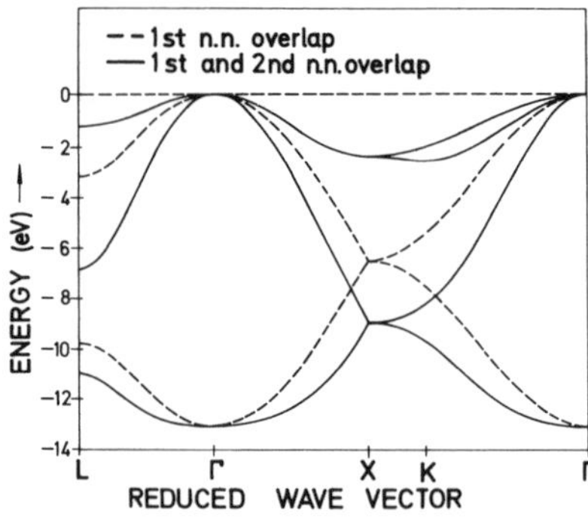

Fig. 2.3. Band structure of germanium calculated with the bond orbital model (BOM). The dashed curve represents the results of a calculation with only nearest-neighbor bond interaction ($\beta-\alpha$). The solid lines represent a BOM calculation which includes interaction between second-nearest-neighbor bonds ($\beta-\beta'$) [2.34]

Once the band structure of a semiconductor has been obtained, it is usually required for the interpretation of photoemission and other measurements to calculate the corresponding density of states

$$N(E)=\sum_{\boldsymbol{k}}\sum_{n}\delta[E-E_n(\boldsymbol{k})] \tag{2.1}$$

where n labels the various bands.

This calculation represents a time consuming integration over the Brillouin zone, which requires the diagonalization of the secular equation at a large number of points. Typical EPM band calculations use 50×50 secular equations [2.30a]. The economy of a calculation based on a 4×4 secular equation like that of the BOM is obvious; we point out again that such work only yields meaningful results for the valence bands. In the BOM Hamiltonian just described the bonding, *occupied* states are decoupled from the empty (antibonding) states. Thus it can be advantageously used to calculate thermodynamic properties such as cohesive energies [2.37b, c]. We show in Fig. 2.4 the density of valence and conduction states obtained by *Herman* et al. with the OPW [2.30b] in comparison with that of the BOM [2.34] for the valence bands. The agreement is reasonable if one considers that the parameters of the BOM (V_1 and V') were not adjusted so as to optimize agreement with the OPW results of [2.30b], but rather chosen so as to fit best the photoemission data [2.34]. The three structures I, II, and III in the density of valence states of Fig. 2.4 can be identified with the two upper p-bands (I), the $p(s)$ band (II), and the lowest $s(p)$ band (III). We note again that the essential assumption of the BOM calculations is the lack of coupling between bonding and antibonding states. Thus the valence bands are *pure* bonding states. The amount of antibonding admixture has been estimated to be 15% for Ge and 9% for Si [2.37b].

Germanium is the prototype of a large number of binary and ternary semiconductors with tetrahedral coordination [2.38]. The simplest and perhaps most important of them are those with zincblende structure, obtained

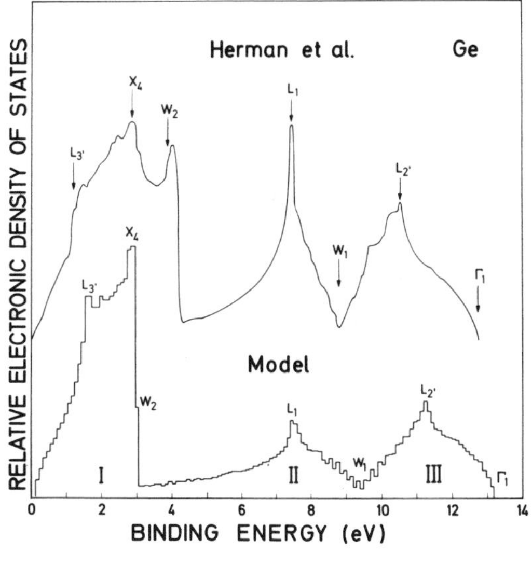

Fig. 2.4. Densities of valence states of germanium calculated with the BOM bands of Fig. 2.3 including second-neighbor bonds (histogram) compared with the results of an OPW calculation by *Herman* et al. [2.30b]. The arrows denote the symmetries of the critical points

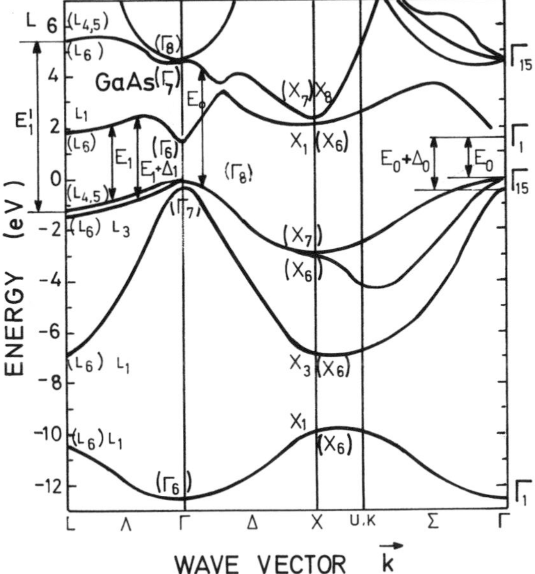

Fig. 2.5. Band structure of GaAs calculated with the nonlocal EPM [2.30a]. The symmetry notation is as in Fig. 2.2. We have indicated the positions of the E_0, $E_0+\Delta_0$, E_1, $E_1+\Delta_1$, E_0' and E_1' critical points which are observed in the reflectance spectra of Fig. 2.21

in the diagram of Fig. 2.1 by replacing atoms 1 and 2 by two different atoms, such that the average number of valence electrons remains four. One thus obtains IV–IV (SiC), III–V's (e.g., GaAs), II–VI (e.g., ZnSe), and I–VII (e.g., CuBr) compounds. Extensive literature exists dealing with the III–V compounds (2.39a–c], and the II–VI compounds [2.40a–c]. The electronic properties of the copper and silver halides (I–VII compounds) have been recently reviewed by *Goldmann* [2.41].

Table 2.1. Compatibility relations between zincblende and diamond space group representations at the Γ, L, and X points [2.42, 43]

Γ-point		L-point		X-point	
Diamond	Zincblende	Diamond	Zincblende	Diamond	Zincblende
$\left.\begin{matrix}\Gamma_1\\ \Gamma_{2'}\end{matrix}\right\}$	Γ_1	$\left.\begin{matrix}L_1\\ L_{2'}\end{matrix}\right\}$	L_1	X_1	$\left\{\begin{matrix}X_1\\ X_3\end{matrix}\right.$
$\left.\begin{matrix}\Gamma_2\\ \Gamma_{1'}\end{matrix}\right\}$	Γ_2	$\left.\begin{matrix}L_2\\ \Gamma_{1'}\end{matrix}\right\}$	L_2	X_2 X_3	X_2 X_4
$\left.\begin{matrix}\Gamma_{12}\\ \Gamma_{12'}\end{matrix}\right\}$	Γ_{12}	$\left.\begin{matrix}L_3\\ L_{3'}\end{matrix}\right\}$	L_3	X_4	X_5
$\left.\begin{matrix}\Gamma_{15}\\ \Gamma_{25'}\end{matrix}\right\}$	Γ_{15}	$L^{\pm}_{4,5}$ $L^{\pm}_6$	$\left\{\begin{matrix}L_4\\ L_5\end{matrix}\right.$ L_6	X_5	$\left\{\begin{matrix}X_6\\ X_7\end{matrix}\right.$
$\left.\begin{matrix}\Gamma_{25}\\ \Gamma_{25'}\end{matrix}\right\}$	Γ_{25}				
$\Gamma^{\pm}_6$	Γ_6				
$\Gamma^{\pm}_7$	Γ_7				
$\Gamma^{\pm}_8$	Γ_8				

We shall first discuss the gross features of the band structures of the III–V and II–VI compounds; those of the I–VII compounds are basically different as a result of the admixture of the *d*-levels of the cations into the valence bands. The band structure of GaAs is shown in Fig. 2.5 to be rather similar to that of germanium (Fig. 2.2). The lowest conduction band minimum is at the Γ point, a rather accidental detail which makes the fundamental edge direct. Other than this, the differences between the band structures of germanium and GaAs are the result of the lack of inversion symmetry as atoms 1 and 2 (Fig. 2.1) become different. The notation of the various space group representations changes as a result of the lifting of the inversion symmetry as shown in Table 2.1 [2.42, 43].

The main qualitative effects of the removal of the inversion symmetry are the splitting of certain states at the X-point of the Brillouin zone and along the X–W line (not shown in Figs. 2.2 and 2.5; for a picture of the BZ of the fcc lattice see [Ref. 2.42, Fig. 7]. As shown in Figs. 2.2 and 2.5 the X_1 points of germanium split for GaAs into X_1 and X_3. We must mention here that the X_1 and X_3 labeling depends on whether the point group under consideration is centered on the Ga or on the As atom; the X_1–X_3 labels reverse in going from a Ga to an As origin. In Fig. 2.5 we have chosen As as the origin which seems to be the most common convention found in the literature [2.30a]. Inconsistencies in this notation are usually encountered in the pseudopotential works, as the symmetry is not obtained directly from the calculation. We note in Fig. 2.5 that the lowest valence band at X has X_1 symmetry, a fact that in our convention means that the wave functions are *s*-like around the As atom (mainly As 4*s* with some admixture of Ga 4*p*). The fact that the second lowest valence band has X_3 symmetry signifies that the wave functions are *p*-like around the As atom or *s*-like around Ga. They are therefore a mixture of As 4*p* and Ga 4*s*. This situation prevails in all III–V and II–VI semiconductors; the X_1–X_3 splitting of the X_1 valence level of Ge is a direct consequence of the ionicity of the material and can be actually used as a measure of that ionicity [2.44] (see Sect. 2.8). The situation concerning the X_1 and X_3 conduction bands is less clear; the splitting is small and it is easy to reverse by changing some of the parameters of the calculation. Most calculations, however, give the lowest conduction band X_1 symmetry (*p*-like around the cation, *s*-like around the anion), a fact which has been confirmed experimentally for GaP [2.45], but seems questionable for AlSb [2.46]. We note that as a result of the ionicity there is a repulsion between states which had different parity in Ge but the same in the zincblende compounds. Hence all gaps tend to increase through the isoelectronic series of group IV, III–V, II–VI, and the I–VII compounds, roughly like the square of the ionicity [2.47]. As already mentioned, the lowest conduction band of GaAs (Fig. 2.5) is at Γ, a fact common to all II–VI compounds and most III–V's. Some III–V compounds, however, have silicon-like minima at or near the X point (e.g., GaP, AlSb).

The spin-orbit splittings of the valence bands of germanium and zincblende-type semiconductors have played a rather important role in the analysis of the optical absorption spectra [2.8]. They have so far, however, played no role in the studies of photoemission for the compounds under consideration. These spin-orbit splittings are basically of atomic nature. That at $\Gamma_{15}(\Gamma_8-\Gamma_7)$ is an average of the splittings of the *p* valence electrons of the two constituent atoms weighted somewhat more heavily towards the splitting of the anion. The splitting at L_3 is roughly two-thirds of the splitting at Γ_{15} while the splitting at X_5 depends on the *difference* of the splittings of the component atoms (it is zero in germanium).

The copper halides (CuCl, CuBr, CuI) and AgI are I–VII compounds and crystallize at room temperature also in the zincblende structure. AgCl and AgBr, however, crystallize in the rock-salt structure. A number of other structural modifications are possible for these materials at higher temperature or under stress [2.41], some of them of technological interest as superionic conductors [2.48]. Some properties of these materials differ drastically from those of their isoelectronic II–VI and III–V counterparts, a fact which is related to the strong hybridization of the valence bands with *d*-electrons of the metal. We show in Fig. 2.6 the valence bands of AgI calculated with the BOM [2.49] without spin-orbit interaction. There are in this figure five more orbital states (the 4*d* states of Ag) than in Fig. 2.3. These states split at Γ under the cubic field into the Γ_{12} doublet and the Γ_{15} triplet, the latter hybridizing strongly with the upper halogen *p*-like band. As a result of this hybridization anomalies in the spin-orbit splitting of the upper Γ_{15} state (observed early in the absorption spectrum) can result; the *d*-admixture yields a negative contribution to the spin-orbit splitting and even produces its reversal: Γ_7 lies above Γ_8 in CuCl

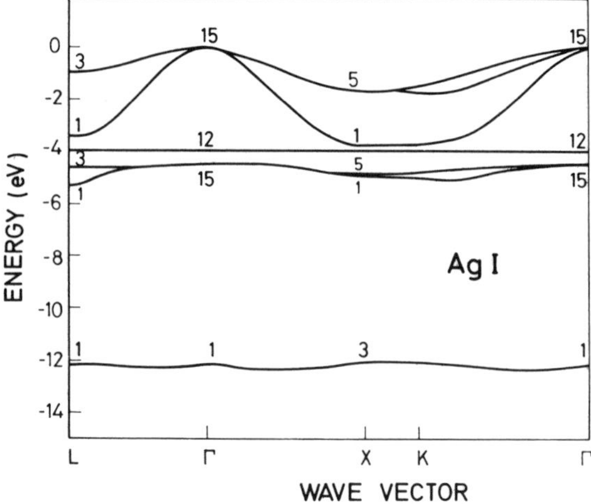

Fig. 2.6. Valence band structure calculated for AgI with the BOM [2.49]

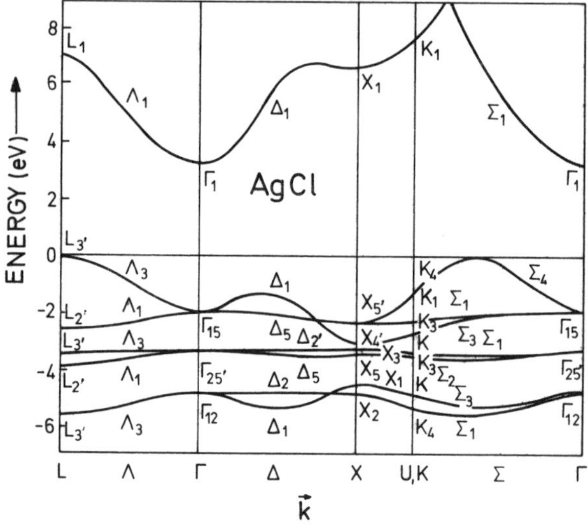

Fig. 2.7. Energy band structure of AgCl calculated with the EPM. Spin-orbit interaction has been neglected [2.51b]

[2.50]. An important contribution of photoelectron spectroscopy has been the elucidation of the halogen-p and metal-d partial contributions to the density of valence states of these materials [2.49].

As already mentioned, AgBr and AgCl crystallize in the rock-salt structure, which contrary to zincblende, possesses a center of inversion. Thus for the states at Γ parity is a good quantum number and $p-d$ admixture is not possible. The p-states have Γ_{15} (odd parity), the d-states $\Gamma_{25'}$ and Γ_{12} symmetry (both even parity). The band structure of AgCl has been calculated by *Bassani* et al. [2.51a] and more recently by *Shy-Yih Wang* et al. [2.51b] (see Fig. 2.7). No spin-orbit-splitting anomalies occur in this structure. Off Γ, however, parity

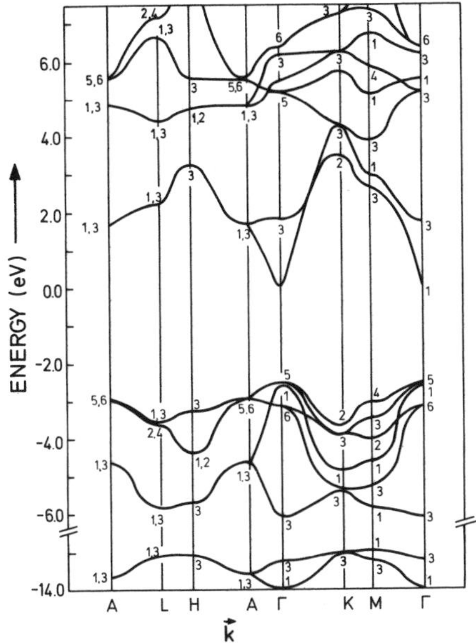

Fig. 2.8. Band structure of wurtzite-type CdS calculated with the OPW method. Spin-orbit splitting neglected [2.53]

does not hold and the top p-states mix strongly with the d-states of the silver. Contrary to what happened in Fig. 2.6 branches of the top valence band move up as a result of the repulsion by the d-states and the absolute maximum of the valence band occurs at L instead of Γ. The fundamental edge is indirect, from L to Γ, with nearly degenerate $\Delta \rightarrow \Gamma$ edges [2.52].

Probably the next most important family of tetrahedral compounds is the wurtzite family. To it belong a few III–V compounds (GaN, AlN) and several II–VI compounds (CdS, CdSe, ZnO, ZnS). Some I–VII compounds (AgI) and SiC can also be found at room temperature in the wurtzite modification. This structure can be considered a variation of the zincblende structure; instead of two interpenetrating cubic closed packed (fcc) lattices one has two *hexagonal* closed packed lattices. Another way of obtaining the wurtzite structure is by rotating atoms 2 by 60° around the 1–2′ axis in Fig. 2.1. This gives rise to the so-called eclipsed configuration, as opposed to the staggered configuration as shown in Fig. 2.1b. The primitive cell of wurtzite has two molecules. Along the hexagonal axis the band structure of a wurtzite crystal is very similar to that of the corresponding zincblende crystal with the bands folded so as to bring the L point of zincblende into Γ of wurtzite to take into account the doubling of the number of atoms per unit cell [2.53], (see Fig. 2.8). At the Γ point, the Γ_{15} states of zincblende split as a result of the hexagonal crystal field. Typical crystal field splittings are between 0.02 and 0.08 eV. Along the other directions of the Brillouin zone the correspondence zincblende-wurtzite is not easy to make.

Another large family of tetrahedral compounds can be derived by performing the following substitution in two molecules of a zincblende compound:

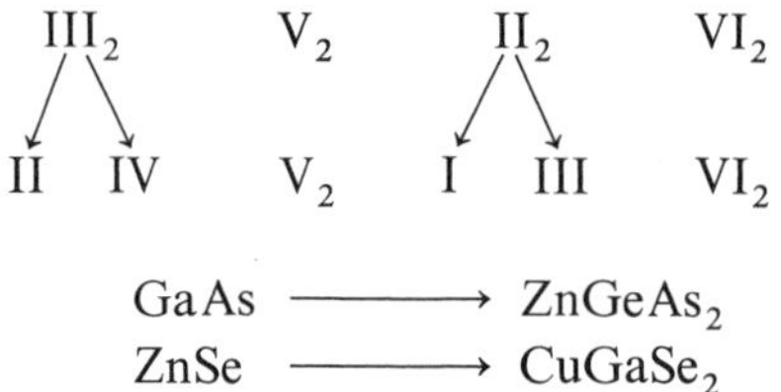

$$\text{GaAs} \longrightarrow \text{ZnGeAs}_2$$
$$\text{ZnSe} \longrightarrow \text{CuGaSe}_2$$

This substitution is made by replacing two of atoms 1-1′ in one of the tetrahedra of Fig. 2.1a by one of the new cations and the remaining two by the other. A chalcopyrite (tetragonal) structure, with 4 atoms per primitive cell, is obtained. These chalcopyrite-type semiconductors have been extensively reviewed by *Shay* and *Wernick* [2.54].

The I–III–VI_2 compounds, like the I–VII binaries, have a valence band structure complicated by the presence of *d*-electrons of the metal. This fact makes band calculations rather unwieldy. A number of calculations have, however, been performed for the simpler II–IV–V_2 compounds [2.54–57]. We show in Fig. 2.9 the band structure calculated in [2.55] for $ZnGeP_2$. The corresponding density of valence states is also presented in Fig. 2.10 [2.56].

A number of binary semiconductors have defect tetrahedral structures, with a tetrahedral network of which some sites are unoccupied. The unoccupied sites can be in regular positions thus giving rise to a superstructure with a larger unit cell, or at random. Many of these systems are described in [2.58]. Among the simplest we consider the "cross products" of the III–V and the II–VI compounds, i.e., the III_2–VI_3 (Ga_2Te_3, In_2Te_3) and the II_3–V_2 (Cd_3As_2, Zn_3As_2, ...). Ga_2Te_3 and In_2Te_3 crystallize in a zincblende structure, with one cation site vacant out of three. An ordered orthorhombic superstructure has been proposed for Ga_2Te_3 [2.58] while In_2Te_3 seems to have a disordered defect zincblende structure at high temperature (β-phase) and an ordered structure (α-phase) at low temperature [2.59]. The II_3–V_2 compounds crystallize in a body-centered tetragonal superstructure with 16 formula units per unit cell [2.60]. This can be considered as an ordered superstructure of an antifluorite structure (fluorite or CaF_2 structure with cations and anions reversed) with one cation out of four missing. The band structure of this defect antifluorite has been calculated in [2.60]. Antifluorite can also be considered as a modified tetrahedral structure with the cations in tetrahedral and the anions in cubic ($\equiv$ two tetrahedral) positions. The Mg_2Si, Mg_2Ge, and Mg_2Sn semiconductors and the semimetal Mg_2Pb crystallize in this structure [2.61]. These materials can be easily derived from their group IV counterparts by replacing one of the group four atoms by two Mg atoms, thus preserving their number of valence electrons per molecule. Their crystal structure is fcc and therefore their band structures and general properties are very similar to those of germanium, except for the slight differences in group theoretical notation.

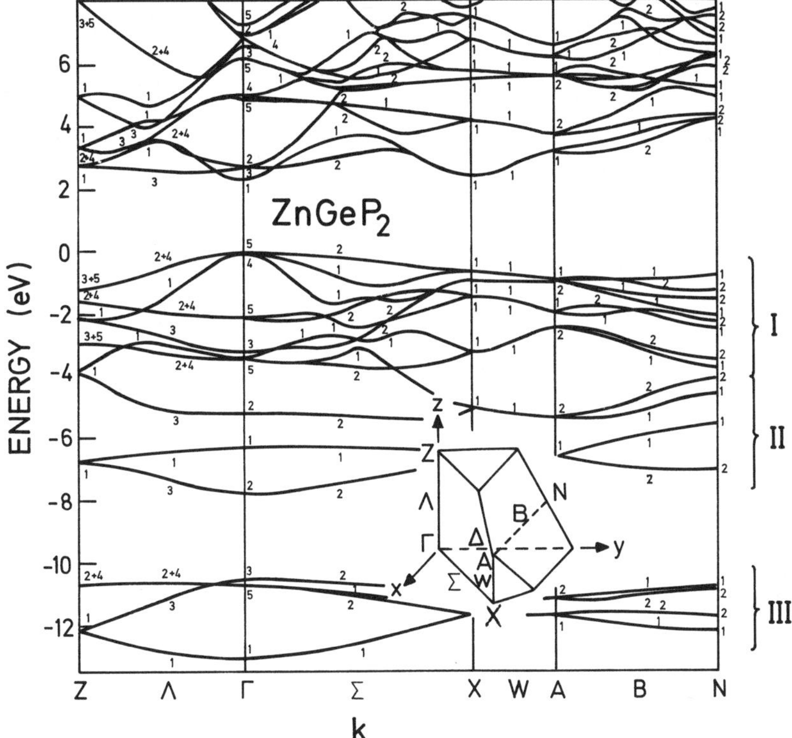

Fig. 2.9. Energy band structure of $ZnGeP_2$ calculated with the EPM. Spin-orbit interaction has been neglected [2.55]

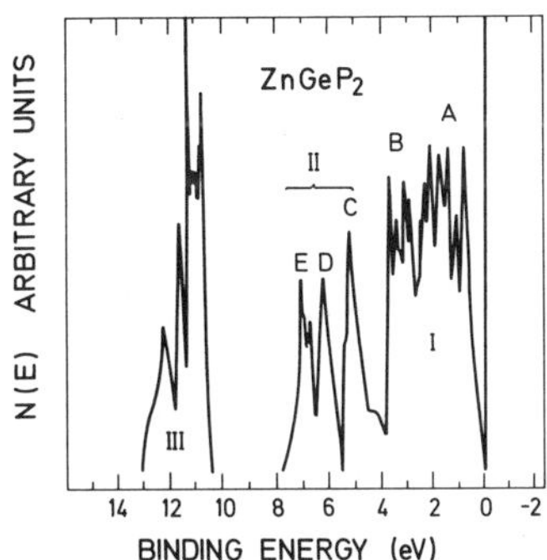

Fig. 2.10. Density of valence states calculated from the energy bands of Fig. 2.9 for $ZnGeP_2$. Spin-orbit interaction [2.56] has been neglected

These differences arise from the fact, that in germanium the center of inversion is between the two atoms while in Mg_2X it lies at the group four atom. Thus the states as X are not all doubly degenerate as was the case for germanium and silicon. The X_1 valence bands of silicon are split in Mg_2Si (see Fig. 2.11) into X_1 and X_4, a fact which could be construed as to represent the ionicity of this

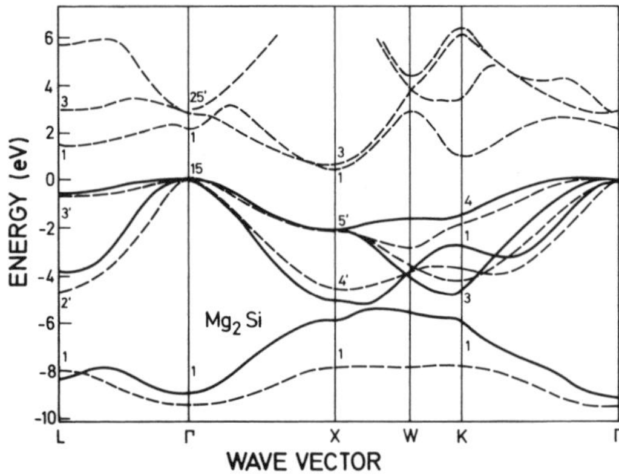

Fig. 2.11. Energy bands of Mg_2Si calculated with the local EPM (dashed) [2.152] as compared with the tight-binding results [2.61] (solid line), which were adjusted so as to fit the experimental XPS spectra (see Fig. 2.35)

material, see Sect. 2.8. The density of valence states of the Mg_2X compounds has been calculated in [2.61] by the empirical tight-binding (LCAO) method. It resembles closely that of a III–V or II–VI compound.

A tetrahedral defect compound which has recently received considerable attention is $CdIn_2Se_4$ [2.62]. The density of valence states calculated for this material [2.62] shows at the top a set of well separated, narrow peaks which are due nearly exclusively to 4*p* functions of selenium. The lowest peak, mainly due to 4*s* functions of Se, is also well separated. The middle peak, however, is broad and has considerable substructure; it corresponds to a mixture of 4*p* electrons of Se, 5*s* of Cd and 5*s* of In.

Another interesting family of compounds which can be regarded as "defect" tetrahedral structures are the III–VI compounds GaS and GaSe [2.63]. These semiconductors have a two-dimensional structure; they can be considered as composed of layers bound together by Van der Waals forces. The basic building block of the layers is a trigonal prism. The atomic arrangement is similar to that in Fig. 2.1a with the 1–2′ bonded atoms being Ga and the bond lying perpendicular to the plane of the layers. The outer atoms of the tetrahedra are in an "eclipsed" configuration like in wurtzite. Several crystal modifications ($\alpha, \beta, \gamma, \delta$) are possible depending on the number of such building blocks per primitive cell and the relative arrangement of the layers. The band structure of GaSe has been calculated by *Schlüter* [2.64] in the β-modification which does not seem to exist in nature [2.63]. The valence bands fall into four groups (I–IV from top to bottom) whose properties are illustrated by the charge distribution diagrams of Fig. 2.12. Group IV is composed of the essentially nonbonding *s* valence electrons of the chalcogens (the nonbonding character of *s* valence electrons of chalcogens is common to many chalcogenides as we shall see later). Group III is mainly composed of bonding p_z orbitals of the gallium (z is the direction perpendicular to the layers). Group II contains bonding orbitals between the Se and the Ga, while group I is composed of Ga–Ga sp_z bonds and of p_z orbitals of the chalcogen.

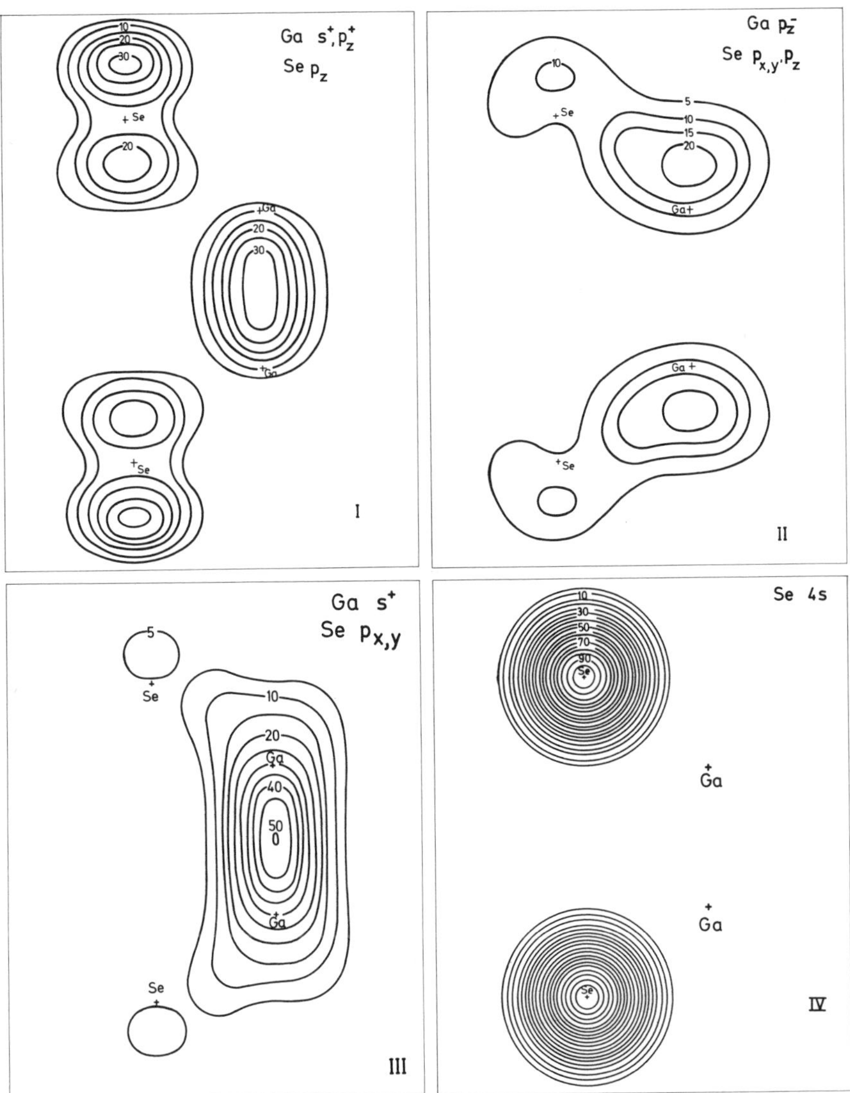

Fig. 2.12. Charge distribution of the four groups of valence bands (I = upper, IV = lowest) found in GaSe, as calculated for the EPM band structure [2.64]

Another large group of tetrahedral semiconductors does not have a well defined crystal structure but appears in amorphous form without long-range order. Most of the group IV and III–V semiconductors just discussed can be prepared in amorphous form with only tetrahedral short-range order. We defer the discussion of their electronic structure to Sect. 2.7.

Table 2.2. Stable modifications of the groups V, IV–VI, and III–VII compounds. Unstable modifications can sometimes be induced by epitaxy [2.578]. The ionicities f_i^{XPS} of the IV–VI compounds (see Sect. 2.8.1) are given in the upper right corner of each corresponding box

Group V Elements	IV–VI Compounds				III–VII Compounds
		Ge	Sn	Pb	Tl
P black ▭	S	0.50 ▭	0.61 ▭ □ Epitaxial	0.67 □ ▭ $p>18$ kbar	F ▭
As semimetal ▱	Se	0.55 ▭	0.69 ▭ □ Epitaxial	0.70 □ ▭ $p>43$ kbar	Cl □
Sb semimetal ▱	Te	0.66 ▱	0.75 □	0.74 □	Br □
Bi ▱		$T>733$ K □	$T<98$ K ▱ $p>40$ kbar	$p>41$–75 kbar ▭	I (□) [*]
▭ D_{2h}^{18}, orthorhombic ▱ D_{3d}^{5}, rhombohedral	▭ D_{2h}^{16}, orthorhombic ▱ C_{3v}^{5}, rhombohedral □ O_h^5, NaCl-Structure				▭ D_{2h}^{16}, [*] D_{2h}^{17} orthorhombic □ O_h^1, CsCl-Structure

2.2.2 Semiconductors with an Average of Five Valence Electrons per Atom

This family of materials is generated from the group V elements (P, As, Sb, Bi. Note: As, Sb, and Bi are semimetals) in a manner similar to that we used to generate the tetrahedral semiconductors from the group IV elements:

$$\begin{array}{ccc} V-V & \qquad & V-V \\ \swarrow \quad \searrow & & \swarrow \quad \searrow \\ IV-VI & & III-VII \end{array}$$

Among the IV–VI compounds we find the rock-salt structure (fcc) lead chalcogenides (PbS). The class of the III–VII compounds comprises basically the thallous halides (TlCl) [2.65]. The alkali trihalogen plumbates ($CsPbCl_3$, $CsPbBr_3$), with perovskite structure, have also been regarded as electronically isomorphic to the lead chalcogenides [2.66a, b]. The cesium atoms are ionized and do not contribute to the valence bands, the halogen atoms have complete valence shells, while the Pb atoms are left with two 5*s* valence electrons, a situation analogous to that of the lead chalcogenides.

Table 2.2 shows the various crystal structures in which the group V, IV–VI, and III–VII compounds occur. The rock-salt structure, characteristic of ionic compounds, is prevalent in the lead chalcogenides. The elemental materials, however, are covalently bonded, a fact which results in a distortion of the rock-salt structure along the (111) axis giving rise to a rhombohedral structure with

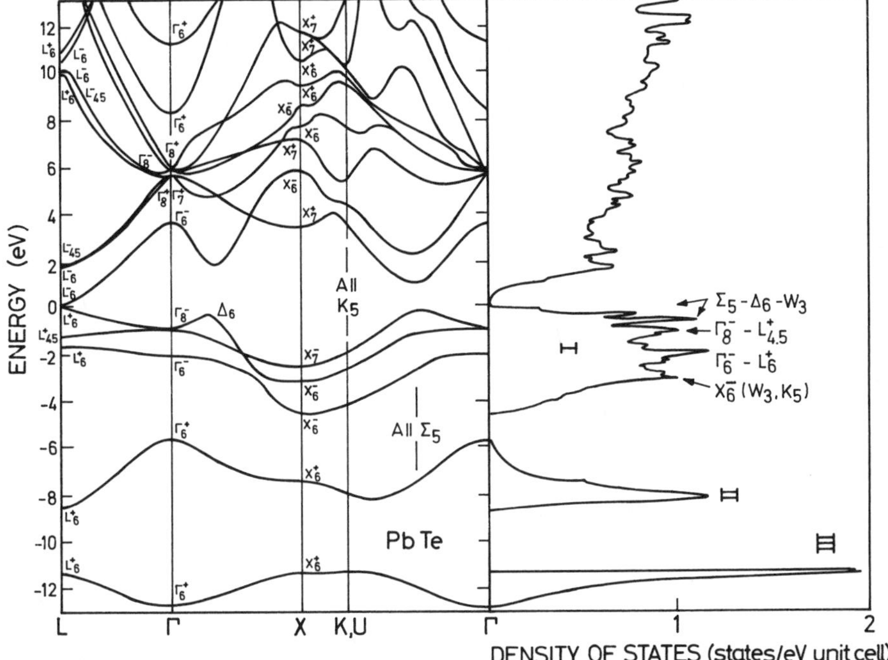

Fig. 2.13. Band structure of PbTe calculated by *Martinez* et al. [2.69] with an empirical pseudopotential method, spin-orbit interaction included. Also, corresponding density of states. We indicate for peak I the critical points possibly associated with the observed substructure (see Sect. 2.4.2)

3-fold coordination. The same structure is found in GeTe and in SnTe at low temperatures. The other Ge and Sn compounds crystallize in the black phosphorus structure, an orthorhombic structure also with 3-fold covalent coordination. This is the structure which the lead chalcogenides exhibit at high pressure, high pressure favoring covalency [2.67a, b], see Sect. 2.8.1. Contrary to the tetrahedral semiconductors, ionicity seems to increase in this family with increasing average atomic number.

The lead chalcogenides have been the object of a number of band-structure calculations [2.68, 69]. Relativistic effects, in particular spin-orbit interaction, must be included in these calculations because of the large atomic number of Pb. We show in Fig. 2.13 the band structure and density of states of PbTe calculated by *Martinez* et al. [2.69] with the relativistic pseudopotential method. Three groups of valence bands appear. The lower ones are nonbonding 5*s* states of Te, the second lowest set 6*s* of Pb with considerable admixture of halogen *p*, and the upper set are halogen *p* bands. We note that the spin-orbit splitting of the upper bands is large, as corresponds to their Te 5*p* character. Strong spin-orbit splitting effects are also seen for the lowest Pb 6*p* *conduction* band. Detailed information on the spin-orbit splitting of the valence bands of

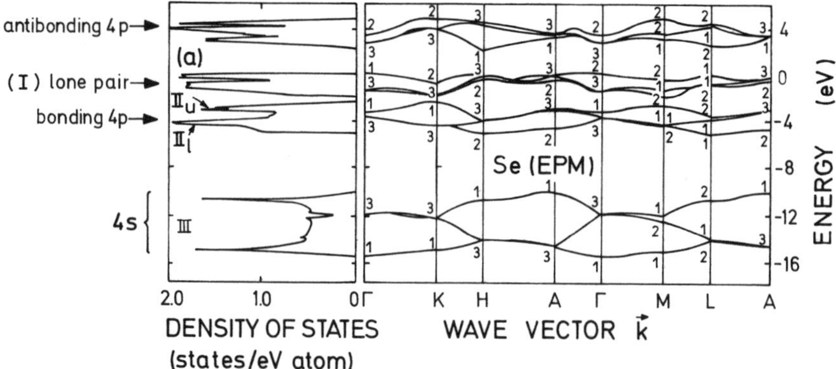

Fig. 2.14. Energy bands calculated for trigonal selenium with the EPM and corresponding density of states [2.77]

these materials has been obtained recently by means of angular resolved photoemission [2.70], as will be discussed in Sect. 2.6.

We note in Fig. 2.13 that the top of the valence band of PbTe does not occur at the Γ point but at L, a fact which is common to all lead chalcogenides and to SnTe [2.68]. This follows from the presence of the 6s-levels of Pb immediately below the 5p-levels of Te. The situation is similar to that described above (Fig. 2.7) for AgCl, where the valence bands were lifted, away from Γ, because of $p-d$ admixture. In the lead chalcogenides the lifting of the L valence bands occurs as a result of $p-s$ admixture which is forbidden at $\boldsymbol{k}=0$.

Many of the materials of this family, in particular the germanium chalcogenides, can be prepared in amorphous form. However, contrary to the tetrahedral semiconductors, the amorphous IV–VI compounds do not have the same short-range order as their crystalline counterparts. This is discussed in Sect. 2.7.2.

2.2.3 Selenium, Tellurium, and the V_2VI_3 Compounds

Selenium [2.71] and tellurium [2.72] crystallize in the trigonal system (space group D_3^4, 3 atoms per primitive cell). Selenium can also be prepared in two different monoclinic structures (α and β, both space group D_{2h}^5, 36 atoms per primitive cell) [2.73]. The monoclinic structures are composed of puckered rings with eight atoms and can be regarded as molecular crystals (Se_8 molecules). Trigonal selenium is composed of parallel helical chains, with a threefold screw axis, and can be regarded as a one-dimensional structure.

A first principles, though nonrelativistic, calculation of the band structure of selenium (self-consistent orthogonal plane waves, SCOPW) has appeared recently [2.74] while a number of semiempirical calculations (pseudopotential, LCAO) have been available for some time [2.75–77]. We show in Fig. 2.14 the valence band structure calculated by *Joannopoulos* et al. [2.77] (LCAO fitted to EPM) and the corresponding density of states. The quasi one-dimensional nature of the material is exemplified by the small dispersion along the $\Gamma - K$ and the $A - H$ lines, perpendicular to the c-axis. The density of states, in particular that of the lowest set of bands (labelled III), also exhibits typical one-dimensional singularities. This one-dimensional aspect is lost in tellurium [2.77], where the chains come closer together and the chain-chain interaction is stronger. The valence bands of Fig. 2.14 fall into three groups, corresponding to the three structures, I, II, and III in the density of states, which are well separated by gaps. An LCAO analysis of the bands [2.76, 77] shows that the lowest group (III) corresponds to nearly pure 4*s*, and peak I and II to nearly pure 4*p* orbitals. The *s*-orbitals are broadened by nearest-neighbor *s*–*s* interaction into bands. The *p*-orbitals split into group I, so-called lone-pair orbitals, and group II, bonding orbitals. The bonding orbitals connect neighboring atoms within a chain, while the lone-pair, nonbonding orbitals are perpendicular to the c-axis; the structure is thus to be regarded as twofold coordinated. The conduction bands shown in Fig. 2.14 are 4*p* antibonding orbitals [2.76]. The bonding orbitals (peak II) split into two groups as a result of chain-chain interaction: electrons in the II_l group are localized in the intrachain bonds while the II_u group has a charge distribution which suggests interchain bonding [2.77].

A number of attempts have been made at calculating the electronic structure of monoclinic selenium, based simply on a molecular LCAO model and neglecting intermolecular interaction [2.77–79]. In [2.78], however, hybridized *sp* orbitals were used. As a result the nonbonding character of the lower *s*-levels was lost. In [2.79] the work was repeated using unhybridized *s*- and *p*-orbitals[1]. A density of states rather similar to that of Fig. 2.14, except for a smearing of the one-dimensional singularities in the band III and a lack of interchain splitting in II, was obtained. A calculation for Se_8 rings with inter-ring interaction was performed in [2.77]. It yielded a density of states rather similar to that of Fig. 2.14 except that peak II_l was considerably weaker than II_u. As we shall see in Sect. 2.7.3, II_l is measured to be weaker than II_u in amorphous Se.

A characteristic feature of selenium and tellurium is the presence of lone electron pairs, a result of the twofold coordination. Lone pairs should also exist in compound chalcogenides with twofold coordinated chalcogens. Such was not the case in the chalcogenides of Sect. 2.2.2 where the chalcogen was either three- or sixfold coordinated. Twofold coordination, however, occurs in a number of V_2VI_3 semiconductors such as As_2S_3. We list these compounds, which can also be prepared in amorphous form (chalcogenide glasses), in Table 2.3 with details about their crystal structures. The compounds with $C_{2n}^{3,5}$ and D_{2H}^{16} space groups have layer structures with twofold coordinated chalcogens and threefold coordinated group V elements. An LCAO band calcu-

1 The angle between bonds in Se is 104°, a fact which is incompatible with pure *p*-bonding orbitals. In [2.78] this problem was solved by using *sp* hybrids. As shown in [2.80], the same thing can be accomplished with very weak *s*- and *d*-admixture.

Table 2.3. Space groups of the stable modifications of the V_3VI_2 compounds. The number of formula units per primitive cell is given in brackets after the crystallographic system

	S	Se	Te
As	C_{2h}^2 monoclinic (4)	C_{2h}^3 monoclinic (4)	C_{2h}^5 monoclinic (4)
Sb	D_{2h}^{16} orthorhomb. (4)	D_{2h}^{16} orthorhomb. (4)	D_{3d}^5 rhombohedral (1)
Bi	D_{2h}^{15} orthorhomb. (4)	D_{3d}^5 rhombohedral (1)	D_{3d}^5 rhombohedral (1)

lation for a simple model of As_2S_3 performed by *Bishop* and *Shevchik* [2.81] shows the existence of lone-pair bands in these compounds. Crystalline As_2Te_3 has a more complicated structure with *threefold* coordinated Te [2.82], thus lone pairs are not expected in this material. Nor are they expected in Bi_2Se_3 and Bi_2Te_3. These compounds, small band gap semiconductors of technological interest as thermoelectric generators [2.83], have a rhombohedral structure with octahedrally (sixfold) coordinated chalcogens [2.84]. Their calculated band structures show no evidence of lone-pair bands [2.85].

We note that throughout the materials of Table 2.3 high coordination number, characteristic of high ionicity, is favored by the compounds with elements of high atomic number, a fact we already noted for the materials with five electrons per atom in Sect. 2.2.2.

2.2.4 Transition Metal Dichalcogenides

The dichalcogenides of the group IV B (Ti, Zr, Hf), V B (V, Nb, Ta), and VI B (Mo, W) transition metals are extremely interesting from the point of view of photoelectron spectroscopy. Some of them (ZrX_2, HfX_2, MoX_2, WX_2) are semiconductors while the compounds of the group V B elements are metals which become superconductors at low temperatures. The Ti-dichalcogenides seem to be semimetals [2.86, 87a, b], although there is no general agreement on this point [2.87c]. We shall nevertheless discuss in this chapter, devoted to *semiconductors*, all of the materials of this family, in order to provide a unified picture. A number of review articles devoted to the transition metal dichalcogenides have been published recently [2.88–90].

The crystals under consideration appear in a number of modifications (polytypes) which can be described as composed of triple sandwich-type layers, a layer of chalcogen atoms, a layer of metal atoms, and another chalcogen layer. These triple layers are held together by van der Waals forces. Thus the crystals cleave easily and, remarkably, the distance between layers can be changed considerably (by as much as 50 Å) by "intercalation" with organic molecules (e.g., pyridine) [2.92]. They can also be intercalated with alkali,

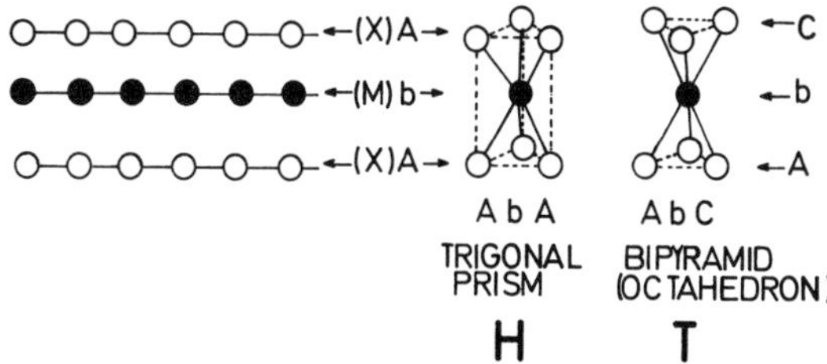

Fig. 2.15. Schematic diagram of the layer structure of the transition metal dichalcogenides MX_2 (e.g., MoS_2). The relative arrangement of the atoms within the triple layers can be either trigonal, prismatic (*H*) or bipyramidal (octahedral coordination, *T*)

Table 2.4. Most common polytypes of the chalcogenides of the group IVB, VB, and VIB transition metals, including their space groups, stacking order, and coordination (see Fig. 2.15)

Polytype	Space group	Stacking	Coordination	Typical examples
1T	D_{3d}^3	AbC–AbC	Octahedral	TiS_2, TaS_2
2H (a)	D_{6h}^4	AcA–BcB	Prismatic	MoS_2, WS_2
2H (b)	D_{6h}^4	AbA–BaB	Prismatic	NbS_2, TaS_2
3R	C_{3v}^5	AbC–BcB–CaC	Prismatic	TaS_2, MoS_2
4H (a)	D_{3h}^4	AbA–BcB–aBa–CbC	Prismatic	$TaSe_2$
4H (b)	D_{6h}^4	AbA–CbA–CbC–AbC	Octahedral-prismatic	$TaSe_2$
4H (c)	D_{6v}^4	AbA–CaC–BaB–CbC	Prismatic	$TaSe_2$
4H (d_1)	C_{3v}^1	AcA–BcA–BcB–CaB	Octahedral-prismatic	$NbSe_2$
4H (d_2)	D_{3h}^1	AcA–BcA–BaB–AcB	Octahedral-prismatic	$NbSe_2$
6R	C_{3v}^5	AbA–BcA–BcB–CaB–CaC–AbC	Octahedral prismatic	TaS_2, $TaSe_2$

alkaline earth, and other metallic atoms [2.93]. The structure of these materials is rather similar (sometimes isomorphic) to that of a number of nontransition metal compounds, such as SnS_2, CdI_2, and PbI_2, which are also of interest in photoelectron spectroscopy. Layer compounds can fully utilize the potential of angular resolved photoemission for the determination of band structures, see [Ref. 2.2, Chap. 6].

The structural modifications of the materials fall into two main categories: those in which the metal is surrounded by a trigonal prism of nearest-neighbor chalcogen atoms, and those in which these nearest neighbors form a trigonal bipyramid (Fig. 2.15) which is equivalent to a slightly distorted octahedron. We thus speak of trigonal or octahedral coordination. Octahedral coordination is characteristic of the IV B materials. As the *d* valence levels of the transition metal atom are filled, the trigonal prismatic configuration is favored, a fact which has been attributed to *d*-electron covalency [2.94]. The group V B compounds can appear, in a variety of both, octahedral and trigonal prismatic configurations, including several in which some triple layers are octahedral and some prismatic [e.g., 4H(b), 4H(d_1), 4H(d_2), 6R of Table 2.4]. The VI B

Table 2.5. Stacking modifications of the layer structure transition metal dichalcogenides. The polytype is indicated by a symbol (e.g., 2H) in which the figure gives the number of X–M–X layers per unit cell and the letter the crystal system (T ≡ trigonal, R ≡ rhombohedral, H ≡ hexagonal)

		S	Se	Te
IV B	Ti	1T	1T	1T
	Zr	1T	1T	1T
	Hf	1T	1T	1T
V B	V	1T	1T	Monoclinic (distorted octahedral)
	Nb	2H (b), 3R	2H (b), 3R, 4H (a), 4H (d_1), 4H (d_2)	Monoclinic (distorted octahedral)
	Ta	1T, 3R, 2H (b)	1T, 2H (b), 3R, 6R, 4H (a), 4H (b), 4H (c)	Monoclinic (distorted octahedral)
VI B	Mo	2H (a), 3R	2H (a), 3R	Monoclinic (distorted octahedral)
	W	2H (a), 3R	2H (a)	Monoclinic (distorted octahedral)

materials are trigonal prismatic except for the tellurides which can have a distorted octahedral structure. The various polytypes listed in Table 2.5 result from several layer arrangements (AbA ≡ prismatic, AbC ≡ octahedral, see Fig. 2.15) and a variety of unit cell sizes along the *c*-axis.

A schematic band structure for the MX_2 compounds (M ≡ transition metal, X ≡ chalcogen atom) was proposed by *Wilson* and *Yoffe* [2.88]. According to these authors the materials are ionic with the chalcogen doubly charged and the metal with configurations d^0, d^1, and d^2 in the IV B, V B, and VI B compounds, respectively. The valence bands are composed mainly of *filled* *s*- and *p*-states of the chalcogens (configuration s^2p^6) and there are *empty* conduction states originating from empty *s*- and *p*-levels of the metal (as we shall see later, these metal and chalcogen orbitals hybridize strongly along the bond directions). Between these two sets of bands, separated by a gap ($\sigma-\sigma^*$ gap) of ~7 eV, the *d* valence levels are located. These levels are split into two sets of bands by the crystal field. The metallic or nonmetallic character of the compounds is then determined in the d^0 configuration by a possible small overlap of these *d*-bands with the σ-valence bands ($TiSe_2$ is a semimetal while ZrX_2 and HfX_2 are semiconductors). The d^2 configurations lead to semiconductors if, and only if, there is a gap between the bands they occupy and the rest of the *d*-bands.

Following these ideas, semiempirical band calculations based on the tight-binding method were performed [2.95]. First principles calculations by *Mattheiss*, based on the nonrelativistic augmented plane wave (APW) method, [2.96], confirmed the essential features of this picture while suggesting a number of detailed modifications. More recent work on the band structure of these compounds includes Xα cluster calculations for MoS_2, NbS_2, and ZrS_2

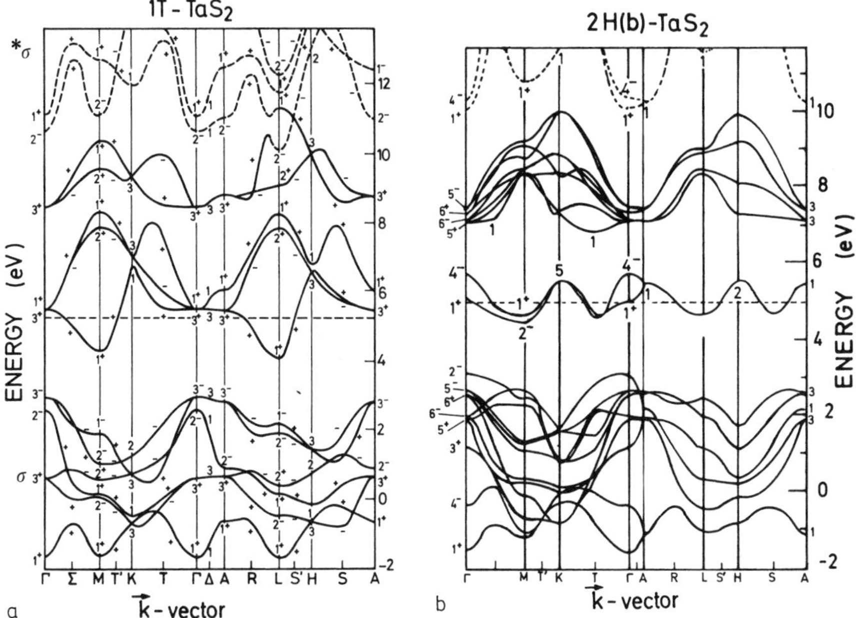

Fig. 2.16a and b. Band structures of $1T$-TaS_2 **a** and 2H–TaS_2 **b** calculated by *Mattheiss* [2.96] with the augmented plane wave (APW) method. Note that the number of bands in 2H–TaS_2 (with 2 formula units per primitive cell) is double that of the $1T$ modification (1 formula unit per unit cell)

[2.97], calculations by *Kasowski* for MoS_2 and NbS_2 using linear combinations of muffin tin orbitals [2.98], the modified KKR work of *Wood* and *Pendry* for MoS_2 [2.99], and the work of *Wexler* and *Woolley* for $NbSe_2$, NbS_2, $TaSe_2$, and TaS_2 using the layer method [2.100, 101]. Empirical pseudopotential work has been recently performed for $NbSe_2$ [2.102].

In spite of this large computational effort, a considerable quantitative uncertainty still exists about the band structure of these compounds and the energies of the gaps involved. We mention, as an example, that following the KKR work of *Myron* and *Freeman* [2.103a] TiS_2 should be a semiconductor with a gap of about 1 eV. The self-consistent OPW work of *Krusius* et al. [2.103b] leads to the same result. According to more recent works, however, this material is a semimetal [2.86, 87a]. The most recent work on this matter [2.87c] favors again semiconducting character for TiS_2, with an indirect gap of ~0.5 eV. According to [2.87b, c] $TiSe_2$, however, is supposed to be a semimetal. This example illustrates again the uncertainties of first principles band-structure calculations unless carefully checked and "touched up" against experimental data.

We shall base our discussion on the work of *Mattheiss* [2.96], which seems to be the most extensive work in the literature, involving 1T-structure HfS_2 and TaS_2, 2H (b) TaS_2 and $NbSe_2$, and 2H (a) MoS_2. The band structures calculated by this author for 1T and 2H TaS_2 are shown in Fig. 2.16. We note that there is indeed a gap of ~ 8 eV between the σ and the σ^* bands. The size of this gap and the lifting of the σ^* bands with respect to the d bands as compared with the pure metal is attributed by *Mattheiss* to chalcogen-p metal-d hybridization; the σ bands are metal-chalcogen bonding bands while the σ^* bands are antibonding.

In comparing Fig. 2.16a with Fig. 2.16b we should keep in mind that the 1T structure (Fig. 2.16a) has one formula unit per unit cell while 2H structures (Fig. 2.16b) have two (twice as many bands). Simple minded crystal field considerations suggest that the *five* d-bands of Fig. 2.16a should split into a triplet (T_{2g}: d_{z^2}, d_{xy}, $d_{x^2-y^2}$) and a doublet (E_g: d_{xz}, d_{yz}), with the triplet being lower. This triplet-doublet splitting can still be recognized in Fig. 2.16a. The T_{2g} triplet should split into a d_{z^2} singlet and a doublet in the crystal field of the 2H prismatic configuration, the d_{z^2} band being lower. This may seem to be the case in Fig. 2.16b where there is a double (2 atoms) lowest d-band well separated from the rest. *Mattheiss*, however, has pointed out that this band, while basically d_{z^2}, is strongly hybridized with d_{xy}, a fact which is actually responsible for the large gap (>1 eV) between this band and the rest of the d-manifold.

If we accept a "rigid band" point of view and fill the bands of Fig. 2.16 according to the number of d-electrons available, the 1T structure would be semiconducting in the d^0 (IV B) configuration and metallic for either d^1 or d^2. The 2H structure would be also semiconducting for d^0 (it does not occur in nature), metallic for d^1 (V B) and semiconducting for d^2. The 1T (octahedral) structure probably does not occur in the d^2 configuration as it would force the material to be metallic; the semiconducting $d_{z^2}-d_{xy}$ hybridization gap which opens in the 2H configuration should stabilize the prismatic phase.

The main reason for the increase in interest during the past years in the transition metal dichalcogenides is the presence of small distortions from the structures of Table 2.4 related to electronic instabilities (charge density waves or CDW [2.91]). The phenomenon is characteristic of the metallic or semi-metallic members of the family (all V B compounds, [2.91, 104], TiX_2 [2.86]); the distortion observed in the structure of most MTe_2 may also be related to CDW's [2.91].

The CDW phenomenon is characteristic of metals of low dimensionality. A phase transition towards an insulating or semiconducting state (Peierls transition) should always take place in one-dimensional metals at low temperatures. A typical example is the Krogmann salt $K_2Pt(CN)_4\,Br_{0.30}\cdot 3H_2O$, undergoes a broad Peierls transition which is at completed at ~ 77 K [2.105a]. The effect can be understood as due to the lowering of the electronic contribution to the cohesive energy which results if a gap opens at the Fermi surface. Such gap will open if the symmetry of the crystals is lowered by formation of an electronic charge density wave with wave vector equal to $2k_F$ (k_F: Fermi momentum). This CDW will obviously also modulate the position of the atomic cores. For a one-dimensional monoatomic metal with one atom per unit cell, one electron per atom, and lattice constant a, k_F equals $\pi/2a$. If the crystal distorts so as to double the size of the unit cell, the Brillouin zone is halved and gaps will appear

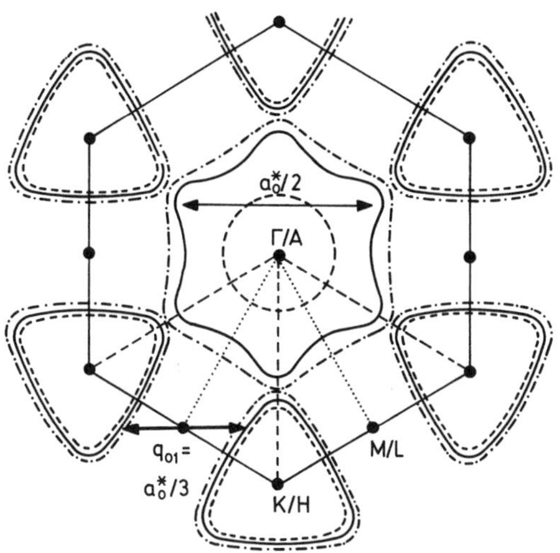

Fig. 2.17. Two cuts through the Fermi surface of $2H–TaS_2$ perpendicular to the c axis. The ΓKM section yields two sheets: (---) lower band and (–·–·–) upper band (see Fig. 2.16b), while for the AHL section both sheets are degenerate (solid line). Indicated are the vectors $q_{01} = a_0^*/3$, which produce observed commensurate charge density waves, and $a_0^*/2$, another possible, though unobserved, source of such phenomena [2.91]

at $k = \pm\pi/2a = k_F$. The periodicity of the distortion (CDW) is indeed $a = \pi/k_F$. As the critical temperature T_d for the formation of the CDW is approached from above, the zone edge phonon which corresponds to the cell distortion becomes soft and its frequency vanishes at T_d.

For two-dimensional solids one expects transitions to CDW states whenever there are flat or nearly flat portions of the two-dimensional Fermi surface. This is the situation which arises quite generally (though fortuitously!) in the transition metal dichalcogenides. We show as an example in Fig. 2.17 the two-dimensional Fermi surface of $2H–TaS_2$. The portions connected by the vectors $k_1 \simeq \pi/a_0$ and $k_2 = 2\pi/3a_0$ are flat and thus are expected to give rise to CDW's with periods $2a_0$ and $3a_0$, respectively. The $3a_0$ distortions have indeed been observed [2.91]. The fact that portions of the Fermi surface nearly match the Brillouin zone of a superlattice is called "nesting". We should mention at this point that the surface reconstructions (superlattice formation) discussed in Sect. 2.9.2 may be interpreted as due to the energy lowering associated with the formation of CDW's in 2-d systems (surfaces) [2.105b].

The charge density waves obtained in the one-dimensional case with half-filled bands are "commensurate" in the sense that their wavelength is a rational number of times the undistorted lattice constant. Under these conditions a larger unit cell can be found which guarantees the translational periodicity of the distorted crystal. The k_1 and k_2 distortions of Fig. 2.17 are at least nearly commensurate. There is, however, in principle no reason why the distortion vectors of Fig. 2.17 and similar cases should be commensurate. In the "incommensurate" case a distortion wave which is not locked to the lattice results. Commensurate CDW's give rise to sharp superstructure points in the diffraction patterns (especially electron diffraction [2.91]). These spots appear blurred if the CDW is incommensurate.

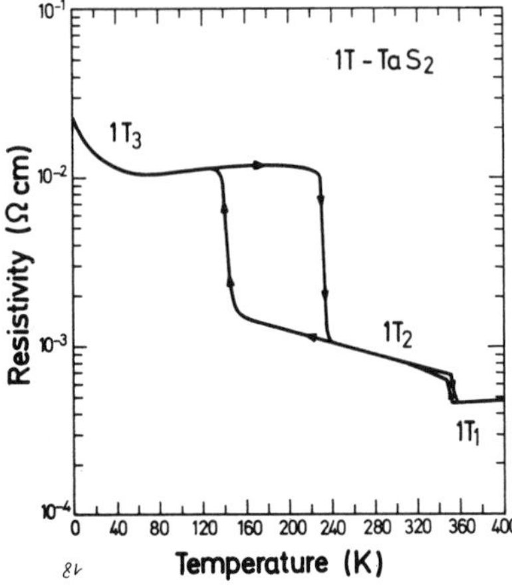

Fig. 2.18. Resistivitiy versus temperature in 1T–TaS_2 exhibiting the phase transitions associated with changes from incommensurate ($1T_1$) to commensurate ($1T_2$) charge density waves. The hysteresis indicates the first-order nature of the phase transitions [2.106a]

A number of other methods can also be used to reveal CDW's. The associated phase transitions were first observed in measurements of resistivity vs temperature. In Fig. 2.18 we show the resistivity of 1T–TaS_2 vs temperature: we see two phase transitions which give rise to three phases. Phase 1 and 2 have been shown to correspond to incommensurate CDW's [2.91] while phase 3 is a commensurate CDW. The undistorted phase, which should appear at higher temperatures, is not seen as the crystal transforms into its 2H-modification (without CDW's). We note that the commensurate CDW phase of this material does not seem to be insulating or semiconducting, as expected for a one-dimensional commensurate CDW, but rather metallic or semimetallic. In the two-dimensional case gaps need not open all around the Fermi surface and some metallic character is always found to remain. Very flat portions of the Fermi surface are rather unlikely to occur in three-dimensional solids. Nearly flat portions which do not suffice to produce CDW's are responsible for the *Kohn* anomalies in the phonon spectrum at $2k_F$ [2.106].

When commensurate CDW's appear the Brillouin zone is folded and the number of *phonon* states at $\boldsymbol{k}=0$ increases. Thus a number of new first-order Raman lines usually appear [2.107], they can be used to study the symmetry and properties of the CDW state. The soft phonons responsible for the transition appear at low frequencies. They are coupled to excitations of the electronic CDW's of which two types are possible: amplitude waves, in which the amplitude of the charge wave is modulated, and phase wave, in which the phase of the CDW is modulated with respect to the atomic lattice [2.108]. Amplitude and phase modes have been recently identified in the Raman spectrum of 2H–$TaSe_2$ [2.109], (see Fig. 2.19).

We should mention here that photoelectron spectroscopy of core levels is an ideal technique to investigate CDW's, provided the amplitude of the distortion is sufficiently large. Upon formation of a superlattice (commensurate CDW) nonequivalent positions appear for a given atom which should give rise to

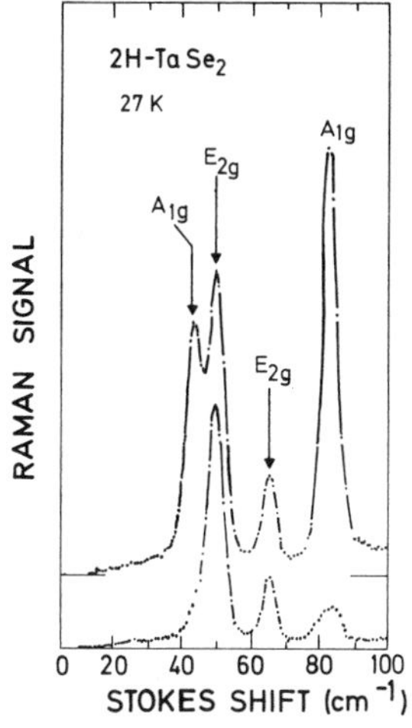

Fig. 2.19. Raman spectrum of 2H–$TaSe_2$ in the commensurate charge density waves (CDW) modification ($T_d = 35$ K), showing four peaks due to amplitude and phase CDW's [2.109]. The upper spectrum was taken for parallel and the lower one for perpendicular incident and scattered fields

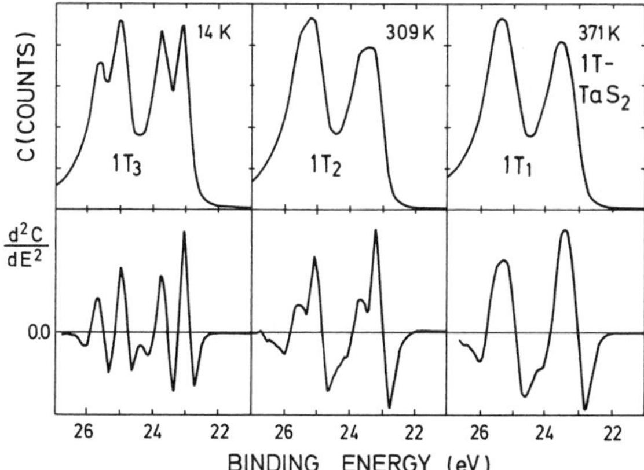

Fig. 2.20. Ta $4f$ lines of TaS_2 showing the splitting produced by commensurate charge density waves ($2T_3$ modification) [2.111]

different core shifts. The splitting pattern of the core shifts below T_d should contain valuable structural information about the number and multiplicity of the nonequivalent sites and about the sign of the electronic charge wave on the given atom: *negative* charge usually giving rise to a decrease in the binding energy of the core level, see [Ref. 2.2, Sect. 1.5.2 and 2.8.2] [2.110, 111]. A semiquantitative determination of the electronic charge on each of the equivalent lattice sites is then possible by comparison with the core shifts observed between the compound under consideration and an isomorphic compound with a different valence electron charge (for TaS_2 for instance $Fe_{1/3}Ta_{2/3}S_2$ [2.110]. Incommensurate CDW's should give rise to a *broadening* but *no splitting* of the core levels.

As an example of this type of analysis we show in Fig. 2.20 the $4f$ core levels of Ta in the $1T_1$, $1T_2$, and $1T_3$ modifications of TaS_2 [2.111]. For better resolution, the second derivatives of these spectra are also plotted. The incommensurate $2T_1$ phase shows the $4f_{7/2}$–$4f_{5/2}$ doublet with an asymmetry towards higher binding energies probably related to the Mahan-Nozières-De Dominicis effect discussed in [Ref. 2.2, Chap. 5], see also [2.110]. A slight splitting of each one of these lines appears in the $1T_2$ phase which was thought to be incommensurate on the basis of electron diffraction. This splitting appears in full in the commensurate $1T_3$ phase.

The slight splitting of the $1T_2$, presumably incommensurate phase, can be interpreted as due to a partial locking with the lattice of the incommensurate CDW, with a small coherence length of the order of the sampling (i.e., escape) depth of the photoelectrons (~ 20 Å). The commensurate phase shows two peaks of nearly equal strength. This splitting pattern is compatible with the structure (lattice constant $\sqrt{13}a_0$) proposed for the $1T_3$ phase on the basis of electron diffraction patterns [2.91]: three nonequivalent Ta sites in the proportion 6:6:1. The *single* nonequivalent site is unresolved and, in view of the sharpness of the low binding energy edges of the core levels of phase $1T_3$ (Fig. 2.20) one may conjecture that its core line would lie on the high binding energy side of the observed doublet. Thus the *single* site should have a defect of electronic charge. Using the calibration procedure mentioned above it has been concluded that the charges around the two nonequivalent sixfold sites differ by about one electron per atom [2.110].

2.3 Methods Complementary to Photoelectron Spectroscopy

Photoelectron Spectroscopy yields information about the occupied electronic states of a semiconductor. A number of other spectroscopic techniques yield related or complementary information. We discuss in the following sections the most important of these techniques: X-ray emission spectroscopy, which also yields information about occupied states with particular emphasis on their angular momentum, optical and electron energy loss spectroscopy, both of which basically yield information on combined valence-conduction density of states, and on density of empty states.

2.3.1 Optical Absorption, Reflection, and Modulation Spectroscopy

Characteristic of semiconductors is the existence of an energy gap which can be observed by optical means: if the semiconductors are pure and at sufficiently low temperatures they are transparent for photon energies below this gap and opaque above. This lowest (sometimes called fundamental) gap can be indirect, if it occurs between a valence band maximum and a conduction band minimum at different points of $\boldsymbol{k}$-space (e.g., germanium, silicon), or direct if the extrema are at the same $\boldsymbol{k}$-point (e.g., GaAs). The investigation of the details of this edge is one of the first steps towards the understanding of the electronic structure of a semiconductor. Discussions of this question can be found in any standard work on semiconductors [2.7, 112–114]. The absorption well above the edge usually becomes strong and can only be directly measured by transmission in very thin films. Measurements on bulk samples are possible by means of reflection techniques [2.7, 8] and also with partial yield photoelectron spectroscopy (see Sect. 6.3.7). The reflection coefficient is a function of the polarization and angle of incidence of the light and of the complex dielectric constant $\varepsilon(\omega) = \varepsilon_1(\omega) + i\varepsilon_2(\omega)$ which determines the linear optical response of the solid. $\varepsilon(\omega)$, usually a tensor, reduces to a scalar for cubic or amorphous materials. We shall discuss here the in which situation $\varepsilon(\omega)$ is scalar. The absorption coefficient α in this case is related to $\varepsilon(\omega)$ through

$$\alpha = \frac{4\pi\kappa}{\lambda} \tag{2.2}$$

with

$$\sqrt{\varepsilon} = n + i\kappa,$$

where n and κ are the refractive and absorption index, respectively, and λ is the wavelength of light in vacuum. Above the lowest direct edge indirect transitions are usually negligible in a *crystalline* solid. The imaginary part of $\varepsilon(\omega)$ is then given by [2.6, 8]

$$\varepsilon_2(\omega) = \frac{A}{4\pi^3\omega^2} \iint_{E_c - E_v = \hbar\omega} \frac{P^2}{|\nabla_{\boldsymbol{k}}(E_c - E_v)|} dS_k \tag{2.3}$$

$$= \frac{A}{\omega^2} \langle P^2 \rangle N_{cv}(\hbar\omega), \tag{2.4}$$

where A is related solely to fundamental constants, E_c and E_v are the energies of empty and occupied states, respectively, P is the matrix element of linear momentum between v and c and $N_{cv}(\hbar\omega)$ is the so-called combined or optical density of states for the transitions from c to v. It represents the number of states which can undergo energy and $\boldsymbol{k}$-conserving transitions for photon frequencies between ω and $\omega + d\omega$. According to (2.3) $\varepsilon_2(\omega)$ will have singularities whenever $\nabla_{\boldsymbol{k}}(E_c - E_v) = 0$, i.e., at the so-called interband critical points which usually occur at high symmetry points or lines of the Brillouin zone. The main object of optical spectroscopy of solids above the fundamental edge is the determination of interband critical points.

Equations (2.3, 4) are valid for *direct* or $\boldsymbol{k}$-conserving transitions. If a strong $\boldsymbol{k}$-conservation violating mechanism is present, like in the case of the amorphous semiconductors, transitions are possible from *any* valence state to *any* conduction state. The imaginary part of ε is then related to a convolution of the density of valence and conduction states N_v and N_c

$$\varepsilon_2(\omega) = \frac{B}{\omega^2} \langle P^2 \rangle \int N_c(\omega - \omega') N_v(\omega') d\omega'. \tag{2.5}$$

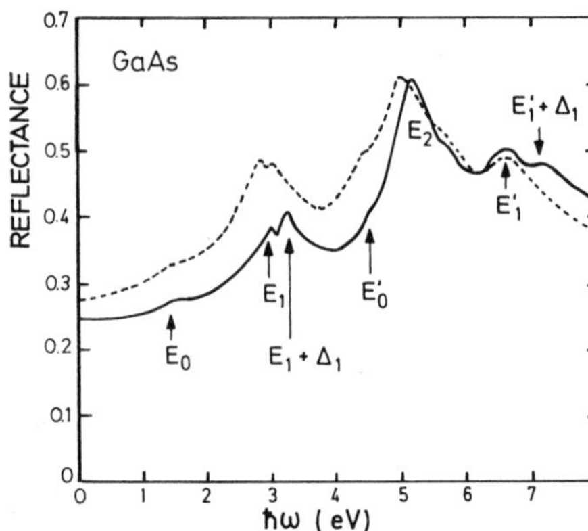

Fig. 2.21. Calculated reflectance spectrum of GaAs (solid line) compared with experimental one at room temperature. The labels indicate the standard notation for the observed critical points [2.30a]

The situation represented by (2.3–5) is similar to that encountered in the expressions for EDC's (direct vs nondirect transitions, see [Ref. 2.2, Sect. 1.5.6]).

Transitions between valence and conduction bands are observed up to about 10 eV photon energies. At higher energies one observes singularities in $\varepsilon(\omega)$ due to transitions from core levels into the conduction bands. The core levels have practically no dispersion: if E_v is replaced by their $\boldsymbol{k}$-independent binding energy E_B (2.4, 5) lead both to the result

$$\varepsilon_2(\omega)=\frac{A}{\omega^2}\langle P^2\rangle N_c(\hbar\omega-E_B)\,. \tag{2.6}$$

As illustrated by (2.6) absorption measurements from core levels yield basically the density of empty conduction states and the corresponding critical points. Angular integrated photoelectron spectroscopy (EDC's) of sufficiently high exciting photon energy yields the density of valence states. Thus the three types of measurements just mentioned, optical spectroscopy of valence to conduction band transitions (combined density of states), optical spectroscopy of core to conduction band transition (density of conduction states) and EDC measurements (density of valence states) are to be regarded as complementary. A band structure fitted to critical points in the combined density of states while representing correctly the difference between the conduction and valence bands, need not describe well the individual valence and conduction bands. It has been suggested that this situation arises in *local* empirical pseudopotential calculations: nonlocal terms have to be added to reproduce correctly the individual bands [2.30a].

As already mentioned $\varepsilon_1(\omega)$ and $\varepsilon_2(\omega)$ can be obtained by means of reflectance measurements. A number of variations of this technique are possible [2.7], the most fruitful one being the normal-incidence reflectance $R(\omega)$ followed by a Kramers-Kronig dispersion analysis in order to obtain both ε_1 and ε_2 from only one spectrum, namely $R(\omega)$. Critical points appear in the $R(\omega)$ spectrum as well as in $\varepsilon_2(\omega)$. As an example we show in Fig. 2.21 the $R(\omega)$

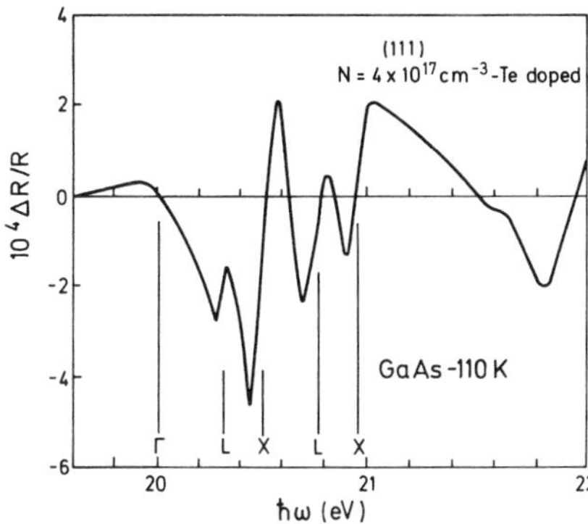

Fig. 2.22. Electroreflectance of GaAs in the photon energy region in which excitations from the Ga 3*d* core electrons to the conduction band take place. From this spectrum the relative position of the Γ, L, and X conduction minima can be determined [2.115]

spectra of GaAs as compared with calculations based on the band structure of Fig. 2.5. The structure (critical point) labelled E_0 corresponds to the lowest absorption edge. A number of other structures (E_1, $E_1+\Delta_1$, E'_0, E_2, E'_1) have been identified in terms of critical points, as shown in Fig. 2.5.

The critical points observed in reflectance spectra are usually not very sharp or well resolved (see Fig. 2.21). Resolution is considerably improved if a derivative spectrum is measured directly by means of one of the many available modulation techniques [2.8]. The most popular one, electroreflectance, measures the modulation of the reflectance spectrum produced by a small electric field sinusoidally varying in time. We show in Fig. 2.22 the electroreflectance spectrum of transitions from Ga 3*d* core levels in GaAs measured by *Aspnes* with synchrotron radiation [2.115]. The lowest conduction minima at Γ, L, and X were identified as labelled in this spectrum (incidentally, these measurements solved a controversy involving the relative order of the X and L minima). The L and X structure appears twice as a result of the $3d^{5/2}$–$3d^{3/2}$ spin-orbit splitting of the core levels. The resolution obtained for this splitting is much better than usually found with photoelectron spectroscopy [2.34].

2.3.2 Characteristic Electron Energy Losses

The term characteristic electron energy losses (CEL) applies to the spectroscopic technique in which monoenergetic (or nearly monoenergetic) electrons are transmitted through or reflected by a sample. The energy spectrum of the outgoing electrons is then measured with an electron energy analyzer. From this CEL spectrum real and imaginary parts of the dielectric constant can be determined. The technique has been reviewed in [2.9]. The energy spectrum of the transmitted electrons is basically proportional to

$$+\frac{1}{\boldsymbol{q}_2}\operatorname{Im}\left\{\frac{1}{\boldsymbol{q}\cdot\overleftrightarrow{\varepsilon}(\omega,\boldsymbol{q})\cdot\boldsymbol{q}}\right\}=\frac{1}{q_2}\frac{\boldsymbol{q}\cdot\overleftrightarrow{\varepsilon}_2\cdot\boldsymbol{q}}{|\boldsymbol{q}\cdot\overleftrightarrow{\varepsilon}\cdot\boldsymbol{q}|^2}\,. \tag{2.7}$$

In (2.7) $\omega = \Delta E/\hbar$, with ΔE the energy loss suffered by the electrons under consideration as measured by the energy analyzer and $\boldsymbol{q}$ the wave change in vector between incident and outcoming electrons. The dielectric constant $\overleftrightarrow{\varepsilon}(\omega, \boldsymbol{q})$ which enters into (2.7) differs from that relevant for optical experiments by its dependence on $\boldsymbol{q}$:$\overleftrightarrow{\varepsilon}(\omega, \boldsymbol{q})$ is, in principle, a nonisotropic tensor even for cubic materials. In the optical case the $\boldsymbol{q}$-dependence is usually negligible since the wave vector of the light is negligible compared with the characteristic solid state wave vectors, i.e., the dimensions of the Brillouin zone.

Most of the CEL measurements which have so far been performed in semiconductors were nevertheless taken for $q \simeq 0$, i.e., for nearly colinear incident and transmitted beams. Hence the results can be compared directly with those of optical measurements. However, it is possible in principle to measure the CEL spectra for large values of $\boldsymbol{q}$. The corresponding dielectric constant must then be calculated from the band structure of the material with a version of (2.4), suitably modified to take care of the fact that $\boldsymbol{q}_c = \boldsymbol{q}_v + \boldsymbol{q}$, rather than $\boldsymbol{q}_c = \boldsymbol{q}_v$ (skew direct transitions). A few measurements under conditions for which $q \neq 0$ have been recently reported for semiconductors, but their quantitative interpretation is not yet available [2.116].

While optical absorption measurements yield basically ε_2, CEL measurements yield $\mathrm{Im}\{\varepsilon^{-1}\}$, (2.7), a fact related to the *longitudinal* nature of the perturbation produced by an electron beam. From the latter quantity (or tensor) $\overleftrightarrow{\varepsilon}$ can be obtained by means of a Kramers-Kronig dispersion analysis [2.9]. The frequency region of maximum accuracy, however, is different for optical and for CEL measurements. Optical transmission measurements are optimal near the fundamental edge, while reflectivity yields accurate values up to the frequency at which ε_2 reaches its maximum. CEL measurements are most sensitive in the region for which $\mathrm{Im}\{\varepsilon^{-1}\}$ is highest, i.e., in the region around the plasma frequency of the valence electrons. This plasma frequency is given by [2.30e]

$$\omega_p = \sqrt{\frac{4\pi N e^2}{m\varepsilon_0}},$$

where N is the valence electron concentration and ε_0 the long wavelength infrared dielectric constant. ω_p lies around 15 eV for most semiconductors, see [Ref. 2.9, Fig. 26]. The plasma frequency is not found as accurately from optical reflectance measurements as from CEL. $\varepsilon(\omega)$ in the region of interband transitions, however, is given rather poorly by CEL spectroscopy (the resolution of CEL spectroscopy is determined basically by the linewidth of the incident electrons, $\sim$0.3 eV for thermionic non-monochromatized electrons used in many of the experiments). On the other hand the region of core transitions can be investigated rather well with CEL as in this region $\varepsilon_1 \simeq 1$, a fact which enhances $\mathrm{Im}\{\varepsilon\}$. So far, however, no measurements of core to conduction transitions with a resolution which compares with that obtained with synchrotron radiation (see Fig. 2.22) has been reported with CEL.

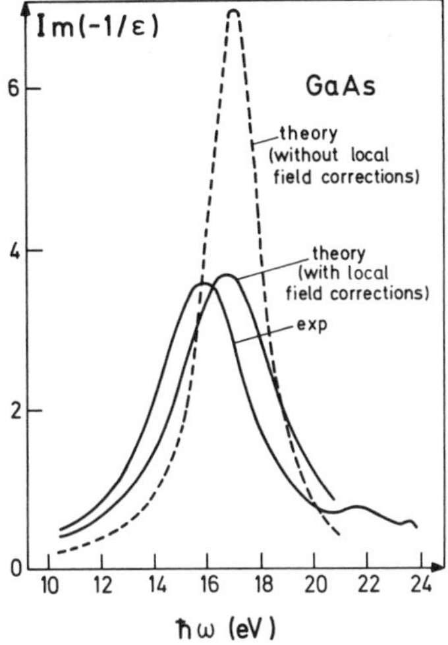

Fig. 2.23. Plasmon characteristic losses measured for GaAs [2.9] and calculations performed by *Sturm* (private communication) with and without local field corrections. The inclusion of local field corrections is found to be vital for the agreement between theory and experiment

The phenomenon of CEL is very closely related to the plasma satellites which appear in photoelectron spectroscopy especially for core levels: the photoexcited electrons on their way to the vacuum suffer characteristic energy losses. These are responsible for the *extrinsic* plasmon losses observed in XPS and discussed for metals in Chap. 7. The plasma frequency ω_p determined from XPS measurements of core levels for many semiconductors (2.34, 49, 117, 118] agrees well with the results of direct CEL measurements. We should mention in closing this section that direct calculations of $\mathrm{Im}(\varepsilon^{-1})$ based on (2.4) do not reproduce the experimental result well. This discrepancy has been recently attributed to the absence of local field corrections in (2.4) [2.119a]. A recent simple pseudopotential calculation for germanium and zincblende-type semiconductors which includes local field corrections has been able to remedy this difficulty [2.119b] (see Fig. 2.23).

2.3.3 X-Ray Emission Spectroscopy

In this technique holes are produced in core levels by means of either electron impact or illumination with X-rays. These holes are filled by electrons from higher occupied levels with emission of energy conserving photons. Of particular interest here is the emission associated with transitions between the valence bands and core levels since it carries information about the density of valence states and their atomic (angular momentum) composition. The excitation can occur, as mentioned, by means of high energy electrons in X-ray tubes or with X-ray photons. The former, which can be made much stronger than the latter, has the disadvantage of heating the sample. Also, the small penetration depth of the electrons mixes the volume information with surface information in a way similar to that discussed for photoelectron spectroscopy.

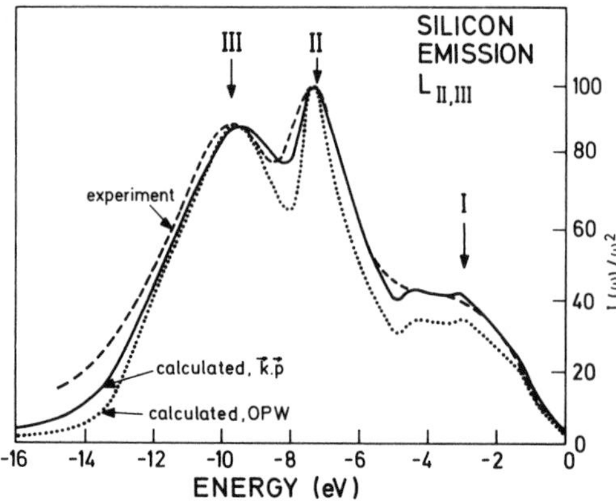

Fig. 2.24. $L_{\text{II}-\text{III}}$ soft X-ray emission spectrum measured in silicon, compared with the results of two calculations: OPW and empirical $\boldsymbol{k}\cdot\boldsymbol{p}$ [2.127, 128]

Synchrotron radiation appears to be promising for photon excited X-ray emission spectroscopy (X-ray fluorescence) [2.120]. A serious difficulty with the technique is poor resolution, especially when the emission is associated with deep $K(1s)$ levels; this poor resolution is due to the natural width of the core levels and the limited resolution of the X-ray spectrometer.

The field of X-ray emission spectroscopy was very active during the 1930's and 1940's. For a review of the early work the reader should consult [2.121, 122]. A large number of emission spectra obtained up to 1955 are given in [2.123], and a discussion of more recent work is found in [2.124–126].

The emission probability $I(\omega)$ is given by a formula similar to (2.4) [2.127]

$$I \propto \omega^2 \iint_{E_\text{v}-E_\text{c}=\hbar\omega} \frac{|P(\boldsymbol{k},E)|^2}{|\nabla_{\boldsymbol{k}} E_\text{v}(\boldsymbol{k})|} ds_k \tag{2.8}$$

where E_c is the energy of the corresponding core level and P the transition matrix element.

We show in Fig. 2.24 the $L_{\text{II},\text{III}}$ spectrum (transitions from valence band to $2p$ core holes) of silicon as observed by *Wiech* [2.127] and as calculated by *Klima* [2.128] from the band structure of this material. These spectra, which agree remarkably well, show three peaks, I, II, III, analogous to those observed in the density of valence states (see Fig. 2.2) but with different strengths. Peak I in the density of states contains two of the four electrons per atom and hence one half of the total density of states. In Fig. 2.24, however, it appears only as a relatively weak shoulder. This fact can be attributed to the p nature of the corresponding core levels which requires s or d valence states if P^2 is not to vanish. The valence bands contain little admixture of d states and their I component hardly any s states, hence it should be weak. Actually, the spectra of Fig. 2.24 can be regarded as the s-like component of the density of valence states.

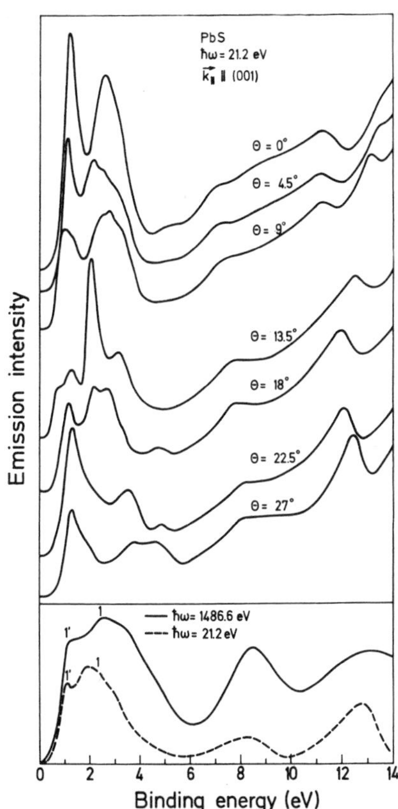

Fig. 2.25. Angle-resolved uv photoemission spectra, angle-integrated X-ray photoemission spectrum, and uv photoemission spectrum (of polycrystalline PbS [2.130]). In the angular resolved data the polar angle of the escaping electrons θ is varied while the azimuth is kept constant. In contrast to Fig. 2.24 the top of the valence band lies in this figure at the left of the horizontal scale. Both types of conventions are used throughout this work as throughout the literature. It would have been impractical to use a unified convention for the large number of figures presented in this work

Likewise spectra due to transitions from valence to *s*-levels should represent the *p*-like component of the density of valence states. Such is the case for the *k* emission bands (valence $\rightarrow 1s$) measured and calculated by *Drahokoupil* and *Simůnek* [2.129] for Ge, GaAs, and ZnSe.

2.4 Volume Photoemission: Angular Integrated EDC's from Valence Bands

The guidelines for the interpretation of EDC's from valence bands are discussed in [Ref. 2.2, Sect. 1.6]. The valence band spectra, although broad (they extend over a typical binding energy range from 0 to 15 eV) are highly structured especially for crystalline semiconductors. Most of the available data were taken in an "angle-integrated" mode.

It has been recently recognized that angular integrated UPS washes out a lot of details which appear in the corresponding angular resolved spectra (see for instance Fig. 2.25, [2.130], and [Ref. 2.2, Chap. 6]). As we shall see later this is easy to understand: angular resolved spectra exhibit basically sharp *one-dimensional* critical points as the component of $\boldsymbol{k}$ parallel to the surface ($\boldsymbol{k}_{||}$) is conserved in the photoemission process. These critical points shift in energy as $\boldsymbol{k}_{||}$ is varied through variation of the escape angle. Thus in an angle-integrated measurement the sharp one-dimensional singularities are washed out giving rise to broader characteristic three-dimensional structures.

Angular resolved photoemission is discussed in detail in [Ref. 2.2, Chap. 6]. We present below in Sect. 2.6 a few specific examples of recent work on the lead chalcogenides using this powerful technique which, in principle, makes possible the determination of the complete energy dispersion relations $E_v(\boldsymbol{k})$ in the case of two-dimensional solids (surfaces, layer structures). The discussion in the present section is mainly confined to angular integrated work.

The formal theory of photoemission was discussed in [Ref. 2.2, Chap. 2]. After a number of drastic assumptions the basic formula for the three-step model of direct transitions was obtained, [Ref. 2.2, Eq. 2.65]

$$\begin{aligned}\langle j_x\rangle = &\frac{2\pi e}{\hbar}\mathscr{A}\sum_{\alpha\alpha}\left(\frac{2A}{2mc}\right)^2\int\frac{2}{(2\pi)^3}d\boldsymbol{k}\,N[E_n(\boldsymbol{k})]l_{\alpha'}T_{\alpha\alpha'}\\ &\cdot\delta[E_{n'}(\boldsymbol{k})-E_n(\boldsymbol{k})-\hbar\omega]\sum_{\mu,\nu}M_\mu M_\nu^*,\end{aligned} \tag{2.9}$$

where $\mathscr{A}$ is the illuminated area, $\boldsymbol{A}$ the vector potential of the radiation $N[E_n(\boldsymbol{k})]$ the density of initial states $(n,\boldsymbol{k})$ $T_{\alpha\alpha'}$ the transmission probability for an electron at the surface to leave the bulk channel α' to go into the vacuum channel α (escape function), $l_{\alpha'}$ the transmission function of the excited electron from the average point of excitation to the surface, $E_n(\boldsymbol{k})$ and $E_{n'}(\boldsymbol{k})$ the energy of initial and final states, respectively and M_μ the dipole matrix elements weighted by $A_\mu/|A|$ (μ represents a coordinate axis). In (2.9) $\langle j_x\rangle$ is the angular integrated photocurrent perpendicular to the surface. The transmission function $l_{\alpha'}$ (labelled T in [Ref. 2.2, Eq. (1.83)]) is given by

$$l_{\alpha'}=\frac{\alpha(\omega)\lambda_{\alpha'}}{1+\alpha(\omega)\lambda_{\alpha'}}, \tag{2.9a}$$

where $\alpha(\omega)$ is the absorption coefficient of the light and $\lambda_{\alpha'}$ the escape depth of the photoexcited electrons.

We now transform (2.9) into the standard three-step model expression for direct transitions [Ref. 2.2, Eq. 1.86] involving only matrix elements of $\boldsymbol{p}$

$$\begin{aligned}&\langle j(E,\omega)\rangle\propto\sum_{nn'}d^3\boldsymbol{k}\,|\langle n'|\boldsymbol{p}\cdot\boldsymbol{A}|n\rangle|^2\\ &\cdot\delta[(E_{n'}(\boldsymbol{k})-E_n(\boldsymbol{k})-\hbar\omega)]\,\delta[(E_{n'}(\boldsymbol{k})-E)],\end{aligned} \tag{2.10}$$

with n and n' the initial and final state, respectively, and E the photoelectron energy. In order to carry out this transformation we must take $l_{\alpha'}$ and $T_{\alpha\alpha'}$ out of

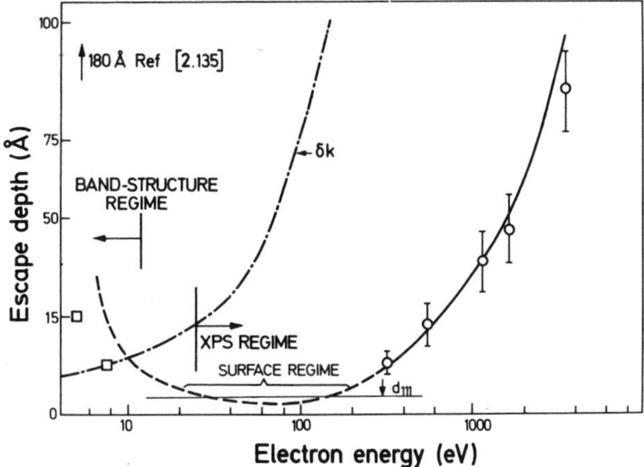

Fig. 2.26. Escape depth of photoelectrons in silicon as a function of their kinetic energy measured with respect to the vacuum level [2.131–135]. We indicate in the curve the three important regimes of photoemission. The surface regime, with surface states yielding a sizeable contribution, XPS-regime, in which the density of bulk valence states is basically investigated, and band-structure regime, with effects of $\boldsymbol{k}$ conservation in the excitation process being dominant.

the integral in (2.9) (i.e., assume they are constant). The consequences of assuming $\lambda_{\alpha'}$ and $T_{\alpha\alpha'}$ constants (isotropic model) will be examined in Sect. 2.4.1.

We have plotted in Fig. 2.26 the escape depth $\lambda(E)$ (averaged for all states with energy E) of silicon as measured by several authors [2.131–135]. This curve seems to be rather universal for most solids (see [Ref. 2.2, Fig. 4.10] and also [Ref. 2.132, Fig. 4]). Thus the discussion which follows should apply to most semiconductors. We see in Fig. 2.26 that for energies between 20 and 200 eV λ is of the order of d_{111}, the lattice constant along [111] (the cleavage surface of silicon is perpendicular to [111]). In this case one would expect the photoemission to be affected strongly by surface effects (e.g., photoemission from occupied surface states), which we shall discuss in Sects. 2.9.1–3. Figure 2.26 shows clearly, however, that the surface regime is reached only marginally and therefore one also expects a substantial bulk contribution even in the 20–200 eV region. This fact is borne out by experiments. It has actually been shown that surface-states structure is strongly dependent on the surface under consideration and can be made to disappear by averaging the spectra of different surfaces [2.136]. Consequently surface-states structure is expected to be strongly washed out in measurements on polycrystalline materials [2.34]. Along the same lines it may be argued it should be absent for amorphous materials unless some preferential cluster orientation occurs at the surface.

The finite $\lambda(E)$ produces a smearing of the component of $\boldsymbol{k}$ perpendicular to the surface, $k_\perp$, which can be interpreted as an imaginary part in $k_\perp = k_\perp^{(1)} + ik_\perp^{(2)}$ with $k_\perp^{(2)} \simeq \lambda_{\alpha'}^{-1}$. $k_\perp$ will cease to be conserved in (2.9, 10). The contribution to $\langle j_x \rangle$ in (2.9) for a given direction of $\boldsymbol{k}_{||}$ yields then under the assumption of constant $l_{\alpha'}$, $T_{\alpha\alpha'}$, and M_{μ} the density of one-dimensional states ($k_{\perp}$ variable) for the $\boldsymbol{k}_{||}$ under consideration. Integration of this density of states over all outgoing directions of $\boldsymbol{k}$ (all $\boldsymbol{k}_{||}$) should then produce the density of valence states. This regime is called the XPS regime (see Fig. 2.26); an estimate of the electron energies at which this regime applies was given in [Ref. 2.2, Sect. 1.6.2, Eq. (1.102)] and in [2.133]. At high electron energies the quasi free-electron bands form a tight mesh in $\boldsymbol{k}$-space as they are folded over into the reduced zone. The separation δk of the energy degenerate bands is given in [Ref. 2.2, Eq. (1.102)] for a typical case which yields for Si $\delta k \simeq 15$Å at 20 eV. The energy dependence of the $(\delta k)^{-1}$ so estimated is also plotted as a dot-dashed line in Fig. 2.26. For $\lambda \ll (\delta k)^{-1}$ $(E > 25\,\mathrm{eV})$ one has the XPS regime while for $\lambda \gtrsim (\delta k)^{-1}$ $(E < 15\,\mathrm{eV})$ one has the band structure regime. The region between 15 and 25 eV is to be considered as a transition region.

We have so far simplified our discussion of the XPS regime by assuming that the matrix elements $\langle n'|\boldsymbol{p}|n\rangle$ can be considered constant and taken out of the summation in (2.10). This will not be the case in general and the EDC's will represent the density of valence states modulated by a factor which corresponds to the average energy dependence of the matrix element. This effect is best discussed by representing the valence states in an LCAO basis: the various atomic wave-function components will have different matrix elements of $\boldsymbol{p}$ to the final state which can be estimated from atomic calculations or experimental cross sections for atoms (see [Ref. 2.2, Chap. 3]). This effect will obviously depend on the orientation of $\boldsymbol{A}$ (i.e., on angle of incidence and polarization of the light) with respect to the emitting surface. This dependence has been discussed for copper and silicon in a recent theoretical paper by *Aleshin* et al. [2.137]. These authors showed that for a single crystal silicon surface

$$\langle j(E,\omega)\rangle \propto \sigma_s(\omega) N_s(E) + \sigma_p(\omega) N_p(E), \tag{2.11}$$

where $\sigma_{s,p}$ are the photoionization cross sections for the 3s and 3p atomic levels of silicon and $N_{s,p}$ the partial densities of valence states. Equation (2.11) coincides with the result for polycrystalline material. The atomic cross sections σ for Al K_{α} excitation are shown in Fig. 4.2.

Equation (2.11), generalized to include more than two atomic orbitals, should remain valid for all polycrystalline materials. It breaks down, however, for single-crystal surfaces whenever d-electrons or more complicated orbitals are present, as shown in [2.137] for copper. Polycrystalline materials have also the advantage of ruling out interference terms between the photoionization cross sections of the constituent atoms [2.138]. An equation of the type of (2.11) can be used to determine the partial densities of states $N_s(E)$, $N_p(E)$, $N_d(E)$, ... if the corresponding σ's are known, from measurements of the EDC's for a series of values of $\hbar\omega$ (see Sect. 2.4.4).

The amount of EDC data obtained for semiconductors in the past twenty years is staggering. In the impossibility of discussing all these data we shall confine ourselves here to giving typical examples in order to illustrate general methods and concepts.

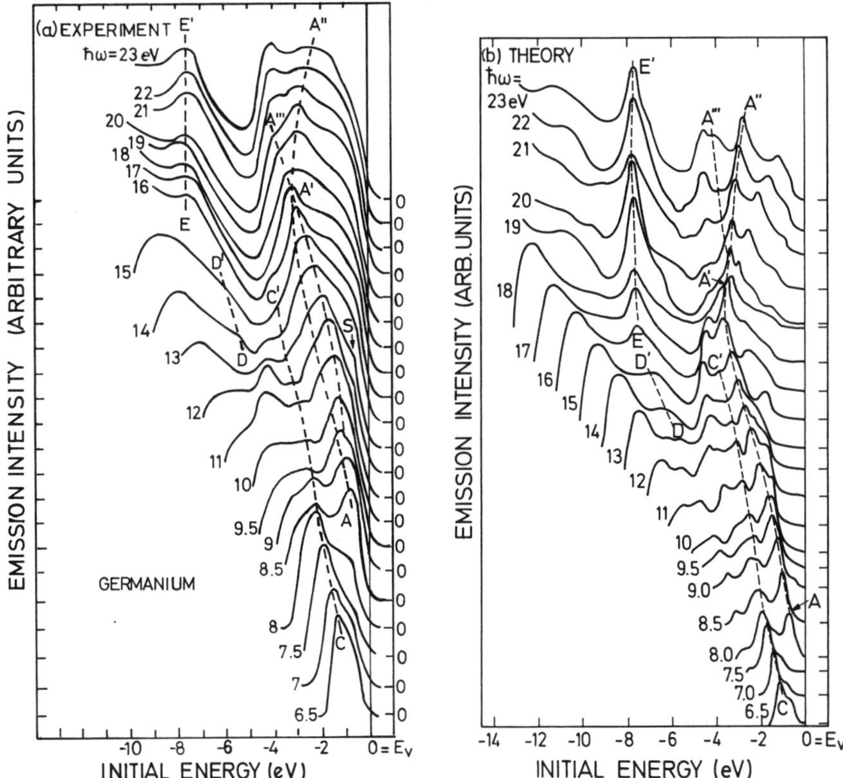

Fig. 2.27. a Experimental EDC's of germanium measured for several photon energies from 6.5 to 23 eV, showing the transition from band structure to XPS regime. **b** EDC's calculated as described in text with a nonlocal EPM band structure [2.133]

2.4.1 Band-Structure Regime: Germanium

Although considerable work has been reported for germanium in the band-structure regime, [2.139, 140] the most complete experimental and theoretical treatment is that of *Grobman* et al. [2.133]. This work was performed on clean Ge surfaces with synchrotron radiation for exciting energies between 6 and 25 eV. By using a cesiated surface the work can be extended to photon energies down to 3.2 eV [2.140]. Figure 2.27a shows the EDC's measured by *Grobman* et al. for several photon energies. Most of the peaks observed shift to higher binding energy with increasing photon energies up to ~ 18 eV. We thus conclude that from 6.5 to 18 eV we are in the band-structure regime. Between 18 and 23 eV the EDC's remain stable and thus we must be already in the XPS regime. The shoulder at ~ -0.8 eV initial energy, labelled *S* on the 13 eV EDC, is due to surface states (see Sect. 2.9.2). We should point out that extremely

Table 2.6. Band energies relative to $E_v(\Gamma_{25'})$ ($L_{25'}$ is the weighted average of the Γ_8^+ and Γ_7^+ valence ban so as to remove the spin-orbit splitting $\Delta_0 = 0.3$ eV). Figures 2.2 and 2.28 define the notation. The results ar for band positions at 300 K. The experimental data are from optical and photoemission work. Error estimate are in parentheses

	Valence bands						Conduction bands							
	Γ_1	$L_{2'}$	L_1	$\Sigma_{1\,\min}$	X_4	$L_{3'}$	L_1	$\Gamma_{2'}$	X_1	Γ_{15}	L_3	$L_{2'}$	L_c^{upper}	X_c^{uppe}
Experiment														
UPS[a]														
(I) From XPS regime	−12.6 (0.3)	−10.6 (0.5)	−7.7 (0.2)	−4.5 (0.2)	−3.15[c] (0.2)	−1.4 (0.3)								
(II) Direct-transition analysis (band structure regime)					−3.2 (0.2)	−1.5 (0.2)	0.8 (0.2)	1.0 (0.1)	1.3 (0.2)	3.25 (0.1)	4.3 (0.2)	7.8 (0.6)	12.6 (1.0)	13.8 (0.6)
XPS[b]	−13.0	−10.3	−7.2											
Indirect gap[c]							0.66							
Theory														
Nonlocal pseudopotential[a]	−12.85	−10.68	−7.72	−4.73	−3.24	−1.52	0.85	0.97	1.08	3.34	4.30	7.97	12.58	13.20

[a] [2.133].
[b] Excitation with monochromatic AlK_α radiation [2.147].
[c] [2.142a].

good vacuum conditions are required for these measurements: a monolayer coverage of oxygene, achieved in a few hours for a vacuum of 10^{-10} Torr, removes the *S*-structure from the spectra. Oxygen coverage also produces a broad peak at initial energies ~ -5 eV which rapidly obliterates the germanium structure for $\hbar\omega \simeq 20$ eV, the cross section σ_p of oxygen being much larger than that of germanium at 20 eV.

Grobman et al. performed an elaborate theoretical calculation of the observed EDC (see Fig. 2.27b) using a nonlocal pseudopotential band structure: it was impossible to fit all of the available optical and photoemission data with a local pseudopotential, a fact which has been noted by a number of other authors [2.30a, 141]. The Schrödinger equation corresponding to the nonlocal pseudopotential is (neglecting spin-orbit interaction)

$$[T + V_L + V_Z + \alpha E_n(\boldsymbol{k})]\,\Psi_{n,\boldsymbol{k}}(\boldsymbol{r}) = E_n(\boldsymbol{k})\,\Psi_{n,\boldsymbol{k}}(\boldsymbol{r}), \tag{2.12}$$

where V_L is the standard local pseudopotential represented by three adjustable parameters v_3, v_8, v_{11} [2.30e], α is a constant which approximates self-energy effects due, at least in part, to the energy dependence of the correlation energy, and V_Z represents the nonlocal angular momentum dependence of the pseudopotential. It is chosen to have only an $l=2$ component, so as to represent the fact that there are $3d$ core levels close to the conduction and valence bands. The function v_2 is approximated by

$$v_2 = \begin{cases} A_2 P_{l=2} & \text{for} \quad r < R_2 \\ 0 & \text{for} \quad r > R_2 , \end{cases} \tag{2.13}$$

where $P_{l=2}$ is the projection operator for the $l=2$ component of $\Psi(r)$.

The calculation of EDC's of *Grobman* et al. in the band-structure (i.e., $\boldsymbol{k}$ conservation or direct transitions) regime is based on (2.9) and represents an attempt to include the dependence of $l_{\alpha'}$ and $T_{\alpha\alpha'}$ on electron energy and direction rather than taking these parameters out of the integral. The dependence of $l_{\alpha'}$ on electron parameters has two sources, which appear clearly as $\lambda = \tau(E) v_{\perp}(\boldsymbol{k})$ is replaced into (2.9a). The energy-dependent scattering time $\tau(E)$ is mainly determined by electron-hole pair production [2.135]. The group velocity $v_{\perp}(\boldsymbol{k}) = dE(\boldsymbol{k})/dk_{\perp}$, which depends strongly on the electron direction, can be easily obtained from the band structure. The surface transmission coefficient $T_{\alpha\alpha'}$ is assumed to be determined exclusively by conservation of energy $\boldsymbol{k}_{\|}$, the component of $\boldsymbol{k}$ parallel to the surface and $k_{\perp}$ for a given plane-wave component of $\Psi_{n,\boldsymbol{k}}$ must be sufficiently large to surmount the potential barrier at the surface. If this is the case the electron is transmitted into the corresponding vacuum channel, otherwise not. If one expands $\Psi_{n,\boldsymbol{k}}$ in plane waves (pseudowave functions) some components will be transmitted and some not, depending on n and $\boldsymbol{k}$. Thus a dependence of $T_{\alpha\alpha'}$ on n and $\boldsymbol{k}$ will result. These EDC calculations also include inelastically scattered electrons (secondaries) computed by the method described in [2.142]. In this manner the curves of Fig. 2.27b were generated: they represent the experimental data better than those calculated on the basis of the isotropic model, i.e., (2.10) (see [Ref. 2.133, Fig. 21]). In order to determine the six parameters of the pseudopotential (v_3, v_8, v_{11}, α, R_2, A_2) *Grobman* et al. noted that several peaks observed in $\varepsilon_2(\omega)$ and in the EDC's while arising from transitions over a large area of the Brillouin zone, occurred always at energies close to high symmetry critical points. The parameters were chosen so as to optimize the agreement between theory and experiment for these peaks. The critical points so determined are listed in Table 2.6.

We discuss now the identification of some of these critical points in connection with Fig. 2.27 and the work at lower photon energies performed on cesiated surfaces [2.132, 140]. We show in Fig. 2.28 the band structure of Ge along lines on the hexagonal face of the Brillouin zone. This figure illustrates the various identified transitions which occur at or near the L-point. The E_1' transitions were early identified in optical spectroscopy [2.7, 8]. They appear in the EDC's as a sharp peak for a photon energy of 5.8 eV coreresponding to an electron energy of ~4 eV (see Fig. 2.29). The energies are measured with respect to the top of the valence band, spin-orbit splitting removed. As shown in Fig. 2.28 the average energy of the excited electrons (4 eV) lies somewhat below the L_3 point. By adding the corresponding shift, taken from the band-structure

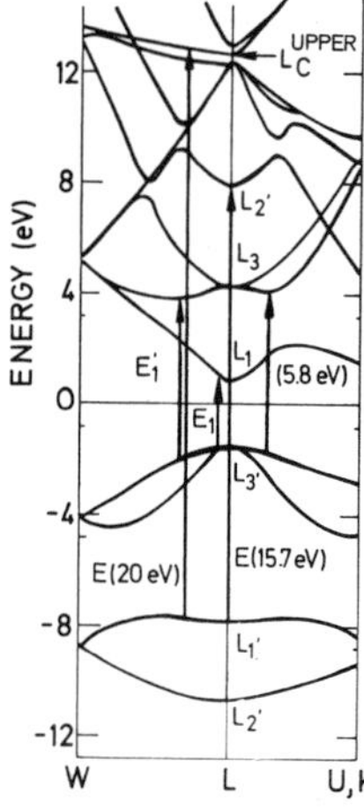

Fig. 2.28. Band structure of germanium (nonlocal EPM) at L and along symmetry lines of the hexagonal face of the Brillouin zone. The main transitions observed in the optical constants and in photoemission around the L point are indicated by arrows

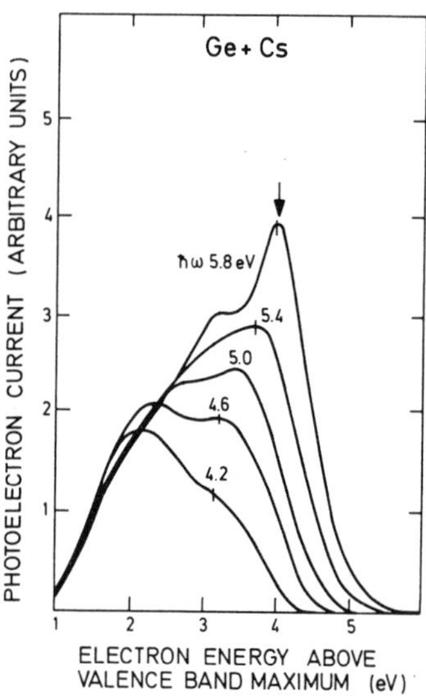

Fig. 2.29. EDC's of cesiated germanium in the low $\hbar\omega$ band-structure regime. The arrow signals the peak which appears for $\hbar\omega = 5.8$ eV ($\simeq E_1'$ gap) for a final state $\simeq L_3$ [2.132]

calculation, we find $L_3 = 4.3$ eV (see Table 2.7). According to Fig. 2.28 one would expect strong structure in the EDC's corresponding to the $L_1 \rightarrow L_{2'}$ transitions since the L_1 valence band is nearly flat. The corresponding peak appears in Fig. 2.27, both a and b, at $\hbar\omega = 16$ eV (E) and remains at the same *initial* energy (dashed line $E - E'$), a consequence of the flatness of the L_1 valence band. From the photon energy for the onset of peak E one finds $E(L_1 \rightarrow L_{2'}) = 15.7$ eV and from the corresponding electron energy $E(L_{2'}) = 7.8$ eV. As shown in both Figs. 2.27a and b the E peak remains nearly

Table 2.7. Ratios of the height of peak I to that of peak III in Fig. 2.30 as found experimentally for Al K_α excitation, calculated with plane waves (PW) [2.143] and orthogonalized plane waves (OPW) final states [2.144]. Also, corresponding ratio of the peaks in the density of states. We have listed the ratio of photoionization cross sections obtained from [2.143–145]

		C	Si	Ge
$\frac{I_I}{I_{III}}$	XPS (experimental)	0.45	1.4	2.8
	XPS (theory, PW)	0.27	1.65	0.56
	XPS (theory, OPW)	—	—	2.4
$\frac{N_I}{N_{III}}$	Density of valence states	2.6	2.3	2.0
	σ_p/σ_s (elements)	0.03	0.5	1.1
	σ_p/σ_s (hydrocarbons)	0.08	0.3	1.0

constant in strength up to ~19 eV to rise again at ~20 eV. The second rise is attributed to the $L'_1 \rightarrow L_c^{\text{upper}}$ transitions of Fig. 2.28.

We conclude this discussion by mentioning that the A-peak, when it starts at $\hbar\omega = 8$ eV in Fig. 2.27, is due to transitions along the [110] (Σ) direction near Γ. It moves to higher k's along Σ as $\hbar\omega$ increases and reaches its maximum initial binding energy at the X_4 valence state (Fig. 2.2). It thus enables the determination of the X_{4v} and the X_c^{upper} energy (X_c^{upper} is a region near the X point where accumulation of nearly-free-electron-like bands occurs, similar to L_c^{upper} of Fig. 2.28).

2.4.2 XPS Regime: Tetrahedral Semiconductors

We show in Fig. 2.30 the EDC's of single-crystal diamond, silicon and germanium obtained in the XPS regime, with AlK_α radiation (1486.6 eV) for the three materials [2.145] and for germanium also the data of [2.133] obtained with $\hbar\omega = 25$ eV. The spectra show a strong similarity to the corresponding densities of valence states (Fig. 2.4) except for a deformation which reduces the ratio of strengths of peak I to peak III in diamond and silicon and enhances it for the 25 eV spectrum of germanium (see Table 2.7). This effect has been attributed to a modulation of the density of states by photoionization cross section which basically depends on the atomic composition of the valence wave functions [see Sect. 2.2.1 and (2.11)]. One may say, in principle, that whether a given cross section is enhanced or suppressed depends basically on whether the oscillations of the atomic radial wave function do or do not match the oscillations in the final-state wave function (determined by the magnitude of $\boldsymbol{k}$). Calculations of this effect have been performed for Al K_α exciation by *Aleshin* et al. [2.143, 144]

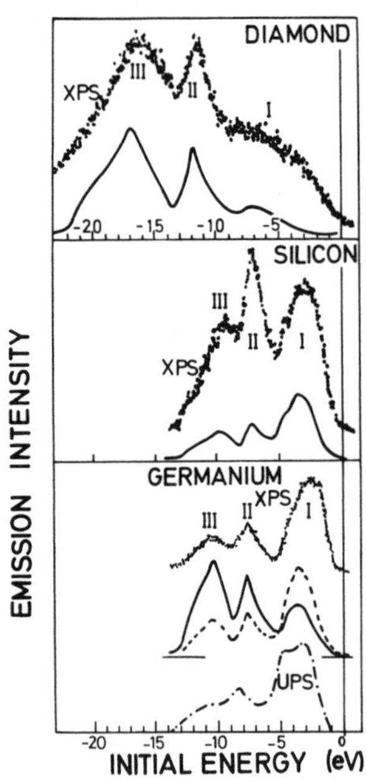

Fig. 2.30. XPS spectra of diamond, silicon, and germanium (dots) compared with the results of tight-binding valence bands calculation with a plane wave final state (solid line) and, in the case of germanium, also with an OPW final state (dashed line), [2.143, 144]. Also, UPS valence spectrum of germanium obtained with synchrotron radiation for $\hbar\omega = 25$ eV (dash-dotted line) [2.133]

using LCAO functions for the valence bands and plane waves (PW) for the final states. The results of these calculations are compared with experiments in Fig. 2.30 and in Table 2.7. For diamond and silicon agreement between the experimental and the calculated curve is good, but not so for germanium. A more elaborate calculation by *Nemoshkalenko* et al. [2.144] using orthogonalized plane waves (OPW) as the final state is able to explain this discrepancy (see dashed curve in Fig. 2.30): contrary to the case of diamond and silicon the $3d$ core levels of germanium contribute sufficiently to the orthogonalization of the final state to render the PW approximation inaccurate. These calculations also yield the partial s and p densities of states N_s and N_p and their contributions to the EDC's I_s and I_p. From these results the ratio of photoionization cross sections

$$\frac{\sigma_p}{\sigma_s} = \frac{I_p}{I_s} \frac{N_s}{N_p}$$

was obtained. It is listed in Table 2.7 together with the ratios σ_p/σ_s measured directly by *Cavell* et al. [2.145] for CH_4, SiH_4, and GeH_4. The agreement between both sets of σ_p/σ_s ratios is reasonable. Semiquantitative agreement with the results of Fig. 4.2 also exists. Table 2.7 documents clearly that the reason for the variation of I_I/I_{III} from C to Ge is solely the change in σ_p/σ_s: peak I is almost purely p-like while peak III is almost purely s-like. The Ge EDC for Al K_α excitation is, fortuitously, a nearly undistorted density of states, as $\sigma_p/\sigma_s \simeq 1$.

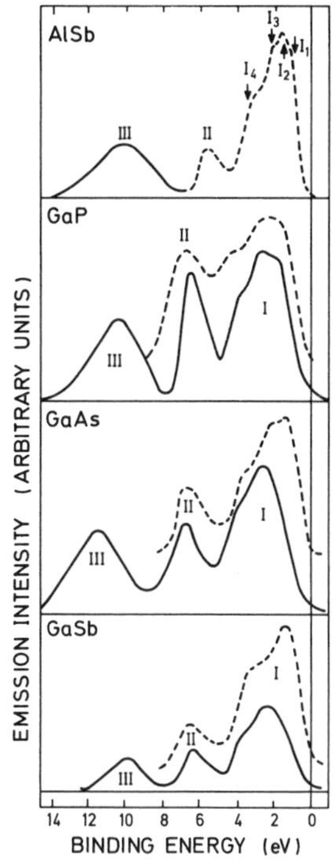

Fig. 2.31. XPS spectra of crystalline GaP, GaAs, GaSb [2.147] and AlSb [2.149] (solid) and UPS spectra (dashed) obtained with He I (21.2 eV) radiation [2.34]. The XPS spectra were excited with monochromatized X-rays with the exception of AlSb

Figure 2.30 also shows the EDC of germanium for $\hbar\omega = 25$ eV after subtracting a surface state peak which occurs between 0 and 2 eV (at $\hbar\omega = 25$ eV we are in the surface regime). Peak III is strongly suppressed in this EDC. This result applies also to silicon and other semiconductors: peak III can hardly be seen with 21.2 eV He I excitation although it is noticeable at 40.8 eV (He II) [2.34]. Failure to see the peak due to the *s* valence electrons for 21.2 eV excitation in semiconductors has led to errors in the interpretation of electronic structures [2.146].

The EDC's of a large number of III–V and II–VI zincblende-type semiconductors have been measured in the XPS regime by a number of authors, the most extensive work being that of *Ley* et al. [2.147], performed with monochromatized Al K_α radiation, *Shevchik* et al. [2.34], with He I and He II excitation, and *Eastman* et al. [2.148] with synchrotron radiation. The work of *Shevchik* et al. was performed on both, polycrystalline and amorphous materials. Figure 2.31 shows the spectra of GaP, GaAs, and GaSb, obtained both with monochromatized X-rays and with uv radiation. The AlSb data in this figure are the composite results of unmonochromatized X-ray excitation (peak III) and uv excitation (peaks I and II).

Figure 2.32 shows X-ray (monochromatized) and uv excited spectra for InP, InAs, and InSb. In the spectra presented in Figs. 2.31, 32 the secondary electron

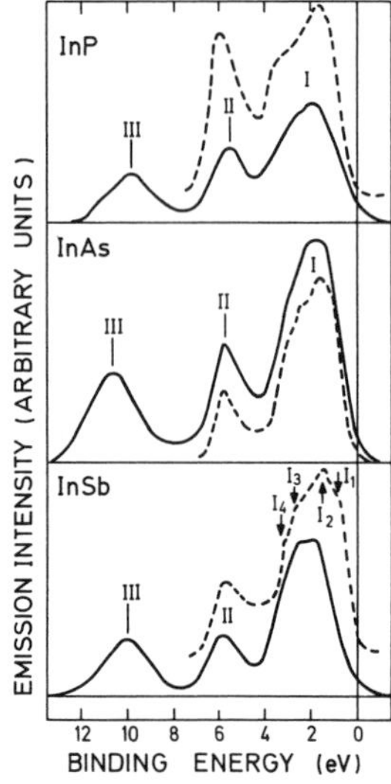

Fig. 2.32. XPS spectra of crystalline InP, InAs, InSb [2.147] (solid) and UPS spectra (dashed) obtained with NeI (16.9 eV) radiation [2.34]

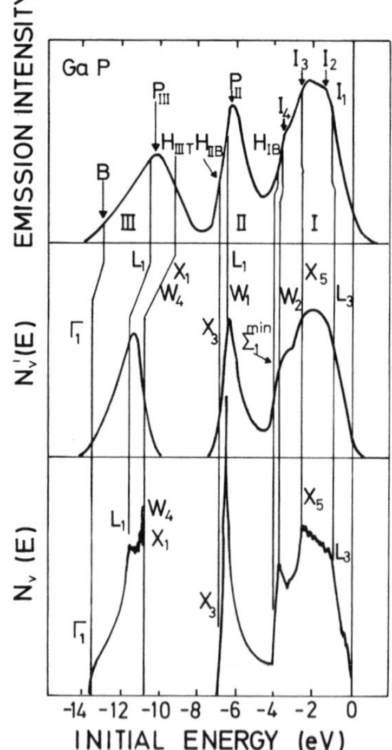

Fig. 2.33. XPS spectrum of the valence band of GaP compared with the calculated (nonlocal EPM) density of valence states $N_v(E)$ and a broadened version of $N_v(E)$ labelled $N_v'(E)$ [2.147]

Table 2.8. Energies (eV) of the various features observed in the EDC's of the III–V compounds in the XPS regime with Al K_α excitation (XPS[a,c]), with uv-excitation (UPS[d]) and with uv-synchrotron radiation excitation (UPS[e]). For comparison we also show the results of nonlocal pseudopotential calculations[f]. For AlSb only local pseudopotential results are available[b]

Structure	I_1	I_2	I_3	I_4	H_{IB}	P_{II}	H_{IIB}	H_{IIIT}	P_{III}	B
Critical point	L_3	L_3	X_5	W_2	Σ_1^{min}	$L_1 - W_1$	X_3	$X_1 - W_4$	L_1	Γ_1
AlSb XPS[a]			2.2			5.8	6.3	8.7	10.0	13.4
AlSb UPS[a]	0.9	1.3	2.0	3.5	3.8	5.8	6.1			
AlSb theory[b]	0.8		1.9		2.7					
GaP XPS[c]	1.7		2.4	3.5	4.0	6.5	7.0	9.1	10.3	13.4
GaP UPS[d]	1.15		2.9	4.3	4.7	6.8	7.7		10.6	
GaP UPS[e]			1.6	3.6	4.3	6.5	7.2		10.2	12.3
GaP theory[f]	1.1		2.7	3.7[e]	4.2	6.8	7.1	9.5	10.6	13.0
GaAs XPS[c]	1.8		2.4	3.8		6.6	7.5	10.0	11.4	14.4
GaAs UPS[d]	1.2		2.2	3.9	4.7	6.7	7.2			
GaAs UPS[e]	0.8				4.1	6.9		10.2	11.2	12.9
GaAs theory[f]	1.20	1.42	2.9	3.5[e]	4.2	6.8	6.9	9.9		12.1
GaSb XPS[c]	1.7		2.1	3.4	3.7	6.4	7.1	9.2	10.0	11.9
GaSb UPS[d]	1.0	1.4	2.1	3.8	4.1	6.5	7.1			
GaSb theory[f]	1.0	1.45	2.5	2.6[c]	3.4	6.25	6.8	9.3	10.2	12.0
InP XPS[c]			1.8	2.7	3.6	5.4	6.2	8.6	9.7	11.6
InP UPS[d]	1.0		2.1	3.5	4.1	6.0	6.7			
InP theory[f]	0.94	1.09	2.1	2.5[c]	3.4	5.8	6.0	8.9	9.7	11.4
InAs XPS[c]	1.7		2.1	3.0	3.5	5.8	6.3	9.4	10.5	12.6
InAs UPS[d]	1.0	1.35	2.3	3.2	3.6	6.0	6.7			
InAs theory[f]	1.0	1.26	2.4	2.7[c]	3.4	6.2	6.6	10.2	10.9	12.7
InSb XPS[c]	2.0		2.5	3.2	3.5	5.9	6.5	8.9	10.0	12.0
InSb UPS[d]	0.9	1.35	2.3	3.0	3.5	6.1	6.4			
InSb UPS[e]	1.05		2.0	3.3	3.65	6.1	6.5	9.0	9.8	11.2
InSb theory[f]	0.96	1.44	2.35	2.8[c]	3.2	5.9	6.4	9.2	9.95	11.7

[a] [2.149]. [c] [2.147]. [e] [2.148].
[b] [2.150]. [d] [2.34]. [f] [2.30a], unless otherwise indicated.

tails were removed in the conventional way [2.147]. Figure 2.33 shows a typical EDC for Al K_α excitation, that of GaP, compared with its density of states, unbroadened $[N_{v'}(E)]$ and broadened $[N_{v'}(E)]$, so as to better simulate the EDC. The three groups of peaks, I, II, III, correspond clearly to the top, middle, and lowest sets of bands, respectively. Some similarity with the EDC's of germanium and silicon (Fig. 2.30) exists, but in the present case peaks II and III are well resolved: the band structure exhibits a gap between the corresponding bands. The critical points $L_3, X_5, W_2, \Sigma_1^{min}, L_1 - W_1, X_3, X_1 - W_4, L_1$, and Γ_1 of $N_v(E)$ correspond to easily identifiable features in $N_v'(E)$ and the EDC. A com-

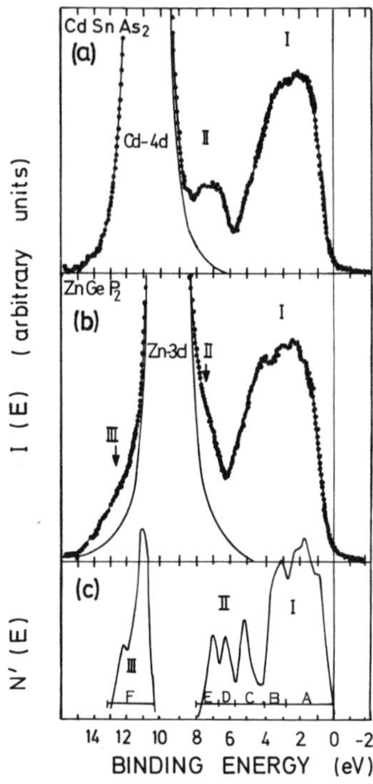

Fig. 2.34. XPS EDC's of $CdSnAs_2$ and $ZnGeP_2$, compared with a broadened version of the density of states of Fig. 2.10. For the origin of the peaks labelled *A–E* see text [2.56]

parison of $N_v(E)$ and $N'_v(E)$, however, shows that the features in $N'_v(E)$ are shifted with respect to the energy of the critical points in $N_v(E)$ by an amount between 0 and 0.5 eV. These shifts can be applied as a correction to the experimental features in the EDC so as to obtain "experimental energies" for the critical points just mentioned. This procedure was applied to the XPS data in [2.147]. In view of the uncertainties involved in this correction, and the scatter in the experimental data, we list in Table 2.8 uncorrected experimental characteristic features obtained with Al K_α [2.147, 149], He I [2.34], and synchroton radiation [2.148]. The theoretical data represent probably the most up-to-date band structures and were obtained with a nonlocal empirical pseudopotential similar to that discussed in Sect. 2.4.1. The labelling of the structures is a combination of the notation of [2.34, 147]: The substructure of peak I is labelled I_1, I_2, I_3, I_4, and H_{IB} (H_{IB} stands for the **H**alf height of the last stretch at the **B**ottom of peak I). Note that peaks I_1 and I_2 are only resolved in the antimonides and in InAs. In view of Fig. 2.33 we assign the I_1-I_2 structure to the L_3 valence band critical points and the I_1-I_2 splittings to the $L_{4,5}-L_6$ spin-orbit splitting of L_3. The experimental values found in Table 2.8 for this splitting agree within experimental error with the splitting of the $E_1-E_1+\Delta_1$ optical transitions [2.8]. We also note in Table 2.8 that the I_1 peak occurs consistently at lower energies in UPS than in XPS. A possible reason for this systematic discrepancy could be the contribution of surface states to I_1 (see Fig. 2.27a). We note, however, that the UPS energies for I_1 are in rather close agreement with the calculated values of L_3 although broadening shifts L_3 towards the position of I_1 in the XPS spectra [2.147]. EDC's for the II–VI compounds of zincblende and wurtzite structure can be found in [2.147, 148, 151].

We show in Fig. 2.34 the XPS spectra of the chalcopyrite compounds $CdSnAs_2$ and $ZnGeP_2$ measured with Al K_α radiation [2.56]. The interpretation of these spectra is hampered by the strong Cd-4*d* and Zn-3*d* core lines around 11 and 10 eV binding energy, respectively. These lines overlap with, and obscure part of, the valence band contribution to the EDC's. We have also shown in this figure (*c*) a suitable broadened version of the density of valence states of $ZnGeP_2$ calculated with a local empirical pseudopotential and presented in Fig. 2.10. In spite of the overlap of the *d*-core lines we can recognize in the experimental spectrum of $ZnGeP_2$ the three classical peaks I, II, and III of the tetrahedral semiconductors. Peak III cannot be resolved in the $CdSnAs_2$ spectrum, in part because the *d*-line has shifted by ~ 1 eV to higher binding energies. Also, peak III should be weaker in $CdSnAs_2$, where it is mainly As-4*s* like, than in $ZnGeP_2$ for which it is P-3*s* like: Arsenic is similar to germanium, phosphorus to silicon. We have shown above (Table 2.7) that the peak III is stronger in Si than in Ge.

Peaks I in Fig. 2.34a, b show considerable fine structure which can be roughly divided into the regions A and B of Fig. 2.34c. Also, the experimental peaks are broader than the calculated ones, a fact known to be related to the use of a local pseudopotential [2.30a]. The experimental peaks I are asymmetric, broader towards the high-binding energy side. This side may hide the C component of peak II in the broadened density of states. *Varea* et al. [2.56] have performed a calculation of the charge distribution associated with *each* of the regions A, B, C, D, E, and F of the valence bands, using pseudowave functions. This work shows that the A–B splitting of peak I corresponds to the difference in binding energies of Ge–P (B) and Zn–P (A) bonds, the Ge–P ones being more covalent and thus more tightly bound. Peak C contains some Zn-4*s* component plus some Ge–P bond-charge. Peak D contains Ge-4*s* plus some Zn–P bonding charge, while E is Ge-4*s* plus some antibonding charge. It is unfortunate that the presence of the *d*-peak prevents the complete observation and assignment of the C–D–E peaks. Peak III (F) is, as usual in tetrahedral semiconductors, anion *s*-like.

As discussed in Sect. 2.2.1, the Mg_2X (X = Si, Ge, Sn, Pb) compounds, with antifluorite structure, are electronically similar to germanium. The band structure of Mg_2Si was shown in Fig. 2.11. We display in Fig. 2.35 the EDC of the valence bands of polycrystalline Mg_2Si measured with monochromatized Al K_α radiation and the density of states which corresponds to the empirical tight-binding (ETB) bands of Fig. 2.11. The tight-binding parameters were adjusted so as to reproduce the bands of [2.152] and then varied so as to obtain the best fit between the calculated density of states and the EDC's. The lowest of the ETB valence bands required for the best fit to the EDC (Fig. 2.35) differs considerably from the pseudopotential results. It is difficult, however, to

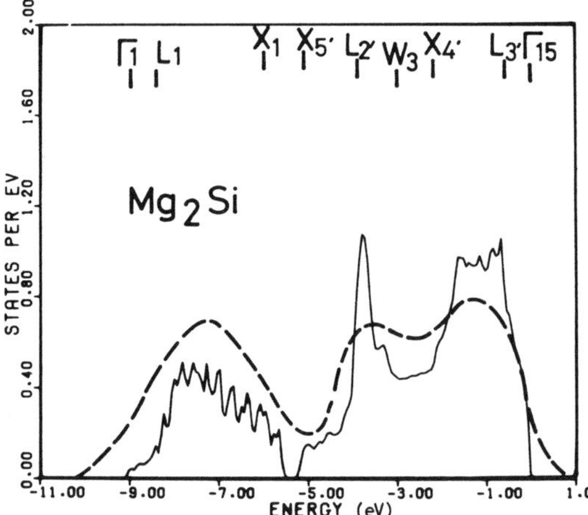

Fig. 2.35. XPS spectrum of the valence bands of Mg_2Si (dashed curve), as compared with the density of states calculated with the tight-binding method (solid line) (see Fig. 2.11). The energies of relevant critical points are also indicated [2.61a]

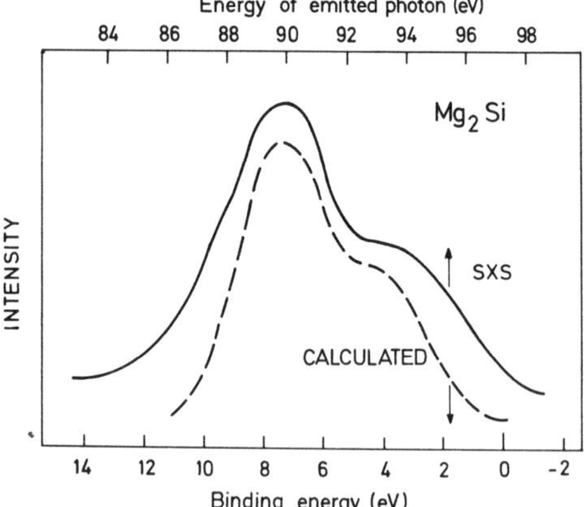

Fig. 2.36. Soft X-ray emission spectrum (Si $L_{II,III}$) of Mg_2Si (solid line) as compared with the calculated partial density of *s* valence states [2.61a]

evaluate the significance of this difference since few other experimental data are available concerning this band. Figure 2.36 shows the Si $L_{II,III}$ soft X-ray emission spectrum of Mg_2Si [2.153] which, as discussed in Sect. 2.3.3, should represent the density of Si *s*-states. The calculated partial density of Si *s* valence states, shown also in Fig. 2.36, agrees well with the experimental data. However, the calculated curve has been Lorentzian broadened by 2.5 eV to account for spectrometer resolution and core level width, a fact which makes the observed agreement a rather insensitive test of the quality of the band structure.

2.4.3 XPS Regime: IV–VI Compounds

The lead chalcogenides have been extensively investigated with photoemission techniques. They possess rock-salt structure and cleave easily, thus providing

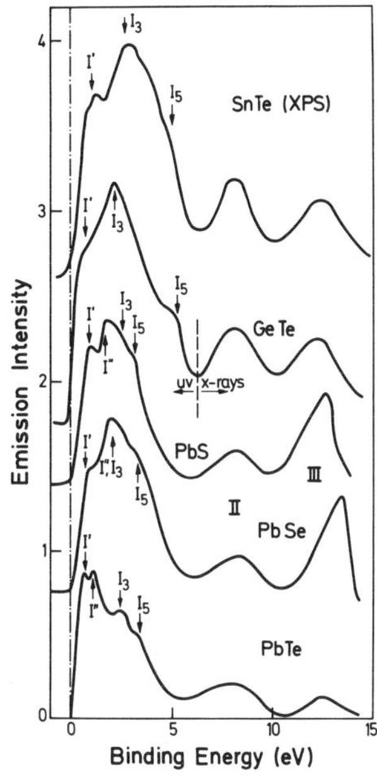

Fig. 2.37. EDC's of the valence bands of PbS, PbSe, PbTe (21.2 eV excitation) SnTe (X-rays), and GeTe (21.2 eV and X-rays) after correction for secondary electrons. The origin of energies has been set at the top of the valence band

good surfaces for photoemission studies. Also, they can be prepared quite easily as epitaxial thin films with either (100) or (111) surface orientation [2.154]. They are not very sensitive to oxygen: typical exposures of $\sim 10^{10}$ Langmuir are required for monolayer coverage [2.155]. Hence they yield trustworthy EDC's even with relatively poor vacuum conditions.

The early work was performed in the "band-structure" regime ($\hbar\omega < 12$ eV) [2.156–158]: initial state energies were observed to shift with photon energy (see, for instance [Ref. 2.158, Fig. 3]). Work with He radiation excitation, at 21.2 and 40.8 eV, [2.159] shows that at these photon energies the initial energies of the peaks observed are stable and thus the XPS regime has been reached. No evidence for effects of surface states has been reported for these materials. The transition region for $\hbar\omega$ from 12 to 21 eV has not been explored, nor has a detailed analysis been performed, similar to that of [2.133] for germanium, of the band-structure regime for EDC's of IV–VI compounds. We thus restrict our discussion here to the XPS regime.

We show in Fig. 2.37 the uv spectra ($\hbar\omega = 21.2$ eV) of PbS, PbSe, PbTe, [2.159], SnTe [2.160], and GeTe [2.161]. With the exception of GeTe all of these materials have rock-salt structure. We shall neglect in our discussion the slight rhombohedral distortion of GeTe and treat it on the same footing as the others. The zero of energy of Fig. 2.37 has been chosen by extrapolation of the steeply rising upper part of the EDC. A comparison with the band structure and the *unbroadened* density of states of Fig. 2.13 suggests that this procedure may lead to some error in determining the top (L_6^+ point) of the valence bands:

because of the small gap associated with this point the corresponding density of states is weak and should appear as a tail above the extrapolation to zero of the steeply rising part. This extrapolation corresponds, in Fig. 2.13, to the Σ_5 maximum of the valence band which lies 0.3 eV below L_6^+. However, as we shall see, the density of states of Fig. 2.13 must be broadened by about 0.25 eV in order to compare it with the upper structure in the EDC: this corrects, at least in part, the error made in the determination of the top of the valence band by the above mentioned extrapolation. An error of about 0.1 to 0.2 eV, however, is excepted to remain. In order to avoid this problem in [2.162] the binding energies were referred to the Fermi level.

The EDC's of Fig. 2.37 show three main features labelled I, II, and III. The structure III shown in Fig. 2.37 for PbS and PbSe seems to be anomalously large: according to the discussion in Sects. 2.2.2 and 2.4.2 this structure, which is due to the *s* valence electrons of the chalcogen, should be very weak. The strong III peaks of PbS and PbSe in Fig. 2.35 are likely due to an improper correction for secondary electrons: they occur very near the maximum of secondary electrons in the uv spectra. This also produces errors in the position of the peaks.

The upper structure I of the EDC's of Fig. 2.37 shows considerable substructure which has been labelled I', I'', I_3, I_5 according to [2.161, 162]. The binding energies at which these structures and the corresponding ones observed in XPS occur [2.162] are listed in Table 2.9. We show in Fig. 2.38 a comparison of the XPS and UPS EDC's of PbTe with the broadened version of the density of states of Fig. 2.13 (the broadening parameter was chosen to increase with photon energy as indicated in the caption). We see in this figure a very good correspondence between the position of the observed and the calculated peaks, in particular for the I', I'', I_3, and I_5 UPS structures. The weakness of peaks II and III in the UPS spectrum with respect to the corresponding peaks in the density of states is what one would qualitatively expect on the basis of the discussion in Sects. 2.2.2 and 2.4.2: peak III is pure Te-4*s* while peak II is Pb-5*s* with some admixture of Te-4*p*. A detailed critical point analysis of the structure in the density of states (i.e., constant energy contours in $\boldsymbol{k}$ space) has not been performed. It is possible, however, by inspection of the large number of published band-structure calculations [2.68, 69] and of Figs. 2.13, 38, to make some reasonable assignments. For this purpose we list in Table 2.9 the position of critical points which are possible candidates for these assignments. They were obtained experimentally from angular resolved photoemission whenever possible [2.70, 163]: these values are considered to be the most reliable ones available in view of the scatter in existing calculations. The remaining critical point energies were taken from typical calculations judged to be reliable.

By examining Table 2.9 and Fig. 2.13 we note that peak I falls close to the upper critical points along $\Sigma(\Sigma_5)$, $W(W_6-W_7)$, and $\Delta(\Delta_6)$. Actually the Δ and W critical points account well for the energy of the I' peaks in all compounds but

Table 2.9. Binding energies (in eV) of the critical points observed in peak I of the EDC's of PbS, PbSe, PbTe, SnTe, and GeTe as compared with the results of angular resolved photoemission (when available) or with several band calculations. All UPS data and the XPS data for SnTe and GeTe are referred to the top of the valence bands. The XPS data for the lead salts are referred to the Fermi energy. The numbers in brackets indicate that the critical point does not seem to correspond to the observed structure under which it is listed. The symmetry notation is that of the double group except for W_3 and K_3

	PbS	PbSe	PbTe	SnTe	GeTe
I' UPS	1.05[a]	0.8[a]	0.65[a]	0.95[b]	0.9[c]
I' XPS	1.3[d]	1.2[d]	0.7[d]	1.2[b]	—
Σ_5	1.0[e]	[0.4[e]]	[0.35[e]]	[0.6[g]]	[0.4[f]]
$W_3(W_6-W_7)$	1.7[e]	1.05[e]	0.75[e]	1.2[g]	0.8[f]
Δ_6	1.6[e]	1.1[e]	0.7[e]	1.3[g]	0.8[f]
I'' UPS	1.8[a]	—	1.1[a]	—	—
I'' XPS	2.4[d]	1.9[d]	—	—	—
Γ_8^-	2.7[e]	2.05[e]	1.4[e]	2.0[g]	1.4[g]
$L_{4,5}^+$	2.7[e]	1.95[e]	1.2[e]	1.8[g]	1.0[g]
I_3 (UPS)	2.65[a]	2.2[a]	2.4[a]	2.7[b]	2.4[c]
$I_3-(I_4)$ (XPS)	2.6 (2.9)[d]	2.3 (3.0)[d]	2.4[d]	2.9[b]	2.3[c]
Γ_6^-	2.95[e]	2.65[e]	2.6[e]	2.8[g]	2.25[g]
L_6^+	2.6[e]	2.15[e]	1.9[e]	2.3[g]	[1.6[g]]
I_5 UPS	3.45[a]	3.4[a]	3.3[a]	5.0[b]	5.4[c]
I_5 XPS	3.7[d]	3.6[d]	—	5.0[b]	—
$W_3(W_6-W_7)$			3.2[f]	4.8[g]	4.4[f]
$K_3(K_5)$	3.6[i]	3.5[h]	3.3[f, h]	5.8[g]	5.5[g]
X_6^-	[4.2[e]]	3.7[e]	3.4[e]	5.4[g]	5.5[g]

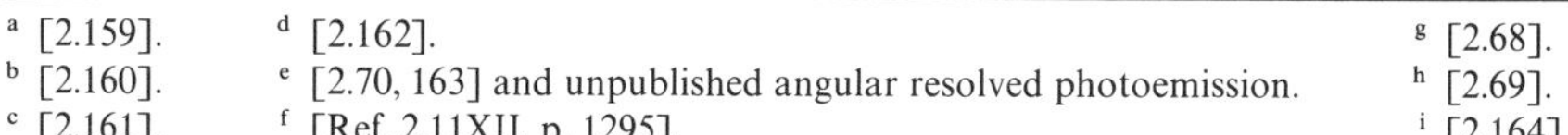
[a] [2.159]. [b] [2.160]. [c] [2.161]. [d] [2.162]. [e] [2.70, 163] and unpublished angular resolved photoemission. [f] [Ref. 2.11XII, p. 1295]. [g] [2.68]. [h] [2.69]. [i] [2.164].

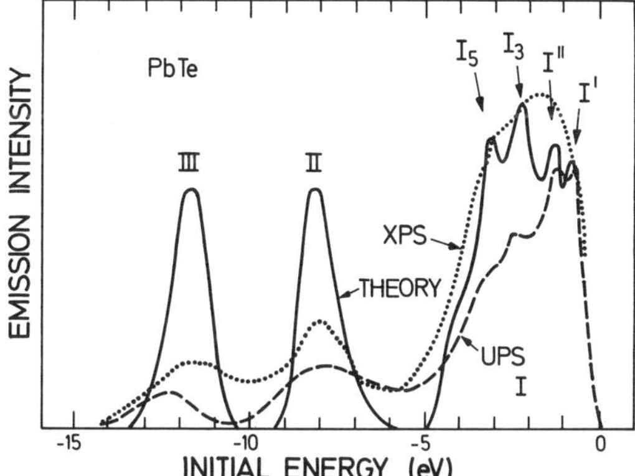

Fig. 2.38. XPS [2.162] and UPS [2.161] spectra of PbTe compared with the density of states of Fig. 2.13, lorenzian broadened with a broadening parameter of 0.25 eV for peaks I, 0.7 eV for peak II and 1.0 eV for peak 3 (solid curve) [2.69]

PbS. If the experimental values of I' in Table 2.10 are correct one would have to assign this peak in PbS to the Σ_5 critical point and thus break the systematic trend of all other compounds of the group. Actually, a density of states calculation for PbS (see [Ref. 2.164, Fig. 13]) yields an I' peak at the energy of the Δ_6 critical point. We therefore suspect that the top of the valence band was

Table 2.10. Binding energies (in eV) of peaks II and III observed with UPS and XPS for PbS, PbSe, PbTe, SnTe, and GeTe as compared with representative values of the calculated energies of the corresponding W_6-K_5 and W_7-K_5 critical lines. The energy references are the same as in Table 2.9

	PbS	PbSe	PbTe	SnTe	GeTe
II UPS	8.3[a]	8.3[a]	8.3[a]	7.7[b]	8.3[c]
II XPS	8.4[d]	8.6[d]	8.2[d]	7.7[b]	8.3[c]
W_6-K_5	6.5[e]	7.3[e]	7.4[e]	7.3[e]	8.3[e]
		8.0[f]	8.0[f]	6.3[g]	7.2[g]
III UPS	12.8[a]	13.4[a]	12.5[a]		
III XPS	12.8[d]	12.9[d]	11.7[d]	11.9[b]	12.4[c]
W_7-K_5	13.8[e]	13.8[e]	10.9[e]	12.1[e]	11.5[e]

[a] [2.159].
[b] [2.160].
[c] [2.161].
[d] [2.162].
[e] [2.68].
[f] [2.69].
[g] [Ref. 2.11XII, p. 1295].

improperly identified in the UPS measurements for PbS, due to the extrapolation procedure mentioned above. We note that the Fermi level was reported in [2.159] to be 0.7 eV above the top of the valence band while the energy gap is only 0.4 eV! While this is, in principle, possible for a very heavily doped *n*-type material (for PbS it would correspond to 10^{21} electrons $\times$ cm^{-3}) an error of $\sim$0.5 eV in the determination of the top of the valence band would reduce the Fermi energy to a more reasonable value. It would also preserve the systematics of the I' assignment.

We note in Fig. 2.13 that the second highest valence band is nearly flat from Γ_8^- to $L_{4,5}^+$. The binding energy of this flat portion agrees reasonably well with the I'' peak observed for PbTe and PbSe although it seems to be somewhat large (2.7 eV) for PbS, even if 0.5 eV are added to the observed I'' energy ($1.8+0.5=2.3$ eV in agreement with the XPS value of 2.4 eV). The third highest valence band in Fig. 2.13 is also flat from Γ_6^- to L_6^+. This portion seems to account for the I_3 peak. The origin of the I_5 peak seems to be the near degeneracy of the W_3, K_3, and X_6^- valence bands as listed in Table 2.9. In particular the bands are nearly flat along the K–W region (side of the square face of the Brillouin zone) [2.69], a fact which accounts for a peak in the density of states.

Peaks III and II, whose energies are listed in Table 2.10, show no substructure and are thus relatively easy to interpret: according to Fig. 2.13 they correspond to the lowest and second lowest valence bands (pure *s*-states of the chalcogen and mainly *s* of the metal), respectively. The peaks should thus occur at the average energy of these bands taken over the Brillouin zone. This average can be approximated by the value of the band energy at the point $\boldsymbol{k}=2\pi a^{-1}$ (0, 1/4, 3/4), the so-called *Baldereschi* point [2.165a]. Unfortunately band calculations for this rather low symmetry point are not available in the literature. We list instead in Table 2.10 an average of the energies at the $W[2\pi a^{-1}(0,1,1/2)]$ and $K[2\pi a^{-1}(0,3/4,3/4)]$ points which are nearly degenerate and relatively close to the *Baldereschi* point. With the exception of peak II for PbS, the agreement between the calculated and the observed energies listed in Table 2.10 is reasonable, especially when one considers that no empirical data have been used to fix the position of the corresponding bands in the band-structure calculations.

The angle-integrated spectra of GeS, GeSe, SnS, and SnSe, IV–VI compounds with orthorhombic structure (see Table 2.2) have also been measured in the XPS regime [2.165b]. Three main structures, equivalent to the peaks I, II, and III discussed above, are also seen. This fact emphasizes the chemical nature of these peaks which reflect essentially the bonding between the two constituent atoms: peak I is mainly anion *p*, peak II cation *s* plus anion *p*, and peak III anion *s*. The details of the fine structure of peak I, closely related to the band structure as seen above, are of course different for the cubic and the orthorhombic compounds. Angular resolved measurements for these materials are reported in [2.165c].

As mentioned above, the IV–VI compounds discussed in this section are closely related to the group V elements As, Sb, and Bi. The rhombohedral structure of these semimetals is basically the same as that of GeTe except that the two atoms per unit cell are equal. Hence one would expect the spectra of the elements to bear a close relationship to those of the IV–VI compounds, a fact borne out by XPS data [2.166]. Peaks II and III, however, are closer together than in the IV–VI compounds, a result of the fact that the *s*-levels contributing to peaks II and III are now the same. Some splitting between peaks II and III remains (see Sect. 2.7.2). The increase in this splitting in going from the covalent element to the ionic compound can be taken as a measure of the ionicity, as will be shown in Sect. 2.8.

We note that, as shown in Table 2.9, the $I''-I_3$ splitting is basically due to spin-orbit splitting of the two upper valence bands ($\Gamma_8^- - \Gamma_6^-$ and $L_{4,5}^+ - L_6^+$ splittings), with the exception of PbS (for this material the observed $I''-I_3$ separation is too large to be due to spin-orbit splitting which is mainly determined by the sulphur atom). Effects of spin-orbit splitting can be clearly seen in peak I for Sb and Bi [2.166].

2.4.4 Partial Density of Valence States: Copper and Silver Halides; Chalcopyrites; Transition Metal, Rare Earth, and Actinide Compounds

As discussed in Sect. 2.2.1 and illustrated in Figs. 2.6, 7, the valence bands of the copper and silver halides contain, beside the traditional four electrons per atom of tetrahedral semiconductors, the 10*d* valence electrons of the metal atom. At $\boldsymbol{k}=0$ the cubic field splits the *d*-levels into $\Gamma_{12}(E_g)$ and $\Gamma_{25'}(T_{2g})$. In the materials with zincblende structure (CuCl, CuBr, CuI, AgI) the Γ_{12} levels mix little with their neighbors and appear as flat core-like bands [2.49] (see Fig. 2.6). This is not the case for the materials with rock-salt structure [2.51a, b].

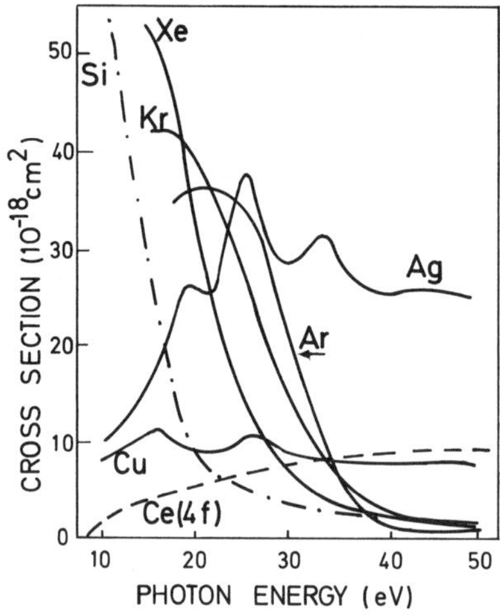

Fig. 2.39. Photoionization cross sections of Si [2.170b], Xe, Kr, Ar [2.170c], Cu and Ag [2.170d], and Ce (from [2.170a]). The data for Si, Ar, Kr, and Xe are basically σ_p cross sections, Ag and Cu essentially σ_d, and Ce σ_f

The UPS spectra of these materials, obtained with He and Ne resonance radiation have been reported in [2.49] for the zincblende and [2.167] for the rock-salt modifications. XPS measurements were reported by *Kono* et al. [2.168] for the copper halides and by *Mason* [2.169] for AgCl and AgBr. All these data fall into the XPS regime in the sense of Fig. 2.26. The most striking feature of the UPS data is the strong change in the shape of the EDC's with exciting photon energy. Similar, but not as striking, changes were discussed in Sect. 2.4.2 in connection with differences in the dependence of the partial photoionization cross section for *s*- and *p*-shells on photon energy, see (2.11). These differences are stronger when *d* or *f* electrons are involved. The general trend of partial photoionization cross sections in the 10–50 eV region is illustrated in Fig. 2.39. The cross section σ_p for photoionization of *p* valence electrons falls rapidly between 20 and 50 eV. The σ_d of *d* valence electrons is nearly constant in this region (except for the peak in Ag at 13 eV) and σ_f increases rapidly up to 30 eV remaining constant above this energy. Strong changes in the EDC's are thus expected in materials with mixed *p*-, *d*-, and/or *f*-like valence bands. These changes can be used to extract the partial *p*, *d*, and *f* densities of states by measuring the EDC's at several photon energies [2.49, 170a, 171].

We should point out that a general discussion of partial photoionization cross sections has been given by *Manson* in [Ref. 2.2, Chap. 3]. Information pertinent to the discussion here can be found in [Ref. 2.2, Figs. 3.2, 11–13]. The behavior of the partial cross sections just discussed is basically related to the atomic potential barrier due to centrifugal potential terms which are particularly strong for *d* and even more so for *f*-electrons (see [Ref. 2.2, Chap. 3]).

The dependence of the EDC on photon energy observed for polycrystalline AgI is shown in Fig. 2.40. The peak between 4 and 6 eV binding energy

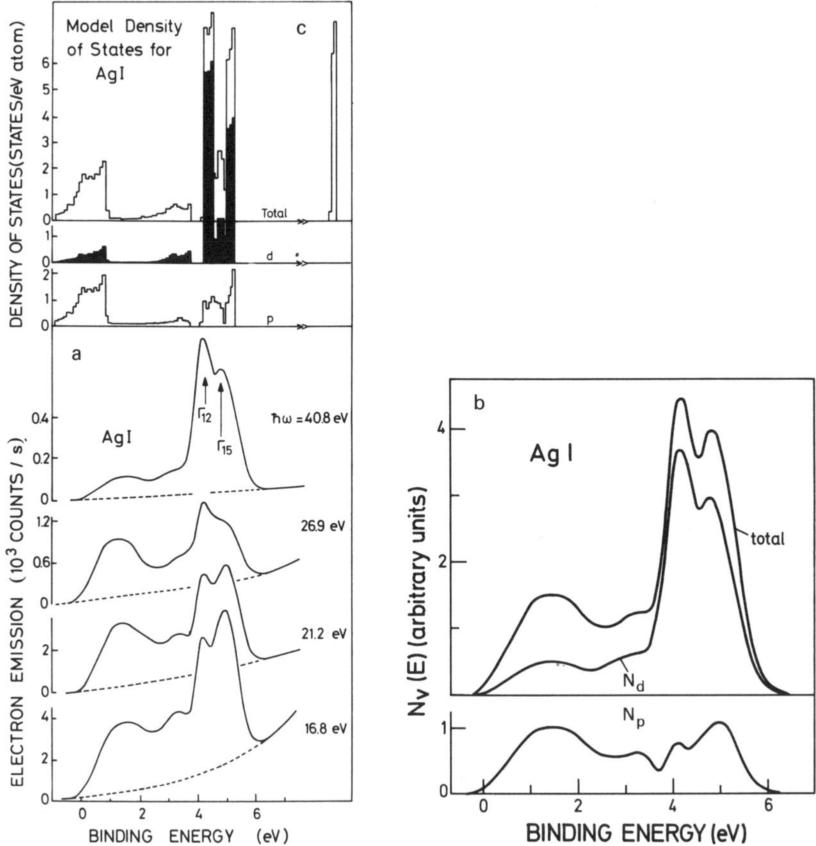

Fig. 2.40. **a** Photon energy dependence of the EDC's of polycrystalline zincblende-type AgI. The dashed lines indicate the background of secondary electrons. **b** Partial densities of states N^{d} and N_{p} and total density of states $N_{\mathrm{v}}(E)=N_{\mathrm{d}}+N_{\mathrm{p}}$ obtained as described in the text from the EDC's. **c** Histograms calculated from the band structure of Fig. 2.6 for these partial and total densities of states [2.49]

increases relative to the structure between 0 and 3 eV as $\hbar\omega$ increases. Hence, according to Fig. 2.39, it must be due mainly to the 4*d* valence levels of silver, while the structure between 0 and 3 eV must be due to the 4*p* levels of iodine. A look at Fig. 2.6 suggests that the *d*-like doublet represent practically the splitting between Γ_{12} and Γ_{15} states at $\boldsymbol{k}=0$.

We now use the equation

$$\langle j(E,\omega)\rangle \propto N_p(E)\sigma_p(\omega)+N_d(E)\sigma_d(\omega)\,, \tag{2.14}$$

analogous to (2.11), to extract the partial densities of states $N_p(E)$ and $N_d(E)$. We use for σ_p that of the rare gases given in Fig. 2.39: the rare gases have the same electronic configuration as the halogen ions neighboring in the periodic table

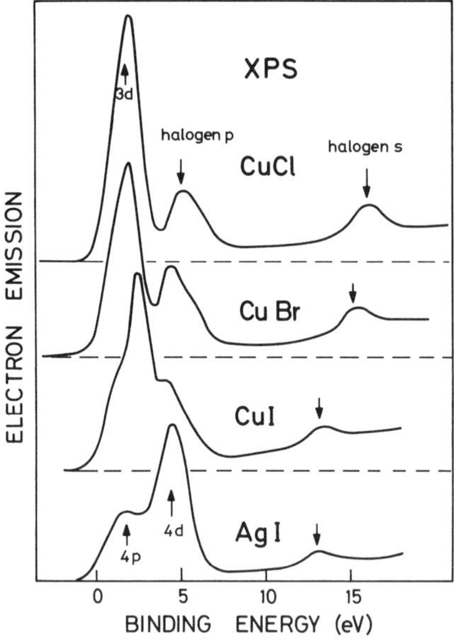

Fig. 2.41. XPS spectra obtained with unmonochromatized Al K_α radiation for CuCl, CuBr, CuI, and AgI. The arrows indicate the main atomic composition of the corresponding peak [2.49]

Note that the *s* valence states of the halogens are neglected; as mentioned in Sect. 2.4.2 they have a small σ_s in the UPS region. The best fit to all of the curves of Fig. 2.40a for AgI is obtained with the N_d's and N_p's of Fig. 2.40b. The unbroadened calculated density of states histograms shown in Fig. 2.40c represent well the experimentally determined densities of states.

We show in Fig. 2.41 the XPS spectra of the Cu-halides and AgI (zincblende). The position of the lower *s*-band can now be seen in all spectra. Also, the *d*-nature of the ~5 eV peak of AgI is apparent in that it is stronger than the *p*-like feature at ~2 eV. As shown in Fig. 4.2 σ_d is larger than σ_p for these materials with Al K_α radiation. The reversal in the intensities of the upper and lower peaks in the copper halides indicates that the low binding-energy peak is mainly *d*-like in these compounds, in contrast to AgI. This qualitative result is borne out by detailed determinations of the partial densities of states from the UPS spectra [2.49].

Similar manipulations of UPS data for several photon energies have been performed for AgCl and AgBr which crystallize in the rock-salt structure [2.167]. We compare in Fig. 2.42 the partial and total densities of states obtained by this method for AgCl with histograms calculated for a tight-binding band structure. Also in this case the predominantly *d* bands are below the *p*-bands. The XPS data of *Mason* [2.169] have been compared with total densities of states for nonlocal pseudopotential band structures in [2.51b].

The technique just discussed can also be applied to other Cu and Ag compounds such as the I–III–VI_2 chalcopyrites (see Sect. 2.2.1). Results for

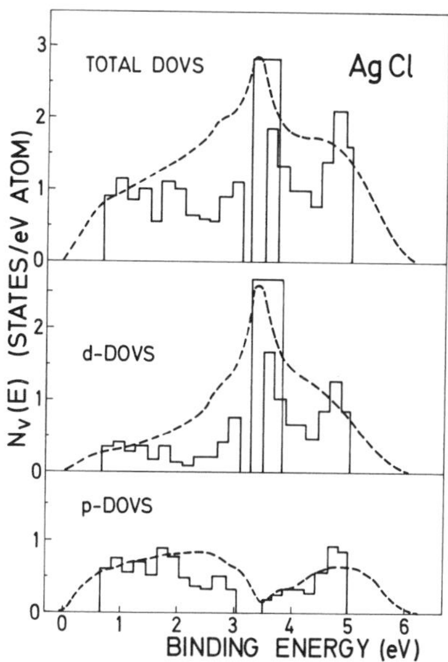

Fig. 2.42. Partial and total densities of valence states obtained from UPS data for AgCl compared with density of states histograms obtained for a tight-binding band structure [2.167]

$AgInTe_2$ and $CuInS_2$ are presented in [2.172]; also in this case the upper bands are mainly d-like for the copper and mainly p-like for the silver compound. In this paper it is also shown that $p-d$ interference terms, which were left out of (2.14), can indeed be neglected for *polycrystalline* samples in the Ne and He radiation region.

Measurements of EDC's in the band-structure regime of Fig. 2.26 ($\hbar\omega \lesssim 12$ eV) have also been performed for the silver halides [2.173] and CuBr [2.174]. Although some dependence of the energy of the initial states on photon energy is seen, the results of the analysis of these data are in basic agreement with those presented above, especially for the high-energy range of $\hbar\omega(\hbar\omega \simeq 12$ eV). A few peaks are observed whose final-state energy E does not change with $\hbar\omega$; they are assigned to structures in the density of *final* states. In AgCl, for instance, such structure occurs 8.1 eV (assignment Cl $3d$) and 9.85 eV (assignment Ag $5p$) above the top of the valence band. Maybe the most interesting part of this work is the measurement of EDC's as a function of temperature (between 77 and 295 K): some peaks broaden considerably and change shape when the temperature is raised. We show in Fig. 2.43 the EDC's of CuBr for $\hbar\omega = 11.8$ eV at 77 and 295 K. The peak labelled F broadens and shifts considerably (~ 10 kT) as the temperature is raised and, to a lesser extent so does B, while D seems to remain unaltered. This rather unusual result was attributed [2.175] to dynamic changes in the $p-d$ hybridization with temperature: As the phonons change dynamically the distance between metal and halogen atoms, the amount of hybridization changes. Peak D is basically the *unhybridized* Γ_{12} peak discussed above, probably the reason why it is not altered by temperature changes.

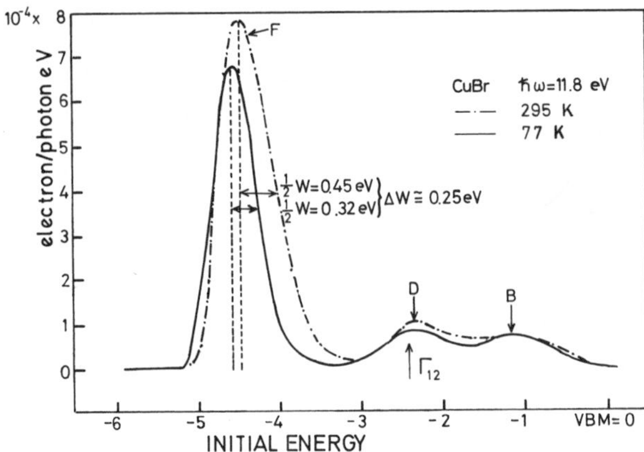

Fig. 2.43. Effect of temperature on the EDC's of CuBr for $\hbar\omega = 11.8$ eV [2.175]

We have so far confined our discussion to $p-d$ hybridization. Measurements as a function of $\hbar\omega$ with the purpose of separating p and d densities of states have been also performed for several transition metal oxides [1.171] and for two transition metal dichalcogenides (2H–$MoTe_2$ and MoS_2) [2.176]: The separation of the chalcogen p and Mo-4d densities of states worked out very well for $MoTe_2$ but not for MoS_2. This fact is probably due to the much larger change in σ_p for Te than for S in the 21–35 eV region of measurement (see for comparison σ_p for Xe and Ar in Fig. 2.39). The method, however, can also be used to separate p and f densities of states in lanthanide and actinide compounds [2.170a] (see Fig. 2.44).

2.4.5 Layer Structures: Transition Metal Dichalcogenides

As mentioned in Sect. 2.2.4, the dichalcogenides of the IV B, V B, and VI B transition metals are rather interesting from the point of view of photoelectron spectroscopy. We already showed in Fig. 2.20 the effect of charge density waves on the core levels of TaS_2. Also, the two-dimensional nature of these layer compounds makes them ideal for studies of the dispersion of the energy bands using angular resolved photoemission (see [Ref. 2.2, Chap. 6]).

In this section we discuss the angular integrated spectra of the valence bands of the group IV B and VI B compounds. The metallic group V B materials will only be mentioned when comparison so requires. The interpretation will be based essentially on the band structures of Figs. 2.16a, b which were calculated by *Mattheiss* for 1T–TaSe and 2H–TaS_2. For the group IV B compounds, with 1T structure (see Table 2.5), we use the band structure of Fig. 2.16a with the d-bands completely empty. For the group VI B compounds, usually with 2H structure, we use Fig. 2.16b with the lowest 2d bands completely filled.

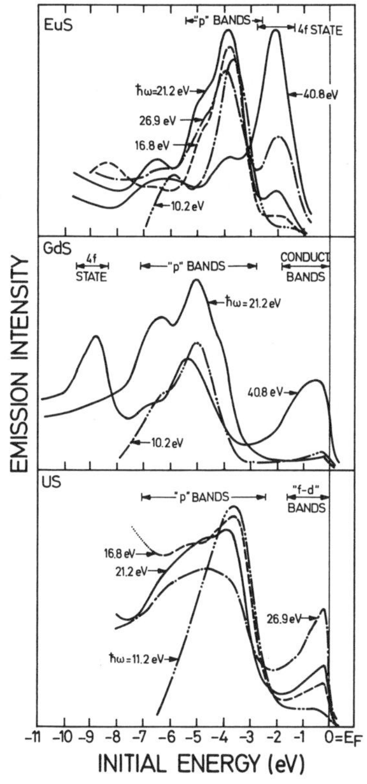

Fig. 2.44. EDC's of EuS, GdS, and US for several photon energies. The upper structures (between 0 and 2.5 eV) are *d*-like as they increase in strength with respect to the lower *p* bands with increasing photon energy (see Fig. 2.39). GdS has an *f*-like structure at -9 eV, corresponding to the Gd^{3+} ions [2.170a]

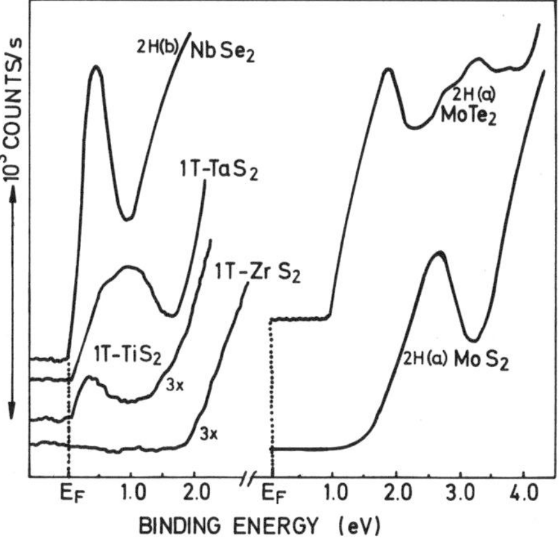

Fig. 2.45. EDC's of several transition metal dichalcogenides taken with $\hbar\omega = 40.8$ eV near the Fermi energy [2.177]

We show in Fig. 2.45 the leading edge of the EDC's of a number of group IV B (TiS_2, ZrS_2), V B (TaS_2, $NbSe_2$), and VI B (MoS_2, $MoTe_2$) dichalcogenides. We note that for ZrS_2 the top of the valence band lies ~ 1.8 eV below the Fermi energy. Hence the material must be a semiconductor (or insulator) as would be expected from Fig. 2.16a if the *d*-bands are empty. Its gap should be $\gtrsim E_F - E_v = 1.8$ eV. Absorption measurements for this material [2.178] do indeed indicate a gap of $\simeq 1.7$ eV. Thus ZrS_2 is an *n*-type, probably degenerate semiconductor. The *n*-character is likely due to a deficiency of sulphur which is easily lost in the growth process (i.e., excess of Zr).

It is reasonable, in view of Fig. 2.16, to attribute the strong peak at the onset of the photoemission seen for $NbSe_2$, MoS_2, $MoTe_2$, and TaS_2 to the filled or half-filled lowest *d*-bands. This assignment has been confirmed by the method

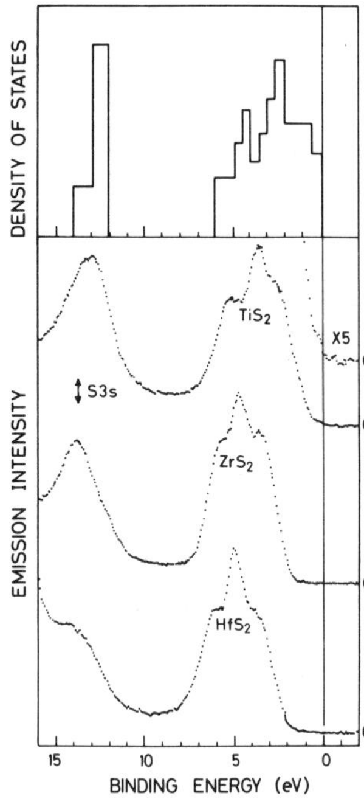

Fig. 2.46. EDC's for TiS_2, ZrS_2, HfS_2, and MoS_2 obtained with AlK_α excitation (XPS) [2.181]. Also, density of states histogram calculated by *Krusius* et al. for TiS_2 [2.103b]

of Sect. 2.4.4 for $MoTe_2$ [2.176] and for MoS_2 by means of angular resolved XPS [2.179]. The width of the *d*-bands, $\simeq 2$ eV for the 2H group VI B compounds is slightly higher than that shown in Fig. 2.16b (≈ 1.3 eV). The half-filled *d*-band of 2H–$NbSe_2$ shows in Fig. 2.45 roughly *half* the width (~ 1 eV) of the corresponding bands of $MoTe_2$ and MoS_2, in agreement with the position of the Fermi level in Fig. 2.16b. For the 1T compound TaS_2 the half-filled *d*-band is wider than for 2H–$NbSe_2$, a fact which also agrees, at least qualitatively, with the wider *d*-bands calculated for the 1T modification of Fig. 2.16a.

We note that 1T–TiS_2 also seems to exhibit a weak *d*-electron peak. Thus this material should be either a semimetal or a very heavily *n*-doped small gap semiconductor. Although the semimetallic character is contrary to the result of band-structure calculations [2.103a, b], it was until recently [2.87c] the accepted picture [2.86, 87]. The Fermi energies $E_F - E_v$ found from Fig. 2.45 for MoS_2 (1.6 eV) and $MoTe_2$ (0.9 eV) agree well with the indirect gaps of these materials (1.6 eV for MoS_2 and 1.0 eV for $MoTe_2$ [2.180]).

The results just discussed are confirmed by XPS work [2.181], as shown in Fig. 2.46. In this figure a Fermi energy 1.7 eV above the top of the valence band is found for ZrS_2 and 2 eV for HfS_2: the optical gap of HfS_2 is 2.6 eV [2.178]. Thus the HfS_2 samples measured are *n*-type but nondegenerate. TiS_2 also shows a weak bump which would be related to *d*-electrons at the Fermi energy if the compound is a semimetal. *Wilson* attributes it to defect induced tailing of the *p*-bands [2.87c].

The width of measured *p*-chalcogen bands ($\sim$6 eV) is somewhat larger than that calculated by *Mattheiss* (4.6 eV, Fig. 2.16), but agrees with the calculations of [2.103b] for TiS_2 also shown in Fig. 2.46. The position of the chalcogen *s*-band is also well reproduced by these calculations. The measured width, however, is much larger than calculated. This fact is common to most deep *s*-bands in semiconductors and is to be attributed to the short lifetime of these states for whose decay a large number of channels is open (Auger, radiative).

The three substructures of the *p*-peak of TiS_2, at 2.2, 3.6, and 5.2 eV, are well reproduced by the calculation of [2.103b]. Similar substructures are also observed for all other 1T materials. They seem to correspond to the upper two *p*-bands, the next lower 3 bands, and the lowest *p*-band of Fig. 2.16a. As expected from Fig. 2.16b this structure is somewhat different for the 2H-type compounds [2.181].

We should mention in closing that the semimetallic nature of TiS_2 has also been concluded by *Fischer* from an analysis of the X-ray emission and absorption spectra [2.182]. Also, a comparison with the sulphur K_β and sulphur L_{III} X-ray emission spectra indicates clearly the Mo-4*d* nature of the first peak (lowest binding energy) in the XPS spectrum of MoS_2 shown in Fig. 2.45: this peak is absent in the sulphur emission spectra which exhibit all other XPS structures (see [Ref. 2.183, Fig. 2.6]).

2.4.6 Layer Structures: SnS_2, $SnSe_2$, PbI_2, GaS, GaSe

To the reasons given above for the interest in photoemission generated by layer structures, one may add the fact that they cleave easily (layers can be peeled off with sticky tape) and it is thus easy to prepare surfaces in situ. Also, the fact that all bonds are saturated keeps these materials from contaminating in a poor vacuum. By replacing the group IV B transition metals by their group IV A counterpart one obtains SnS_2 and $SnSe_2$ which crystallize in the same structure (octahedral 1T of Table 2.5) as the IV B counterparts. The band structures of these compounds [2.184, 185a] should also bear some relationship to that of the IV B chalcogenides (see Fig. 2.16a). The removal of the *d*-electrons in the $\sigma-\sigma^*$ gap of Fig. 2.16a, however, results in a closing of this gap (2.1 eV for SnS_2 and 1.0 eV for $SnSe_2$ [2.185a, b]) and in a large change in the ordering of the valence bands. If one compares these compounds with SnS and SnSe one would expect the 5*s*-electrons of the metal to be transferred to the anion, assuming for the compounds the ionic configuration $Sn^{+4}(S^{-2})_2$. One would thus expect the Sn 4*s*-like peak II of the IV–VI compounds discussed in Sect. 2.4.3 to be absent in SnS_2 and $SnSe_2$. In a tight-binding picture the *p*-electrons of the two chalcogen atoms per unit cell can occupy even and odd p_z orbitals (p_z^+ and p_z^-) and also p_x^+, p_x^-, p_y^+, and p_y^- orbitals (*z* is perpendicular to the layers). The p_z^+ orbital will mix with the Sn-4*s* orbital and form bonding and

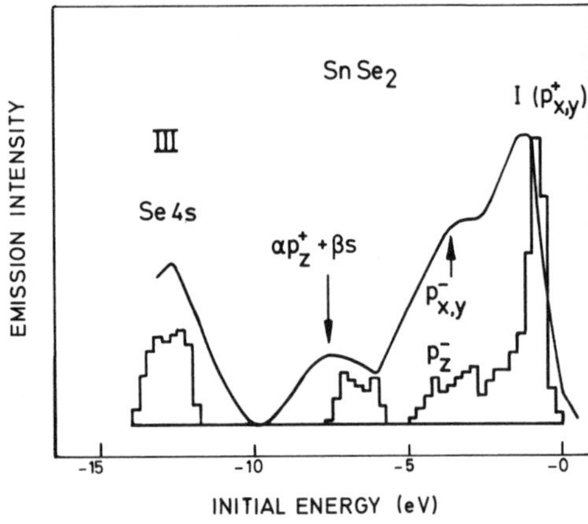

Fig. 2.47. EDC's of $SnSe_2$ excited with unmonochromatized AlK_α radiation, compared with a density of states histogram [2.185a, b]

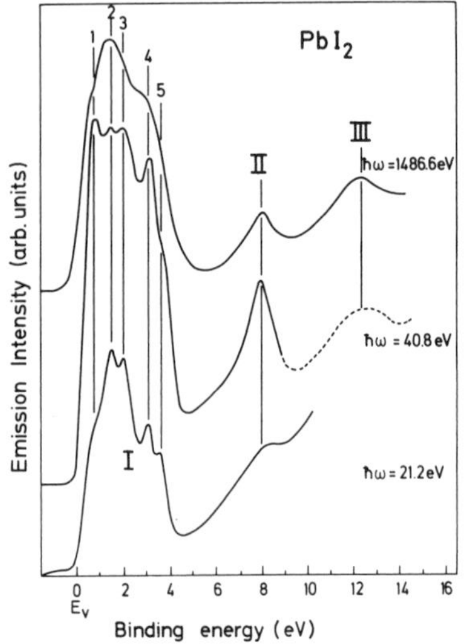

Fig. 2.48. EDC's of PbI_2 [2.187a]. The dashed line was obtained after correction for the Pb 5d lines produced by 48.4 eV radiation

antibonding orbitals, the amount of mixture determining the deviation from the purely ionic configuration. The $\alpha p_z^+ + \beta s$ bonding orbital will be lowered in energy and occupied while the $\beta p_z^+ - \alpha s$ antibonding orbital will be raised and empty.

The EDC obtained with unmonochromatized Al K_α radiation for $SnSe_2$ is shown in Fig. 2.47 and compared with a calculated density of states histogram. The peak at ~ 8 eV binding energy corresponds to the $\alpha p_z^+ + \beta s$ state. The peak at 1.5 eV corresponds mainly to the $p_{x,y}^+$ states while the broad shoulder at 3.5 eV contains, according to the calculations of [2.185a], the $p_{x,y}^-$ and the p_z^- states of Se. High resolution measurements on $SnSe_2$ have recently been reported [2.185b, c].

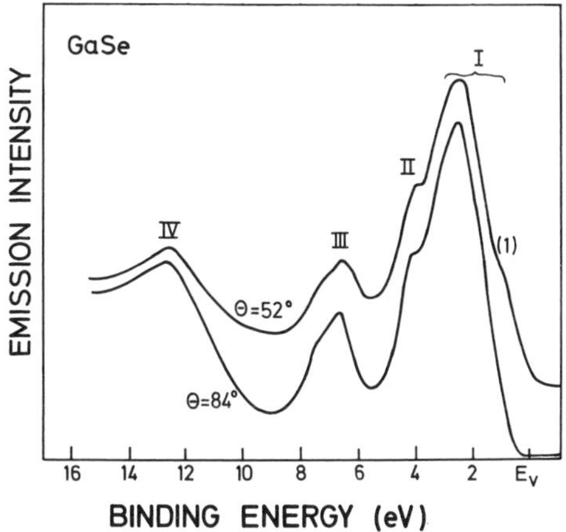

Fig. 2.49. XPS spectra of GaSe for two different electron take-off angles with respect to the normal θ [2.179]

PbI_2 has the same symmetry as SnS_2 and $SnSe_2$ but two more electrons per unit cell. Thus one would expect in a rigid-band type of model the $\beta p_z^+ - \alpha s$ states, empty in SnS_2, to be occupied. Band calculations yield a gap of ~ 2 eV between this s-like top valence band and the remaining valence bands which should be analogous to those of $SnSe_2$. If this gap exists, the absorption edge of PbI_2 would correspond essentially to a one-electron excitation of the Pb^{+2} ion from $6s$ to $6p$ [2.186]. XPS and UPS measurements speak against this conclusion [2.187a]. The XPS and UPS spectra of PbI_2 (see Fig. 2.48) are very similar to those of SnSe [2.165b] as one may guess from the similar electronic configuration. Peak I does not show any evidence for a well separated Pb-$6s$ component while peak II has nearly the same dependence on $\hbar\omega$ as for SnSe [2.165b] and is thus expected to be mainly Pb-$6s$ like. These conclusions have been confirmed in a recent band-structure calculation by *Robertson* [2.187b]: the uppermost valence band is separated only by a small gap from the lower valence bands; it is approximately 40% Pb-$6s$ and 60% I-$5p$.

BiI_3, a layer compound with an electronic configuration similar to PbI_2, has also been investigated with XPS [2.188]. In this compound peak II (Bi-$6s$) has moved to higher binding energies and is nearly degenerate with peak III (I-$5s$).

As discussed in Sect. 2.2.1 GaS and GaSe are also interesting layer structure materials. With an odd number of electrons per molecule they owe their semiconducting character to the fact that they possess an even number of molecules per primitive cell (at least two for the simplest modifications). As layer-type crystals they have also generated a considerable amount of photoemission work, both XPS [2.188, 179] and UPS [2.189]. The XPS spectra of GaSe for two different take-off angles Θ of the electrons (relative to the surface normal) are shown in Fig. 2.49. As discussed in Sect. 2.2.1 and shown in Fig. 2.12, the valence bands of this compound fall into four groups I–IV (top I-bottom IV) which are responsible for the corresponding structure shown in

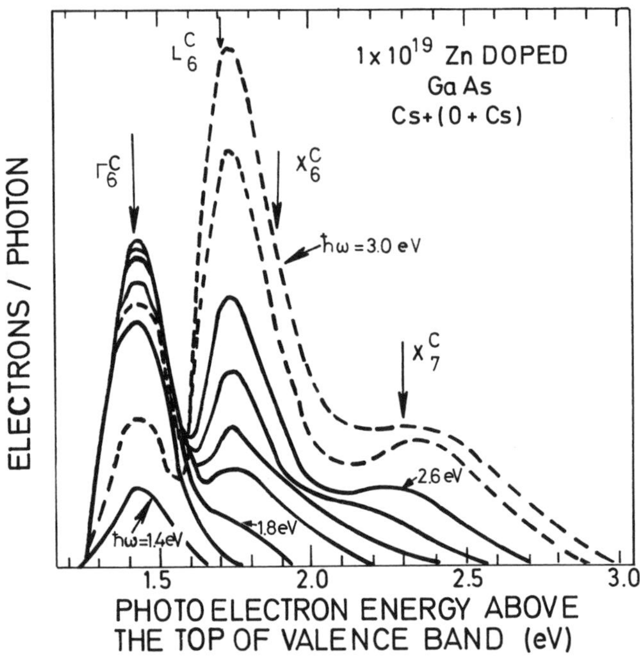

Fig. 2.50. EDC of a cesiated GaAs surface in the region of secondary electron emission. According to [2.115] the structure observed is due to the Γ_6, L_6, and X_7 conduction band minima. The experimental data are from [2.190]

Fig. 2.50. Peak I has a high energy shoulder (1) which disappears for large take-off angle Θ. This fact enabled *Williams* et al. to conclude [2.179] that shoulder $I(1)$ is due to the p_z – Se component of the charge density of Fig. 2.12I.

2.5 Photoemission and Density of Conduction States

As discussed above the angular integrated EDC's yield in the XPS regime information about the density of valence states, while in the band-structure regime information about a combined EDJDOS [Ref. 2.2, Sect. 1.6.1] is obtained if $\boldsymbol{k}$-conservation is assumed. Under some circumstances it is, however, possible to obtain information on the density of *conduction* states from photoemission experiments. If nondirect transitions apply, at least in part, [Ref. 2.2, Eq. (1.93)] holds

$$\langle j(E,\omega)\rangle \propto N_v(E-\hbar\omega)N_c(E)\,. \tag{2.15}$$

Thus for $E-\hbar\omega=$constant the spectra reproduce the density of conduction states $N_c(E)$. The technique of measuring $\langle j(E,\omega)\rangle$ while keeping $E-\hbar\omega$ constant, i.e., changing the *photon* energy continuously as the *electron* energy is being swept, is known as constant initial-state spectroscopy (CIS). Such work is usually carried out with synchrotron radiation: It is thus discussed in Sect. 6.3.4 of this volume.

Equation (2.15) applies to an *elastic* photoemission process. The photo-excited electrons can also suffer energy losses before being emitted and dribble down in the band structure and pile up at regions of high density of states. Thus peaks in the EDC which correspond to conduction band extrema or flat bands can appear. We have already mentioned such results obtained by *Lin* et al. [2.174] for the silver halides in Sect. 2.4.4. Whenever a peak in the EDC does not change E while changing $\hbar\omega$, changes are it is due to such mechanisms.

2.5.1 Secondary Electron Tails

We discuss here an example of secondary electrons which carry information about the density of conduction states. The lowest conduction band of GaAs has three sets of minima [(000), {111}, {100}, see Fig. 2.5] at which photo-excited secondary electrons should remain metastable a sufficiently long time to be emitted without further losses. Unfortunately these states lie below the vacuum level for clean GaAs surfaces. Cesiation of a p-type sample, however, brings them above the vacuum level (see negative affinity, [Ref. 2.2, Fig. 1.10]) and enables their emission into the vacuum. The EDC's of the corresponding electrons are shown in Fig. 2.50. The structures are observed at electron energies independent of the exciting photon energy. Their assignment is that of [2.115] in agreement with the present ideas about the ordering of the conduction bands in GaAs.

2.5.2 Partial Yield Spectroscopy

The term refers to the measurement of an integrated portion of the EDC's as a function of photon energy $\hbar\omega$

$$Y(\omega, E_{\mathrm{m}}, E_{\mathrm{M}}) = \int_{E_{\mathrm{m}}}^{E_{\mathrm{M}}} \langle j(E, \omega)\rangle \, dE\,. \tag{2.16}$$

One usually chooses for the domain of integration the low-energy region in which mainly secondary electrons are emitted, with E_{m} = vacuum level and E_{M} a few electron volts above it. With a typical $\alpha(\omega) \leqq 10^5\,\mathrm{cm}^{-1}$ and, according to Fig. 2.26, $\lambda \lesssim 50$ Å (2.9) yields the transmission function

$$l \simeq \alpha(\omega) \langle \lambda(E) \rangle\,, \tag{2.17}$$

where $\langle \lambda(E) \rangle$ is averaged between E_{m} and E_{M}. If one makes the reasonable assumption that the secondary electrons have lost all memory of the states they came from, the yield spectra will be simply proportional to the transmission function of (2.17), i.e., for a fixed electron energy window to $\alpha(\omega)$. Thus the partial yield method enables one to measure the spectrum of $\alpha(\omega)$ to an arbitrary scaling factor. The method is particularly useful to investigate

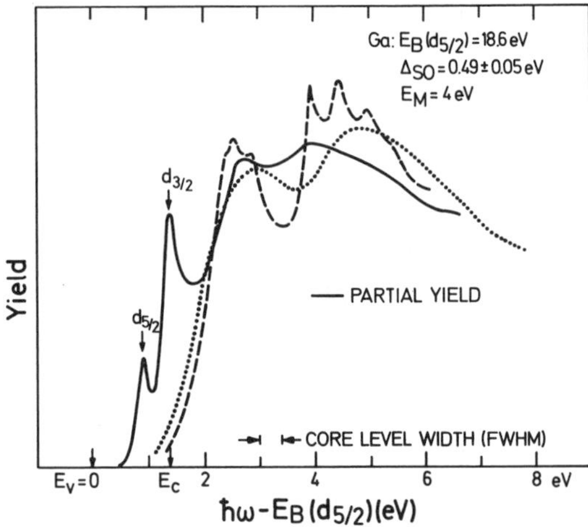

Fig. 2.51. Partial yield spectrum of the Ga $3d \to$ conduction transitions in GaAs (solid line). For comparison we also show the absorption spectra of the material as measured on thin films (dotted line) [2.193] and the density of conduction states convoluted with the density of core levels (dashed line) [2.192]

transitions from core levels as it gets around the preparation of thin films required for standard transmission measurements. The measurements are usually performed with synchrotron radiation and a discussion of the method with a few examples of applications is given in Sects. 6.3.7, 8 and 6.4. We show in Fig. 2.51 the partial yield spectrum of GaAs as compared with the absorption coefficient measured by transmission and the density of conduction states [2.191, 192]. The yield spectrum does bear a close resemblance to the two other spectra *above* E_c but shows two striking lines, separated by 0.5 eV, the $3d_{5/2} - 3d_{3/2}$ spin-orbit splitting of gallium, below and near E_c. In order to interpret these lines we must realize that, while according to (2.17) the yield spectrum measures $\alpha(\omega)$, this is the $\alpha(\omega)$ in a surface layer of depth $\langle\lambda\rangle \simeq 25$ Å. This "surface" absorption coefficient can differ from that of the bulk if surface states are present. The fact that the $d_{5/2}$ peak occurs below E_c was used as evidence that these peaks are due to empty surface states [2.193]. While these states exist in the gap of silicon and germanium, (see Sect. 2.9.2) they lie above the bottom of the conduction band in GaAs [2.194]. The explanation of the surface absorption peaks must then shift to surface core *excitons* with binding energies sufficiently large to bring the excitations into the continuum surface states below E_c. It is well known that two-dimensional excitons have a binding energy much larger ($\simeq \times 4$) than that of their three-dimensional counterparts [2.195].

2.6 Angular Resolved Photoemission from the Lead Salts

Considerable more information than from angular integrated spectra can be obtained through angular resolved photoemission spectroscopy (ARPES) from single-crystal surfaces. Smith has dealt with this topic in detail in [Ref. 2.2,

Chap. 6] and has given examples for the application of this technique to layered structures and GaAs. The essential point is that $\boldsymbol{k}_{\|}$, the component of the electron momentum parallel to the sample surface, is conserved in the photoemission process. This is a consequence of the translational symmetry of the system parallel to the well-ordered single-crystal surface. $\boldsymbol{k}_{\|}$ can easily be determined from the measured kinetic energy E of the photoelectrons and the angle θ between their trajectory and the surface normal (in a.u.)

$$k_{\|} = \sqrt{2E}\,\sin\theta\,. \tag{2.18}$$

Any structure in the EDC can thus be identified with valence states confined to a line in reciprocal space defined by $\boldsymbol{k}_{\|}=$constant. The "width" of this line is of course determined by the angular resolution $\Delta\theta$ of the analyzer and the energy E of the electrons. For $E \sim 15$ eV and $\Delta\theta = 3°$, $\boldsymbol{k}_{\|}$ is defined to within better than 1/10 of the size of a typical Brillouin zone and by varying θ it is possible to measure the dispersion of structure in the EDC as a function of $\boldsymbol{k}_{\|}$. The range of θ necessary to cover a Brillouin zone is approximately 40° at these energies. To obtain a similar resolution in $\boldsymbol{k}$ space with electron energies around 1400 eV, as they occur typically in XPS, would require acceptance angles as small as 0.1°. This would be accompanied by an unacceptable loss in intensity. It has been recently shown by *Shevchik* that even with good angular resolution $\boldsymbol{k}_{\|}$ is smeared by phonons (Debye–Waller factors) in XPS photoemission [2.196]. Angular resolved photoemission for the study of energy dispersion is therefore the domain of UPS. The high selectivity of angular resolved photoemission is demonstrated in Fig. 2.25 for the (100) surface of PbS [2.130]. Angular resolved spectra for a number of angles θ are compared with an angle-integrated spectrum taken with the same photon energy and with an XPS spectrum. The angular resolved spectra show considerably more and sharper structure. Peaks as narrow as 0.3 eV can be observed in some instances and peak positions and intensities vary rapidly with θ. The integration over a wide range of angles in the conventional spectra washes out almost all of this structure. A word of caution seems in order here. Interpreting conventional UPS spectra in terms of a density of states or an energy distribution of a joint density of states (see [Ref. 2.2, Chap. 1]) tacitly assumes that a complete sampling of the whole Brillouin zone takes place. This may not be the case in work on single crystals unless the acceptance angle of the analyzer is $\sim 40°$ which is hardly ever realized. As mentioned in Sect. 2.4, the measurement of polycrystalline samples helps in the uniform sampling of the Brillouin zone.

Peak positions referred to initial state energies as a function of $\boldsymbol{k}_{\|}$ are summarized in Fig. 2.52 for the (100) surface of PbS according to the work of *Grandke* et al. [2.70]. Two geometries were used: $\boldsymbol{k}_{\|}$ parallel to the (001) and $\boldsymbol{k}_{\|}$ parallel to the (011) direction, respectively. The photon energy was 21.2 eV and the angle of incidence of the photons was kept fixed around 50°. Because the Brillovin zone of the lead salts has mirror symmetry with respect to the [001] and the [011] planes, the points in Fig. 2.52 are

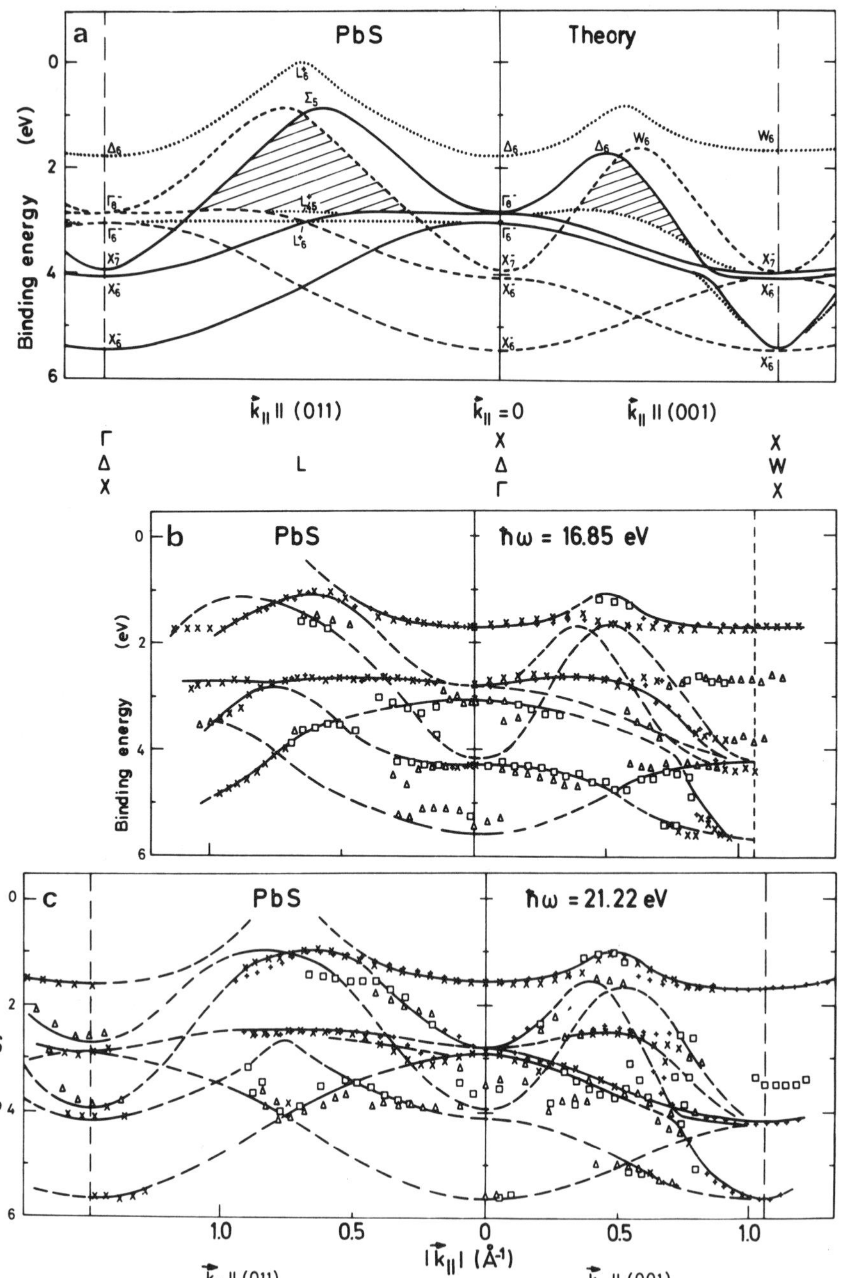

Fig. 2.52. a Energies of critical points in the one-dimensional density of states plotted vs $\boldsymbol{k}_{\parallel}$. The solid lines refer to critical points at $\boldsymbol{k}_{\parallel}=0$, the dashed lines to critical points at $\boldsymbol{k}_{\parallel}=2\pi/a$ and the dotted lines to critical points at some intermediate value of $\boldsymbol{k}_{\parallel}$

b Peak position vs $\boldsymbol{k}_{\parallel}$ the reduced momentum component parallel to the (100) surface of PbS for $\hbar\omega=16.85$ eV. The experimental points have been connected by lines to give energy vs momentum curves similar in shape to those of a [2.70]

c Same as Fig. 2.52b; $\hbar\omega=21.22$ eV

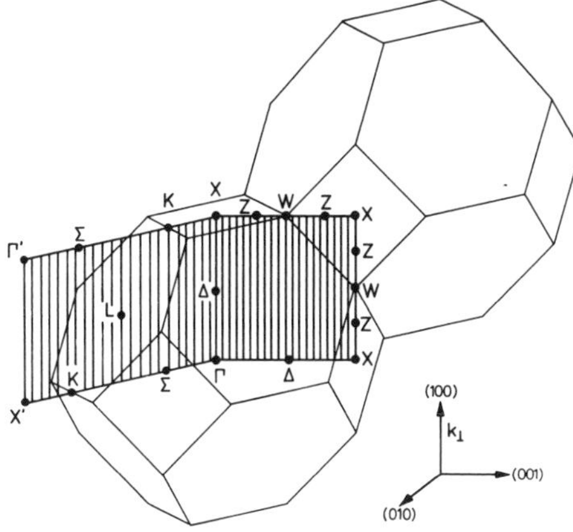

Fig. 2.53. Extended zone scheme of PbS. The measurements correspond to $\boldsymbol{k}$ vectors lying in the $\Gamma-X-X-X$ plane ($\boldsymbol{k}_{||}$ parallel [001] and in the $\Gamma-X-K-L-K$ plane ($\boldsymbol{k}_{||}$ parallel [011]). One-dimensional densities of states were calculated along $\boldsymbol{k}_{||}$=constant, as indicated by vertical lines

derived from spectra taken both at $\theta<0°$ and $\theta>0°$ for each direction of $\boldsymbol{k}_{||}$. For a layer compound such a plot of E vs. $\boldsymbol{k}_{||}$ would give a complete experimentally determined band structure of the material under study: the band dispersion is in this case negligible along $k_{\perp}$ (see [Ref. 2.2, Chap. 6]). For a material with a truly three-dimensional band structure some information about the as yet unknown quantity $k_{\perp}$ is necessary to locate each state unambiguously in the Brillouin zone and to make contact with band-structure calculations. The translational symmetry perpendicular to the sample surface is broken due to the short escape depth of the photoelectrons. The final states are therefore no longer eigenstates of $k_{\perp}$ and a conservation of this quantity is no longer necessary. *Grandke* et al. [2.130] carried this consideration to its extreme and assumed that all transitions along a line $\boldsymbol{k}_{||}$=constant would be equally allowed. Structure in the ARPES is then due to singularities in the one-dimensional density of states calculated along $k_{\perp}$ for the appropriate $\boldsymbol{k}_{||}$.

The positions of $\boldsymbol{k}_{||}$ and the directions of $k_{\perp}$ relevant to these experiments are shown in Fig. 2.53 superimposed on the extended zone scheme of PbS. The upper half of Fig. 2.52 traces the positions of singularities in the one-dimensional densities of states versus $\boldsymbol{k}_{||}$. Most singularities coincide with points of high symmetry along the lines $\Gamma-X$ and $\Gamma-K$ ($k_{\perp}=0$) or $X-K$ and $X-W$ ($k_{\perp}=2\pi/a$) (compare the band structure of PbTe in Fig. 2.13). Two singularities, (dotted lines in Fig. 2.52), however, occur at a general point in the Brillouin zone. The uppermost one corresponds to the local maximum in $E(k_{\perp})$ along the line $\Gamma-X(k_{||}=0)$. It merges for $k_{||}=2\pi a^{-1}$ (0, 1/2, 1/2) *with the* L-point and marks the top of the valence bands. The experimental E vs $\boldsymbol{k}_{||}$ curves in the same figure reproduce the general pattern set by the singularities quite well. Energies of critical points have been extracted from these data with a precision of a tenth of an eV and some of them are compared in Table 2.9 with other experimental values.

The authors [2.70, 197] have extended their work to the other lead chalcogenides. Figure 2.54 shows three spectra for special values of θ such that

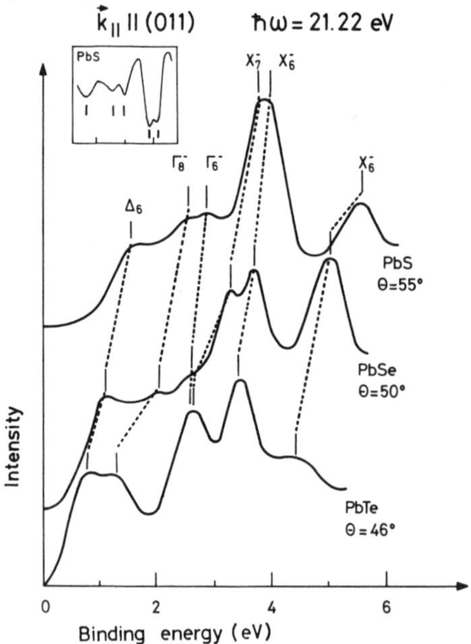

Fig. 2.54. Angular resolved photoemission spectra of PbS, PbSe, PbTe. The dotted lines connect peaks corresponding to states of high energy so as to indicate the increase in spin-orbit splitting [2.197]

Table 2.11. Spin-orbit splitting (eV) of valence states at Γ and X for the lead salts as determined by angular resolved photoemission [2.197]

Photon energy $\hbar\omega$ [eV]	$\boldsymbol{k}_{\|}$	PbS		PbSe		PbTe	
		$\Delta_{so}(\Gamma)$	$\Delta_{so}(X)$	$\Delta_{so}(\Gamma)$	$\Delta_{so}(X)$	$\Delta_{so}(\Gamma)$	$\Delta_{so}(X)$
16.85 (Ne I)	(000)	—	—	0.5_2	—	1.1_5	—
	(001)	—	—	—	0.5_6	—	0.9_5
21.22 (He I)	(000)	—	—	—	—	(0.9_7)	—
	(001)	—	—	—	(0.6_4)	—	0.9_5
	(011)	0.3_0	0.2_2	0.6_7	0.5_2	1.2_0	0.9_0
Recommended value		0.3	0.2	0.6	0.5	1.15	0.9
APW calculation [2.68]		0.24	0.23	0.60	0.44	1.10	0.69

singularities at X and Γ mark the increase in the spin-orbit splitting at X and Γ from PbS to PbTe. Accurate peak positions obtained from the second derivative of these and other spectra are set out in Table 2.11 and are compared with relativistic APW calculations [2.68]. The spin-orbit splitting increases with increasing atomic number of the anion as expected for predominantly anion-p derived states. The relativistic APW calculations [2.68] appear to reproduce these splittings well except for the X point in PbTe.

Let us briefly return to Fig. 2.52. It is apparent from it that not all singularities expected do show up as peaks in the spectrum. In particular the top of the valence bands is never actually reached in PbS. This might be different for other directions of $\boldsymbol{k}_{\|}$ but, as mentioned in Sect. 2.4.3 the top of the valence bands in PbS as determined from conventional spectra in [2.159, 162] could be in error because the possibility of negligible emission from the uppermost levels was not taken into account.

The interpretation of the ARPES data on the lead salts presented so far rests on the assumption of a complete relaxation of $k_{\perp}$ selection rules. The other extreme, direct $k_{\perp}$ conserving transitions, has also been adopted successfully for the interpretation of angular resolved photoemission for a number of metals [2.198–201] and for GaAs [2.202] despite the fact that the mean free path in these materials is not substantially longer than that in the lead salts. Both interpretations must indeed be regarded as limiting cases of a model developed by *Feibelman* and *Eastman* [2.203] which replaces the strict $k_{\perp}$ selection rule of the bulk photoemission model (compare [Ref. 2.2, Chap. I]) by a relaxed $k_{\perp}$ conservation such that transitions from an initial state with a given $k_{\perp}^{\mathrm{i}}$ are possible into a range of final states centered around $k_{\perp}^{\mathrm{i}}$. Transition matrix elements are weighted by their distance from $k_{\perp}^{\mathrm{i}}$ according to a Lorentzian with a width that is approximately equal to twice the inverse of the electron mean free path λ. In the limit of $\lambda \to \infty$ we regain the model of bulk optical transitions with conservation of $k_{\perp}$ and $\boldsymbol{k}_{\|}$, and for $\lambda \to 0$ the model of $k_{\perp}$ nonconserving transitions applies. For photoelectrons of about 15 eV energy λ is ~ 20 Å. This corresponds to an uncertainty in $k_{\perp}$ of about 10% of the radial dimension of the Brillouin zone of PbS and we would therefore expect at least remnants of direct transitions in the spectra of PbS. The lack of some structures calculated on the basis of singularities in the one-dimensional density of states and peaks that do not coincide with these singularities (see Fig. 2.52) do indeed confirm this conjecture. Most convincing evidence for $k_{\perp}$ conserving transitions is, however, a slight shift (of the order of $\gtrsim 0.2$ eV) in most structures when $\hbar\omega$ is changed from 21.2 eV to 16.9 eV (Figs. 2.52b, c). An analysis of the normal emission spectra in terms of the *Feibelman* and *Eastman* formalism [2.197] confirms, however, that most transitions tend to fall at or very close (± 0.1 eV) to singularities in the one-dimensional density of states: away from those singularities the density of states is small and no transitions are observed. Whether quasi-direct or quasi-indirect transitions dominate depends on the band structure of the material, the electron mean free path, the photon energy and the take-off angle θ. Away from $\boldsymbol{k}_{\|}=0$ more final states are usually available to couple to an outgoing plane wave than at $\boldsymbol{k}_{\|}=0$ and thus a spectrum resembling the one-dimensional density of states becomes more likely.

2.7 Amorphous Semiconductors

A characteristic feature of the more covalently bonded semiconductors and insulators is the relative ease with which they can be prepared in amorphous form. Glasses solidify from the melt without crystallization and remain in their metastable amorphous state for a period of time that may be regarded as infinite for all practical purposes, provided their temperature does not exceed the crystallization temperature. Glasses are most readily prepared from compounds containing one of the chalcogens O, S, Se, and Te. Examples are the oxides of Si and Ge (SiO, SiO_2, GeO_2); the chalcogenides of As (As_2S_3, As_2Se_3, As_2Te_3) and the alloys of Ge and Te: Ge_xTe_{1-x}. Selenium can also be quenched from the melt to form a glass. The quenching rates obtainable from the liquid phase are often not sufficient to bypass crystallization. Higher quenching rates are obtained by quenching from the vapor phase onto a substrate held at sufficiently low temperature. Preparation methods along this line include evaporation, sputtering, and decomposition of gaseous compounds by dc or rf discharges [2.204]. These techniques are applicable to virtually all materials that form an amorphous solid phase and since they lend themselves readily to the preparation of thin films in vacuum they are widely used in photoelectron spectroscopy of amorphous materials (see the following sections). Crystalline solids can also be made amorphous by irradiation with energetic ions (usually rare gases), which renders a highly damaged surface layer with many of the properties of the amorphous state.

Amorphous materials are characterized by the loss of the translational periodicity found in the crystal. This lack of "long-range order" is reflected in the loss of sharp X-ray diffraction structures which are replaced by a few broad halos. An analysis of these halos in terms of the probability distribution of the distances [the radial distribution function (RDF)] and coordination numbers of the atoms (also given by the RDF) reveals, however, that a certain degree of "short-range order" is preserved in the amorphous state. This short-range order relates in particular to the first coordination sphere. The RDF's of the 4-fold coordinated solids Si, Ge, SiC, GaP, GaAs, GaSb, and InSb, for example, show a first peak whose area is equivalent to 4 ± 0.1 nearest neighbors and a nearest-neighbor distance that is only between 0.02 and 0.07 Å larger than that in the corresponding crystals [2.205]. A second coordination sphere at a distance comparable to that in the crystalline form can be distinguished in a number of cases, but beyond that the RDF's of the amorphous modifications bear little resemblance with those of their crystalline counterparts.

A comparatively new method for the structure determination of amorphous solids is EXAFS (extended X-ray absorption fine structure) [2.206]. The fine structure of an X-ray absorption edge contains information about the first coordination sphere of an atom. It has the further advantage that the absorption can be tuned to a particular element. These properties make EXAFS ideally suited for the investigation of the local structure in amorphous compounds. The method has been applied to the Ge and As chalcogenide glasses [2.206, 207].

A number of structural models have been proposed that give X-ray diffraction patterns and RDF's in reasonable agreement with the experimental results [1.205]. In the microcrystal approach the amorphous solid is thought to consist of crystallites of about 100 atoms imbedded in a completely disordered matrix. In the perturbed crystal model defects, voids and distortions are introduced into the crystalline phase until the RDF of the model matches that of the real crystal. This approach has been adopted to a-Se [2.208], a-Ge [2.209], and a-Si [2.209]. The model of a continuous random network (CRN) is mainly applied to the tetrahedrally coordinated elements and compounds. An infinite nonperiodic network of atoms is built (manually or by computer) in which the short-range bonding requirements of each atom are fulfilled according to the characteristics of the crystal. The continuous random network for a-Ge and a-Si by *Polk* [2.210] has found wide acceptance as a model for the average valence-four amorphous solid. We refer the reader to the articles by *Grigorovici* [2.205] *and Turnbull* and *Polk* [2.211] for a critical review of the structures of amorphous semiconductors.

It turns out that neither the diffraction methods, which work so well in the crystalline case nor EXAFS are sufficient to distinguish unambiguously between the various models proposed for the structure of amorphous semiconductors. In this situation the interrelation between the vibrational [2.212] and electronic properties of a system and its geometrical structure plays a crucial role in further elucidating the atomic arrangement in amorphous solids. Photoemission of the valence bands made valuable contributions to our present knowledge of the structure of a variety of amorphous systems. As we shall see in the following sections, it is the short-range order extending primarily up to the third coordination sphere that correlates most strongly with the shape of the valence density of states (DOS). Amorphous semiconductors are all primarily covalent and their electronic structure is therefore most sensitive to disorder. Simple metals where electrons interact only weakly with the ions through small pseudopotentials are free-electron-like with a DOS determined mainly by the density and hence the kinetic energy of the electrons rather than by the detailed arrangement of the atoms. At the other extreme of insulating materials with large gaps and narrow bands, the electrons are well localized around the ions and are therefore again rather insensitive to order. The covalent semiconductors correspond to the intermediate cases of maximal sensitivity of electronic structure to atomic structure and composition [2.213]. Comprehensive reviews of the structural, electronic and transport properties are given in the proceedings of the conferences on amorphous semiconductors [2.13] and in a number of monographs covering the topic [2.214–217].

2.7.1 Tetrahedrally Coordinated Amorphous Semiconductors

a) Amorphous Si and Ge

The XPS spectra of the valence bands of amorphous silicon (a-Si) and amorphous germanium (a-Ge) obtained by *Ley* et al. [1.218] are compared with those of their crystalline counterparts in Fig. 2.55. The amorphous samples

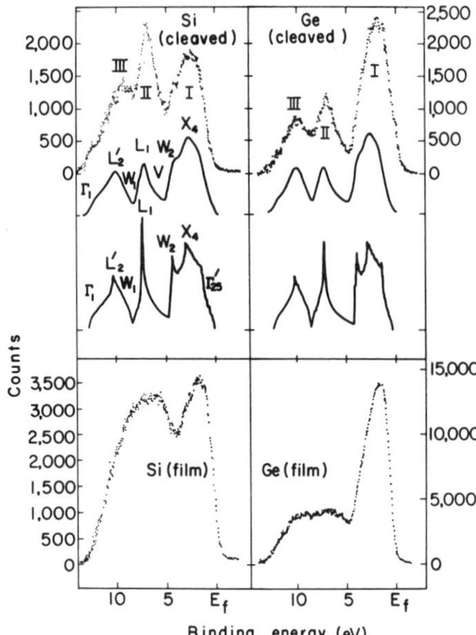

Fig. 2.55. Valence band spectra of crystalline and amorphous silicon and germanium excited with monochromatized Al K_α X-rays. Also shown are theoretical densities of states, unbroadened (lower solid lines) and broadened, in order to facilitate comparison with the experimental spectra [2.218]

were prepared by evaporation onto substrates held at room temperature, and the crystalline samples used were cleaved single crystals.

The valence bands of crystalline silicon (c-Si) and crystalline germanium (c-Ge) exhibit three peaks which correlate well in shape and position with corresponding peaks in the density of states, taking into account the finite resolution and the lifetime broadening of the hole states, which add up to a total of $\sim$0.7 eV (full width, half maximum, FWHM). The area under these peaks in the DOS corresponds to 1 electron each per atom for the two lower peaks and 2 electrons for the topmost peak. These intensity ratios are distorted in the XPS spectra due to differences in the atomic cross sections of the outermost *s*- and *p*-orbitals, as discussed in Sect. 2.4.2. The relationship between band structure, density of states, and the photoemission spectra has also been discussed in detail in Sect. 2.4.2 and will not further be elaborated upon. The results for amorphous Si and Ge can be summarized as follows:

(I) The overall appearance of the DOS is similar for the amorphous and crystalline materials. We find in particular that the total width of the valence bands remains unchanged as well as the valley between the region of predominantly *s*-derived states and the *p*-like peak at the top of the DOS.

(II) The two lower lying peaks (corresponding to the L_1 and L'_2 critical points in the crystals) merge into one hump. Since the edges of this hump do not broaden, the authors conclude that the one-hump structure occurs as the result of states that fill the dip between L_1 and L'_2 rather than through a mere broadening of the two peaks.

(III) The centroid of the p-like peak shifts towards the top of the valence bands by 0.4 eV in Si and 0.5 eV in Ge. This results in a distinct steepening of the leading edge in the DOS of the amorphous samples.

These observations are corroborated by low-energy photoemission [2.219–222] and soft X-ray emission spectra involving transitions from the valence bands of a-Si [2.223].

Depending on the deposition conditions and annealing treatments, the reported densities of a-Ge and a-Si films range from 75 to virtually 100% of the crystalline density [2.224–228]. Generally, the most dense films are those which have been deposited or annealed at the highest temperatures possible, short of crystallization [2.228] or at the slowest sputtering rates [2.229]. The rather low density obtained for a-Ge by *Clark* [2.224] (3.9 gcm^{-3} compared to 5.35 gcm^{-3} for c-Ge) led *Herman* and *van Dyke* [2.230] to consider the electronic structure of a dilated Ge crystal as a first theoretical model to explain the blue shift of peak I in the photoemission spectrum.

The small-angle scattering studies on a-Si by *Moss* and *Graczyk* [2.231, 232] indicate, however, that films deposited onto room temperature substrates contain small voids. These voids that have been observed directly by *Donovan* and *Heinemann* [2.233] in a-Ge and by *Ohdomari* et al. [2.234] in a-Si using electron microscopy are responsible for the density deficit in films of a-Ge and a-Si. Annealing studies [2.232, 234] show indeed that the voids disappear as the density of the films increases. The voidless amorphous material has therefore a density virtually identical to that of the crystals and the dilated crystal model by *Herman* and *van Dyke* cannot be applied. This conclusion is substantiated by the plasmon energy that does not change between crystalline and amorphous Si or Ge [2.218]. The plasmon energy depends on the electron density and therefore on the density of a material (see Sect. 2.3.2). The regions of high density even in a film with voids are apparently big enough to sustain undisturbed plasma oscillations and the densities of states measured by photoemission are expected to be representative for the ideal dense material.

Let us therefore start by considering the effect of the loss of long-range order on the electronic structure of a-Si and a-Ge. The arrangement of the atoms in a solid enters the Hamiltonian of the electrons through the structure factor $S(\boldsymbol{q})$. This factor relates the crystal pseudopotential $\langle \boldsymbol{k}|V|\boldsymbol{k}+\boldsymbol{q}\rangle$ (we use the plane-wave representation) to the pseudopotential form factor $\langle \boldsymbol{k}|v|\boldsymbol{k}+\boldsymbol{q}\rangle$ of an individual atom according to

$$\langle \boldsymbol{k}|V|\boldsymbol{k}+\boldsymbol{q}\rangle = S(\boldsymbol{q})\langle \boldsymbol{k}|v|\boldsymbol{k}+\boldsymbol{q}\rangle . \tag{2.19}$$

$S(\boldsymbol{q})$ reduces to a series of δ functions centered at the positions of reciprocal lattice vectors $\boldsymbol{G}_i$ for the periodic arrangement of atoms in a crystal (one atom per primitive cell is assumed here). $\langle \boldsymbol{k}|V|\boldsymbol{k}+\boldsymbol{q}\rangle$ can be regarded as a coherent scattering matrix element that mixes only waves differing by reciprocal lattice vectors, with the well-known result that the electron wave functions are Bloch states with the reduced wavevector $\boldsymbol{k}$. Deviations of the periodic arrangement introduces nonvanishing structure factors for all $\boldsymbol{q}$'s and the conventional band-structure concept breaks down. For small deviations of the atoms from their positions in the crystal nonvanishing values of $S(\boldsymbol{q})$ are concentrated around the positions of lattice wave vectors $\boldsymbol{G}_i$ and drop off rapidly away from $\boldsymbol{G}_i$ [2.235]. The additional incoherent scattering matrix elements $\langle \boldsymbol{k}|V|\boldsymbol{k}+\boldsymbol{q}\rangle$ $(\boldsymbol{q} \neq \boldsymbol{G}_i)$ can under these circumstances be considered as perturbations on an otherwise perfect crystal band structure analogous to the scattering matrix elements introduced by dislocations or phonons in a crystal and states may still be classified according to their wave vector $\boldsymbol{k}$. Following this line of reasoning *Brust* [2.236, 237] concluded that the band structure of amorphous Ge was just a broadened version of the crystalline band structure. Each eigenstate $E(\boldsymbol{k})$ of the crystalline band structure is no longer stationary but lifetime broadened as a result of the scattering probabilities which carry $E(\boldsymbol{k})$ into a number of states $E(\boldsymbol{k}')$. Assuming an energy and $\boldsymbol{k}$ vector independent scattering matrix element M_{scatt} and a phase-space factor equal to the density of states $N(E)$ he obtained for the width $\Gamma(E)$ of a level E

$$\Gamma(E)=\pi|M_{\text{scatt}}|^2 N(E)\,. \tag{2.20}$$

M_{scatt} was adjusted to reproduce the $\varepsilon_2(\omega)$ spectrum of a-Ge as measured by *Donovan* et al. [2.220] and the level widths $\Gamma(E)$ were found to lie between ~ 0.5 and 1.5 eV [2.236].

Maschke and *Thomas* [2.238] and *Kramer* [2.239–240] used the ansatz (2.19) to calculate the band structure of a number of amorphous materials using a generalized pseudopotential formalism based on Green's function techniques. The scattering contributions of the pseudopotential in (2.19) for $\boldsymbol{q} \neq \boldsymbol{G}_i$ introduce an imaginary part into the electron self-energy which leads to nonstationary eigenstates with complex energies. The imaginary part of the energy can be interpreted as the lifetime of the state in analogy to the discussion given above. *Kramer* et al. were able to obtain from these complex band structures densities of states and the imaginary part of the dielectric function $\varepsilon_2(\omega)$ for a-Si, a-Ge, a-Se, and a number of amorphous III–V compounds [2.240–243]. The calculations have the character of model calculations in that the authors use a simplified form of the structure factor $S(\boldsymbol{q})$. The squared amplitude of the structure factor is the Fourier transform of the pair distribution function of the atoms in a solid. As such, experimental structure factors obtained, e.g., from X-ray diffraction, could be used directly in the band-structure calculation. Experimental structure factors have been used by *Shaw* and *Smith* [2.244] to calculate the band structure of liquid metals in second-order perturbation theory.

Kramer [2.239, 243] chose for computational reasons the form factor $S(\boldsymbol{q})$ so that the δ functions of the crystalline form factor are replaced by Gaussian functions centered at the reciprocal lattice sites of the crystals with

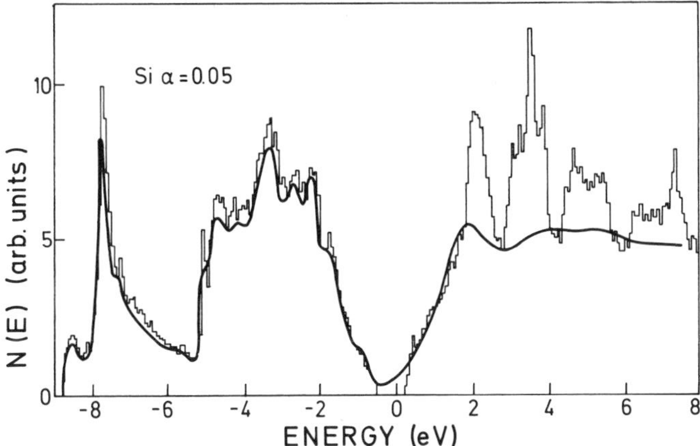

Fig. 2.56. Theoretical density of states of crystalline (histogram) and amorphous (solid line) silicon. The density of states for a-Si is based on the complex band structure model [2.240]

widths proportional to the magnitude of the reciprocal lattice vector $\boldsymbol{G}_i$. The corresponding two-particle correlation function is

$$P_2(\boldsymbol{x}) = \delta(\boldsymbol{x}) + \sum_{l \neq 0} \frac{1}{\pi^{3/2}} \frac{1}{a|\boldsymbol{x}|^3} \exp\left[-\frac{(\boldsymbol{x}-\boldsymbol{R}_l)^2}{\alpha^2|\boldsymbol{x}|^2}\right]. \tag{2.21}$$

It places two atoms such that the probability to find them separated by a lattice vector $\boldsymbol{R}_l$ decreases proportional to their distance $|\boldsymbol{x}|$. The parameter α can be used to define a short-range order region, the radius of which is

$$R_0 = \frac{a_0}{2\alpha} \quad (a_0\text{: nearest-neighbor distance}).$$

The many-atom correlation function is taken as a product of two-atom correlation functions.

This is a global description of the amorphous structure in the spirit of the distorted crystal model with short-range order and long-range disorder but no defects or voids. The densities of states calculated for c-Si and for a-Si with a disorder parameter $\alpha = 0.05$ by *Kramer* [2.240] are compared in Fig. 2.56. The calculations omit the lowest valence band (compare Figs. 2.54, 55) they are nevertheless instructive for our purpose. As anticipated by *Brust* [2.236, 237] the long-range disorder produces a broadening in the density of states. Unlike *Brust*, *Kramer* found, however, that the broadening is strongest near critical points and is not simply weighted by the density of states. A sizable contribution of states that tail into the gap is the most obvious result. It is apparent from Fig. 2.56 that the broadening is considerably stronger for the

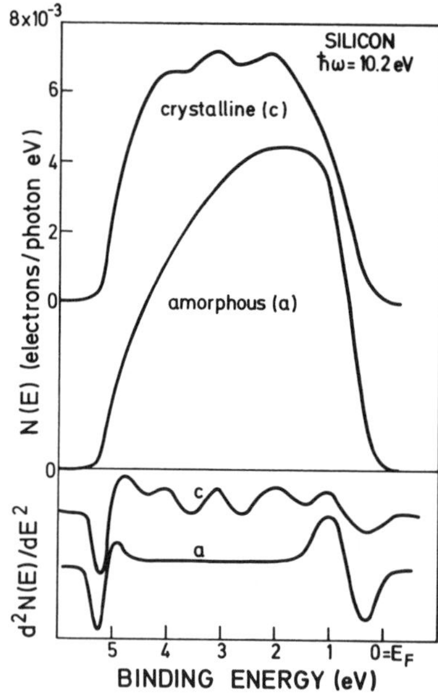

Fig. 2.57. UPS spectra ($\hbar\omega = 10.2\,\text{eV}$) of crystalline and amorphous silicon showing the loss of fine structure in a-Si. This structure is emphasized in the first derivative spectrum in the lower half of the figure [2.221]

conduction bands than for the valence bands. In fact, the model of *Kramer* gives a good representation of the rather flat (except for a hump at the onset) and structureless density of conduction states observed by *Brown* and *Rustgi* [2.245] and by *Eastman* et al. [2.246]. This result is not surprising because the conduction bands, as essentially free-electron-like states, are expected to be more susceptible to changes in long-range order than the electrons in the valence bands, which are rather well localized. It is therefore reasonable to attribute the loss of fine structure in low-energy photoemission from a-Si and a-Ge [2.221, 247] compared to their crystalline counterparts see Fig. 2.57, mainly to the relaxation of $\boldsymbol{k}$-conservation in the photoemission process and the unstructured density of final states. While the disorder parameter $\alpha = 0.05$ ($R_0 = 20\,\text{Å}$) may have been chosen somewhat too small in Fig. 2.56, later calculations by the same author with $\alpha = 0.09$ [2.248] (for this α the pair distribution function resembles more that of a-Ge) confirm that the broadening introduced in the valence bands is hardly sufficient to wipe out fine structure in the density of occupied states completely, let alone to reproduce the characteristic blue-shift of peak I and the filling of the valley between peaks II and III inherent to the amorphous state.

The conclusion to be drawn from these calculations is therefore that the lack of long-range order is certainly a necessary requirement for the structure of a-Si and a-Ge but not sufficient to explain their density of states.

It is interesting to note that a structureless density of conduction states and in particular the relaxation of the $\boldsymbol{k}$-selection rule is sufficient to reproduce the optical spectrum of a-Ge and a-Si (Fig. 2.58) quite well [2.236, 248, 259] and

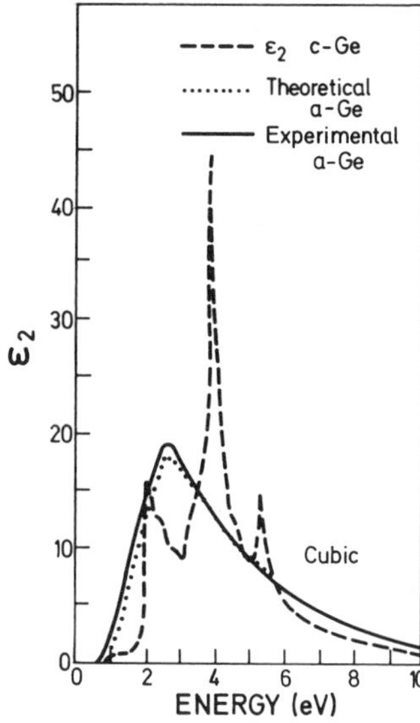

Fig. 2.58. Imaginary part (ε_2) of the dielectric constant for crystalline and amorphous germanium. The theoretical curve was calculated using the complex band structure model [2.248]

that structural models which do much better on the valence density of states give essentially no improvement over the treatment discussed so far [2.249]. This bears witness to the integrating nature of the optical properties as soon as direct transitions are no longer required.

Weaire and *Thorpe* investigated the electronic properties of nonperiodic tetrahedrally coordinated structures using a bond orbital model (BOM) for the Hamiltonian (see Sect. 2.2.1) [2.250–253]

$$H = V_1 \sum_{i,j,j'} |ij\rangle\langle ij'| + V_2 \sum_{i,i',j} |ij\rangle\langle i'j| \,. \tag{2.22}$$

The bond orbitals are denoted by $|ij\rangle$ where i labels the atom and $j(j=1,\ldots,4)$ the four sp^3 hybrids centered on each atom. The first term gives $4V_1$ for the $s-p$ splitting $\Gamma_1 - \Gamma_{25'}$ when (2.22) is used to calculate the band structure of a diamond-type crystal (compare Sect. 2.2.1). The second term by itself results in a bonding-antibonding splitting of $2V_2$ between valence and conduction bands. The complete Hamiltonian thus describes the solid as a giant covalently bonded molecule with the basic tetrahedral units of the diamond lattice as building blocks. *Thorpe* and *Weaire* [2.250] were able to show that the Hamiltonian (2.22) would give a band gap between unoccupied and occupied states regardless of the amount of disorder for values of $V_1/V_2 < 1/2$ as long as the tetrahedral coordination is preserved. The ratio V_1/V_2 is about 1/3 for a reasonable reproduction of the Ge valence band structure. Thus they gave a rigorous proof of the empirical *Ioffe–Regel* rule [2.254] which states that a semiconductor remains a semiconductor independent of its structure, as long as its coordination number remains unchanged. This result emphasizes the

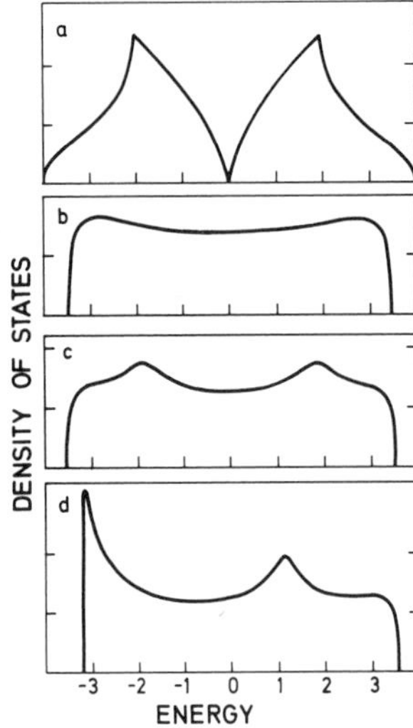

Fig. 2.59a–d. Model densities of states using the one-parameter Hamiltonian of *Thorpe* and *Weaire*.
a For the diamond (*FC*-2) structure; **b** for the Bethe lattice; **c** for the Husumi cactus with sixfold rings; **d** for the Husumi cactus with fivefold rings [2.251, 252]

chemical nature of the fundamental gap in semiconductors as opposed to its Bragg character as a result of the coherent scattering of electrons in a crystal. A contradiction between the result of *Weaire* and *Thorpe* and the existence of tail states (Sect. 2.7.4) does not necessarily exist because fixed values of V_1 and V_2 imply basically the absence of bond length and bond-angle variation in the model calculations.

As we pointed out earlier (Fig. 2.3) the BOM with only the V_1 and V_2 interactions gives a dispersionless p-band at the top of the valence bands. Structural information that would be reflected in a dispersion somewhere between that of a free electron and that of a core level is obviously contained in the s-part of the model alone. *Thorpe* and *Weaire* proceeded therefore to transform (2.22) into an even simpler Hamiltonian that had only one matrix element V between bond-orbitals on adjacent atoms [2.252]. Since the position of the atoms does not longer enter in such a Hamiltonian it is the connectivity of the network which determines the electronic structure. Consequently they investigated different structures of fourfold coordinated atoms; the results for the densities of states of four such structures are shown in Fig. 2.59. The top panel gives the density of states of a diamond lattice with the by now well-known two-peaked structure of the s-like valence bands.

The next calculation is for a so-called Bethe lattice, (Fig. 2.60) which is an infinitely branching tree-like structure of fourfold coordinated atoms [2.255]. The Bethe lattice has no closed loops of bonds. The density of states is essentially flat with well-defined upper and lower bounds. The densities of states of Figs. 2.59c, d are for "Husumi cactus lattices" [2.256]. They are similar to the Bethe lattice but are made up entirely of six- and fivefold rings of atoms respectively, so that exactly two rings go through each atom (Fig. 2.60). The

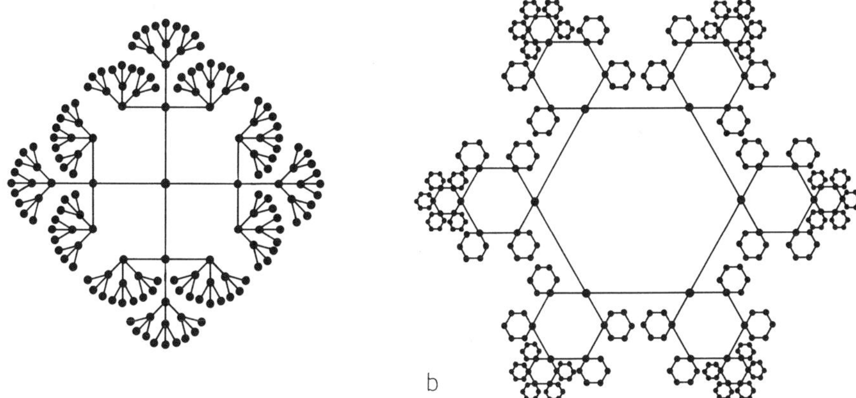

Fig. 2.60. a Bethe lattice and **b** Husumi cactus structure

sixfold Husumi cactus gives a density of states with two humps, comparable in position with the two singularities of the diamond lattice, whereas the five-ring cactus has one peak near the center of the band and another at its botton. These pseudolattices are relevant to amorphous solids because they are nonperiodic and can be used to illustrate the role of the ring structure in determining the spectrum. The main feature to notice in Fig. 2.59 is that all nonperiodic structures tend to fill the valley between the two *s*-like peaks in the density of states of the diamond structure. This effect, however, is minimized for the Husumi cactus with sixfold rings (Fig. 2.59c).

The Husumi cacti just discussed have only either five- or six-fold rings. For real lattices we define as a ring the shortest path we must follow to return to a given atom after leaving it along a given bond. A sixfold ring of the chair-like configuration in the diamond lattice is shown in Fig. 2.61a. On the basis of the results of Fig. 2.59 it has been argued that the disappearance of the valley between peaks II and III in the spectra of a-Ge and a-Si is due to the presence of rings other than sixfold as occur in the *Polk* model [2.210].

It might be surprising at first that the ring aspect in the highly connected diamond lattice, with 12 sixfold rings passing through each atom can be paralleled with the much simpler Husumi cactus structure. *Joannopoulos* and *Cohen* [2.249] have given a plausibility argument along the following lines. Consider a number of isolated sixfold rings with one orbital on each atom and a one-parameter Hamiltonian of the kind used by *Weaire* and *Thorpe* to describe the interaction. The eigenvalues E_n of any ring of order N with that Hamiltonian are

$$E_n = -2V\cos(2\pi n/N) \qquad n = 0, 1, \ldots, N-1\,, \tag{2.23}$$

where V is the intra-ring interaction. For a sixfold ring the resulting density of states is sketched in Fig. 2.62b. The density of M-isolated sixfold rings is, of course, just the M-fold degenerate density of states of a single ring. As the rings

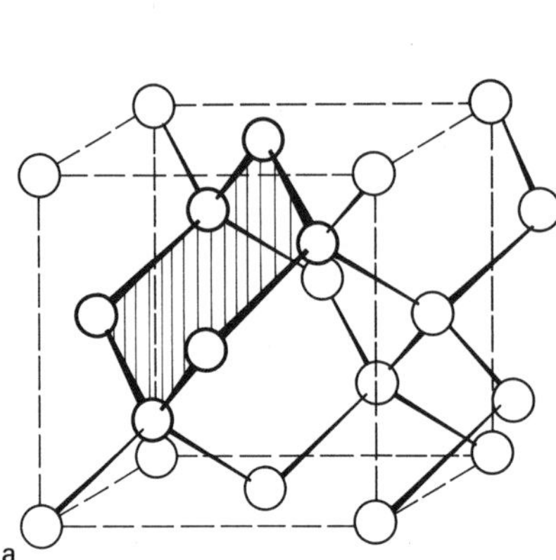

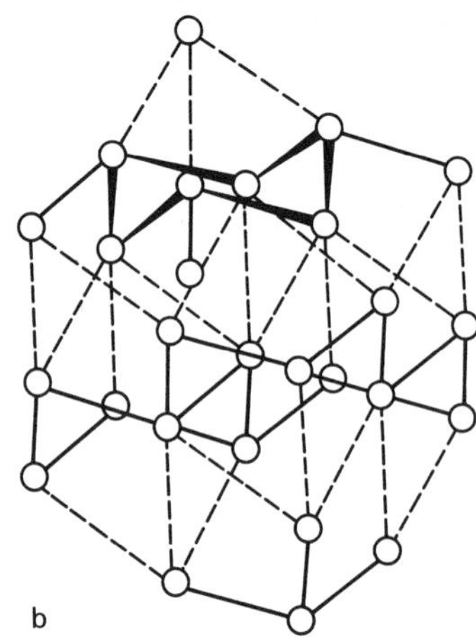

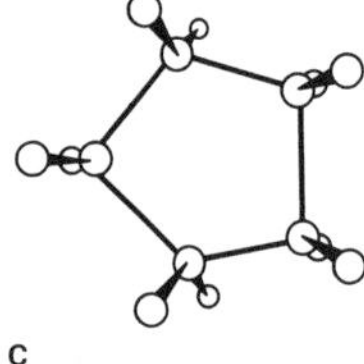

Fig. 2.61a–c. Sixfold ring in the diamond structure **a**, and in the A7 structure of arsenic, antimony, and Bismuth **b**. Also shown is a planar five-membered ring which can be formed with only a slight adjustment of the tetrahedral bond angle **c**

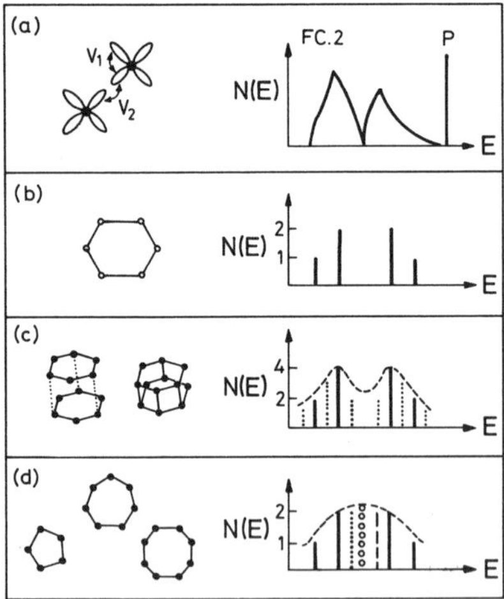

Fig. 2.62a–d. Model densities of states using the one-parameter Hamiltonian of *Thorpe* and *Weaire.*

a For the diamond (*FC*-2) lattice; **b** the degeneracy and energy levels of a sixfold ring; **c** the degeneracy and energy levels of two noninteracting sixfold rings (solid lines) and two interacting sixfold rings brought together to make a total of 5 sixfold rings (dotted lines). The dashed line represents the expected shape of the spectrum when a very large number of rings is brought together; **d** the contributions of fivefold (dashed line), sevenfold (dotted line) and eightfold (open circles) rings of bonds to the spectrum of a sixfold ring system. The trend of other than sixfold rings is to fill in the valley region as shown by the dashed curve [2.249]

are brought together they will interact and the *M*-fold degenerate state will split. The important point is now, that it is necessary to bring rings together no closer than the intra-ring distances to form the large number of rings necessary in the diamond lattice. In Fig. 2.62c two sixfold rings are brought together so as to make a total of five sixfold rings. The central gap in the eigenvalue pattern of

Fig. 2.62b is not removed by the splittings produced while bringing the rings together. Thus even the M-ring system will have a spectrum of two humps with a valley in between. With the presence of five-, seven-, and eighfold rings of atoms, new states would be introduced in the valley region as sketched in Fig. 2.62d.

It should be noted here that *Weaire* and *Thorpe* were not the first to point out that fivefold rings appeared to be a necessary and distinguishing ingredient in the structure of a-Si and a-Ge compared to the diamond lattice, *Richter* and *Breitling* [2.257] noticed as early as 1958 that the maxima in the RDF at 4.7 and 7.1 Å which are missing in the RDF of a-Ge correspond to those interatomic distances which change drastically when neighboring tetrahedra are rotated about their common bond (Fig. 2.1). This observation led *Coleman* and *Thomas* [2.258], and *Grigorovici* and *Mănăilă* [2.259] to explore the possibilities of building small three-dimensional clusters from atoms linked tetrahedrally to each other, but not solely in the staggered configuration (Fig. 2.1b) that is exclusively present in the diamond lattice. If only eclipsed bonds (Fig. 2.1b), known to exist in the wurtzite lattice, are used to link atoms together, fivefold planar rings are formed as shown in Fig. 2.61c. The bond angle of 108° differs only slightly from the tetrahedral angle of 109°28′ and this configuration is therefore likely to occur in amorphous networks. The presence of such rings has been detected by *Mader* [2.260] in small Ge particles deposited by electron beam evaporation on NaCl substrates.

An important step towards a more realistic calculation of the spectrum of a-Si and a-Ge was the investigation of polytypes of these materials with complex structures compared to the diamond lattice. A number of crystalline phases of Ge and Si that were discovered in high pressure work are metastable under normal conditions [2.261–264]. They are listed in Table 2.12 together with some of their relevant properties. Si II is a wurtzite-like structure with 4 atoms per unit cell ($2H$–4). The atoms are in perfect tetrahedral arrangement and the dihedral angles form both eclipsed and staggered configurations. Si III (Ge in the same structure is called Ge IV) is body-centered cubic with eight atoms in a primitive cell (BC-8). Bond lengths differ up to $\pm 2\%$ from those in Si and there are two types of bond angles: $\sim 118°$ and $\sim 100°$, respectively. Ge III, finally is simple tetragonal with twelve atoms per unit cell (ST-12). Bond lengths are all nearly the same and about 1% shorter than in the diamond structure. The bond angles vary between -20% to $+25\%$ from the ideal tetrahedral angle of 109°28′. A unique feature of the ST-12 structure is the presence of 5, and 8 membered rings not found in any of the other structures. In addition a polytype of SiC has also been considered. This is a hexagonal structure with eight atoms in the unit cell [2.261].

Weaire and *Williams* [2.262] suggested that the increasing complexity of these structures with atoms in configurations that deviate from the perfect tetrahedral arrangement in the diamond lattice would make these crystals suitable objects to study the influence of short-range disorder. As crystals they have the advantage to be amenable to a realistic treatment by proven band-

Table 2.12. Some characteristics of structures used for model amorphous tetrahedrally coordinated solids. Densities relative to the density of the material in the diamond structure (ϱ/ϱ_0), average bond length fluctuation ($\Delta R/R$), bond-angle distortions ($\Delta\Theta$), and the distribution of n-fold rings are given [2.249, 276]

	Name[a]	Structure	Number of atoms per primitive cell	$\frac{\varrho}{\varrho_0}$	$\frac{\Delta R}{R}$ [%]	$\Delta\Theta$	Average number of n-fold rings, $n=$ 5	6	7
Periodic	Diamond, FC-2	fcc	2	1.00	0	0°	0	12	7[b]
	Wurtzite, Si II, 2H-4	hcp	4	1.02	0	0°	0	12	0[b]
	Si III, Ge IV, BC-8	bcc	8	1.10	2	9°	0	9	0[b]
	Ge III, ST-12	st	12	1.11	1.4	16°	3.3	2	4.7[b]
	Henderson [2.286]	CRN[d]	61	1.03	3.8	12.3°	0.44	0.8	0.51[c]
Finite clusters	*Polk-Boudreaux* [2.282]	CRN[d]	519	0.995	1.1	7.1°	0.38	0.93	1.04[c]
	Connel–Temkin [2.288]	CRN[d]	238	0.99	1.2	11.5°	0	2.43	0[c]

[a] Names under which the structure appears in the literature.
[b] Average number of rings passing through one atom.
[c] Average number of rings per atom.
[d] Continuous random network.

structure methods. The influence of variations in bond length and bond angle can thus be studied and not just the topological disorder as was the case for the one-parameter BOM. Band-structure calculations were performed by *Alben* et al. [2.263] (Ge III, Si III) using a tight-binding formalism, and by *Ortenburger* and *Henderson* [2.261, 264, 265] and by *Joannopoulos* and *Cohen* [2.249, 266–270] using the empirical pseudopotential method (EPM). The results of *Joannopoulos* and *Cohen* [2.267] for three polytypes of germanium, which differ only slightly from those obtained by *Ortenburger* et al., are summarized in Fig. 2.63. The large p-bonding peak (I) is there for all structures. Its top is shifted, however, to higher-binding energy as the bond-angle distortions increase from FC-2 over BC-8 to ST-12. Electrons in this peak are localized in the bond region between atoms. As the bond angles are changed away from the tetrahedral direction the Coulomb interaction between the bond charges changes as well. Because the tetrahedral arrangement is the energetically most favorable one any change in the Coulomb interaction results in a net increase in the energy of the electrons. The s-region is more strongly affected. In the BC-8 (Si III) structure (Fig. 2.63c) the gap between the two peaks survives, even though the peaks are modified. In ST-12 (Ge III), however, the position of the valley is taken by a single broad hump in agreement with the discussion given above. After the fine structure, a result of the Bragg gaps in the crystal, is removed, the density of states resembles the photoemission spectrum of a-Ge of Fig. 2.55 quite closely. The density of conduction states is rather featureless also in agreement with experiment. These results, when compared with the complex band structure approach of *Kramer* et al. (Fig. 2.56), stress the importance of the *short-range* disorder for the electronic properties of a-Si and a-Ge. The lack

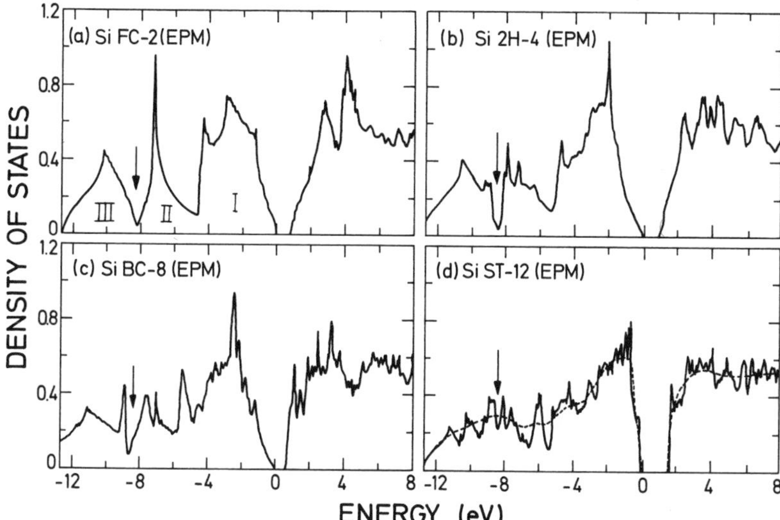

Fig. 2.63. Densities of states of polymorphs of silicon using the empirical pseudopotential method. The polymorphs model the bond length and the bond-angle distortions of the amorphous phase (see Table 2.12).

The gap between regions II and III (arrow) disappears only for the *ST*-12 structure which contains five- and sevenfold rings of bonds. The dashed line indicates the likely density of states of Si *ST*-12 in the absence of long-range order [2.249]

of long-range order is clearly of secondary importance for the densities of valence states.

In recent years a number of methods [2.271–281] have been developed to deal with continuous random networks (CRN's) [2.210, 282–288] which are at present the best representation of tetrahedrally coordinated amorphous materials. Aside from the *Henderson* model, [2.285, 286] which is a quasi-periodic lattice with 61 [2.286] atoms per unit cell and can thus be treated like a very complex crystal [2.274, 275], the main computational difficulty in a CRN arises from the surface atoms of these finite clusters. Even for the *Polk–Boudreaux* model [2.282] with 519 atoms, about 25% of the atoms reside on the surface and have thus at least one dangling bond. The corresponding "surface states" show up mainly at the top of the valence bands [2.272, 273]. Densities of states for finite clusters are therefore calculated as "local densities of states" confined to atoms well inside the cluster [2.275–281]. It is particularly interesting to compare the densities of states for the relaxed (with respect to its elastic energy) *Polk* model [2.283, 284] with that for the *Connell–Temkin* model [2.288]. Both models are very similar (Table 2.12) and their RDF's compare equally well with that measured for a-Ge [2.289]. Also the distributions of dihedral angles between neighboring tetrahedra are almost identical [2.288]. They differ, however, in that the *Connell–Temkin* model has no fivefold rings. The density of states calculated for this model does show a distinct dip between the *s*-like peaks which is absent in the *Polk* CRN [2.276, 277, 281]. This result is important in view of the doubts that have been raised [2.271, 290–293] as to the significance of ring statistics vs. bond length and bond-angle deformations for the electronic and vibrational properties of a-Si and a-Ge.

b) Amorphous III–V Compounds

The RDF's of amorphous III–V compounds are very similar to those of their elemental counterparts as we have mentioned earlier. It would therefore appear reasonable to apply the structural models that worked so well for a-Si and a-Ge also to the a-III–V's. The odd membered rings present in these models require, however, that ~10% of the bonds be between like atoms, e.g., Ga–Ga and As–As bonds in a-GaAs. Much of the work on a-III–V's has focused on the question of whether these "wrong" bonds do actually exist and the *Connell-Temkin* model [2.288] was designed to proof that a random network can be built that has all the structural characteristics of the amorphous valence 4 semiconductors without the necessity to invoke odd membered rings. Aside from a slight broadening in the first peak of the RDF of a-GaP and a-GaSb compared to the crystalline RDF's, which seems to indicate the presence of Ga–Ga, P–P, and Sb–Sb pairs [2.205], one has to look again at nonstructural information to answer whether there are wrong bonds or not. The pressure and temperature dependence of the fundamental gap in a-GaAs and a-GaP cannot be explained in terms of suitably averaged crystalline states as it may be done for a-Si and a-Ge [2.294, 295] and *Connell* offered the presence of wrong bonds as a possible explanation [2.295]. A more direct way to observe wrong bonds is possible by using the Raman spectra of amorphous semiconductors [2.296–299]. Anion pairs have been identified by their characteristic vibrational frequencies in a number of flash evaporated samples by *Wihl* et al. [2.297]. Some of the same compounds have later been studied by *Lannin* [2.298, 299] with a negative result using sputtered samples whereas he finds evidence for wrong bonds in sputtered GaP, InP, and InSb. The results of these and other measurements concerning wrong bonds are listed in Table 2.13. The apparent contradictions are to a considerable extent due to the difficulty in preparing amorphous compound semiconductors that fulfill the requirement of stoichiometry. It is not even clear if like bonds, when they are observed, are due to segregated microclusters or to wrong bonds in the CRN. Particularly the flash evaporated films of *Wihl* et al. [2.297] did show variations in stoichiometry and possibly signs of recrystallization [2.298].

The effect of like atom bonds on the density of states and the charge density has been studied in considerable detail by *Joannopoulos* and *Cohen* [2.300] and by *Yndurain* and *Joannopoulos* [2.301] using the complex crystal models already referred to in connection with their work on a-Ge and a-Si. Taking GaAs as a prototype they placed Ga and As at the atomic positions of these polytypes so as to form stoichiometric complex III–V crystals with varying

Table 2.13. Evidence for like atom bonds in III–V amorphous semiconductors (+represents positive evidence, – evidence against)

a–III–V	From temperature and pressure dependence of gap	From Raman spectra	
		Flash evaporated	Sputtered
GaAs	+[2.293, 294]		– [2.295]
GaP	+[2.293, 294]	– [2.296]	+ [2.298]
GaSb			+ (2–4%)[2.298]
InP		+ [2.296][a]	+ [2.298][a]
InAs			– [2.298]
InSb		(+)[2.296][b]	(+)[2.296][b] – [2.297]
AlSb		+ [2.296][a]	

[a] Likely due to nonstoichiometry.
[b] No clear evidence.

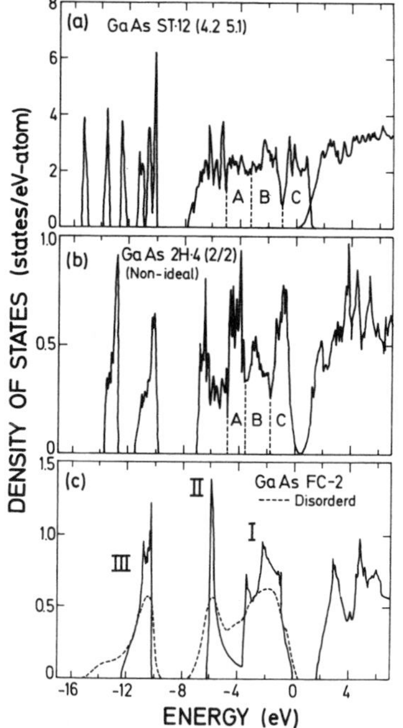

Fig. 2.64a–c. Densities of states of model structures of GaAs using the EPM. The *ST*-12 structure in **a** leads to an overlap between valence and conduction bands. In **b** the lattice constant has been adjusted so as to remove the overlap of valence and conduction bands, which is also found for the 2H-4 (2/2) structure. **c** Estimated density of states of a-GaAs containing 10% wrong bonds (dashed line) superimposed on the density of states for c-GaAs [2.249]

percentages of like-atom bonds, different clustering configurations of like atoms and topological variations of clusters of like atoms [2.300]. Two representative results of densities of states in the presence of like-atom bonds are compared with the density of states of normal zincblende GaAs (FC-2 in the notation of *Joannopoulos* and *Cohen*) in Fig. 2.64. The calculations were

performed using the empirical pseudopotentional method (EPM). GaAs 2H-4 (2/2) refers to the structure 2H-4 (see Table 2.12) in which the atoms are in perfect tetrahedral arrangement with both eclipsed and staggered configurations of the tetrahedral units; (2/2) characterizes the like-atom bonds. We have two chains, each one with two cations and the other with two anions in the primitive cell of this particular model. Similarly GaAs ST-12 (4, 2/5, 1) refers to the ST-12 structure which has bond-angle and bond-length variations and chains of 4 cations, 2 cations, 5 anions, and one anion in the unit cell, respectively. This corresponds to 33.3% like-atom bonds for GaAs ST-12 (4, 2/5, 1) and 25% for GaAs 2H-4 (2/2). The region of the *p*-bonding peak (peak I in Fig. 2.64) at the top of the valence bands is broadened considerably. This is a direct result of like-atom bonds and *Joannopoulos* and *Cohen* were able to assign spectral regions to particular bonds on the basis of charge density calculations. Region A corresponds mainly to anion-anion bonding states, region B to anion-cation bonds and region C to cation-cation bonds. The character of states at the top of the valence bands changes therefore presumably from cation-anion *p*-bonding in zincblende III–V's to cation-cation bonding in amorphous III–V's in the presence of like atom bonds. This lends some support to the interpretation of the unusual pressure and temperature dependence of the fundamental gap in a-GaAs and a-GaP in terms of wrong bonds by *Connell* [2.295]. The broadening tends to reduce the fundamental gap again in agreement with the ~0.8 eV reduction observed by *Connell* and *Paul* [2.294] both for a-GaAs and a-GaP. The model calculation apparently overestimates this broadening as the gap disappears completely (GaAs ST-12) in Fig. 2.64 unless a slight adjustment in the bond lengths is applied (GaAs 2H-4 in Fig. 2.64).

The splitting of the anion *s*-peak (peak III in Fig. 2.64) into two substructures in GaAs 2H-4 (2/2) reflects directly the clustering of anions into pairs of two atoms. In ST-12 (4, 2/5, 1) we find a total of 6 peaks in the same region, five as a result of the presence of five-membered chains and one for the single anion. Thus the anion *s*-region in the density of states signals most clearly the presence *and* also the topology of like-atom bonds in amorphous compound semiconductors. Similar arguments hold for the cation *s*-peak (peak II in Fig. 2.64), but here the overlap between peaks I and II obliterates any detailed analysis in this region. The heteropolar gap between regions II and III is unaffected by disorder and wrong bonds. It is a consequence of the difference in atomic potentials between the elements of group V and group III and as such related to the ionicity of the bonding (see Sect. 2.8.1). For a heteropolar amorphous compound with 10% wrong bonds, realized in the form of two-membered chains only, we would expect a spectrum as indicated by the dashed line in Fig. 2.64. The model density of states is broadened to account for the loss of long-range order according to the complex band-structure calculations of *Kramer* et al. [2.302].

Photoemission spectra of a number of amorphous III–V compounds have been measured by *Shevchik* et al. [2.34] using uv excitation and by *Ley* et al.

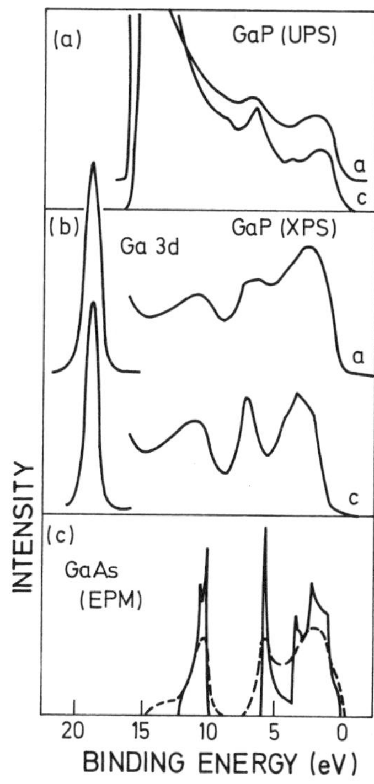

Fig. 2.65a–c. EDC's of the valence bands and the Ga 3*d* core levels of c-GaPs and a-GaPs. **a** UPS; **b** XPS; **c** theoretical densities of states of c-GaAs (solid line) and a-GaAs (dashed line) with 10% wrong bonds [2.304]

[2.303] using monochromatized Al K_α X-rays. The amorphous samples were in both cases prepared by sputtering. In the latter experiments the stoichiometry could be checked by comparing the core level intensities of the films with those of cleaved single crystals. The XPS spectrum of a-GaP in Fig. 2.65 is typical for the results obtained. The dip between peaks I and II is filled in and the *p*-bonding peak appears to have gained in intensity. Fine structure due to van Hove singularities that was present in peak I for c-GaP (Fig. 2.33) has disappeared in a-GaP. A blue shift (e.g., a shift towards the top of the valence bands) of peak I like that in a-Si or a-Ge has not been observed [2.34, 303]. The heteropolar gap between peaks II and III is retained in accord with the arguments given above. New structure in the region of the anion *s*-band could not be resolved in any of the samples by *Ley* et al. [2.303] nor in the XPS spectrum (nonmonochromatized) of a-AISb measured by *Shevchik* et al. [2.34]. In the UPS spectra the bottom of the valence bands is obscured by a steeply rising background of secondary electrons. The 4*d* core levels of indium in a-InAs and InSb have also been studied [2.34] in an attempt to identify wrong bonds through the accompanying chemical shift (see [Ref. 2.2, Sect. 1.5.2] and 2.8.2). No shifted components have been observed for stoichiometric compounds.

These results appear to argue strongly against the presence of like-atom bonds in amorphous compound semiconductors. Upon closer scrutiny the evidence is, however, insufficient to rule out the presence of 5–10% wrong bonds. As pointed out earlier, the double-peaked structure in the anion *s*-region of the valence bands is only expected if like-atoms occur in chains of two. Like-atom chains longer than two will distribute the already small (~10%) intensity of additional structure over several peaks (compare Fig. 2.64a) which might no longer be discernible under the much more intense *s*-like band due to "right" bonds [2.304]. In order to estimate the expected chemical shifts under conditions of mixed bonding, it is helpful to invoke the concept of the ligand chemical shift which is successful in predicting 3*d* core level shifts in, e.g., arsenic compounds [2.305]. From the average 0.3 and 0.6 eV shift in the In-4*d* levels between In metal and InSb and InAs, respectively, [2.34] we obtain a shift of 0.08 and 0.15 eV per wrong bond. One would certainly be hard pressed to identify a 10% component shifted by that amount in a line that is about 0.6 eV wide. More promising in this respect might be GaP. Here the Ga-3*d* shift per wrong bond is about 0.5 eV. We see that the question of wrong bonds or not is still largely unresolved for the amorphous heteropolar semiconductors. The only indication of wrong bonds might be the filling of the valley between peaks I and II. The complex band-structure calculations of *Kramer* et al. do not show enough broadening to account for that effect [2.302]. They do, however, reproduce the optical spectra of a-III–V's as measured by *Stuke* and *Zimmerer* [2.306] rather nicely.

2.7.2 Amorphous Semiconductors with an Average of Five Valence Electrons per Atom

As pointed out earlier (Sect. 2.2.2) As, Sb, and Bi are semimetals when crystallized in the rhombohedral A7 structure with two atoms per unit cell. The A7 structure may be regarded as a distorted cubic structure produced by an elongation along one of the (111) directions. This leaves puckered layers containing six-membered rings of covalently bonded atoms [2.307] (see Fig. 2.61b). Within the layer each atom has three nearest neighbors separated by 2.51 Å (As), 2.87 Å (Sb), and 3.10 Å (Bi), respectively. The bonding between layers is accomplished by two weaker long bonds which results in the anisotropy of these materials with easy cleavage along the planes. The coordination number of three is in agreement with the Bradley and Hume-Rothery "8-*N* rule". This rule states that covalent bonding in compounds with an average of *N* valence electrons implies 8-*N* nearest neighbors. The bond angles between short bonds are 97° for As and close to 95.5° for Sb and Bi, sufficiently different from 90° to require some *s*-hybridization in an otherwise pure *p*-bonding crystal [2.308]. From As over Sb to Bi there is a tendency to equalize the two different bond lengths. Their ratio drops from 1.25 in As to 1.17 in Sb and to 1.12 in Bi. Despite their semimetallic character the bonding in these materials is predominantly covalent. The semimetallic conductivity stems from a slight overlap of valence and conduction electrons in certain areas of the Brillouin zone which leads to localized electron and hole pockets [2.307]. There is, however, a tendency towards metallic bonding in the series that increases with atomic weight. The structure of Bi is only slightly distorted from simple cubic with a coordination number of six. The same tendency is observed in the group four elements: tin comes in a semiconducting (grey tin) and a metallic (white tin) modification, and lead is, of course, a metal. This is believed to be the result of the "relativistic dehybridization" [2.309] which takes place for the heaviest elements as a result of the relativistic lowering of the s-electrons as they penetrate the heavy atom core. $s-p$ hybridization necessary for a three- or fourfold coordination is therefore increasingly difficult to achieve.

Amorphous arsenic can be prepared as a bulk glass [2.310, 311]. Films of a-As can also be obtained by electrodeposition [2.312], evaporation onto substrates held at or below room temperature [2.313], by sputtering [2.314, 315], and by a glow discharge decomposition of AsH_3 [2.316, 317]. In a-As there is generally a density deficit of $\sim 12\%$, which cannot be attributed mainly to voids. The first peak of the radial distribution function of a-As remains, however, at 2.49 Å [2.312, 318] and corresponds to three atoms. Amorphous arsenic is therefore not merely an isotropic dilation of crystalline As. *Greaves* and *Davis* [2.319] have shown that the RDF of a-As can be reproduced quite well through a continuous random network model of 533 atoms in which the short-bond distance of 2.5 Å and the threefold coordination is maintained. The average bond angle is 97° with a variation of $\pm 10°$. The structure is no longer two-dimensional and the density deficit must be related to a weakening of the metallic interlayer bond [2.320, 321]. The corresponding increase in covalency is reflected in the semiconducting properties of a-As and a-Sb [2.313]. The band gap of a-As is between 1.0 and 1.1 eV, as measured by electrical or optical means [2.316, 322–324]. The optical properties of crystalline and amorphous As have been studied by several groups [2.323–328]. The Raman spectra of crystalline As, Sb, Bi, and of a-As and a-Sb have been reported by *Lannin* et al. [2.329, 330]. Photoemission spectra of the valence bands of all three elements as measured by *Ley* et al. [2.331] are set out in Fig. 2.66. The spectra for crystalline samples were obtained from cleaved single-crystal surfaces and the amorphous samples were prepared by prolonged argon-ion bombardment of the same surfaces. The valence band spectrum of a-As in Fig. 2.66 is virtually identical to that obtained by *Shevchik* [2.332] from a-As prepared by sputtering onto substrates held at room temperature. All spectra are divided into two regions: a p-like bonding peak (peak 3(4) in Fig. 2.66) at the top and a region of s-like bands (1 and 2 in Fig. 2.66) at the lower end of the valence bands. The intensity ratio of nearly 2:3 (s:p) for arsenic is in accord with the atomic configuration s^2p^3: like for germanium the cross sections for s- and p-electrons should be equal (see Sect. 2.4.2). This and the clear separation between the two regions makes their identification with the atomic s- and p-electrons more valid than in the group IV elements. It is further evidence for a reduced hybridization. The p-peak is split into two components for Bi as a result of the 2.16 eV spin-

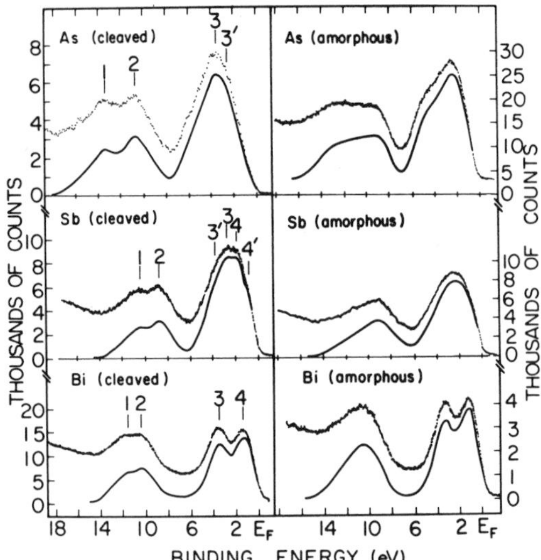

Fig. 2.66. Valence band spectra of crystalline and amorphous arsenic, antimony, and bismuth excited with monochromatized X-rays. The original spectra (dots) and the spectra corrected for a background of ineleastically scattered electrons (solid line) are shown [2.331]

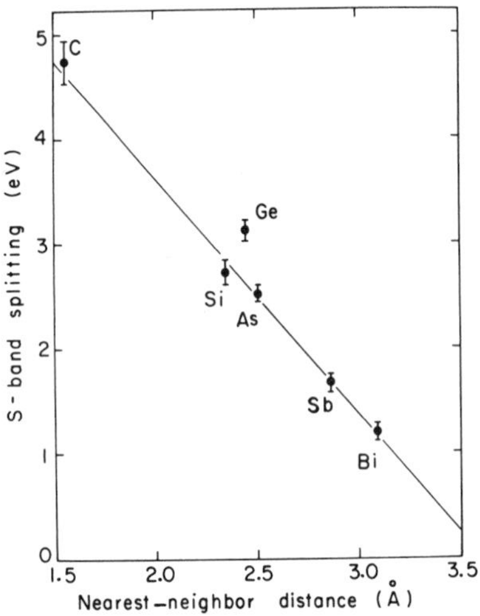

Fig. 2.67. Splitting of the *s*-like peaks (II and III for C, Si, Ge; 1 and 2 for As, Sb, Bi) versus the nearest-neigbor distance in group IV A and V A elements [2.331]

orbit splitting of the 6*p* level in atomic Bi. A similar splitting has been observed in Pb [2.333, 334].

A number of band-structure calculations have been performed for the group V elements [2.335–341]. They are in generally good agreement with the measurements; the energies of the calculated critical points are compared in detail with the experimentally observed fine structure in [2.331]. A noteworthy feature of the valence band spectra of c-As, c-Sb, and c-Bi is the bifurcation of the *s*-peaks noted by *Ley* et al. [2.331]. This bifurcation was attributed to the presence of six-membered rings in the A7 structure in analogy to the arguments given above for Si and Ge. The analogy is supported by three additional facts: First, the positions of the two peaks are well described by the energies of the T_1 and T_2' critical points of the A7 band structures. The T point of the rhombohedral Brillouin zone corresponds to L in the (diamond) structure. Thus the T_1–T_2 splitting of arsenic can be made to correspond to the L_1–$L_{2'}$ splitting of the group IV elements (see Fig. 2.55). Second, when plotted as a function of interatomic distance the separations of the s-peaks in group IV and group V elements fall on the same line (see Fig. 2.67). And most important in the present context, the bifurcation disappears in the density of states of the disordered samples in agreement with the presence of odd-numbered rings in the continuous random network model of *Greaves* and *Davis* [2.319]. The importance of the ring structure in A7 materials for the shape of the s-peak is corroborated through calculations by *Robertson* [2.320] and by *Kelly* and *Bullett* [2.321].

In addition to the by now well-documented loss of fine structure in the spectra of amorphous materials, we observe a shoulder at the high-binding energy side of peak 3 in a-As (see Fig. 2.66) that is not observed in the spectrum of rhombohedral As. *Bullett* [2.342] has suggested that this shoulder is the result of orthorhombic structure elements in a-As. Black phosphorus, the group V element above As, crystallizes in the orthorhombic structure and As can also be prepared in that structure [2.343, 344] though only in the presence of ~1 to 2% Hg. Structural studies of both crystalline modifications and of a-As based on X-ray diffraction [2.318, 344] and inelastic neutron scattering experiments [2.345] have indeed indicated that the atomic configuration in a-As may be related more closely to the metastable orthorhombic than to the rhombohedral structure of arsenic. The band structure of orthorhombic As has been calculated by *van Dyke* [2.346] and by *Bullett* [2.341]. Despite differences in detail, both calculations give a p-bonding peak at the top of the valence bands that would yield a doublet after some broadening to account for configurational averaging, whereas the rhombohedral structure yields a single, rather structureless hump. Fine structure in the low-energy reflectivity spectrum of orthorhombic and a-As observed by *Greaves* et al. [2.324] supports this interpretation. *Shevchik* [2.332] reported little difference in the top part of the valence bands between sputtered a-As films and films that had been deposited at 220 °C. He suggested that the differences in this region between amorphous and rhombohedral As observed by *Ley* et al. [2.331] arose from directional effects (see [Ref. 2.2, Chap. 6]). This is unlikely because the cross sections for all possible bond directions were about equal in the geometry used. A possible explanation is that *Shevchik*'s films may have crystallized to the orthorhombic rather than the rhombohedral form [2.341].

Among the binary compounds with an average of 5 valence electrons (see Sect. 2.2.2) the Ge-chalcogenides have attracted some attention: the short-range order in the amorphous modification of these compounds differs appreciably from that of the distorted NaCl structure of the crystals. For

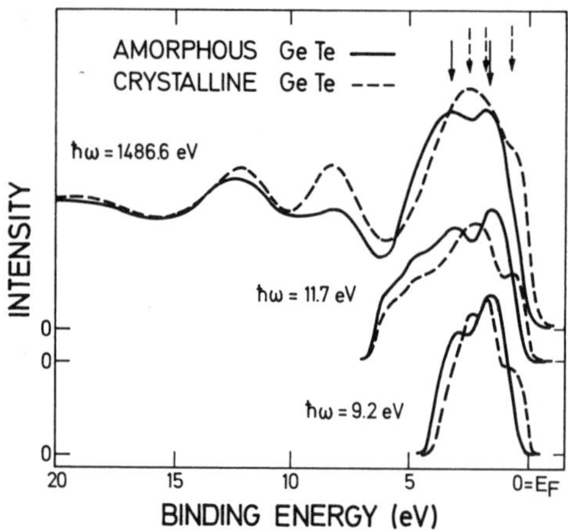

Fig. 2.68. Valence band spectra of crystalline and amorphous GeTe using a range of exciting photon energies [2.363]

instance, the RDF of crystalline GeTe, with three nearest neighbors at 2.86 Å and three next-nearest neighbors at 3.16 Å, shows an unresolved peak corresponding to six atoms at an average distance of 3.0 Å. In a-GeTe this peak is shifted down to 2.7 Å and reduced in strength so as to represent only three neighbors [2.347, 348]. This reduction in bond length from values comparable to the sum of ionic radii to values that are well described by covalent radii [2.349] is accompanied by an opening of the band gap from 0.2 in c-GeTe to 0.8 eV in a-GeTe [2.350]. The corresponding changes in the electronic structure have been well documented by now for GeTe (see Fig. 2.68) using high resolution photoemission [2.351–353]. Two structural models have been proposed for the vitreous Ge chalcogenides [2.349, 354] none of which can be ruled out on the basis of X-ray data alone [2.355]. The apparent increase in the covalency of the a-IV–VI compounds bears a certain resemblance with a-As. Exploiting this similarity *Bienenstock* [2.349] has proposed a model for the amorphous Ge chalcogenides in which each Ge atom is surrounded by three chalcogens and vice-versa in a structure that bears a certain resemblance to that of black phosphorous. The model yields RDF's in good agreement with the measured ones and is plausible in terms of its analogy to a-As. A weak point of all threefold coordinated models of binary systems is, however, that they require an average of $8-3=5$ valence electrons per atom and are thus susceptible to phase separation away from stoichiometry. The search for phase separation in nonstoichiometric Ge–Te and Ge–Se glasses was, however, unsuccessful [2.356, 357]: Raman [2.358–360] and photoemission [2.352] studies of Ge–Te and Ge–S glasses gave no indication of discontinuous changes in the vibrational or electronic properties near the GeS or GeTe composition.

The second structure proposed is the random covalent model [2.355] in which each atom is coordinated according to its bonding requirements in the elemental semiconductor: the Ge atoms have four neighbors and the chalcogens two. This local bonding arrangement can support amorphous alloys

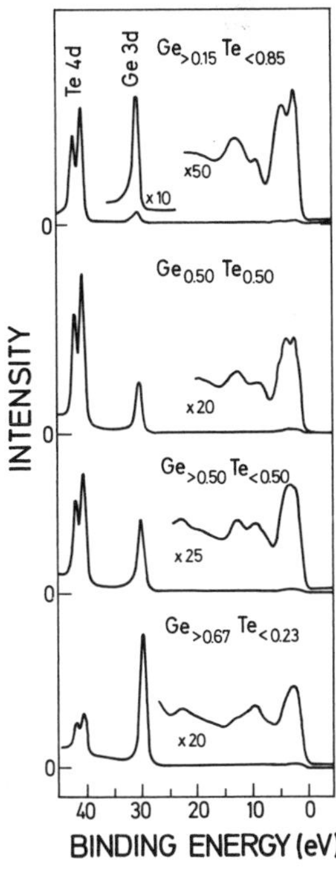

Fig. 2.69. Valence and outer core level spectra for a series of glassy Ge_xTe_{1-x} alloys excited with monochromatized X-rays [2.363]

over a wide range of concentrations. The average coordination is three for the glasses with a 1:1 composition. The radial distribution functions of all alloys, within the system Ge–S–Te are predicted with great accuracy by this model which can be regarded as an extension of the proposal by *Mott* [2.361] that atoms in amorphous semiconductors are individually coordinated according to the 8-*N* rule. This suggestion was originally based on the observed difficulty to change the conductivity of amorphous semiconductors through doping by impurities. A feature in the photoemission spectra which supports this model is the leading peak in the Ge–Te alloy spectra of Fig. 2.68 around 1.8 eV below the top of the valence bands. This peak cannot be identified in the spectrum of c-GeTe (Fig. 2.68). Its intensity scales with the Te concentration as can be seen by comparison with the Te-4*d* core levels in Ge_xTe_{1-x} alloy (Fig. 2.69). It is therefore a likely candidate to represent the two nonbonding *p*-electrons (lone pair) of a twofold coordinated Te atom (valence configuration s^2p^4) [2.352]. The bonding Te *p*-electrons, together with the 4 Ge bonding *p*-electrons, are mainly responsible for the broader second peak in Figs. 2.68, 69. The valence *s*-electrons are at 8 eV (Ge) and 12 eV (Te) in Fig. 2.68.

This rather indirect evidence for the random covalent model would be insufficient to rule out other models were it not for strong indications of $GeTe_4$ and GeS_4 units in the vibrational spectra of Ge–Te and Ge–Se glasses [2.362, 363] which support the fourfold coordination of Ge in these alloys.

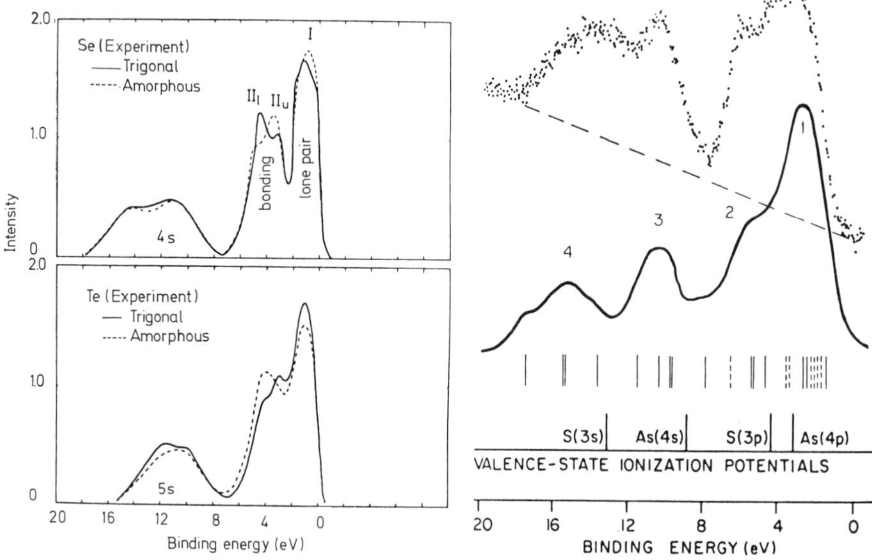

Fig. 2.70. Experimental densities of states from monocrystalline and amorphous selenium and tellurium [2.371]

Fig. 2.71. XPS valence band spectrum (monochromatized Al K_α) of As_4S_4 (dots) compared with a density of states obtained by broadening molecular orbitals (MO's) with a 1 eV wide Gaussian (solid line). The MO's were calculated self-consistently; bonding orbitals are indicated as solid lines and nonbonding (lone-pair) orbitals as dashed lines [2.364]

The identification of the leading peak in the spectra of Figs. 2.68, 69 as due to the Te lone-pair p-electrons is based on the analogy with the spectrum of elemental Te (Fig. 2.70). This has to be done with caution, as demonstrated in Fig. 2.71. The figure shows an XPS spectrum of As_4S_4 and the result of a self-consistent molecular orbit calculation for the As_4S_4 molecule both due to *Salaneck* et al. [2.364]. As_4S_4 belongs to the class of As-chalcogen glasses. This glass is a molecular solid (see Chap. 5) in the sense that the bonding is strong within the molecular unit (here As_4S_4) and weak between the units. The electronic properties of the molecular unit dominate therefore the electronic structure of the solid. There are virtually no changes in this structure as witnessed by photoemission [2.363, 364] between the crystalline and amorphous modification, because the molecular units provide the building blocks in both instances. As_4S_4 is built according to *Mott*'s rule: each S atom is bonded to two As atoms and each As atom to one As and two S atoms. The positions of the molecular orbital (MO) energies as calculated are given by vertical bars in Fig. 2.71: solid lines for bonding orbitals and dashed lines for nonbonding orbitals. These MO's when broadened, give a good representation of the measured valence-band spectrum allowing for some differences in cross sections. Two things are worth noting:

(I) there is a rather thorough mixing of As-4*p* and S-3*p* electrons in the top two peaks (1 and 2 in Fig. 2.71) with peak 1 more S-3*p*-like.

(II) Orbitals with nonbonding character are concentrated at the top of the spectrum but they are also found throughout peaks 1 and 2. The highest occupied orbital is an As–As bonding orbital instead of a lone pair! The reason for this greater complexity in the spectrum of a compound compared to that of an element is, of course, due to the higher number of nondegenerate atomic levels (4 as compared to 2) from which the MO spectrum derives.

A closely related class of materials is the group *V*-trichalcogenides. Photoemission spectra of As_2S_3, As_2Se_3, As_2Te_3 [2.365], and Sb_2Se_3 [2.366] have been measured both for the crystals and the glasses. The spectra of the amorphous modifications exhibit a "lone-pair" peak at the top of the valence bands thus signaling the twofold coordination of the chalcogens; arsenic is threefold coordinated. The lone-pair peak is preserved in the crystals for all materials. This is to be expected for the sulfides and selenides which retain their $3-2$ coordination numbers in the crystalline state. Crystalline As_2Te_3, however, has a complicated monoclinic chain structure (see Table 2.3) with the tellurium atoms threefold coordinated and the arsenic occupying both tetrahedral and octahedrally coordinated sites. Nevertheless its EDC is also very similar to that of the amorphous modification, thus casting severe doubts on the usefulness of the lone-pair concept for the identification of short-range order in chalcogenides. A simple explanation of this fact has been proposed by *Shevchik* and *Bishop* [2.367]. This explanation is based on the assumption that in As_2Te_3 only interactions between nearest-neighbor (As–Te) $p-p$ orbitals are important. Since there are 6 such orbitals of As and 9 of Te per molecule it is possible to choose three *linear combinations* of the tellurium orbitals which do not interact with their nearest-neighbor arsenics. These orbitals play essentially the same role as the lone pairs.

2.7.3 Amorphous Group VI Semiconductors

The electronic structure of trigonal selenium and tellurium has been discussed in Sect. 2.2.3. Photoemission measurements have been performed on Se and Te in their trigonal as well as in their amorphous modification by a number of workers [2.117, 146, 368–373]. Sulfur and selenium can also be prepared in an orthorhombic (S) or monoclinic (Se) structure, respectively containing eight-membered puckered rings of atoms [2.71, 374]. Photoemission spectra of the valence bands of these allotropes have been measured [2.375–379]. Low resolution XPS valence band measurements for amorphous S have been reported by *Fisher* [2.363]. High resolution overviews of the valence-band spectra of amorphous and crystalline Se and Te are shown in Fig. 2.70 [2.117, 371]. The a-Se sample was obtained by sputtering Se onto a substrate held at room temperature; the film was subsequently crystallized at 130 °C [2.117]. No check on the structure of either of the two films was performed. Amorphous Se is known to exist in two distinctly different modifications: black amorphous Se, prepared by vapor deposition, and the red modification produced by quenching of the melt [2.380]. These amorphous modifications contain Se_8 rings similar to those of monoclinic selenium. The solubility in CS_2 suggests that red a-Se is composed entirely of Se_8 rings while black a-Se is composed of rings and chains in the approximate ratio of 2 to 3 [2.380]. According to Raman data the amount of rings vs. chains in both modifications is, however, controversial [2.381, 382].

Starting material for the spectrum of trigonal Te in Fig. 2.71 was a single crystal of Te cleaved in vacuum. It was argon-ion bombarded for 2 h to make the surface amorphous [2.371]. The peaks in Fig. 2.70 are labelled according to the discussion in Sect. 2.2.3 and Fig. 2.14. There is a systematic decrease in the separation between the lone-pair orbitals (peak I) and the *p*-bonding bands (peak II) from 3.5 eV in S (not shown, [2.363]) to 2.7 eV in Se and about 2 eV in Te. This reflects the decreasing anisotropy in the bonding properties of the group VI elements with increasing atomic number. Polonium, the element following Te in the series, is cubic and the distinction between bonding and nonbonding *p*-electrons ceases to exist.

The spectra of the amorphous modifications are in many respects very much like those of their crystalline counterparts. They retain the clear separation in three regions; a broadening of the amorphous with respect to the crystalline spectra is not observed on the scale of the resolution of Fig. 2.70 ($\sim$0.5 eV). Only the high resolution uv-induced spectra of *Laude* et al. [2.370] for a-Se show a disorder induced washing out of fine structure in the leading peak that reflects the loss of long-range order according to the calculations of *Kramer* [2.383]. The most distinct changes occur in peak II: the intensity ratio of II_l and II_u is reversed in the amorphous phase for both Se and Te even though the trends are opposite in Se and Te. From the analysis of the bonding character of peaks II_l and II_u in trigonal Se by *Joannopoulos* et al. [2.77], *Schlüter* et al. [2.371] conclude, that the weakening of II_l corresponds in amorphous Se to a decrease of the number of pure intra-chain bonding states. Thus there are in a-Se more electrons occupying states which are partially localized outside the chains. In an attempt to correlate this observation with structural models for a-Se those authors [2.77, 371] calculated the density of states for model structures containing only either six- or eightfold rings. Both models reproduce the strengthening of II_u in agreement with experiment. The model containing six-membered rings best represents the experimental density of states: six-membered rings yield a two hump *s*-peak as discussed in Sect. 2.7.1, whereas eight-membered rings have a central *s*-peak [see (2.23) and Fig. 2.62] that tends to fill the dip in peak III of a-Se rather than deepen it, as observed experimentally. This line of reasoning seems to be supported by photoemission spectra of orthorhombic sulphur which consists entirely of S_8 molecules. The data of *Richardson* and *Weinberger* [2.376] and of *Salaneck* et

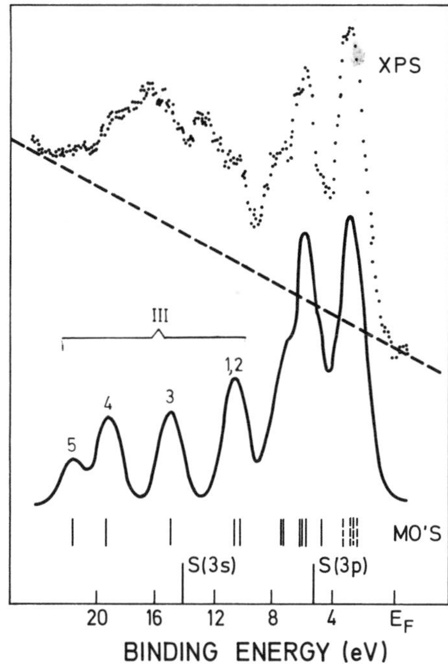

Fig. 2.72. Valence band spectrum of S_8 excited with monochromatized Al K_α radiation. Also shown are the molecular orbitals (MO's) for a single S_8 ring which have been broadened by a 1 eV wide Gaussian to facilitate comparison with experiment [2.377]

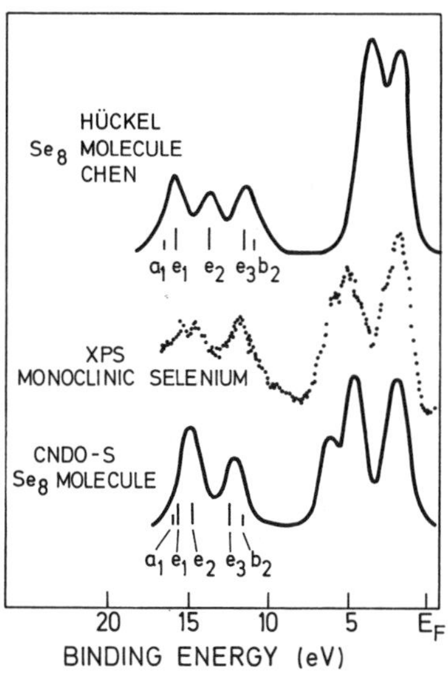

Fig. 2.73. XPS valence band spectrum of monoclinic Se (dots) for monochromatized Al K_α excitation compared with two molecular orbital (MO) calculations [2.378]

al. [2.377] both show a region of *s*-electrons with 5 peaks spaced approximately according to (2.23) and the MO calculation of *Chen* [2.384] with a central peak where there should be a minimum (Fig. 2.72). Recent photoemission measurements on monoclinic Se (Se_8 units) by *Salaneck* et al. [2.378], however, reveal a very distinct splitting of peak III (see Fig. 2.73). This is the result of a redistribution of the five levels of (2.23) into a lower group of three and an upper one of two. Thus the presence of eight membered rings in a-Se cannot be excluded.

A strengthening of peak II_u relative to trigonal Se is observed both for the monoclinic and the amorphous modifications. The origin of this phenomenon has been investigated by *Shevchik* [2.117] using a single-chain approximation for Se. He finds that a distortion of the dihedral angle from 102° in the trigonal form to $\sim 80°$ in the amorphous form reproduces the reversal in intensity of II_u and II_l. A dihedral angle of $\sim 80°$ is also appropriate for the Se_8 ring. Thus a sizeable admixture of Se_8 rings in addition to broken Se chains appears to give a reasonable explanation of the observed electronic structure. The situation for a-Te, in contrast, is far less satisfactory. The strengthening of

II_1 would be consistent with an increase in the covalency of Te in the amorphous phase. The information obtained from the *s*-like states is controversial: *Shevchik* et al. (2.117] find an increase in the dip, whereas *Schlüter* et al. [2.371] observe a piling up of states in the dip region (see Fig. 2.70). A solution of this puzzle must obviously await further experimental input.

2.7.4 Gap States in Amorphous Semiconductors

As we saw in previous sections, the loss of long-range order has a relatively minor effect on the density of valence states; it does affect the electrical conductivity, however, rather drastically. The energy gaps in the band structure of a crystal are a consequence of the periodic boundary conditions imposed on the wave functions of the electrons moving in the regular array of ion cores. Relaxation of these boundary conditions in disordered systems can lead to tails of band states extending into the gap [2.385–388]. Additional gap states are usually introduced through the presence of vacancies, chain ends, dangling bonds and microvoids in the amorphous semiconductor [2.387–389]. The concentrations of these gap states vary by many orders of magnitude depending on the material used and the way it was prepared. Densities of gap states as high as $10^{20}\,\mathrm{eV}^{-1}\,\mathrm{cm}^{-3}$ at the Fermi energy and as low as $10^{16}\,\mathrm{eV}^{-1}\,\mathrm{cm}^{-3}$ have been observed in a-Si with different methods of preparation [2.389–391b]. The gap states are localized and the amorphous semiconductor remains a semiconductor in so far as the conductivity at high temperatures is governed by carrier thermally excited into extended states [2.385, 392].

The *extended* valence and conduction states are separated by the mobility gap which corresponds approximately to the fundamental gap in crystalline semiconductors. The gap states determine the transport properties either through hopping conductivity between the localized states of amorphous semiconductors or through their influence on the position of the Fermi level [2.393].

The influence of these gap states on the fundamental absorption edge has been reviewed by *Tauc* [2.394]. There is ample evidence for a broadening and tailing of the absorption edge in many amorphous semiconductors. It is, however, not possible to ascribe these effects unambiguously to localized gap states because other broadening mechanisms (e.g., Urbach tails) and changes in the optical transition matrix element near the mobility edges [2.394] could also contribute to an absorption edge that differs from that of the crystal. The widest variety of absorption edges is observed in a-Ge and a-Si depending on the preparation technique and the thermal history of the sample. The extreme positions are illustrated by two measurements on a-Ge. A sharp absorption edge with an onset at 0.6 eV was observed by *Donovan* et al. [2.395] for a well annealed sample evaporated and measured in ultrahigh vacuum. In the opinion of the authors such a sample represents a-Ge that comes closest to the ideal of the defect-free random network model of *Polk* [2.282]. The a-Ge films, deposited at high rates onto substrates held at room temperature, on the other hand, show considerable tailing and it is difficult to define an optical gap [2.396]. A systematic investigation of the deposition conditions and the optical properties of a-Ge has been reported recently [2.397]. The absorption edge of a-Si has been studied as well [2.398, 399]. The onset of absorption ($\alpha \simeq 100\,\mathrm{cm}^{-1}$) varies from 0.6 to 1.2 eV with increasing annealing temperature [2.398, 399]. *Brodsky* et al. [2.398] have shown that the degree of tailing of the absorption correlates with the spin density believed to be representative of the number of dangling bonds in a-Si. Their measurements extrapolate to a pseudogap of ~1.7 eV for a vanishing electron spin resonance (ESR) signal [2.398b]. Even if one were to assign features in the optical absorption spectrum to transitions involving gap states it would remain nearly impossible to decide whether the initial state or the final state or both are gap states. Photoemission offers some advantages in this situation in addition to the obvious fact that the tail states observed are the occupied ones:

(I) The higher photon energies reduce the variations in the matrix element between extended and localized states because the cross sections tend to approach atomic cross sections (see [Ref. 2.2, Chap. 3]). One is therefore less likely to mistake a "matrix-element edge", separating undetected localized states from "normal" extended states, for a sharp valence-band edge with no tail states.

(II) It is possible to determine the position of the Fermi level E_F relative to the valence band of the semiconductor, because the kinetic energy of electrons originating at E_F is a known parameter of the spectrometer. This offers the opportunity to define the mobility edge E_c in the EDC if one derives $E_F - E_c$ from the temperature dependence of the exponential part of the conductivity $\sigma(T)$ according to

$$\sigma(T) = \sigma_0 \cdot \exp\left(\frac{(E_F - E_c)}{T}\right).$$

These advantages have to be balanced against the serious drawback of photoemission spectroscopy: its finite sampling depth, which limits the concentration of states that can be observed to some finite value that depends on the electron mean free path or escape depth (see Fig. 2.26), and the noise of the spectrum. Optical absorption spectroscopy is in principle free from this limitation, since a given absorption can always be maintained by increasing the thickness of the sample, such that the total number of states that contribute to the absorption is constant. A number of attempts have however been made to observe band tailing and gap states directly in photoemission from a-Ge [2.400–402], a-Si [2.403–406], and a-Te [2.407].

The mean free path of electrons as a function of energy has been discussed in connection with Fig. 2.26. In order to increase the mean free path (i.e., the sampling depth) it is advantageous to work at electron energies as low as possible. It is therefore not surprising that the first evidence for gap states was presented by *Peterson* et al. [2.403] using photon energies of 6.05 and 6.53 eV,

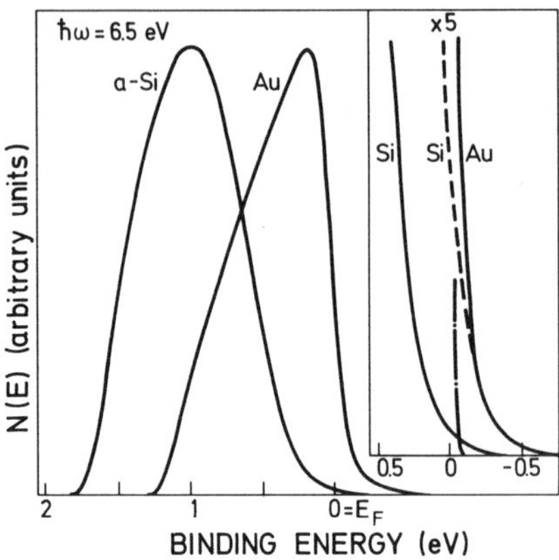

Fig. 2.74. Low-energy photoemission spectrum of a-Si and of gold used to investigate band tailing in a-Si. In the insert the high-energy edges of both spectra are superimposed in order to show that no gap states need to be invoked to explain the tailing of the a-Si spectrum. The dashed-dotted line represents the shape of the Fermi distribution

rather close to the photoemission threshold. For an a-Si film evaporated onto a substrate held at liquid nitrogen temperature those authors found a tail in their EDC's that extends up to E_F. The intensity of this tail is reduced after an unspecified "mild heat treatment". *Pierce* and *Spicer* [2.404], however, repeated these experiments on samples evaporated also at 100 K and attributed the observed tailing to the finite resolution (~0.25 eV FWHM) of their measurements as determined by the spectrum of the Fermi edge of gold. This procedure together with the spectrum of *Peterson* et al. is shown in Fig. 2.74. It is evident that in addition to the finite resolution the assignment of the position of the valence-band edge E_v plays a crucial role ascribing any portion of the spectrum to tail states. *Peterson* et al. [2.403] assumed E_v to coincide with the inflection point of the high-energy slope of their EDC, whereas *Pierce* and *Spicer* [2.404] find reasons to believe that E_v lies at the point where the linear extrapolation of the leading edge of the EDC crosses the baseline. Both choices are on equally shaky grounds.

Fischer and *Erbudak* [2.405, 406] bypassed this problem rather elegantly by using photon energies just below the threshold for photoemission from "band states". Any electrons emitted under these conditions must therefore be attributed to gap states. We reproduce in Fig. 2.75 a spectrum obtained by *Erbudak* and *Fischer* [2.406] on an a-Si sample covered with a monolayer of cesium. The cesium lowers the work function of a-Si from ~4.2 eV to about 1.8 eV and with it the photoemission threshold to 2.9 eV. The mean free path of the photoelectrons emitted at threshold thus increases by about two orders of magnitude (compare Fig. 2.26). This enhances the sensitivity for the detection of gap states by the same amount and at the same time alleviates any problem in the interpretation of the EDC's that might be connected with the cesiation of the a-Si surface. It is apparent from the spectrum taken at $\hbar\omega = 3.02$ eV in Fig. 2.75 that the identification of a hump between E_v and E_F is rather insensitive to the exact position of E_v which was chosen by the authors to lie 1.1 eV below E_F.

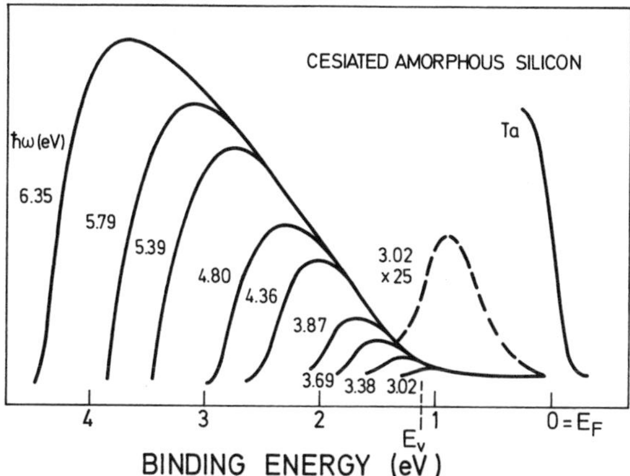

Fig. 2.75. A series of low-energy photoemission spectra of cesiated amorphous silicon. The position of the Fermi level, E_F, is determined by the sharp onset of a tantalum spectrum (Ta). The top of the valence bands extrapolates to a position marked E_v. The emission intensity between E_v and E_F is attributed to gap states [2.406]

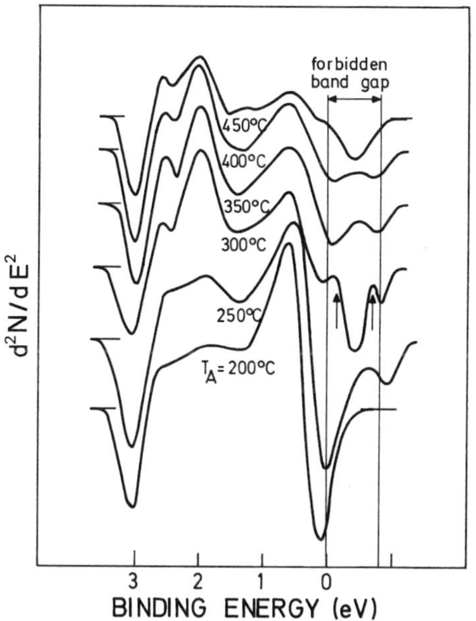

Fig. 2.76. Second derivative of the EDC's of a-Ge for a series of annealing temperatures T_A. The arrows indicate features that have been attributed to gap states [2.401, 402]

Laude et al. [2.401, 402] studied the effect of the preparation parameters of a-Ge on the distribution of gap states using the second derivative of the EDC's which greatly enhances any structure that is present in the spectra. Films that were prepared so as to develop considerable internal stress exhibit structure in the second derivative of the EDC above E_v upon annealing at $\sim 300\,°C$ (see

Fig. 2.76). This structure was attributed to localized states due to defects that form during the thermally activated rearrangement processes that take place as the strained amorphous network of the films begins to order. It is worth noting that the energy distribution of gap states in Fig. 2.76 resembles rather closely that obtained for a-Si by *Spear* and *LeComber* [2.408a] by analyzing conductivity and photoconductivity measurements. From these examples we can draw two conclusions: (I) photoelectron spectroscopy has a considerable potential for the investigation of gap states in amorphous semiconductors, in particular if one combines photoemission near threshold (possibly lowering the threshold through cesiation) with a derivative technique to enhance the sensitivity for the detection of such states in the EDC. (II) It is mandatory to combine photoelectron spectroscopy with auxiliary measurements that clearly establish the dividing line between extended and localized states and thus eliminate the ambiguities that go into the determination of the mobility edge E_v in the examples given. A rather straightforward way for *p*-type materials would be to determine the activation energy $(E_v - E_F)$ from the temperature dependence of the conductivity and combine this result with the position of E_F in the EDC to obtain E_v [2.408b].

2.8 Ionicity

We probably all have an intuitive understanding of what ionicity means. Consider in the present context the isoelectronic series of semiconductors and insulators Ge, GaAs, ZnSe, and KBr which share 8 valence electrons. The ionicity increases in this series from zero in Ge to a maximum in KBr. Germanium is covalently bonded with the valence electrons distributed symmetrically in the central bonding region between nearest neighbors. In GaAs the bonding charge is displaced from the center between atoms towards the As atoms. This displacement increases as the difference in the number of atomic valence electrons increases until all 8 valence electrons are concentrated around Br leaving a net charge of unity and opposite sign on both partners. Charge distributions in agreement with these expectations have been calculated by *Walter* and *Cohen* [2.409] for a number of semiconductors using pseudowave functions. The driving force behind the displacement of the bonding charge is the greater ability of the nonmetals to the right of germanium to attract electrons compared to the metallic atoms on the left in the periodic table. The bonding characteristics change at the same time from the highly directional bonding in Ge to the almost purely electrostatic attraction between charged spheres in the alkali halides which lacks all directionality and favors a maximum number of neighbors with opposite charge.

The balance between ionic and covalent bonding determines a number of electronic and mechanical properties of semiconductors. For instance, the fundamental gap increases with ionicity, the static dielectric constants $\varepsilon(0)$

decreases, the shear moduli decrease and the splitting between longitudinal-optical (LO) and transverse-optical (TO) phonons at the zone center increases with increasing polarity of the bonds. A detailed discussion of the relation between these and other properties with ionicity is given by *Phillips* [2.410].

A quantitative measure of ionicity should be a useful concept to unify a host of experimental data. A number of ionicity scales have indeed been devised with this intention. They assign an ionicity f_i between 0 and 1 to each material such that the most covalent ones have small values of f_i and for the most ionic f_i is close to unity. Homopolar solids like Ge, Si, etc. have by definition $f_\mathrm{i} \equiv 0$. Any property that is related to the charge transfer between cation and anion may be used to define the ionicity f_i. One of the oldest and still most widely used scales is that of *Pauling* [2.411a]. It is based on the observation that the heat of formation $H_\mathrm{f}(\mathrm{AB})$ for a compound AB exceeds the average heat of formation $[H_\mathrm{f}(\mathrm{A})+H_\mathrm{f}(\mathrm{B})]/2$ of the two constituent solids A and B. This extra ionic heat of formation $-\Delta H_\mathrm{f}(\mathrm{AB})$ can be parametrized according to

$$-\Delta H_\mathrm{f} = g N_\mathrm{r}(X_\mathrm{A} - X_\mathrm{B})^2 \tag{2.24}$$

for compounds containing no oxygen or nitrogen. N_r is the number of resonating bonds per atom pair, which equals one as long as we are dealing with coordination numbers in accord with the 8–N rule. (For further details concerning the treatment of N_r see [2.411b]). The constant g, with the dimension of an energy, equals about 1 eV or 27 kcal/mol. X_A and X_B are the electronegativities of atoms A and B, respectively. *Pauling*'s definition of the ionicity of a single bond is

$$f_\mathrm{i}^P = 1 - \exp[-(X_\mathrm{A} - X_\mathrm{B})^2/4]. \tag{2.25}$$

The electronegativities defined through (2.24) are obtained from thermochemical data. They are given in the periodic table on the inside cover of this volume.

Other ionicity scales express the ability of atoms to attract electrons through the mean value of electron affinity and ionization potential [2.412]. *Coulson* et al. [2.413] define ionicity through the probabilities to find a valence electron on either of the two atoms using trial wave functions based on atomic orbitals. Two more recent approaches confine themselves to compounds with an average of four valence electrons. *Phillips* and *van Vechten* [2.414–416] allocate covalent and ionic contributions to the mean separation E_g between valence and conduction bands as obtained from the dielectric constant (dielectric theory, DT) E_g corresponds approximately to the average energy of the optical transitions between valence and conduction bands. *Harrison*'s approach, similar to that of *Coulson* et al. [2.413], relates the covalent and ionic contributions to the bonding in $\mathrm{A}^N\mathrm{B}^{8-N}$ compounds to matrix elements in an LCAO description of the valence bands [2.37, 417, 418]. The relative merits of all these ionicity scales are critically reviewed in [2.410, 416, 417]. In the next section we shall describe an ionicity scale that is derived from the photoemission spectra of the valence bands of compound semiconductors and insulators.

Ionic charges play obviously a decisive role in the physics of heteropolar compounds and yet they are quantities difficult to assess in a consistent and meaningful way. Even if we know the distribution of the valence charge in a semiconductor, obtained, e.g., from pseudopotential calculations, the ionic charges are arbitrary, depending on the way we divide the space assigned to individual atoms. In the context of bond orbital models [2.37, 417, 418] the ionicity, or polarity α_p, as it is called in [2.37, 417, 418], is directly related to the admixture coefficients of the cation and anion hybrids $|h^c\rangle$ and $|h^a\rangle$ which combine to form a bond $|b\rangle$

$$|b\rangle = \frac{1}{\sqrt{2}}(1-\alpha_p)^{1/2}|h^c\rangle + \frac{1}{\sqrt{2}}(1+\alpha_p)^{1/2}|h^a\rangle . \tag{2.26}$$

A simple definition of the charge on each atom is obtained by interpreting the square of the admixture coefficients in (2.26) as the probability to find the 8 valence electrons that are shared between cation and anion on either atom in an A^NB^{8-N} compound. Subtracting from this charge the number of electrons necessary to make the ions neutral we obtain for the ionic charge of the anion q_i [2.418]

$$q_i = -8\cdot\tfrac{1}{2}(1+\alpha_p)+(8-N) = -4\alpha_p - N + 4 . \tag{2.27}$$

Equation (2.27) is sometimes also used to derive ionic charges for an ionicity defined in a framework other than the bond orbital model. Ionic charges using the polarity α_p obtained by *Harrison* and *Ciraci* (q_i^{BOM}) [2.418] and the ionicity f_i^{DT} derived by *Phillips* and *van Vechten* (q_i^{DT}) [2.416] are given in Table 2.14 for a number of tetrahedrally coordinated compounds. We find that the charges are quite different indeed for the two ionicity scales. It is not easy to relate the static q of (2.27) to an experimentally observable quantity and thus to obtain, within this model, an experimental value for the ionicity f_i.

The atomic form factor that determines the intensity of Bragg peaks in X-ray scattering depends on the charge distribution of the constituent atoms. Thus in principle the measurement of scattering intensities yields information on the ionic charges. Unfortunately this works well only for materials composed of light elements: the large number of core electrons reduces the accuracy with which q can be determined. Effective ionic charges of 1.7 ± 0.5 have been determined for AlN and GaN by this method [2.419]. The binding energy of core levels as measured in photoemission is also sensitive to the valence charge distribution as shown in some detail in [Ref. 2.2, Sect. 1.5.2]. Chemical shifts observed for the core level binding energy of one element in different compounds have therefore been interpreted in terms of ionic charges as will be discussed in Sect. 2.8.2.

Besides the "static" charge q just discussed it is also possible to define dynamical charges related to the infrared absorption (e_T^*) [2.420] and the piezoelectric effect (e_p^*) [2.421]. These charges, however, differ from each other

Table 2.14. Core level shifts ΔE and effective cation charges q_i for a number of binary semiconductors. The experimental shifts are relative to the binding energy in the corresponding elemental solid. Reference level is in both cases the vacuum level. Shifts ΔE(calc) are calculated using effective charges q_i obtained with (2.27) from *Harrison's* polarity α_p [2.418] and replacing α_p by the *Phillips van Vechten* ionicity f_i^{DT} [2.414]. The last column gives the shifts calculated by *Hübner* and *Schäfer* [2.454] (for details see text). All energies are given in eV

Crystal	ΔE(exp)[a]	f_i^{DT}	q_i^{DT}	ΔE(calc)[a]	α_p	q_i^{BOM}	ΔE(calc)	ΔE(calc)[d]
AlSb	−1.8	0,426	0.70	−1.64	0.54	1.16	−2.7	—
GaP	−1.8[d]	0.374	0.50	−1.90	0.52	1.08	−4.1	−1.8
GaAs	−1.7	0.310	0.24	−0.90	0.50	1.00	−4.5	−1.7
GaSb	−1.1	0.261	0.045	−0.20	0.44	0.76	−3.4	−0.9
InP	−1.6	0.421	0.69	−1.12	0.58	1.32	−2.1	−1.5
InAs	−1.0	0.357	0.43	−0.93	0.53	1.12	−2.4	−1.0
InSb	−0.7	0.321	0.28	−0.69	0.51	1.04	−2.6	−0.8
ZnSe	−1.65	0.676	0.70	−2.48	0.72	0.88	−3.1	
ZnTe	−1.19	0.546	0.18	−0.67	0.72	0.88	−3.3	
CdS	−1.56	0.685	0.70	−1.65	0.77	1.08	−2.6	
CdSe	−1.34	0.699	0.80	−2.1	0.77	1.08	−2.8	
CdTe	−1.44	0.675	0.70	−2.2	0.76	1.04	−3.3	
CuCl	+1.1 (5)[b]	0.746	0.02	−0.07[b]	0.75	0.00	±0.0[c]	
CuBr	+1.5 (5)[b]	0.735	0.06	−0.31[b]	0.79	−0.16	+0.8[c]	
CuI	+0.8 (5)[b]	0.692	0.23	−1.23[b]	0.78	−0.12	+0.7[c]	
AgI	+0.9 (5)[b]	0.770	−0.08	+0.35[b]	0.83	−0.32	+1.4[c]	

[a] [2.34]. [c] [2.452]
[b] [2.49]. [d] [2.454].

and bear no direct relationship to the static charge. Simple expressions for e_T^* and e_p^* based on the bond orbital model have been given for tetrahedral semiconductors by *Harrison* and *Ciraci* [2.418]. A microscopic calculation of e_T^* based on the pseudopotential band structure has been published recently [2.422].

2.8.1 An Ionicity Scale Based on Valence Band Spectra

An ionicity scale that uses the splitting of the low-lying *s*-like peaks in the density of states of semiconductors has been developed by *Kowalczyk* et al. [2.423]. As shown in Fig. 2.77 the separation ΔE_s of these two peaks increases monotonically in the isoelectronic series Ge, GaAs, ZnSe, and KBr. Thus ΔE_s must be related to the ionic character of the bonding. The splitting ΔE_s does, however, not vanish for Ge. *Kowalczyk* et al. proceeded therefore to decompose ΔE_s into its covalent (ΔE_s^c) and ionic contributions (ΔE_s^i)

$$\Delta E_s = \Delta E_s^c + \Delta E_s^i . \tag{2.28}$$

The covalent gap ΔE_s^c was identified with the splitting in the elemental semiconductors. As we have seen in Fig. 2.67 ΔE_s^c follows a simple linear

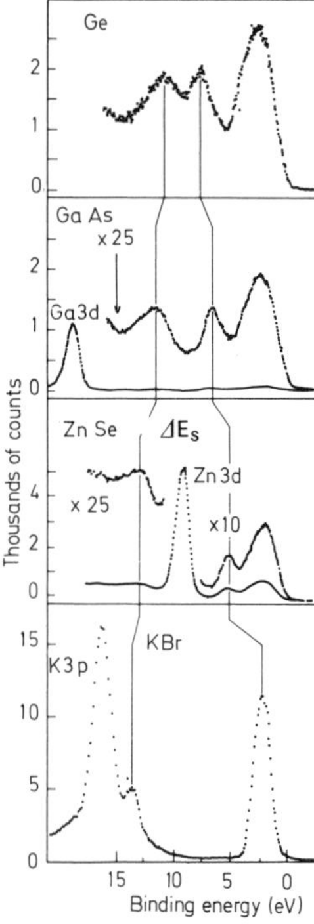

Fig. 2.77. Photoemission spectra of a series of isoelectronic octet crystals showing the increase in the splitting ΔE_s of the s-derived bands with increasing ionicity [2.423]

relationship as a function of the interatomic distance d

$$\Delta E_s^c = 8.0 - 2.2d\,, \tag{2.29}$$

with ΔE_s^c measured in eV and d in Å.

The dependence of ΔE_s^c on d is qualitatively what one would expect for the bonding-antibonding splitting of two states interacting through the overlap of their orbitals (compare with the discussion of the *Weaire* and *Thorpe* model in Sect. 2.7.1). This makes ΔE_s^c a good measure of the covalency of the bonds. For heteropolar compounds ΔE_s^c is calculated according to (2.29) and the ionic gap ΔE_s^i is obtained as the difference between the observed splitting ΔE_s and ΔE_s^c.

An ionicity f_i^{XPS} is now defined as

$$f_i^{\mathrm{XPS}} = \frac{\Delta E_s^i}{\Delta E_s} = \frac{\Delta E_s^i}{\Delta E_s^i + \Delta E_s^c}\,. \tag{2.30}$$

A "covalency" could be defined similarly by replacing E_s^i by E_s^c. This symmetrical treatment reflects adequately the balance between ionic and covalent contributions to the bonding properties of a heteropolar crystal. The ionicity of *Pauling* for instance lacks this symmetry. He used only the *ionic* contribution to the heat of formation to define f_i. The definition of f_i due to *Phillips* and *van Vechten* [2.414, 416] is very similar: they defined the average separation E_g of valence and conduction bands as the pythagorean sum of an ionic part C and a covalent part E_h:

$$f_i^{DT} = \frac{C^2}{E_h^2 + C^2} = \frac{C^2}{E_g^2}. \tag{2.31}$$

The energy E_g is obtained from the low-frequency dielectric constant $\varepsilon(0)$ in the framework of the *Penn* model [2.424].

The ionic energy C is obtained from E_g and the gap E_h of the corresponding isoelectronic homopolar compound (e.g., Ge for GaAs). The homopolar gap E_h scales with interatomic distance d-like $d^{-2.5}$. A list of ionicities f_i^{XPS} based on photoemission data is given in Table 2.14. A distinct advantage of the XPS derived scale is that it is directly applicable to average valence 5 compounds and ternary compounds as well (see Table 2.14). In the latter case a properly weighted average of interatomic distances has been taken to calculate E_s^c with the help of (2.29). The Phillips–van Vechten scale has also been modified to be applicable to other than A^NB^{8-N} compounds [2.425, 426].

The relationship between structure and ionicity has been of particular interest since the pioneering work of *Mooser* and *Pearson* [2.427]. These authors were able to separate fourfold coordinated and sixfold coordinated A^NB^{8-N} crystals into two distinct groups when plotted in a coordinate system with *Pauling*'s electronegativity difference as abscissa and the average period number $\bar{n} = (n_A + n_B)/2$ as ordinate. The compounds with rock-salt structure fall in the upper right half of the plot and those with zincblende or wurtzite structure in the lower left half with only about 8 cases on the wrong side of the boundary. The interatomic distance d of an A^NB^{8-N} compound is a monotonically increasing function of $\bar{n}$. d, in turn, decreases monotonically with increasing covalency (2.29). Thus $\bar{n}$ may be taken to represent the "lack" of covalency. The upper right half of *Mooser* and *Pearson*'s plot corresponds therefore to compounds with low covalency or high ionicity and for the lower left half the situation is reversed. Had Mooser and Pearson plotted the compounds on a one-dimensional diagram as a function of electronegativity difference alone, there would have been considerably more mixing between the structures. This result emphasizes the importance of the *balance* between covalent and ionic bonding in determining the structure that is adopted by a material.

Since the XPS ionicity scale takes this balance into account, we might hope that compounds can be separated into high and low coordinated ones on the basis of their ionicity f_i^{XPS} alone. That this is indeed the case, is shown in Fig. 2.78. A critical ionicity $F_i^{XPS} = 0.68$ separates 41 compounds into the two groups. PbS, with its NaCl structure and an ionicity of 0.67, is a borderline case. Also shown in Fig. 2.78 are 36 compounds arranged according to their dielectric ionicities. The critical ionicity is 0.71 in this case and two compounds

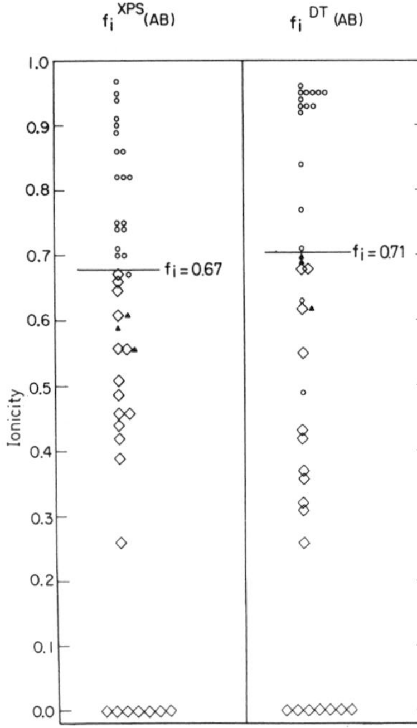

Fig. 2.78. Octet compounds plotted according to their ionicity in the XPS derived scale (f_i^{XPS}) and the scale based on the dielectric theory. The diamonds and triangles stand for the four- and threefold coordinated structures, respectively; the circles indicate compounds in the NaCl or CsCl structures. The horizontal lines indicate the critical ionicities that separate high from low coordination [2.423]

fall on the wrong side. The ordering of compounds is in general similar in both scales. It appears, however, that the XPS scale works better in the region of the very ionic alkali halides. They span a range of ionicities between ~0.8 and nearly one, whereas f_i^{DT} is around 0.95 for all of them. A range of ionicities smaller than 1 for the alkali halides suggests a nonvanishing covalent contribution to their bonding in contrast to the model of closed shell ions interacting only through their electrostatic forces. This covalency is reflected in the valence band spectra as has been observed by a number of workers [2.428–430]. *Poole* et al. [2.430] in a series of investigations of the alkali halides using uv excitation (He I and He II) have noticed a distinct narrowing of the uppermost valence band (corresponding to the halide *p*-level) with increasing interatomic distance. The influence of band-structure effects on the valence bands of the alkali halides has been discussed by *Kowalczyk* et al. [2.429]. The top peak of the sodium and potassium halides from [2.429] is reproduced in Fig. 2.79 together with the corresponding value of f_i^{XPS}. For compounds with the same halogen, i.e., for the same atomic origin of the valence bands, the peak in the sodium halide shows more structure than that of the potassium halide. This signals a higher degree of "banding" in the latter, in agreement with the lower ionicity of the sodium halides.

Of some concern to *Kowalczyk* et al. [2.423] was the fact that the ionicities of compounds with the same cation but different anions were opposite to those

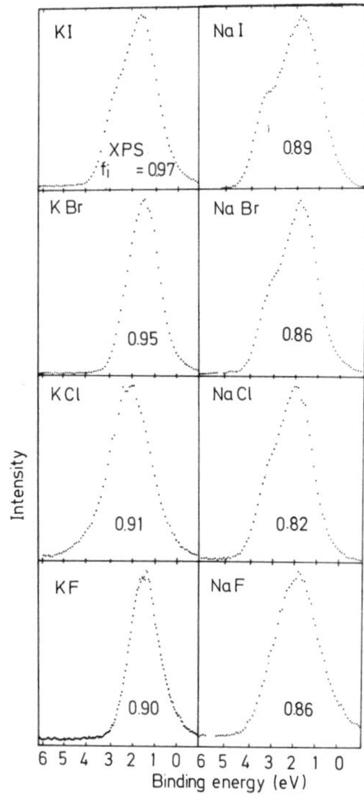

Fig. 2.79. The anion-derived top peak in the XPS spectrum of the valence bands of potassium and sodium halides. Also given are the corresponding ionicities f_i^{XPS} of the compounds [2.429]

of Phillips and van Vechten's or Pauling's scale. For the series of Fig. 2.79 f_i^{XPS} decreases slightly from the iodides to the fluorides, whereas f_i^P and f_i^{DT} have the opposite trend. The reason for that inversion was conjectured to lie in a size factor of the anions [2.423]. *Kemeny* et al. [2.431] have shown that such an "inverted" order of f_i^{XPS} as compared with f_i^{DT} [2.432, 433] is also found for the IV–VI compounds. The ionicities f_i^{XPS} of nine IV–VI compounds are shown in Table 2.2 together with their crystal structures. The noncubic structures have a coordination of 3+3, i.e., each atom has three nearest neighbors and three next-nearest-neighbors at a somewhat larger distance. These distances are 2.86 and 3.16 Å for GeTe, 2.58 and 3.34 Å for GeSe, and 2.48 and 3.25 Å for GeS [2.453].

Kemeny et al. [2.431] regarded these compounds as threefold coordinated: the ionicities obtained with the shorter interatomic distance are all smaller than the critical value 0.68. PbS and SnTe are borderline cases, a fact which is borne out in particular by SnTe which can be prepared in the NaCl structure as well. Also listed in Table 2.2 are the pressures at which some of the compounds undergo a phase transition from the NaCl to the orthorhombic structure. This phase transition from high coordination to low coordination can be understood in terms of the ionicity under the following assumptions [2.431].

(I) The bond length of the NaCl structure is reduced under pressure: this leads to an increase of the covalent contribution ΔE_s^c to the s-band splitting according to (2.29). (II) The ionic splitting ΔE_s^i, which reflects the difference in the atomic potentials through the differences in the binding energies of the s-like bands, is unaffected by pressure. As a consequence f_i^{XPS} decreases with pressure until one reaches the critical ionicity which favors the low-coordinated or orthorhombic structure. The transition pressures do indeed increase with increasing ionicity f_i^{XPS} for the lead and for the tin salts: this trend is reversed if f_i^{DT} (2.432, 433] is used! Using the compressibilities of the materials, *Kemeny* et al. were able to calculate transition pressures of 43 kbar for PbSe and 63 kbar for PbTe, in more than qualitative agreement with experiment. The connection of these phase transitions with the concepts of ionicity and metallicity are discussed in [2.426, 432].

The relationship between the splitting of the s-like valence bands at the X-point of the Brillouin zone ($X_1 - X_3$, see Fig. 2.5) and their ionic character has already been mentioned in Sect. 2.2.1. *Grobman* et al. [2.434] have actually proposed that this splitting is a direct measure of the ionic energy C of *Phillips*' theory for octet compounds (GaAs, NaCl). *Chadi* et al. [2.435] have shown that this $X_1 - X_3$ splitting can be expressed in terms of certain Fourier coefficients of the antisymmetric part of the pseudopotential. Experimental values for this gap have been given by *Ley* et al. [2.147] and by *Grobman* et al. [2.434]. *Unger* and *Neumann* [2.436] combined this gap with the total width of the valence bands to define an ionicity scale. Band edges are, however, usually less well defined than peak positions; thus it seems physically more reasonable to choose regions of high densities of states to define a "chemical" property like the ionicity than edges which carry no spectroscopic weight.

2.8.2 Binding Energy Shift and Charge Transfer

The connection between the shift in the binding energy of core levels and the chemical environment has been one of the central problems of XPS since its inception. Much of the early work and the basic ideas are discussed in the book by *Siegbahn* et al. [2.437]. The principles that govern chemical shifts in solids, including the problem of relaxation and the choice of a proper reference level, have been formulated by *Fadley* et al. [2.438] and applied to shifts measured for a number of europium and iodine compounds. A rather qualitative correlation between chemical shifts in a series of II–VI compound semiconductors and ionicities based on different scales has been established in an exploratory study by *Vesely* and *Langer* [2.439]. A linear dependence between chemical shifts of Ge core levels, corrected for a Madelung term and Pauling charges has been found by *Hollinger* et al. [2.440] for seven germanium compounds (see [Ref. 2.2, Fig. 1.21]).

The relationship used to derive the "static" charge q_i on an atom i from the binding energy shift $\Delta E(q_i)$ of one of the core levels of the same atom is invariably derived from an expression of the form (in atomic units)

$$\Delta E(q_i) = \frac{q_i}{\bar{r}} + \sum_{j \neq i} \frac{q_j}{r_j} \tag{2.32}$$

(see also [Ref., 2.2, Eqs. (1.51, 52)]. In (2.32) ΔE is the balance of two electrostatic potentials of opposite signs acting on the core electron. The first term is the potential set up by the valence charge q_i of the atom under consideration, which is assumed to be spread evenly over a sphere of radius $\bar{r}$, the valence charge radius. The second term is the potential due to the charges q_j centered at the position of all other atoms in the solid. For binary compounds $q_i = -q_j$ and the lattice potential can be expressed in terms of the Madelung constant α [2.437, 441]

$$\sum_{j \neq i} \frac{q_i}{r_j} = -\frac{q_i \alpha}{R}, \tag{2.33}$$

where R is the distance between cation and anion. The Madelung constant α has values of 1.75, 1.76, and 1.64 for the sodium chloride, cesium chloride, and zincblende structures, respectively [2.442]. Combining (2.32) and (2.33) we obtain

$$\Delta E(q_i) = q_i \left(\frac{1}{\bar{r}} - \frac{\alpha}{R} \right). \tag{2.34}$$

In this expression it is obvious that the direct valence term and the lattice sum (Madelung) nearly cancel each other as $\bar{r}$ and $R/1.71$ are usually about equal The shifts observed typically are indeed only of the order of 1 to 3 eV despite the fact that each term in (2.34) contributes 10–15 eV per unit charge to ΔE. The correlation between q_i and $\Delta E(q_i)$ is therefore very sensitive to the exact choice of $\bar{r}$.

The potential model is a rather gross simplification of the two-electron integrals that contribute to ΔE. This is discussed in detail in [Ref. 2.2, Sect. 1.5.2]. Equation (2.34) can, however, be very useful if regarded as a parametrization scheme with $\bar{r}$ as effective valence charge radius chosen so as to represent best the Coulomb interaction between core and valence electrons. A contribution to the chemical shift that cannot be parametrized in the form of (2.34) is the relaxation or polarization shift (see [Ref. 2.2, Sects. 1.52, 4.2]). The relaxation energy is the energy imparted to the photoemitted electron by the polarization of surrounding electrons that screen the suddenly created hole. *Fadley* et al. [2.438] used the polarization corrections introduced by *Mott* and *Gurney* [2.443] in connection with the energy of a vacancy in an ionic crystal. This form of polarization correction has become standard in work on the binding energy of the alkali and alkaline earth halides [2.444–446]. The polarization energy E_p so obtained amounts to 1–3 eV for a typical insulator. Differences in E_p that would affect $\Delta E(q_i)$ are usually $\sim$0.5 eV but values as large as 1.4 eV have been reported [2.438]. An alternative approach to the calculation of polarization corrections is that of *Kowalczyk* et al. [2.447] and *Wagner* and *Biloen* [2.448]. These authors use the combination of Auger electron shifts and core level shifts to obtain E_p. *Shalvoy* et al. [2.433] have applied this formalism to the relaxation shifts of Ge compounds and obtained values of ΔE_p relative to elemental Ge of

$-0.35\,\text{eV}$ (GeS), $-0.2\,\text{eV}$ (GeSe), and $+0.1\,\text{eV}$ (GeTe), respectively, all small compared to the observed chemical shifts of 1.6, 1.9, and 0.95 eV for the same compounds. Polarization shifts are consequently neglected in most investigations.

Chemical shifts are usually measured with respect to the elements in solid form. In view of the small chemical shifts found in semiconductors and insulators the problem of the reference level for binding energies, both for the compound and the corresponding elements, warrants particular attention. The proper level is the vacuum level, i.e., the binding energy of an electron that is removed to infinity. The quantity directly measured with a photoelectron spectrometer is the binding energy relative to the Fermi level E_F of the spectrometer which, for a sufficiently conducting sample, coincides with the Fermi level of the sample. The Fermi level of a metal is a well defined quantity that is separated from the vacuum level by the work function ϕ. In a semiconductor E_F can, in principle, lie anywhere within the fundamental gap. It can be pinned at the surface by surface states. In this case band bending will change the position of E_F further from its position in the bulk [2.449]. The Fermi level is therefore not a useful internal reference in order to obtain meaningful chemical shifts. The problem can be aggravated if the sample is a bad conductor: it charges up and the equivalence of spectrometer and sample Fermi level is lost [2.147]. Masking partly the sample surface with a layer of gold or carbon and using the Au $4f$ or C $1s$ line as secondary references is a widely used technique to correct for the effect of charging [2.450]. The conducting layer is supposed to float at the potential of the charged surface. This technique is, however, not without problems, due to changes in band bending and possibly chemical reactions between the masking material and the substrate [2.451].

An internal reference that avoids these problems is the top of the valence bands, a point that can be defined with some reliability in most EDC's taken with monochromatized X-rays or uv-sources [2.34, 147]. The difference between the top of the valence band and the vacuum level is the photoemission threshold E_T, which is directly related to the total width W of a photoemission spectrum and the photon energy $\hbar\omega$

$$E_T = \hbar\omega - W. \tag{2.35}$$

W is the difference between the top of the valence band and the low-energy cutoff of the spectrum provided E_T is greater than the work function of the spectrometer. *Shevchik* et al. [2.34] and *Goldmann* et al. [2.49] determined in this way chemical shifts for the cation of a number of tetrahedrally coordinated semiconductors. These authors parameterize $\bar{r}$ in (2.34) in the form

$$\bar{r} = r_m / A(\Gamma),$$

where r_m is the interatomic distance for the cation in elemental metallic form and $A(\Gamma)$ a geometrical factor calculated under the assumption that the charge

q_i is evenly distributed over a shell with an outer radius r_m and an inner radius $\Gamma \times r_m$. By comparing the measured shifts with those calculated using (2.34) and the ionicities of *Phillips* and *van Vechten* they found that $\Gamma \simeq 0.5$, i.e., $A(\Gamma) = 2.6$, gave the "best" agreement between theory and experiment. The results are listed in Table 2.14; it is obvious that this agreement is far from satisfactory. For the Cu-halides [2.49] the calculated shifts have even a sign opposite to that observed! It has been shown by *Pantelides* and *Harrison* [2.452] that the correct sign could be obtained in this case if the charges are calculated according to (2.27) using the polarities based on the bond orbital model rather than those of the dielectric model. The results for ΔE calculated using the ionic charges of the bond orbital model are also listed in Table 2.14. Aside from the copper halides, this model seems to give shifts that exceed the observed shifts by about a factor of two and are therefore far inferior to those obtained with the "dielectric" charges. We have to keep in mind, however, that the geometrical factor Γ was chosen in [2.34] so as to fit the q_i^{DT} best: a different choice of Γ could bring the shifts calculated using q_i^{BOM} in comparably good agreement with experiment. This example illustrates the sensitivity of the present approach to geometrical considerations.

A substantial improvement for the calculated shifts of the III–V compounds has been obtained by *Hübner* and *Schäfer* [2.453, 454]. These authors take explicitly into account the bonding charge which in the more covalently bonded compounds is *shifted* from its central position with increasing ionicity (see e.g., [Ref. 2.410, Fig. 6.6]). Consequently they divide the potentional seen by the core electrons of the atom into three parts: (I) the potential of the effective atomic charge q_i^*, (II) the potential of the bond charge q_B, and (III) the Madelung potential of the lattice. The effective atomic charges in an $A^N B^{8-N}$ compound are:

$$q_i^*(\text{cation}) = N - 8\left[1 - \frac{1}{\varepsilon(0)}\right]\left(\frac{1-f_i}{2}\right), \tag{2.36}$$

$$q_i^*(\text{anion}) = (8-N) - 8\left[1 - \frac{1}{\varepsilon(0)}\right]\left(\frac{1+f_i}{2}\right). \tag{2.37}$$

The bond charge per unit cell (four bonds) q_B is given by [2.410]

$$q_B = -\frac{8}{\varepsilon(0)}, \tag{2.38}$$

where $\varepsilon(0)$ is the static dielectric constant.

Equation (2.37) corresponds to (2.27), generalized so as to remove the bonding charge: it transforms into (2.27) for $\varepsilon(0) \to \infty$. The atomic charges (2.36, 37) satisfy together with the bond charge (2.38) the requirement of charge neutrality. The contributions of the atomic charges to the chemical shift are evaluated by a procedure analogous to that described above. The bond charge is shifted from its central position by an amount $\Delta\tau$ towards the anion [2.454]

$$\Delta\tau = \frac{2}{\pi}\frac{R}{2}\arcsin(f_i^{1/2}), \tag{2.39}$$

where R is the interatomic distance. The displacement $\Delta\tau$ contributes to the chemical shift an energy $\Delta E(q_B)$

$$\Delta E(q_B) = \frac{8}{\varepsilon(0)(R/2+\Delta\tau)}. \tag{2.40}$$

The bond charges of more distant bonds are considered to be shared by the ions in the usual way. Combining all contributions to the chemical shifts, *Hübner* and *Schäfer* [2.454] obtain the values for the chemical shifts of the cation listed in Table 2.14. These values are in truly excellent agreement with the measured shifts. Chemical shifts of a somewhat different kind have been investigated in alkali and alkali-earth halides [2.444–446]. These are the shifts of the outermost levels between the free atom or free ion and the ionic solid. The binding energies for atoms and ions are obtained from optical data; if ionic data are used, the problem of determining $\bar{r}$ in (2.34) is avoided. The implicit assumption in using free ion values is, of course, that the effective ionic charges in the solid are integer multiples of the electron charge. The motivation behind these investigations is therefore not to determine q but to measure the specific solid state contributions to the binding energy and to compare them with the theoretical models. It turns out that the Madelung term and the polarization corrections reproduce binding energies (or rather binding energy differences between cation and anion so as to avoid the charging problem) very well. Contributions of the repulsive interactions between overlapping valence electrons on neighboring ions proved to be unimportant. Thus, as stated succintly by *Poole* et al. [2.446], "the individual electron energy levels of strongly ionic solids are apparently insensitive to the wave function overlap, whereas the width and structure of the valence bands bear witness to the existence of this overlap". We might add that this conclusion is only true in the framework of the model which assumes a priori integer effective charges on the ions disregarding the fact that the ionicities of these compounds are actually smaller than 1.

2.9 Photoemission Spectroscopy of Semiconductor Surfaces

In the remaining portion of this chapter we turn our attention from photoemission spectroscopy studies of bulk solid properties to those specifically designed to probe the electronic structure of solid surfaces. The investigation of surface electronic structure is in a considerably more rudimentary stage of development than that of the bulk, however recent advances in both experimental and theoretical techniques have generated wide interest in this area of study which is presently active and growing. Our objective here will be to present an outline of the role photoemission spectroscopy is playing in the current understanding of semiconductor surfaces and to introduce the reader to the photoemission literature. Comprehensive reviews of the many experimental results and theoretical methods can be found in the literature [2.455]. Although progress is being made toward a microscopic understanding of semiconductor surfaces, there are a large number of new results and interpretations, many of which are still being tested.

2.9.1 Semiconductor Surface States

One can intuitively imagine that the wave functions at the surface of a solid will differ from those inside the bulk because of the reduced coordination of the atoms at the surface. Also it is not surprising that atoms at the surface that are missing one or more stabilizing nearest-neighbor bonds often have stable crystal arrangements different from that in the bulk. What is not immediately obvious is the fact that the solid-vacuum interface often produces wave functions that are *localized* at the surface. These states are commonly called *surface states.* There are also *surface resonances* with wave functions that have a large amplitude at the surface and, unlike surface states which decay, they have smaller oscillations into the bulk.

The presence of surface states was predicted by *Tamm* [2.456] in 1932. Using a model calculation *Tamm* showed that solutions to the Schrödinger equation, which were forbidden in the infinite solid because they diverge to infinity (band gap solutions), could be matched at the solid-vacuum interface onto allowed vacuum states that decay away from the surface into the vacuum. This composite wave function does not diverge and is thus allowed at the surface at energies within the band gap of the bulk. These states are localized at the surface by being trapped between the forbidden bulk and forbidden vacuum. Thus the two basic properties which characterize surface states were already defined: (I) they are localized near the surface and (II) their eigenvalues fall in bulk band gaps.

In 1939 *Schockley* [2.457] specified the existence conditions for surface states in bonding-antibonding band gaps and pointed out the importance of crystalline symmetry and the missing bonds at the surface and thus introduced the connection between chemical properties and surface states. The "broken bond" orbitals at the surface became known as *dangling-bonds* which in the case of semiconductors were expected to fall in the primary gap and thus constitute surface states. *Davisson* and *Levine* [2.458], *Henzler* [2.459], *Forstmann* [2.460], and *Jones* [2.461] review the quantum mechanical model calculations appearing in the literature which define the mathematical properties of surface states.

The problem of describing the electronic structure at a surface differs from that in the bulk in several important ways. The origin of most of these differences is the reduced crystal symmetry at the surface and the concomitant variation of potential on going from the bulk through the surface into the vacuum. In addition, the arrangement of atoms which is stable in the bulk often no longer is stable in the surface region where the coordination and potential are different. Consequently lower free energy states are expected to exist with surface layers moved inwards or outwards (relaxation) or surface atoms rearranged into two-dimensional surface unit cells differing from a simple truncation of the bulk (reconstruction). Often, depending upon the method of preparation, several different surface reconstructions are obtained for the same crystal face. For example, the natural cleavage face of Si is the (111) face and exhibits a 2×1 reconstruction at room temperature which transforms to 1×1 above 840 °C and then to 7×7 upon cooling back to room temperature. The notation $n \times m$ (e.g. 7×7) signifies the relationship of the two-dimensional symmetry at the surface with respect to the projection of the bulk structure at the surface. It also is used to describe the ordered arrangement of adsorbed species with respect to the bulk unit cell. Because of the various possible surface structures and their sensitivity to vacuum conditions, it is experimentally very important to produce, maintain, and identify a single phase of a given surface to insure reproducible electronic properties [2.462]. Low-energy electron diffraction (LEED) patterns signal when surfaces are reconstructed and yield the size of the surface unit cell. The positions of the atoms within the unit cell can sometimes be deduced by trial and error fits of experimental LEED spot intensities versus the incident electron beam kinetic energy using model dynamical LEED calculations [2.463]. The results of such a procedure however are limited by the availability as well as validity of a hypothetical surface structure which can be used to fit the model calculation [2.464].

Experimentally until about fifteen years ago the existence of surface states was only inferred from indirect experimental methods. Shortly after *Bardeen* [2.465] in 1947 postulated the existence of surface states in semiconductors a great number of field effect, photoelectric, work function, and surface recombination techniques were used to deduce the properties of surface states indirectly from studies of the space charge layer in semiconductors [2.466, 467]. Recently several methods have been devised to probe surface states directly. They include surface photoconductivity [2.462] which yields a measure of the empty surface states, field emission [2.468, 469], and ion neutralization spectroscopy (INS) [2.470] which both yield a measure of the filled surface states, infra-red internal reflection [2.471], and electron energy loss (ELS) [2.472, 473] spectroscopies which both probe joint densities of states, and photoemission spectroscopy which presently is the most widely used technique for obtaining the density of filled surface states. Confidence in assignment of photoemission structure as emission from filled surface states of both semiconductors and metals began to appear in the literature around 1972. More recently, angle-resolved photoemission techniques [Ref. 2.2, Chap. 6], which can be used to

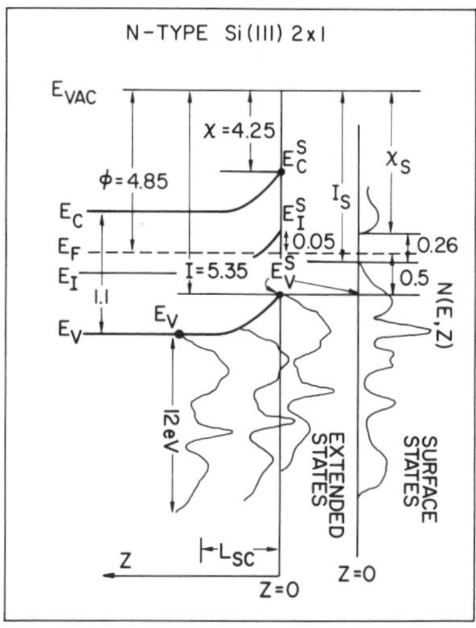

Fig. 2.80. Schematic energy level diagram for lightly doped n-type Si(111) 2×1. The density of states at the surface ($Z=0$) is decomposed into extended and surface states which are shown separately.

E_{VAC}: Vacuum level; E_C: Bottom of bulk conduction bands; E_I: Intrinsic level in the bulk; E_V: Top of the valence bands; E_C^S: Bottom of the conduction bands at the surface; E_I^S: Intrinsic level at the surface; E_V^S: Top of the valence bands at the surface; ϕ: Work function; I: Ionization potential; χ: Electron affinity; I_S: Surface ionization potential from the filled surface states; χ_S: Electron affinity of empty surface states; Z: Distance into bulk from the surface; L_{SC}: Thickness of space charge layer

measure the dispersion of surface state energy bands, have been developed. In addition the various yield spectroscopies (see Sects. 6.3.7, 6.3.8, 6.4) which are closely related to photoemission spectroscopy, help complete the picture by probing the density of empty states. Experimental results from such techniques, which directly probe the electronic structure of surface states, have become a testing ground for a relatively new generation of realistic electronic structure calculations [2.474] which attempt to determine the properties of specific surfaces accurately enough to interpret experimental results. These detailed surface electronic structure calculations started to appear in the literature also around 1972.

2.9.2 Silicon Surface States

Silicon surfaces have received a great deal of attention undoubtedly because of the technological importance of that semiconductor and also because its bulk electronic structure is rather well known. A schematic view of the properties relevant to photoemission spectroscopy of the natural (111) cleavage face of lightly doped n-type silicon is shown in Fig. 2.80. Inside the crystal for $Z>L_{sc}$ where L_{sc} is the space charge layer thickness, the position of the Fermi level E_F within the gap is determined by the impurity concentration and upon continuously varying the doping from moderately p- to n-type will move from near the valence band edge E_V through the intrinsic level $E_I \sim (E_V + E_{gap}/2)$ to the conduction band edge E_C. The bulk potential is defined as $(E_F - E_V)$ and varies by $\sim E_{gap}$ for this range of doping. At the surface ($Z=0$) a different behavior is observed. The surface potential $(E_F - E_V^S)$ is relatively independent of the dopant level while the bulk potential varies by E_{gap}. E_V^S is the energy at the top of the valence band at the surface position $Z=0$. This "pinning" of E_F in the gap at the surface is manifested by the relatively small variation of the work function ϕ and ionization potential I with doping. *Bardeen* [2.465] postulated the existence of surface states in the gap near the Fermi level, which would form a space charge layer L_{sc} by collecting impurity carriers at the surface, and by leaving immobile ionized impurities behind between the surface and the bulk. This process continues until a dipole forms which is equal in magnitude and opposite in direction to the difference between the bulk and surface potentials and which aligns E_F at the surface and in the bulk. In Fig. 2.80 the local density of extended states are plotted at three locations: at the surface, in the middle of the space charge layer, and in the bulk. The local density of filled and empty surface states at the surface position ($Z=0$) have been schematically plotted separately on the right side. The ionization potential of an atomically clean surface is I_S and the electron affinity is χ_S. These quantities, which are properties of the surface state density in the gap, and not their bulk counterparts I and χ, govern the surface chemistry on clean surfaces. Extreme care in an ultrahigh vacuum environment ($\sim 10^{-11}$ Torr) is necessary to obtain intrinsic surface states. Surface reactions, for example oxidation, may stabilize the dangling-bond states down to energies below E_V^S. The surface states are thus removed from the gap and the Fermi level pinning is greatly reduced. Any residual pinning is now attributed to interface states at the silicon-silicon oxide junction. An interesting case which we will discuss below is the Schottky barrier formed at a silicon-metal interface where significant pinning remains after a metal layer is deposited onto the silicon. Photoemission spectroscopy can directly probe all the electronic states discussed above: extended, surface, and interface states. Photoelectron escape depths λ, depending on $\hbar\omega$ are for the usual experiments ~ 5 to 20 Å (see Fig. 2.26) and therefore, for all but degenerate semiconductors, the space charge thickness L_{sc} is much larger than λ and the photoemission spectrum will reflect the electronic structure at the surface potential end of the space charge layer near $Z=0$.

Chemical reaction rates, activation energies, and Schottky barriers all directly depend upon the position of valence and conduction band edges. Band edges are observed in photoemission spectra, but usually can be determined with higher accuracy by other techniques. Results from some of these techniques are shown in Fig. 2.80 for comparison with the photoemission spectra presented below. The work function ϕ is 4.85 ± 0.05 eV for all but degenerately doped samples because of the Fermi level pinning. Kelvin probe contact potential difference measurements [2.475, 476] as well as high sensitivity total photoemission yield threshold measurements [2.477] agree on this value. In the total photoemission yield experiment, the total photoemission current per incident photon is measured as a function of increasing photon energy. The value of $I = E_{vac} - E_V^S = 5.35 \pm 0.02$ eV was also obtained from the total photoemission yield measurements of *Sebenne* et al. [2.477]. We believe that this value is more accurate than the earlier widely quoted value of 5.15 ± 0.08 eV of *Allen* and *Gobeli* [2.475] which was also obtained from photoemission yield spectra. *Sebenne* et al. carefully removed a significant surface state emission from their yield spectra before determining E_V^S thus obtaining a larger value for the ionization potential I. This problem of determining E_V^S, which is often masked by surface states, is even more difficult for photoemission kinetic energy spectra (EDC's) where the resolution is between 0.1 and 0.5 eV compared with 0.01 and 0.02 eV for the total photoemission yield technique. The existence of a gap between the filled and empty surface state bands (Fig. 2.80) is supported by evidence from infrared absorption [2.471] and electron energy loss spectroscopy [2.478]. However there are still questions about the interpretation of these experiments and the existence of the gap is not yet firmly established (see below).

A growing number of photoemission energy distribution spectra now exist which contain many examples of spectral peaks and bands which are extremely surface sensitive and consequently have been identified with *Bardeen*'s surface states. Current efforts are directed toward understanding both the nature of this surface emission as well as its relation to semiconductor surface states.

a) Photoemission from Si(111) 2 × 1 and 7 × 7 Surfaces

The experimental problems of producing atomically clean (111) silicon surfaces with reproducible photoelectric properties were solved in a series of very careful experiments by *Allen* and *Gobeli* [2.475, 479, 480]. They measured the total photoemission yield ($\hbar\omega < 6.18$ eV) in UHV (10^{-10} Torr) from cleaved crystals with a wide range of doping. *Allen* and *Gobeli* observed a tail in the low-energy region ($5 < \hbar\omega < 5.5$ eV) of the yield spectrum of n-type crystals which they interpreted as direct emission from surface states [2.479]. Several years later they measured the photoemission kinetic energy distribution curves (EDC's) excited with several different photon energies ($\hbar\omega < 6.18$ eV) from a cleaved Si(111) sample which is known from LEED [2.462] to have a 2×1 reconstruction at 25 °C. They also reported spectra from the same samples after annealing at 1000 °C and cooling to room temperature [2.480]. This treatment is known to produce a 7×7 reconstruction. A spectral peak observed between E_V^S and E_F in the spectrum from the annealed surface was assigned to surface states. In retrospect, much of the observed structure from the cleaved surface was also from surface states. However, at the time no structure distinguishing it from valence band emission was observed, because the 2×1 dangling-bond surface state emission has a higher binding energy than that from the 7×7 surface and at the low photon energies used the surface states masked the top of the valence band edge at the limit of their spectral range (2 eV binding energy).

Experimentally progress was made with the introduction of higher photon energy light sources such as rare gas resonance lamps and synchrotrons which probe higher binding energies. These source allow study of surface state emission not just between E_V^S and E_F but overlapping much of the valence bands. The experimental challenge became, and still is, to distinguish surface state photoemission from bulk state emission [2.481, 482] and the theoretical

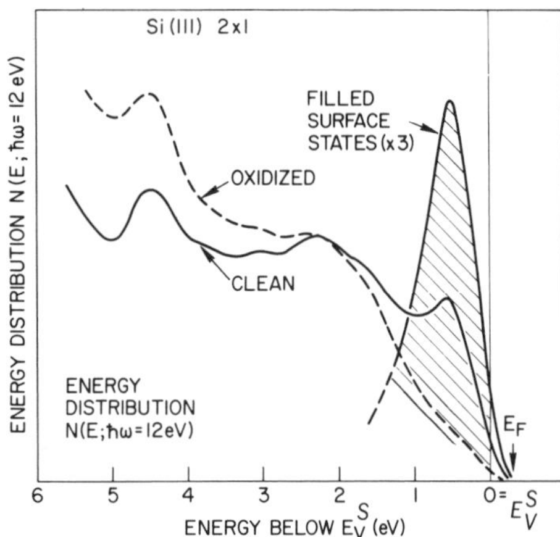

Fig. 2.81. Angle-integrated photoelectron energy distributions obtained with a cylindrical mirror analyzer from clean and oxidized Si(111) 2×1 surfaces. The difference curve of these spectra is a measure of the optical density of intrinsic surface states [2.484]

challenge to interpret the surface state emission in terms of the electronic structure [2.474, 483]. Although much work is active in both areas both tasks have just begun.

In 1972 *Eastman* and *Grobman* [2.484], and *Wagner* and *Spicer* [2.485] obtained EDC's from cleaved Si(111) 2×1 in which a band overlapping E_V^S could be distinguished from bulk emission because of (I) the sensitivity of the spectrum to surface contamination and (II) its insensitivity to transition matrix-element modulation at different photon excitation energies. This band is selectively attenuated by at least 90% when one monolayer of oxide is grown on the surface [2.486]. The spectra from a clean and oxidized Si(111) 2×1 surface are shown in Fig. 2.81 [2.484]. The difference curve is a measure of the filled band of surface states. A number of groups have reported EDC's from Si(111) 2×1 and 7×7 surfaces obtained with photon energies between 5.5 and 11.8 eV. These spectra are compared in [Ref. 2.481, Fig. 15]. Small disagreements, which are not presently well understood, are discussed in terms of differences in experimental conditions and uncertainties, such as differences in doping, in cleavage dependent surface morphology, in the ratio of surface to bulk emission versus photon energy, in the effective spectrometer acceptance angle, or in the angle of incidence of the exciting light.

The approximate densities of surface states near the band gap which have been derived from photoemission measurements are summarized in Fig. 2.82 after *Rowe* et al. [2.478] for bulk Si (top panel) and the 2×1 cleaved (middle panel) and 7×7 annealed (bottom panel) surfaces. The zero of energy in Fig. 2.82 corresponds to E_V^S in Fig. 2.80. It is clear that the dangling-bond surface state structure denoted with the letter A is different for the two surface reconstructions of the Si(111) surface. In the 2×1 density of states the dangling-bond peak is centered $\sim$0.8 eV below E_F and a tail of states extends up to E_F while in the 7×7 density of states the dangling-bond band occurs at E_F with a metal-like edge and two peaks. Although there have been many plausible

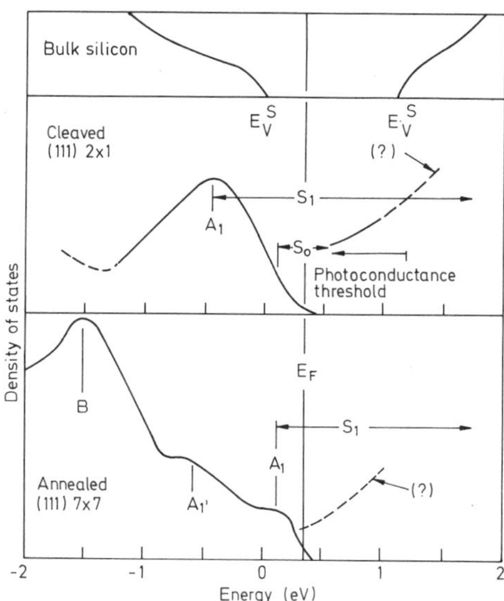

Fig. 2.82. Approximate density of states for bulk silicon and for the cleaved and annealed surfaces. The energy zero is taken at the bulk valence band maximum. The relative energy positions from one curve to another are uncertain to ± 0.25 eV while the positions of the structure within each curve are accurate to ± 0.1 eV. The position of the Fermi level is $+0.35 \pm 0.15$ eV for both surfaces. S_0 and S_1 are assignment for electron energy loss transitions. The dashed portions of the empty conduction band regions denoted with question marks are estimated [2.478]

model calculations, the origin of these differences in terms of the electronic and surface crystal structure (which is also not yet known) is still not well established.

Wagner and *Spicer*'s EDC's, taken with $\hbar\omega = 11.8$ eV from degenerate p (p^{++}: 0.0014 Ω-cm, 8×10^{19} holes × cm^{-3}), degenerate n (n^{++}: 0.001 Ω-cm, 8×10^{19} electrons × cm^{-3}) and lightly doped n-type (n^-: 25 Ω-cm, 10^{14} electrons × cm^{-3}) cleaved Si(111) 2×1 are shown in Fig. 2.83a. The n^- spectrum in Fig. 2.83a is very similar to the spectrum shown in Fig. 2.81 which was obtained from 5 Ω-cm n-type Si. The bands are quite flat throughout the electron escape depth for both of these lightly doped samples and no manifestation of the band bending is expected. Only the bulk states at the surface potential end of the space charge layer are observed. However, the space charge thickness becomes of the order of the photoelectron escape depth for the degenerate p^{++} and n^{++}-type samples and the states between 2 and 6 eV are observed to shift toward the respective bulk potentials. The movement of the bulk features is measured to be 0.7 eV rather than the expected 1.1 eV bulk potential swing from p^{++} to n^{++} doping because the spectra from the degenerately doped materials are a composite of emission from the surface potential region where E_V^S is pinned, the space charge layer, and the bulk potential region (see Fig. 2.80). The small 0.2 eV shift of the surface state band centered near 0.8 eV is believed to be related to the change in surface state occupation (Fig. 2.83b). The observed shift towards the bulk potential of the extended states (2 to 6 eV region) and not of the surface state band is evidence that the surface states do not extend deeply into the space charge layer (12 Å) of the degenerate materials, but must be *localized* within the first few atomic layers

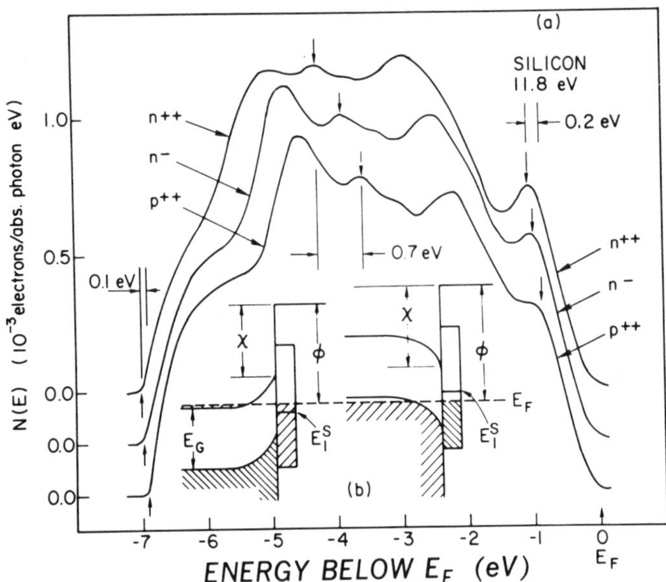

Fig. 2.83. **a** Energy distribution curves for n^{++}, n^{-}, and p^{++} silicon are shown referenced to the Fermi energy E_F. The high-energy (right-hand) peak is due to electrons in surface states. **b** The pinning effect of a high density of surface states for n^{++} and p^{++} semiconductors [2.486]

of the surface. Similar observations have been made recently by measuring the EDC's of the $2p$ core levels of silicon with synchrotron radiation [2.449].

When Si is cleaved along its (111) natural cleavage face not every atom will be at a terrace site [see Fig. 2.87 for (111) atomic arrangement] unless the surface is perfect. A common defect is the surface step between two terrace planes which are on the average separated by one double atom layer distance of 3.14 Å [2.487]. Atoms at step-edges have a different coordination than those at terrace sites. On surfaces with high step densities ($\sim 10\%$) this is reflected in the electronic structure as extrinsic surface state emission peaks surrounding the intrinsic surface state emission originating from terrace atoms [2.488, 489]. Some of the discrepancies among earlier Si surface state photoemission measurements may be attributed to differences in the surface morphology.

If a surface state feature in the photoemission spectrum can be distinguished from bulk features, it is possible, using angle-resolved photoemission, to measure the dispersion (E vs $\boldsymbol{k}$) of this surface state band (see [Ref. 2.2, Chap. 6]). During photoemission the component of the electron wave vector parallel to the surface $\boldsymbol{k}_{\parallel}$ is conserved within a surface reciprocal lattice vector and is easily determined from the measured photoelectron kinetic energy E_{ke} and emission angle θ defined with respect to the surface normal. The azimuthal angle of emission around the surface normal determines the line in the two-dimensional surface Brillouin zone along which the dispersion is determined. For a given azimuth the simple relation between E_{ke}, θ, and $\boldsymbol{k}_{\parallel}$ is given by (2.18)

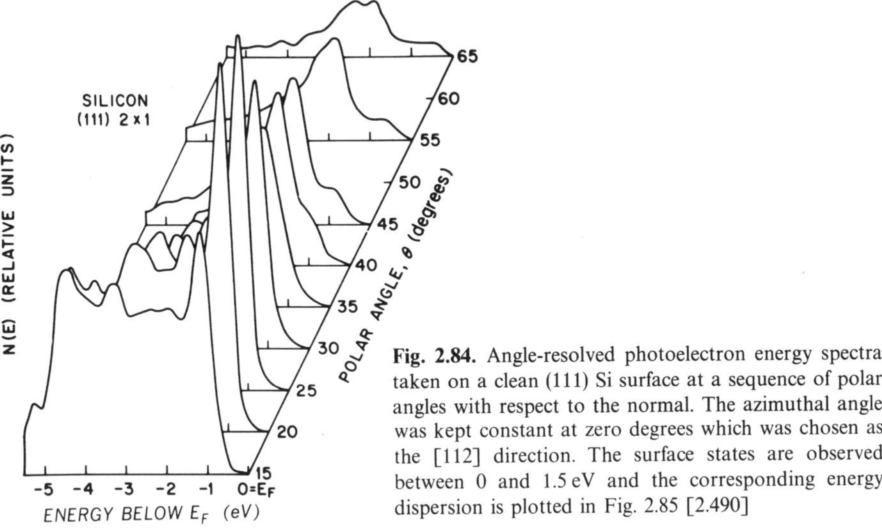

Fig. 2.84. Angle-resolved photoelectron energy spectra taken on a clean (111) Si surface at a sequence of polar angles with respect to the normal. The azimuthal angle was kept constant at zero degrees which was chosen as the [112] direction. The surface states are observed between 0 and 1.5 eV and the corresponding energy dispersion is plotted in Fig. 2.85 [2.490]

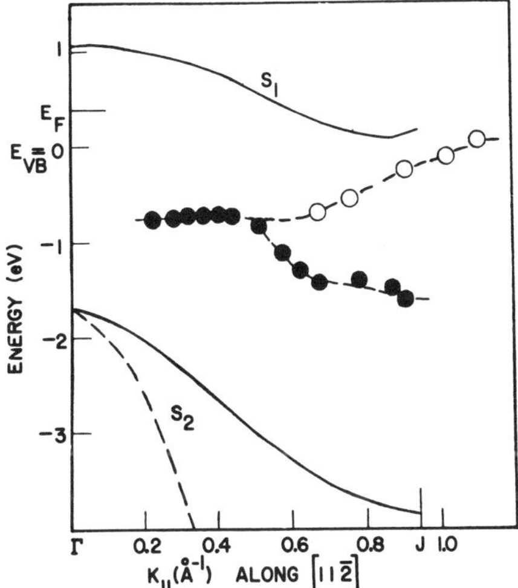

Fig. 2.85. Energy versus parallel wave vector for the experimental surface state peak (filled circles) and the weaker split-off peak (open circles). The dashed curve has been drawn to connet the circles smoothly. Also shown are the calculated S_1 and S_2 surface states (solid lines) and S_2 surface resonance (broken line) of [2.497]. The center of the two-dimensional zone edge has been labelled *J*. E_F is the Fermi level and E_{VB} is the top of the bulk valence bands [2.490]

with $E = E_{ke}$. Thus the dispersion can be obtained by measuring the kinetic energy of the surface state peak as a function of either polar angle or photon energy.

Traum et al. [2.490] measured the angular distribution of photoelectrons from the cleaved Si(111) 2×1 surface and found a strong dependence of the energy and intensity of the dangling-bond surface state upon polar angle of emission, their results are shown in Fig. 2.84. From the angle and kinetic energy of the surface state peaks near E_F, they deduced the two-dimensional dangling-bond band structure shown in Fig. 2.85 using (2.18). Note that at large angles the surface band splits. In addition they measured the azimuthal dependence of

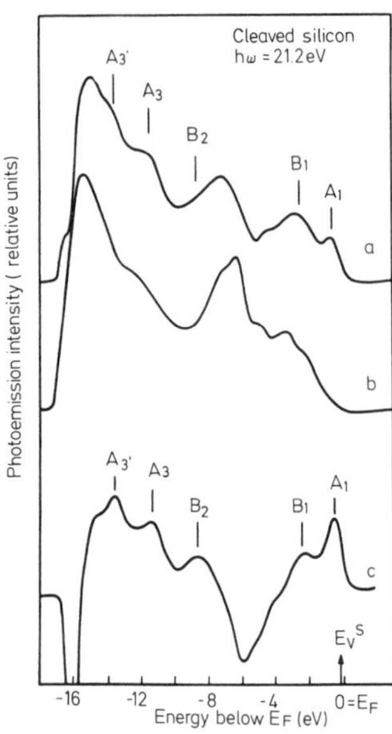

Fig. 2.86. UPS for clean low-step-density surface (Curve *a*), after adsorption of ~1 monolayer of atomic *H* (Curve *b*), and the difference curve (Curves *a* and *b*) with the vertical scale times 1.5 (Curve *c*). The features B_1, B_2, A_3, and A'_3 are due to back-bonding surface states below the valence band maximum [2.488]

the emission intensity between E_F and -1.7 eV at a polar angle of 30° and found a threefold symmetric angular distribution. The three emission lobes coincided with the directions of the three bonds of the surface atom to the second layer atoms. It was suggested that the surface state photoemission bands near E_F may consequently have a significant *back-bonding* as well as dangling-bond character.

Model electronic structure calculations [2.483] show that in addition to the dangling-bond states there are surface states and surface resonances mostly associated with the back-bonds that extend to energies as far down as the bottom of the valence bands. Experimentally it is more difficult to distinguish photoemission from these states because they often occur in relative band gaps and overlap strong emission from extended states. The most promising technique to probe the back-bonding surface states is angle-resolved photoemission; directions (regions of $\boldsymbol{k}$ space) can be chosen where bulk emission is weak and surface emission strong. Detailed angle-resolved studies of the back-bonding surface states of Si have not been reported, however evidence for such states has been obtained indirectly from angle-integrated measurements [2.491]. Figure 2.86 shows the EDC obtained by *Rowe* et al. [2.488] with $\hbar\omega = 21.2$ eV from (a) the valence band of Si(111) 2×1, (b) the same surface after adsorption of ~1 monolayer of atomic hydrogen, and (c) the difference curve. It is assumed that the adsorption of hydrogen attenuates the surface state structure so that the difference spectrum will give some *indication* of the position of the surface states. A_1 labels the dangling-bond band and the remaining structures B_1, B_2, A_3, and A'_3 are assigned to back-bonding surface bands.

b) Electronic Structure Theory of Si(111) Surfaces

There have been several theoretical studies of the ideal (truncation of the bulk) Si(111) surface [2.492–500] and a few attemps to model the 2×1 reconstruction [2.495, 496–502]. The latter calculations however suffer from the same problem encountered in dynamical LEED calculations. The results are only as sound as the ad hoc choice of the type and magnitude of the atomic displacements used to model the surface reconstruction. Nevertheless the trends and general features of the calculated results are proving very useful in understanding the electronic structure of specific surfaces and the dependence on surface crystal structure. Since LEED intensity data and dynamical LEED structure calculations are not available for many surfaces, the atom positions within the unit cell must be deduced from other data. *Haneman* reviews the structural models that have evolved for semiconductor surfaces and discusses the surface energy associated with different atom arrangements [2.503].

The procedure for determining the *bulk* electronic structure entails; (1) measuring the crystal structure using X-ray diffraction, (2) calculating the potential for the measured atomic arrangement, (3) solving the Schrödinger equation for the wave functions and energy levels, (4) and if carried to self-consistency, calculating a new potential from the new electronic arrangement specified by the wave functions obtained in step 3, (5) iterating steps 3 and 4 until the potential no longer varies, and finally (6) comparing the results of such calculations with photoemission spectra and other experiments to evaluate the choice of parameters and approximations. This method has been extensively developed for bulk solids during the 30 years.

The theoretical treatment of the electronic structure of solid surfaces however has not progressed to the degree that the techniques used for the bulk have, for several reasons: (1) there is presently no technique available to accurately determine the atomic positions in the surface region that corresponds to the bulk X-ray diffraction techniques, (2) theoretical techniques have only recently been developed to account for the surface potential barrier (variation of the potential perpendicular to the surface from the bulk into the vacuum see [Ref. 2.2, Sect. 1.3]) as well as changes parallel to the surface, and (3) even assuming factors 1 and 2 are adequately treated, often the inherent reduction in symmetry at the surface, together with large two-dimensional unit cells generally parallel to the surface (observed by LEED), make the problem mathematically so complex that in many cases only simple model calculations can be performed. X-ray diffraction techniques cannot distinguish the first few atomic layers from the bulk. Because of the lack of surface crystallography information, it is often necessary to model or guess the surface atomic positions.

While bulk band-structure theory was well established when bulk photoemission data became widely available during this decade, the uncertainties discussed above made the surface electronic structure theory unmanageable. The rather recent experiments, including photoemission, which directly probe the surface electronic structure, have provided guidance in choosing surface

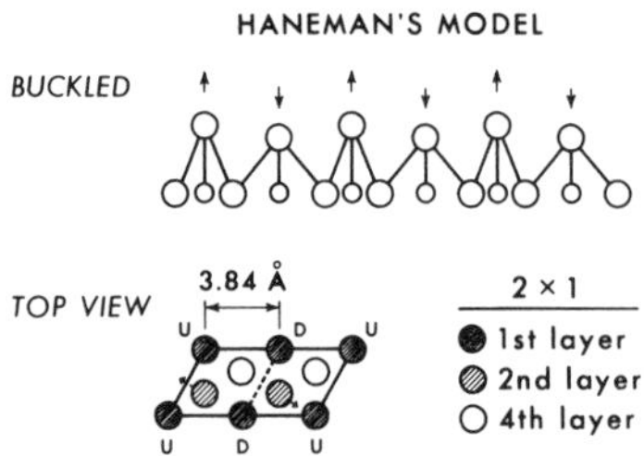

Fig. 2.87. Schematic representation [2.499] of the ideal and (2 × 1) reconstructed Si(111) surface. Reconstruction is done according to *Haneman*'s model [2.503] and leaves the surface buckled as indicated by the arrows. Alternating rows of atoms are raised by 0.18 Å and lowered by 0.11 Å. Slight lateral shifts of the second-layer atoms are also indicated by arrows in the top view

structural models. Until accurate surface crystallography techniques are developed, photoemission will thus play a more essential role in the difficult surface problem than it does in the study of the bulk. The sensitivity of the electronic structure at the surface to the particular crystal structure makes photoemission spectroscopy, together with realistic electronic structure calculations, a technique which may possibly become as useful or even more than LEED analyses for determining the surface crystallography [2.495].

LEED shows that Si as well as Ge(111) cleaved at room temperature exhibit a 2 × 1 surface reconstruction with two atoms per unit cell. The atomic structure within the unit cell has however not been conclusively determined. While several models for this structure have been presented, the most widely used model is that shown schematically in Fig. 2.87 which was proposed by *Haneman* and is designed to explain experimental LEED, EPR, and crystal mating results. It is based on simple chemical bonding arguments [2.503]. *Haneman* hypothesized that the hybridization of the dangling-bonds on the ideal (111) surface was likely to differ from bulk sp^3 hybridization, because on the vacuum side their environment is free-atom-like and should therefore induce them to dehybridize toward purer atomic *s*- or *p*-type. A purer *p*-type dangling-bond would leave the back bonds more sp^2-like which would tend toward a planar geometry and thus pull the surface atom downwards, presumably spreading the second-layer neighbors. Assuming that this inward movement occurred for half the surface atoms and that the second-layer spreading pushed alternate rows of surface atoms outward, the back-bond angles of the outward moving atom would be reduced, which is consistent with a rehybridization where the dangling-bond states on the outward moving atoms loose sp^3 and gain *s*-type character, while the back bonds gain *p*-type character. The dangling-bonds with increased *s* character on the atoms which move outward d_{out} will have lower energy than the *p*-type dangling-bonds on the atoms that move inward d_{in} since the atomic *s* orbital is more stable than the atomic *p* orbital. The *Haneman* model thus results in a splitting of the dangling-bond energy band into a band of filled and a band of empty states

[2.495, 499]. LEED intensity data have just been reported for the cleaved Si(111) 2×1 surface [2.504]; it will be interesting to see whether dynamical LEED calculations will fit this data using the *Haneman* model.

After a specific arrangement of atoms has been *chosen* (e.g., the *Haneman* model) the next problem is to determine the electron distribution for that given structural model so that the total potential can be calculated. To account for the reduced symmetry at the surface, the electron wave functions are expanded in a representation that consists of two-dimensional plane waves describing the periodicity parallel to the surface, which are multiplied by functions that depend upon the remaining spatial coordinate z perpendicular to the surface. The z dependent functions used differ among the various groups working in this field depending upon how they treat the boundary conditions in this direction [2.499]. It is more difficult to predict a priori the electron distribution in the vicinity of the surface-vacuum interface [2.458] than it is in the bulk. For this reason the first self-consistent calculations by *Appelbaum* and *Hamann* (AH) of the potential and electronic charge density of a discrete lattice at the surface of both a metal [2.505] in 1972 and a semiconductor [2.494] in 1974, heralded a wave of new realistic surface calculations. In AH's calculation for the Si(111) surface [2.494] the only adjustable parameter was the surface geometry which they *chose* using empirical structural chemistry rules. AH argued that the surface atom back bond-order increases from 1 to 4/3 when the surface is formed and using *Pauling*'s formula for bond order versus bond length they found that the last atomic plane should move 0.33 Å inwards. Also they assumed that reconstruction would be a small perturbation and calculated the relaxed but unreconstructed 1×1 surface. In addition to the surface charge density and potential AH were able to predict the ionization potential $I=5.4\,\text{eV}$ and the work function $\phi=5.1\,\text{eV}$ which can be compared with the experimental values $I=5.35\,\text{eV}$ and $\phi=4.85\,\text{eV}$ shown in Fig. 2.80 for the Si(111) 2×1 surface. A band of surface states associated with the dangling-bonds was found in the lower part of the energy gap. For the three geometric models that AH used as initial guesses they found that I and ϕ were much less sensitive to surface relaxation than were the positions of the dangling-bond bands.

Pandey and *Phillips* [2.495–497] approached the problem of determining the electron distribution at the surface from a more chemical point of view. They assumed that the chemical nature, i.e., the bonding properties of the surface atoms, will be preserved; thus, rather than starting with a lattice of silicon ions to which electrons are added and distributed using a self-consistent calculation, they started with silicon atom charge densities and used a *non*-self-consistent tight-binding scheme to calculate the energy bands. The surface atoms do show changes in hybridization and charge densities through the changes in orbital interactions during relaxation and reconstruction. They found excellent agreement with the self-consistent calculations indicating that the changes in chemical bonding accompanying surface relaxation and reconstruction can account for the surface electron distribution rather well.

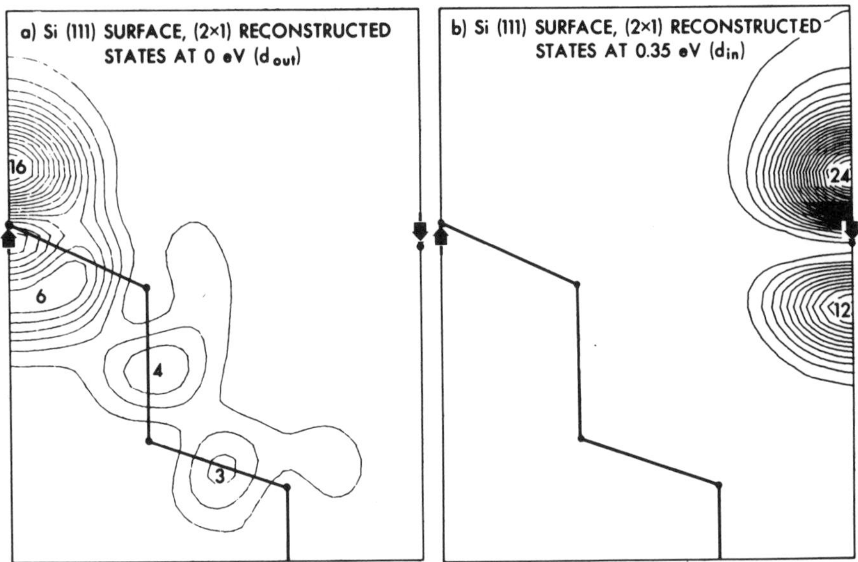

Fig. 2.88a and b. Charge density contour plots for the dangling-bond states d_{out} **a** and d_{in} **b** of (2 × 1) Si(111). Charge is plotted in a (210) plane of (2 × 1) Si which corresponds to the (110) plane of (1 × 1) Si. The raised and lowered atoms are marked by arrows [2.499]

Theory yields band structures $E(\boldsymbol{k})$, densities of states $N(E)$, and charge densities $\varrho(\boldsymbol{r})$. Because of the variation in the electronic structure near the surface it is instructive to decompose $N(E)$ into a local density of states function $N(E, \boldsymbol{r})$ which can be integrated over a region (3 to 10 atomic layers) near the surface. Such a local density of states function is useful for separating the states at the surface from those in the bulk and comparing theory with experiments that probe a characteristic depth such as photoemission. Charge density distributions near the surface for small energy intervals over the width of the valence bands are also calculated because they are useful for distinguishing surface from bulk states. States localized at the surface along with the nature of their orbitals are clearly seen in such charge density plots, Fig. 2.88 [2.499].

Chiarotti et al. [2.471] using internal reflection infra-red spectroscopy on a cleaved Si(111) 2 × 1 surface, observed a peak with an onset at 0.26 eV and maximum at 0.45 eV, which they assigned to optical transitions between filled and empty surface states within the band gap. A similar peak was observed in electron energy-loss experiments [2.478]. Because theoretical calculations do not find a splitting of the dangling-bond band unless the 2 × 1 reconstruction is included in the model structure, these experimental results have been interpreted as evidence for a super-lattice energy gap between the empty and filled dangling-bond bands [2.495]. Density of states calculations based on ideal Si(111) structural models [2.492–500] give a surface state band in the middle of the gap as exemplified by the solid $N(E)$ curve for the unrelaxed

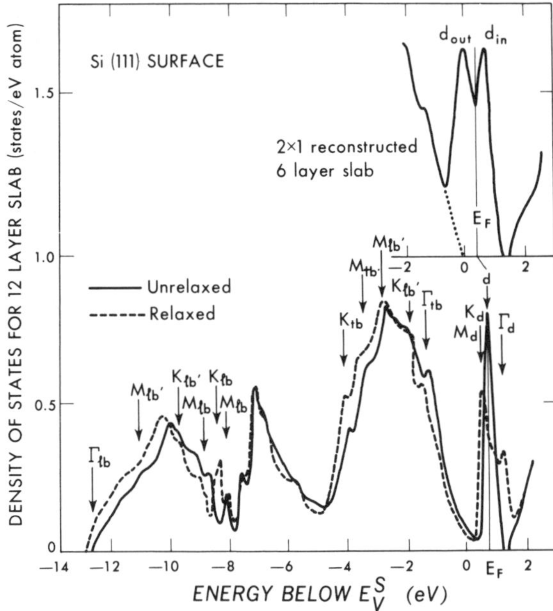

Fig. 2.89. Density of states curves for the self-consistent pseudopotential method results on 12-layer films for the relaxed (broken line) and unrelaxed (solid line) surface geometry. Surface states are indicated by arrows. Inserted is the density of states in the vicinity of the fundamental gap for a 6-layer (2 × 1) reconstructed surface model [2.499]

(ideal) surface in Fig. 2.89 [2.499]. When the outermost atomic layer is relaxed inward 0.33 Å as AH argued, the dangling-bond band shows an increased dispersion parallel to the surface (dashed curve in Fig. 2.89). Only when the Haneman [2.499, 502] or similar models [2.495] are used to describe the 2 × 1 reconstruction does a splitting of the surface dangling-bond band occur (insert) in Fig. 2.89). This splitting increases and opens a gap when a more detailed treatment of the periodicity (12 layer slabs, see [2.499]) perpendicular to the surface is used. The lower filled energy band d_{out} was found localized around the outward moving atom and the upper empty band d_{in} was found localized around the inward moving atom. This is clearly seen in Fig. 2.88 from the charge density plots for energies at the respective centers of the split d_{in} and d_{out} bands.

Haneman's model for the structure of Si(111) 2 × 1 has been used by several groups as the starting point of their electronic structure calculation. The atomic positions in the model have been adjusted to obtain better agreement between the location of the dangling-bond band in theory and in the photoemission experiments where it is observed overlapping the top of the valence bands and centered 0.5 eV below E_V^S (see Fig. 2.82). This method has however resulted in significantly different structural parameters among the different groups who

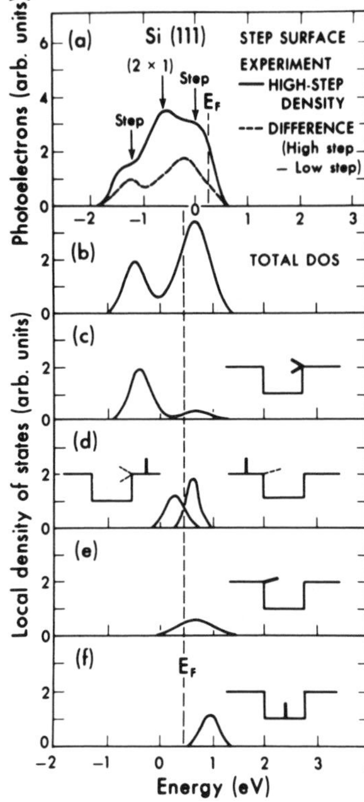

Fig. 2.90a–f. UPS data **a** for the surface dangling-bond states near the valence-band maximum [2.488]. Calculated total **b** and local **c–f** densities of states for six dangling-bonds on the (4 × 1) stepped surface model. The positions of the dangling-bonds relative to the step edges are indicated schematically [2.489]

have calculated the electronic structure of the Si(111) 2 × 1 surface [2.495, 499, 502].

Schlüter et al. [2.489] have studied unreconstructed Si(111) model structures containing steps and have found additional dangling-bond surface states associated with the atoms at and near the steps. Their results are summarized in Fig. 2.90. There are six types of step-edge sites: (Fig. 2.90c) edge double-dangling (dd) bonds with two neighbors, (Fig. 2.90d) upper terrace dangling-bonds, (Fig. 2.90e) edge single-dangling (sd) bonds with three nearest neighbors, (Fig. 2.90f) lower terrace dangling-bonds. The (dd) edge bonds are experimentally observed while the (sd) edge bonds are not. The latter are included in the calculation because of computational limitations requiring a geometry which includes (sd) edge sites. The authors argue that the contribution of the (sd) density of states (Fig. 2.90e) does not appreciably affect the total density of states (Fig. 2.90b). The calculations indicate that the (dd) edge bonds are more stable than the terrace bonds and the step-edges therefore attract electrons from the surrounding terraces. The calculation does not consider reconstruction which may change the picture. The dangling-bond energy region of the experimental photoemission spectrum from a high-step density (10%)

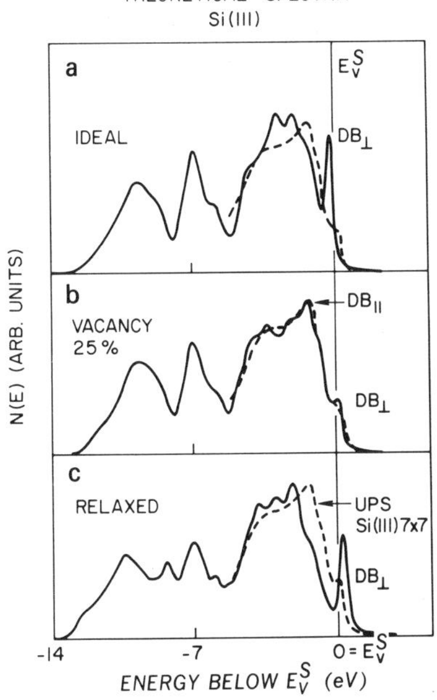

Fig. 2.91a–c. A comparison of the UPS spectra (dashed curve) for Si(111) 7×7 surface with the theoretical spectra based on: **a** the ideal Si(111) 1×1 surface, **b** Si(111) 2×2 (25% vacancies plus relaxation), and **c** a relaxed Si(111) 1×1 surface. $DB_\perp$ and $DB_{||}$ respectively denote density from dangling-bond sites pointing perpendicular to the surface and those next to vacancies pointing parallel to the surface [2.506]

surface [2.488] is shown as a solid line in Fig. 2.90a. A measure of the contribution of the steps to this spectrum is shown as the dashed curve in Fig. 2.90a which was obtained by subtracting from it the dangling-bond spectrum from a low-step density (3%) surface. This difference curve is accounted for rather well by the total theoretical density of states in Fig. 2.90b.

One may expect that surface defects like steps will exhibit characteristic electronic structure. *Pandey* [2.506] has modelled the Si(111) 7×7 surface with 25% surface atom vacancies after the model of *Lander* [2.507]. The calculations were performed for a 2×2 structure which approximates the 7×7 structure up to about the fifth-nearest neighbor. It was assumed that the longer-range ordering of the 2×2 subcells is a much smaller effect than the short-range intracell ordering. The vacancies, in a similar manner to steps, produce a characteristic distribution of surface states. *Pandey*'s results for ideal Si(111) 1×1, Si(111) 2×2 (25% vacancies plus relaxation), and relaxed Si(111) 1×1 are compared in Fig. 2.91 with the ultraviolet photoemission spectrum from Si(111) 7×7 obtained with $h\nu = 21.2$ eV [2.478]. $DB_\perp$ and $DB_{||}$ respectively denote densities of states from dangling-bonds pointing perpendicular to the surface and those next to vacancies pointing parallel to the surface. Experimental photoemission structure is observed at energies corresponding to both of these states. The vacancy model shows best agreement with the UPS results.

2.9.3 Surface States on Group III–V Semiconductors

It is well established for cleaved Si that Fermi level pinning $(E_F - E_V^S) \sim 0.4$ eV occurs at the surface and it is attributed to filled and empty bands of surface states near E_F (middle panel of Fig. 2.82). Cleaved III–V semiconductors with the zincblende structure do not exhibit similar Fermi level pinning. *Van Laar* et al. [2.508] found for cleaves with low defect densities that the contact potential differences between moderately doped *n*- and *p*-type GaAs ($E_{gap} = 1.43$ eV) was equal to the bulk potential difference (1.37 eV for their specific doping values) proving that the Fermi level could be swept through the gap and that there were neither filled nor empty surface states in the gap. However, they did find that the Fermi level position was pinned in the gap for samples which had large defect densities which probably induced extrinsic surface states in the gap [2.509]. They found that GaSb ($E_{gap} = 0.70$ eV), InAs ($E_{gap} = 0.36$ eV), and InP ($E_{gap} = 1.35$ eV) also did not have surface states in the gap [2.507]. For GaP ($E_{gap} = 2.26$ eV) they however did find Fermi level stabilization. Because photoemission spectra do not show filled dangling-bond surface states in the GaP gap, one can conclude from the contact potential difference measurement of 1.7 eV between moderate *n*- and *p*-type material that the Fermi level will be stabilized only on *n*-type GaP as a consequence of an empty surface state band with its lower edge in the gap 1.7 eV above the valence band maximum [2.508].

The natural cleavage face of III–V zincblende semiconductors is the (110) face which is nonpolar with one cation and one anion per surface unit cell. Other high symmetry surfaces such as the polar (111) and (100) surfaces where the top layer can be either all cation or all anion can be produced by molecular beam epitaxy or by spark cutting followed by vacuum cleaning. The nonpolar (110) surface of GaAs is schematically shown in Fig. 2.92. There are two dangling-bonds per surface atom. LEED analysis tells us that although there is no reconstruction, relaxation from their ideal position of the one Ga and one As atom within the (1×1) unit cell is necessary to explain the LEED intensity data [2.510]. The model that is commonly used to describe the relaxation within the unit cell is characterized by the tilt angle θ_T shown in Fig. 2.92 between the surface plane and the plane through the outward moving As atom and the inward moving Ga atom. The structural chemistry argument explaining the driving force for the relaxation follows similar lines to that given above for the *Haneman* model. The As and Ga bonds upon rehybridization gain *s*- and *p*-character respectively, stabilizing the filled As dangling-bond states and destabilizing the empty Ga dangling-bond states [2.511, 512]. In addition to this relaxation-splitting of the dangling-bond states, the III–V compounds have ionic as well as covalent contributions to the thermal gap which affect the separation of the filled and empty surface states. Loss of nearest neighbors at a surface reduces the covalent part of the interaction responsible for the gap, but not the ionic part, which is largely an atomic property. Since the surface states of the III–V's are observed outside the thermal gap the relaxation splitting apparently compensates for the loss of the covalent bonding-antibonding splitting from the

ATOMIC STRUCTURE OF GaAs(110) 1x1 SURFACE

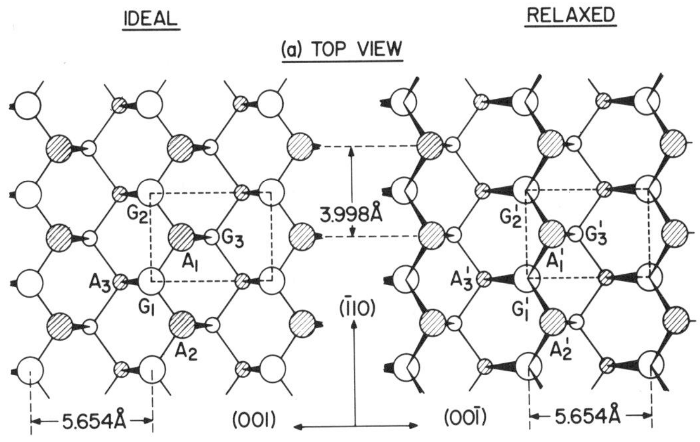

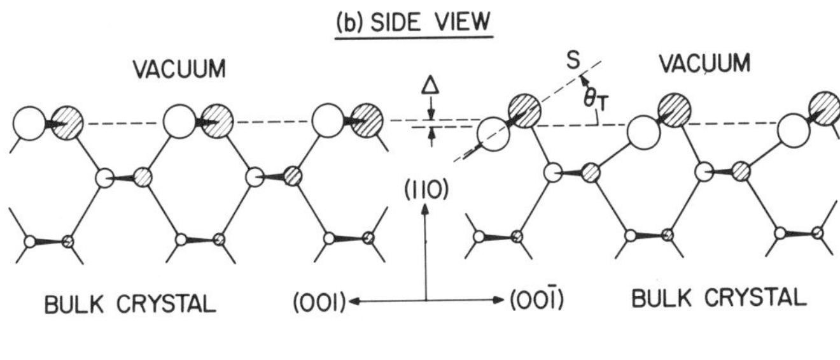

Ga ATOMS: ○ - SURFACE LAYER ○ - SECOND LAYER ○ - THIRD LAYER

As ATOMS: ◍ - SURFACE LAYER ◍ - SECOND LAYER ◍ - THIRD LAYER

θ_T = TILT ANGLE

Fig. 2.92a and b. Atomic structure of ideal and relaxed GaAs(110) 1×1 surfaces. All atomic coordinates are completely specified by the single parameter θ_T, the tilt angle. It is defined as the angle between the (110) surface and the plane S passing through the nearest-neighbor surface Ga and As atoms. The two-dimensional surface unit cell is shown by dashed lines. Note that as a result of relaxation the average interlayer spacing between the surface and the second layer decreases [2.506]

two missing surface atoms and thus leaves the filled anion dangling-bond density of states overlapping the valence bands and the empty cation dangling-bond surface density of states overlapping the conduction bands outside the gap. This overlap makes observation of the dangling-bond surface states on the III–V's as difficult as the back-bonding surface states on Si.

Lubinsky et al. [2.510] LEED analysis of the GaAs (110) surface yielded a relaxation characterized by a tilt angle $\theta_T = 35°$. *Pandey*'s [2.506] tight-binding

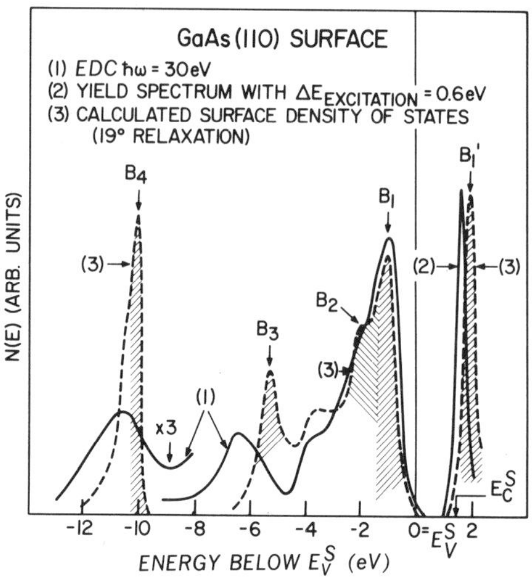

Fig. 2.93. Comparison of the UPS spectrum of the GaAs(110) 1×1 surface [2.512] with the theoretical calculation of *Pandey* for the ideal, 19° relaxed model [2.506]

model density of states shown in Fig. 2.93 however agrees best with photoemission results [2.512] when he uses a tilt angle $\theta_T \sim 19°$. With the help of angle-resolved photoemission spectroscopy *Knapp* and *Lapeyre* [2.513] identified peaks B_1 and B_2 as surface states because of their sensitivity to hydrogen adsorption. By plotting the shift of B_1 as a function of polar angle of emission they also obtained the dispersion of B_1 over part of the surface Brillouin zone [2.513]. The empty dangling-bond band B_1' has been observed by several methods, including characteristic energy loss [2.514] and partial yield spectroscopy techniques [2.482] (see also Sect. 6.4) which effectively measure the energy necessary to excite an electron from the localized Ga $3d_{5/2}$ core level with a binding energy of 18.6 eV to the empty surface state band B_1'. In several of the earlier publications the position of the empty dangling-bond bands was derived directly from this one-electron transition energy, which placed the empty dangling-bond bands of almost all the III–V's near the middle of the band gap. This is ruled out by *van Laar* et al. [2.508] CPD (contact potential difference) measurements. For the one exception, GaP, *van Laar* et al. determined the actual position of the surface state band inside the band gap at 1.7 eV above E_V^s while *Gudat* and *Eastman*'s [2.512] yield measurement placed it at 1.2 eV. This discrepancy of 0.5 eV has been attributed to an "excitonic shift" or relaxation resulting from the Coulomb attraction of the Ga $3d_{5/2}$ hole acting upon the excited electron in Ga derived dangling-bond states. The localized nature of this excitation has also been exhibited by the fact that excitations from the As $3d$ core levels to empty surface states are not observed [2.512] because this would involve charge transfer from the As to Ga dangling-bonds. The validity of the interpretation of the ir absorption results of *Chiarotti* et al. as evidence for a super-lattice band gap on Si(111) 2×1 is now being questioned for the same reason; the transitions would involve a charge transfer between $d_{\text{out}}^{\text{Si}}$ and

d_{in}^{Si} [2.471, 499]. *Forstmann* [2.474] carries the relaxation question further in his recent review of surface electronic structure theory where he concludes that it is still an open question to what extent single-particle energies or densities of stationary states can at all agree with photoemission densities of states. He stresses that the importance of relaxation during photoemission from *localized* surface states must still be evaluated. Operationally at present relaxation is assumed insignificant for photoemission from filled valence band and surface states and an approximate excitonic correction of the ~0.5 eV based on the result for GaP is made for partial yield spectroscopy measurements. This is exemplified by the experimental B'_1 peak which has been shifted up by 0.6 eV in Fig. 2.93.

2.9.4 Surface Chemistry of Semiconductors – Si(111):H and Si(111):SiH_3

Presently the best understood chemisorption system is hydrogen bonded to atomically clear silicon. *Sakurai* and co-workers tested the simple chemical argument that H atoms should saturate dangling orbitals available at the surface of silicon to form Si–H covalent bonds similar to those in SiH_4 and Si_2H_6 [2.515–519]. This series of papers demonstrates the power of close collaboration between theoretical and experimental efforts. They used LEED to characterize the surface structure, Auger spectroscopy to measure surface cleanliness, and UPS to measure the photoemission energy distributions ($\hbar\omega = 21.2$ eV) from clean and hydrogen covered Si(111) [2.515, 516], Si(110) [2.517), and Si(100) [2.519] surfaces, which have one, one, and two dangling-bonds per surface atom respectively. A LEED observation common to all surfaces was, that upon atomic hydrogen adsorption the surface reconstruction on the clean Si(111) 7×7, Si(110) 5×1, and Si(100) 2×1 surfaces would revert for a H saturated surface to a 1×1 pattern, as evidenced by the gradual disappearance of the nonintegral LEED spots. From the excellent agreement between their theoretical density of states calculation and experiment they concluded that the monohydride –SiH (Fig. 2.94) formed on the (111) and (110) surfaces where there was one dangling-bond and the dihydride $-SiH_2$ formed on the (100) surface where there were two dangling-bonds per surface atom. In addition, on the (110) surface, which has a rather open structure, they found a second weakly bound phase, which desorbs at much lower temperatures than the temperature required to desorb the hydrogen in the monohydride phase. Similar types of bonds have been recently observed in the UPS of *bulk* hydrogenated amorphous silicon [2.520].

It is possible to obtain yet another metastable Si(111) surface reconstruction, the Si(111) 1×1 structure, by heating the (7×7) surface to 800 °C for 10 min and quenching to room temperature [2.521]. Atomic hydrogen adsorption again produces the monohydride phase, however with prolonged exposure the UPS spectrum changes. *Pandey* et al. [2.515] tried several theoretical model calculations to explain the new structure and found the best

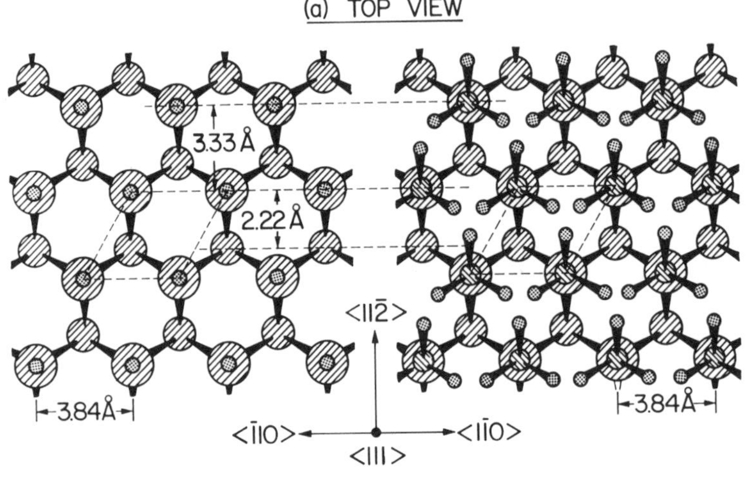

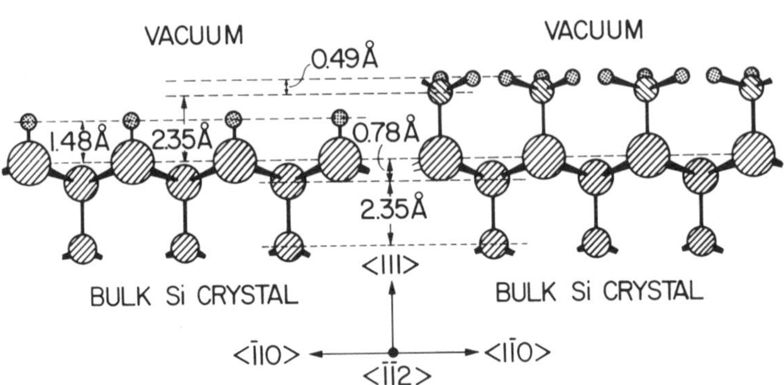

Fig. 2.94a and b. Structural models for the low- and high-coverage phases of H saturated Si(111) 1×1 surfaces

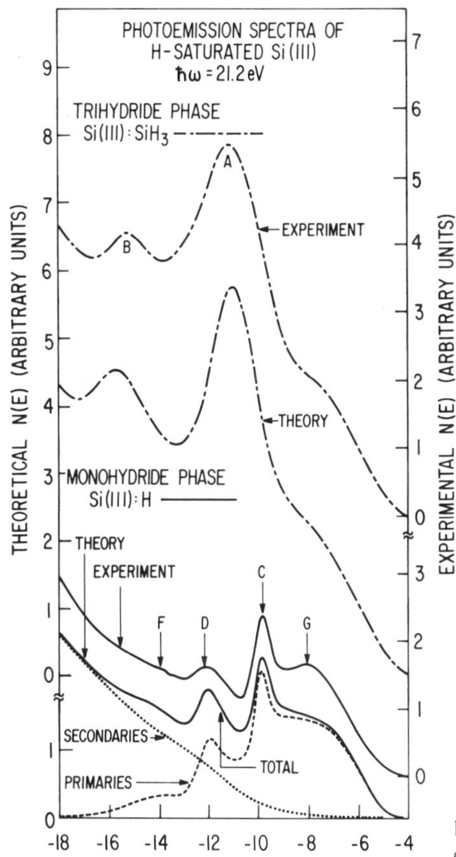

Fig. 2.95. Comparison of the theoretical and experimental spectra for the two hydrogen saturated phases of Si(111) 1×1 [2.515]

agreement for a monolayer of $-SiH_3$ radicals bonded to the dangling-bonds of Si(111). This model structure is shown in Fig. 2.94. Since the Si(111) surface normally has one dangling-bond per surface atom and not the three required by SiH_3, the surface must chemically reconstruct. This may occur with the removal of one Si layer by the desorption of SiH_4. This brings the second layer to the surface and the three back-bonds now become dangling-bonds. *Pandey* et al. believe that a random arrangement of vacancies on the Si(111) 1×1 surface makes it susceptible to chemical reconstruction, while an ordered arrangement of vacancies responsible for the symmetry of the (7×7) surface protects it from chemical attack [2.515]. The photoemission spectra and theoretical densities of states of the monohydride Si(111):H and the trihydride Si(111):SiH_3 phases shown in Fig. 2.95 are in excellent agreement.

The challenges in the fields of surface chemistry and heterogeneous catalysis awaiting the new surface physics tools which include photoemission spectroscopy have recently been reviewed by *Fischer* [2.522]. Most experiments to date have dealt with strongly chemisorbed species [2.468, 486, 512, 523, 524], because they comprise stable static systems which have been experimentally convenient for the initial study of chemisorption and the related electronic properties of the interacting solid and gas molecules. Future areas where surface physics and chemistry will most surely contribute include chemical reaction rates, reaction intermediates, activated adsorption, reactions between coadsorbed gases, modifications of adsorption properties of one species by coadsorption of another, mechanisms of catalyst poisoning and promotion, and the relationship between the catalytic properties of solids and their physico-chemical properties [2.522].

2.9.5 Interface States: Metal-Semiconductor Electrical Barriers

Semiconductor-metal contacts have been known since 1874 to pass current preferentially in one direction [2.15]. A large body of phenomenological data exists for this effect [2.525], however the physics at the metal semiconductor interface (MSI) is still quite uncertain and controversial. This phenomenon has been explained by *postulating* that a certain energy was necessary to take an electron from the Fermi energy of a metal and place it at a point of the conduction band of a semiconductor immediately close to the metal. This energy is called the interfacial barrier energy ϕ_B or more commonly the Schottky barrier. The barrier energies on ionic insulators have been observed to vary strongly with the work function of the metal while in the case of covalent semiconductors the barrier energies are relatively independent of the metal [2.526] with a value of $\phi_B \sim 2/3$ the band gap of the semiconductor [2.527].

Because of the lack of microscopic information about the MSI the theory has been limited to mostly macroscopic and some very idealized microscopic models. The basic question is to understand how the picture in Fig. 2.80 for the semiconductor-vacuum interface changes when a layer of metal makes intimate contact with the surface of the semiconductor. On an atomically clean Si surface the space charge layer adjusts the potential in the bulk so that the Fermi level in the bulk aligns with the Fermi level defined at the surface by the surface states. Similarly some form of dipole layer must build up at the MSI to align the Fermi level of the metal and that in the semiconductor. A microscopic understanding of the electronic states responsible for this dipole should lead to an understanding of the phenomenological behavior described above.

Bardeen [2.465] postulated that surface states pinned the Fermi level at the MSI just as they did on the clean surface. *Heine* [2.528] however, argued that the semiconductor surface states would no longer exist once the interface was formed, since they would match to the extended metal states. He postulated that the tails of the metal Bloch states however, would extend into the semiconductor, and they would act as the source and sink of electrons that would pin the Fermi level. *Louie* et al. [2.529] have extended these ideas to specific systems (Si, GaAs, and ZnS in contact with Al or jellium) using a method involving self-consistent pseudopotentials. They found states which

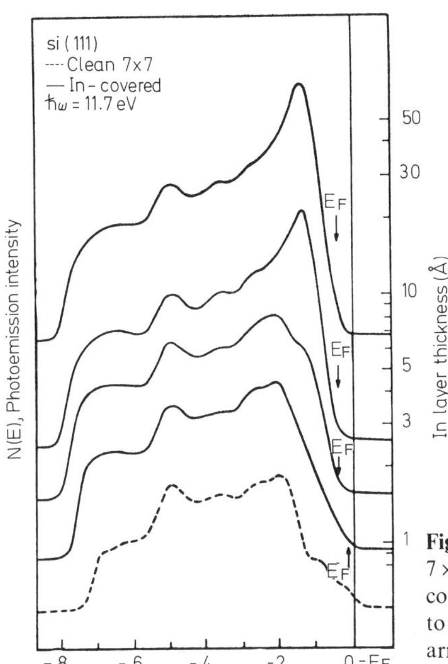

Fig. 2.96. Photoelectron spectra for a clean Si(111) 7×7 surface (dashed line) and for increasing In coverages (solid lines). The curves have been shifted to align the bulk-silicon feature at ~ -4.8 eV, and arrows indicate the position of the Fermi level for the metal-covered surfaces [2.531]

they named metal induced gap states (MIGS) which extend across the semiconductor gap and decrease in state density $N(E)$ with increasing ionicity. Recently *Varma* and *Pandey* [2.530] suggested a model where there was an atomic rearrangement at the interface resulting from the chemical interaction of the metal and semiconductor atoms. This reconstruction produces interface states responsible for the pinning. They proposed that for insulators and the more ionic semiconductors the chemical reconstruction would not be energetically favorable and the pinning states would not be formed.

Our discussion here will be limited to illustrating how photoemission spectroscopy is for the first time beginning to supply information about the "local character" of the MSI. We will again use Si as an example. *Margaritondo* et al. [2.531] have studied the formation of Schottky barriers between argon sputter cleaned and annealed Si(111) 7×7 surfaces and the group III metals Al, Ga, and In using UPS. The spectra from Si covered with In layers between 0 and 6 Å are shown in Fig. 2.96.

Margaritondo et al. concluded that; 1) the band bending $(E_C - E_F)$ shown in Fig. 2.97 and work function $(E_V - E_F)$ are modified for low metal coverages between one and four monolayers and 2) the "intrinsic" filled surface states observed between 1 and -1.5 eV on the clean surface (dashed curve in Fig. 2.96) are replaced with "extrinsic" metal-related interface states betwen 0 and -3 eV which are different for the different metals studied. The disappearance of the dangling-bond states at the beginning of the barrier formation was interpreted as an indication that the first layer of metal adatoms saturated these orbitals. The new metal-related surface states, which subsequently grow with greater

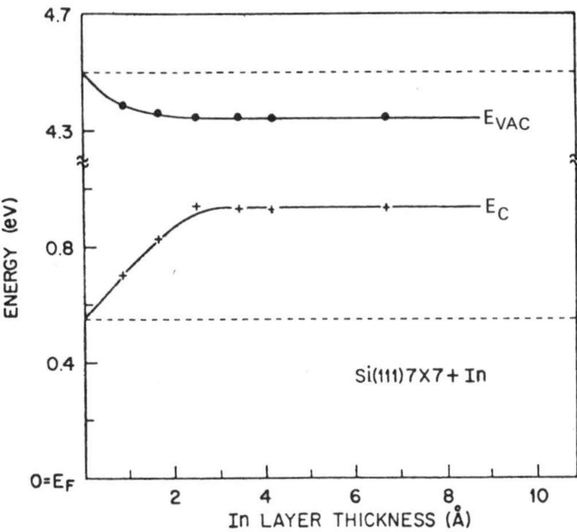

Fig. 2.97. Position of the conduction-band minimum E_C (crosses) and the vacuum level E_{VAC} (dots) vs In coverage. The dashed lines correspond to the clean surface position. The vertical scale is reffered to the Fermi level. The position of E_C has been evaluated subtracting the clean surface value of $E_F - E_V = 0.55$ eV and the measured shift of E_F from the energy gap, $E_C - E_V = 1.1$ eV. The uncertainties are ~ 0.1 eV for the energy and $\sim 15\%$ for the thickness [2.531]

coverage, tail into the gap causing the pinning of E_F in a new position and saturating its shift. The Schottky barrier height ϕ_B is determined by the new space charge induced by the new distribution of interface states at the MSI. We will call this interpretation of *Margaritondo* et al. the "one-electron picture" since ϕ_B can be read directly from the one-electron diagram in Fig. 2.97 giving $\phi_B = |E_C - E_F|$. These results give a step-by-step description of the local density of states during the barrier formation.

The above one-electron interpretation yields values for ϕ_B consistent with those obtained by other experimental methods (e.g., current-voltage plots, photoconductivity) which directly measure the excitation energy across the barrier. We however point out once again the weakness of photoemission to measure the surface potential accurately and will now describe how this uncertainty can affect the one-electron interpretation. The determination of ϕ_B in the one-electron picture given above depends directly on the accuracy of the evaluation of E_V^s and E_{gap} at the MSI. An error of 0.1 to 0.2 eV or even more is possible because, first the valence band edge has to be determined by extrapolating through the surface state emission from the clean surface, and second the assumption that the band gap remains equal to the clean surface value during the barrier formation may not be valid. *Inkson* [2.532] for example proposed, that changes in the electron exchange and correlation potential are related to collective excitation at the interface of a metal-semiconductor junction, and that these interactions change discontinuously across the gap and thus reduce the effective band gap. An additional factor that must be accounted for is the observation in Fig. 2.96 that the metal-semiconductor bonds at the interface produce orbitals that differ from both the bulk metal and the bulk semiconductor states. The model of

[a] Energy loss spectroscopy.
[b] Angular resolved EDC.
[c] Partial yield.
[d] Constant final-state spectroscopy.
[e] Constant initial-state spectroscopy.

Table 2.15. Experimental studies of semiconductor surfaces

Surface	Method	Photon energy [eV]	Remarks	Ref.
Expt. review			Semicond. surfaces	[2.533]
Expt. review			Surface states	[2.534]
Si(100) 2×1				[2.491, 538, 542]
Si(100) 2×1				[2.519]
Si(100) 2×1	ELS[a]			[2.543]
Si(100) 5×1	EDC	21.2	H chemisorp.	[2.517, 518]
Si(111) 1×1	EDC	21.2	Quenched from 800C	[2.515]
Si(111) 2×1	EDC	10, 12, 20	Surface states	[2.484]
Si(111) 2×1	EDC	6.2 to 11.8	Surface states	[2.485, 486]
Si(111) 2×1				[2.537]
Si(111) 2×1	EDC	21.2	Surface states	[2.478, 491, 538]
Si(111) 2×1	EDC	21.2	Surface steps	[2.488]
Si(111) 2×1				[2.538]
Si(111) 2×1	AREDC[b]			[2.490, 539]
Si(111) 2×1	EDC	$<$ 90	Oxidation	[2.546]
Si(111) 2×1	EDC, PY[c], AREDC	85		[2.561]
Si(111) 7×7	EDC	21.2		[2.478, 491, 540]
Si(111) 7×7				[2.486]
Si(111) 7×7				[2.538]
Ge(100) 2×1	ELS			[2.548]
Ge(100)/In	EDC, ELS	21.2	Interface states	[2.535]
Ge(110)/In	EDC, ELS	21.2	Interface states	[2.535]
Ge(111) 2×1				[2.543]
Ge(111) 2×1				[2.547]
Ge(111) 2×1	EDC	12		[2.484]
Ge(111) 2×1	EDC	10.2	Cleaved	[2.544]
Ge(111)/In	EDC, ELS	21.2	Interface states	[2.514, 535]
Ge(111) 2×8	EDC	10.2	2×1 annealed	[2.544]
Ge(111) 2×1	EDC	21.2	Cleaved	[2.473, 545]
Ge(111) 2×1	EDC	$<$130	*n*-type, oxidation	[2.468]
Ge(111) 8×8	ELS			[2.548]
GaAs(100)	ELS			[2.472]
GaAs(110)	AREDC, CFS[d], CIS[e]	$<$ 30		[2.470, 553]
GaAs(110)	EDC	25	*n*-type	[2.484]
GaAs(110)	EDC	$<$100	Oxidation	[2.553, 555]
GaAs(110)				[2.556]
GaAs(110)	EDC, CIS	10–100	Empty surf. states	[2.512, 557–559]
GaAs(111) As	ELS			[2.472]
GaAs(111) 2×2	AREDC, EDC	21.2	Polar Ga face	[2.561, 562]
GaAs(111)/In	EDC, ELS		Interface states	[2.514, 535]
GaAs(111) As	AREDC, EDC	21.2		[2.561, 562]
GaAs(111) As	ELS			[2.472]
GaP(110)	EDC, CIS	10–100		[2.512]
GaP(111) P	ELS			[2.550]
GaSb(110)	EDC, CIS	10–100		[2.512]
GaSb(110)	EDC	$<$160	Oxidation	[2.553]
InAs(110)	EDC, CIS	10–100		[2.512]
InP(110)	EDC, CIS	10–100		[2.512]
InP(110)	EDC	$<$100	Oxidation	[2.553]
InSb(110)	EDC, CIS	10–100		[2.512]
ZnS(100)	EDC	85		[2.561]
ZnSe(110)	EDC	85		[2.561]
ZnTe(110)	EDC	85		[2.561]
CdTe(110)	EDC	85		[2.561]

Table 2.16. Theoretical studies of semiconductor surfaces

Surface	Method	Remarks	Ref.
Review articles			[2.474, 483, 562, 564]
Si(100)	Tight binding	Ideal	[2.566]
Si(100)/H	SCF	H bonded to Si	[2.568]
Si(100) 2 × 1	Tight binding		[2.569]
Si(100) 2 × 1	SCF	Ideal, relaxed, reconstructed	[2.567]
Si(111)	Tight binding		[2.563]
Si(111)	Tight binding	Ideal	[2.566]
Si(111) 1 × 1	Semiempirical tight binding	$E(k)$ and $N(E)$: unreconstructed relaxed, and reconstructed models	[2.495–497]
Si(111) 1 × 1, 2 × 1			[2.498–500]
Si(111) 1 × 1	SCF	Ideal, unreconstructed and relaxed	[2.492–494]
Si(111) 2 × 1	Bond orbital model	Reconstructed Haneman, extension of Haneman model	[2.501, 502]
Si(111)/Al, Jellium	SC-EPM	Interface states: $N(E)$, $\varrho(E,\boldsymbol{r})$	[2.529, 578]
Ge(111) 1 × 1			[2.570]
Ge(111) 1 × 1	Tight binding		[2.497]
Ge(111) 1 × 1	Tight binding	Unreconstructed	[2.571]
Ge(111) 2 × 1	Nonlocal EPM	Anisotropic direct transition model	[2.563]
GaAs(100)	SCF	Ga surface, unreconstructed, relaxed	[2.576]
GaAs(110)		Ideal and relaxed	[2.573]
GaAs(110)		Ideal	[2.574]
GaAs(110)		Ideal	[2.575]
GaAs(110)		Ideal and relaxed	[2.506]
GaAs(111) and GaAs($\bar{1}\bar{1}\bar{1}$)	Tight binding		[2.572]
GaAs(111) and GaAs($\bar{1}\bar{1}\bar{1}$)	Tight binding	Unreconstructed	[2.571]
ZnS(110)	Tight binding	$E(\boldsymbol{k})$, $N(E)$	[2.577]
ZnSe(110)	Tight binding	$E(\boldsymbol{k})$, $N(E)$	[2.577]
ZnTe(110)	Tight binding	$E(\boldsymbol{k})$, $N(E)$	[2.577]
CdTe(110)	Tight binding	$E(\boldsymbol{k})$, $N(E)$	[2.577]

Varma and *Pandey* [2.530] contains both the formation of new bonds and collective effects at the interface. There presently is considerable theoretical and experimental activity addressing the Schottky barrier problem and many new interesting results should be appearing shortly. We conclude this chapter with tables listing experimental photoemission (Table 2.15) and theoretical (Table 2.16) studies of surface states in semiconductors.

References

2.1a P. Görlich: Z. Phys. **101**, 335 (1936)

2.1b see R.L. Bell: *Negative Electron Affinity Devices* (Clarendon Press, Oxford 1973)

2.2 M. Cardona, L. Ley (eds.): *Photoemission in Solids I, General Principles*, Topics in Applied Physics, Vol. 26 (Springer, Berlin, Heidelberg, New York 1978)

2.3 W.E. Spicer, R.E. Simon: Phys. Rev. Lett. **9**, 385 (1962)

2.4 F.G. Allen, G.W. Gobeli: Phys. Rev. **127**, 141 (1962)

2.5 D. Long: *Energy Bands in Semiconductors* (J. Wiley and Sons, New York 1968)
2.6 J.C. Phillips: *Solid State Physics*, Vol. 18, ed. by F. Seitz, D. Turnbull (Academic Press, New York 1966) p. 1
2.7 D.L. Greenaway, G. Harbeke: *Optical Properties of Semiconductors* (Pergamon Press, Oxford 1968)
2.8 M. Cardona: *Modulation Spectroscopy* (Academic Press, New York 1969)
2.9 J. Daniels, C.V. Festenberg, H. Raether, K. Zeppenfeld: Springer Tracts in Modern Physics, Vol. 54 (Springer, Berlin, Heidelberg, New York 1970), p. 77
2.10 G.L. Pearson, W.H. Brattain: Proc. IRE **43**, 1794 (1955)
2.11 Proceedings of the International Conferences on the Physics of Semiconductors
I. *Semiconducting Materials*, ed. by K.H. Henish (Butterworths Scientific Publications, London 1951)
II. Phys. **20** (1954)
III. *Halbleiter und Phosphore*, ed. by M. Schön, H. Welker (Vieweg, Braunschweig 1958)
IV. J. Phys. Chem. Solids **8** (1958)
V. Proc. of the Int. Conf. on Semiconductor Physics (Publishing House of the Czechoslovak Academy of Sciences, Prague 1961)
VI. Rpt. on the Int. Conf. on the Physics of Semiconductors, ed. by A.C. Strickland (The Institute of Physics and the Physical Society, London 1962)
VII. Physique des Semiconducteurs, ed. by M. Hulin (Dunod, Paris 1964)
VIII. J. Phys. Soc. Jpn. 21 Suppl. (1966)
IX. Int. Conf. on the Physics of Semiconductors, Moscow, ed. by S.M. Ryvkin (Nauka, Leningrad 1968)
X. Proc. of the 10th Int. Conf. on the Physics of Semiconductors, ed. by S.P. Keller, J.C. Hensel, F. Stern (US Atomic Energy Commission, Washington D.C. 1970)
XI. Int. Conf. on the Physics of Semiconductors Proceedings (PWN-Polish Scientific Publishers, Warsaw 1972)
XII. *Physics of Semiconductors*, ed. by M.H. Pilkuhn (Teubner, Stuttgart 1974)
"Physics of Semiconductors", in Proc. of the 13th Int. Conf., ed. by F.G. Fumi (Tip. Marves, Rome 1976)
XIII. Edinburgh, Sept. 1978 (to be published)
2.12 M. Cardona: In [Ref. 2.11 XII, p. 1339]
2.13 Proc. of the Int. Conf. on the Physics of Amorphous and Liquid Semiconductors
I. Prague 1965 (proceedings unpublished)
II. Bucharest, 1967 (proceedings unpublished)
III. J. Non-Cryst. Solids **4** (1970)
IV. J. Non-Cryst. Solids **8–10**, 172 (1972)
V. *Amorphous and Liquid Semiconductors*, ed. by J. Stuke, W. Brenig (Taylor and Francis, London 1974)
VI. *Structure and Properties of Non-Crystalline Semiconductors*, ed. by B.T. Kolomiets (Nauka, Leningrad 1976)
VII. *Amorphous and Liquid Semiconductors*, ed. by W.E. Spear (Center for Industrial Consultancy and Liaison, University of Edinburgh 1977)
VIII. Cambridge, Massachusetts, 1979
2.14 M. Faraday: *Experimental Researches in Electricity*, Vol. I, (B. Quaritch, London 1839) p. 122
2.15 F. Braun: Ann. Phys. Chem. **153**, 556 (1874)
2.16 W. Schottky, F. Waibel: Ann. Phys. **17**, 501 (1933)
2.17 C. Wagner: Z. Phys. Chem. B **11**, 181 (1933)
2.18 A.H. Wilson: Proc. R. Soc. London A **133**, 458 (1931); A **134**, 273 (1931)
2.19 J. Bardeen: Phys. Rev. **71**, 717 (1947)
2.20 J. Bardeen, W.H. Brattain: Phys. Rev. **74**, 230 (1948)
2.21 G.K. Teal, J.B. Little: Phys. Rev. **78**, 647 (1950)
2.22 W.G. Pfann: Trans. Am. Inst. Min. Metall. Pet. Eng. **194**, 747 (1952)
R.A. Smith: *Semiconductors* (Cambridge Uni. Press, Cambridge, 1964)

2.23 F. Herman: Phys. Rev. **93**, 1214 (1954)
2.24 B. Lax, H.J. Zeiger, R.N. Dexter: Physica **20**, 818 (1954)
G. Dresselhaus, A.F. Kip, C. Kittel: Phys. Rev. **98**, 368 (1955)
2.25 G.L. Pearson, C. Herring: Physica **20**, 975 (1954)
2.26 G.G. MacFarlane, V. Roberts: Phys. Rev. **97**, 1714 (1955)
2.27 H.R. Philipp, E.A. Taft: Phys. Rev. **113**, 1002 (1959)
2.28 J. Tauc, A. Abraham: In [Ref. 2.11 V, p. 375]
2.29a B.O. Seraphin, R.B. Hess: Phys. Rev. Lett. **14**, 138 (1965)
2.29b M. Cardona, W. Gudat, B. Sonntag, P.Y. Yu: In [Ref. 2.11 X, p. 106]
2.30a J.R. Chelikowsky, M.L. Cohen: Phys. Rev. B **14**, 556 (1976)
2.30b F. Herman, R.L. Kortum, C.D. Kuglin, J.P. Van Dyke: In *Methods in Computational Physics*, Vol. 8 (Academic Press, New York 1968) p. 193
2.30c D.J. Stukel: Phys. Rev. B **3**, 3347 (1971)
2.30d D. Auvergne, J. Camassel, H. Mathieu, M. Cardona: Phys. Rev. B **9**, 5168 (1974)
2.30e M. Cardona: In *Atomic Structure and Properties of Solids*, ed. by E. Burstein (Academic Press, New York 1972) p. 514
2.31 G. Dresselhaus, M.S. Dresselhaus: Phys. Rev. **160**, 649 (1967)
2.32 K.C. Pandey, J.C. Phillips: Phys. Rev. B **13**, 750 (1976)
2.33 S.T. Pantelides, W.A. Harrison: Phys. Rev. B **11**, 3006 (1975)
2.34 N.J. Shevchik, J. Tejeda, M. Cardona: Phys. Rev. B **9**, 2627 (1970)
2.35 D. Stocker: Proc. R. Soc. London **270**, 397 (1962)
2.36 M.F. Thorpe, D. Weaire: Phys. Rev. B **4**, 3518 (1971)
2.37a D. Weaire, M.F. Thorpe: In *Computational Methods for Large Molecules and Localized States in Solids*, ed. by F. Herman, N.W. Dalton, T. Koehler (Plenum Press, New York 1972)
2.37b R.N. Nucho, J.G. Ramos, P.A. Wolff: Phys. Rev. B **17**, 1843 (1978)
2.37c W.A. Harrison: Phys. Rev. B **8**, 4487 (1973)
2.38 N.A. Goryunova: *The Chemistry of Diamond-like Semiconductors* (MIT Press, Cambridge, Mass. 1965)
2.39a R.K. Willardson, A.C. Beer (eds): *Semiconductors and Semimetals* (Academic Press, New York) Vols. 1–12, 1966–(continuing)
2.39b O. Madelung: *Physics of III–V Compounds* (Wiley and Sons, New York 1964)
2.39c C. Hilsum, A.C. Rose-Innes: *Semiconducting III–V Compounds* (Pergamon Press, Oxford 1961)
2.40a *The Physics and Chemistry of the II–VI Compounds*, ed. by M. Aven, J.S. Prener (North-Holland, Amsterdam 1967)
2.40b II–VI Semiconducting Compounds, ed. by D.G. Thomas (Benjamin, New York 1967)
2.40c Rev. Physique Appl. **12** (1977)
2.41 A. Goldmann: Phys. Status Solidi B **81**, 9 (1977)
2.42 G.F. Koster: *Space Groups and their Representations*, (Academic Press, New York 1957)
2.43 R.H. Parmenter: Phys. Rev. **100**, 573 (1955)
2.44 W.D. Grobman, D.E. Eastman, M.L. Cohen: Phys. Lett. A **43**, 49 (1973)
2.45 T.N. Morgan: Phys. Rev. Lett. **21**, 819 (1968)
2.46 L. Laude, M. Cardona, F.H. Pollak: Phys. Rev. B **1**, 1436 (1970)
2.47 M. Cardona: In [Ref. 2.39a, Vol. III, p. 125]
2.48 K. Funke: Prog. Solid State Chem. **11**, 345 (1976)
2.49 A. Goldmann, J. Tejeda, N.J. Shevchik, M. Cardona: Phys. Rev. B **10**, 4388 (1974)
2.50 M. Cardona: Phys. Rev. **129**, 69 (1963)
2.51a F. Bassani, R.S. Knox, W.B. Fowler: Phys. Rev. **137**, A 1217 (1965)
2.51b S. Shy-Yih Wang, M. Schlüter, M.L. Cohen: Phys. Status Solidi b **77**, 295 (1976)
2.52 R.A. Stullen, G. Ascarelli: Phys. Rev. B **13**, 5501 (1976)
2.53 R.N. Euwema, T.C. Collins, D.G. Shankland, J.S. de Witt: Phys. Rev. **162**, 710 (1967)
2.54 J.L. Shay, J.H. Wernick: *Ternary Chalcopyrite Semiconductors: Growth, Electronic Properties and Applications* (Pergamon Press, Oxford 1975)
2.55 C. Varea de Alvarez, M.L. Cohen, S.E. Kuhn, Y. Petroff, Y.R. Shen: Phys. Rev. B **10**, 5175 (1974): $ZnSeP_2$

2.56 C. Varea de Alvarez, M.L. Cohen, L. Ley, S.P. Kowalczyk, F.R. McFeely, D.A. Shirley, R.W. Grant: Phys. Rev. B **10**, 596 (1974): $ZnGeP_2$
2.57 L. Pasemann, W. Cordts, A. Heinrich, J. Monecke: Phys. Status Solidi b **77**, 527 (1976): $ZnSiP_2$
2.58 P.C. Newman, J.A. Cundall: Nature **200**, 876 (1963)
2.59 A.I. Zaslavskii, N.F. Kartenko, Z.A. Karachentseva: Sov. Phys.-Solid State **13**, 2562 (1971)
2.60 P.J. Lin-Chung: Phys. Rev. **188**, 1272 (1969)
2.61 J. Tejeda, M. Cardona: Phys. Rev. B **14**, 2559 (1976) and references therein
2.62 A. Baldereschi, F. Meloni, F. Aymerich, G. Mula: Solid State Commun. **21**, 113 (1977)
2.63 A. Kuhn, A. Chevy, R. Chevalier: Phys. Status Solidi a **31**, 479 (1975)
2.64 M. Schlüter: Nuovo Cimento B **13**, 313 (1973)
2.65 R.Z. Bachrach, F.C. Brown: Phys. Rev. B **1**, 818 (1970)
2.66a Yu.I. Ravich, B.A. Efimova, I.A. Smirnov: *Semiconducting Lead Chalcogenides* (Plenum Press, New York 1970)
2.66b K. Heidrich, H. Künzel, J. Treusch: Solid State Commun. **25**, 887 (1978)
2.67a P.C. Kemeny, J. Azoulay, M. Cardona, L. Ley: Nuovo Cimento B **39**, 709 (1977)
2.67b D. Schiferl: Phys. Rev. B **10**, 3316 (1974)
2.68 F. Herman, R.L. Kortum, I.B. Ortenburger, J.P. van Dyke: J. Phys. (Paris) Colloq. C **4**, 29 (1968) and references therein
2.69 G. Martinez, M. Schlüter, M.L. Cohen: Phys. Rev. B **11**, 651 (1975) and references therein
2.70 T. Grandke, L. Ley, M. Cardona: Solid State Commun. **23**, 897 (1977)
2.71 *The Physics of Selenium and Tellurium*, ed. by W.C. Cooper (Pergamon Press, Oxford 1969)
2.72 P. Grosse: *Die Festkörpereigenschaften des Tellur*, Springer Tracts in Modern Physics, Vol. 48 (Springer, Berlin, Heidelberg, New York 1969)
2.73 G.B. Abdullayev, Y.G. Asadov, K.P. Mamedov: In [Ref. 2.71, p. 179]
2.74 P. Krusius, J. von Boehm, T. Stubb: Phys. Status Solidi b **67**, 551 (1975)
2.75 R. Sandrock: Phys. Rev. **169**, 642 (1968)
2.76 N.J. Shevchik: J. Phys. C **8**, 3767 (1975)
2.77 J.D. Joannopoulos, M. Schlüter, M.L. Cohen: Phys. Rev. **11**, 2186 (1975)
2.78 I. Chen: Phys. Rev. B **7**, 3672 (1973)
2.79 I. Chen: Phys. Rev. B **11**, 3976 (1975)
2.80 J.D. Joannopoulos, M. Kastner: Solid State Commun. **17**, 221 (1975)
2.81 S.G. Bishop, N.J. Shevchik: Phys. Rev. B **12**, 1567 (1975)
2.82 J. Cornet, D. Rossier: J. Non-Cryst. Solids **12**, 85 (1973)
2.83 J.R. Drabble: *Progress in Semiconductors*, Vol. 7 (Heywood, London 1963) p. 45
2.84 J.R. Drabble, C.H.L. Goodman: J. Phys. Chem. Solids **5**, 142 1(958)
2.85 S. Katsuki: J. Phys. Soc. Jpn. **26**, 58 (1969)
2.86 F.J. DiSalvo, D.E. Moncton, J.V. Waszak: Phys. Rev. B **14**, 4321 (1976)
2.87a A.H. Thompson, F.R. Gamble, C.R. Synnon: Mater. Res. Bull **10**, 915 (1975)
2.87b M.M. Traum, G. Margaritondo, N.V. Smith, J.E. Rowe, F.J. di Salvo: Phys. Rev. B **17**, 1836 (1978)
2.87c J.A. Wilson: Phys. Status Solidi b **86**, 11 (1978)
2.88 J.A Wilson, A.D. Yoffe: Adv. Phys. **18**, 193 (1969)
2.89 A.D. Yoffe: *Festkörperprobleme* XIII (Pergamon/Vieweg, London–Braunschweig 1973) p. 7
2.90 A.D. Yoffe: In [Ref. 2.11 XI, p. 611]
2.91 J.A. Wilson, F.J. di Salvo, S. Mahajan: Adv. Phys. **24**, 117 (1975)
2.92 F.R. Gamble, J.H. Osiecki, M. Cais, R. Pisharody, F.J. di Salvo, T.H. Geballe: Science **174**, 493 (1971)
2.93 C.B. Carter, P.M. Williams: Philos. Mag. **26**, 393 (1972)
2.94 R. Huisman, R. de Jonge, C. Haas, F. Jellinek: J. Solid State Chem. **3**, 56 (1971)
2.95 R.B. Murray, A.D. Yoffe: J. Phys. C **5**, 3038 (1972) and references therein
2.96 L.F. Mattheiss: Phys. Rev. B **8**, 3719 (1973)
2.97 R.A. de Groot, C. Haas: Solid State Commun. **17**, 887 (1975)
2.98 R.V. Kasowski: Phys. Rev. Lett. **30**, 1175 (1973)
2.99 K. Wood, J.B. Pendry: Phys. Rev. Lett. **31**, 1400 (1973)

2.100 G. Wexler, A.M. Woolley: J. Phys. C **9**, 1185 (1976)
2.101 B. Riccó: Phys. Status Solidi B **77**, 287 (1976)
2.102 C.Y. Fong, M.L. Cohen: Phys. Rev. Lett. **32**, 720 (1974)
2.103a H.W. Myron, A.J. Freeman: Phys. Rev. B **9**, 481 (1974)
2.103b P. Krusius, J. von Boehm, H. Isomäkki: J. Phys. C **8**, 3788 (1975)
2.104 A.H. Thompson: Phys. Rev. Lett. **34**, 520 (1976); J.E. Smith, Jr., J.C. Tsang, M.V. Shafer: Solid State Commun. **19**, 283 (1976)
2.105a H.R. Zeller: *Festkörperprobleme* XIII (Pergamon/Vieweg, London–Braunschweig 1973) p. 31
2.105b E. Tosatti: In [Ref. 2.13 XIII, p. 21]
2.106 W. Kohn: Phys. Rev. Lett. **2**, 393 (1959)
2.106a A.H. Thompson, R.G. Gamble, J.F. Revelli: Solid State Commun. **9**, 981 (1971)
2.107 E. Steigmeier, G. Harbeke, H. Auderset, F.J. di Salvo: In [Ref. 2.13 XIII, p. 369]
2.108 W.L. McMillan: Phys. Rev. B **12**, 1187 (1975)
2.109 J.A. Holy, M.V. Kline, W.L. McMillan, S.F. Meyer: Phys. Rev. Lett. **37**, 1145 (1976)
2.110 G.K. Wertheim, F.J. DiSalvo, S. Chiang: Phys. Lett. A **54**, 304 (1975)
2.111 H.P. Hughes, R.A. Pollak: Philos. Mag. **34**, 1025 (1976)
2.112 T.S. Moss: *Optical Properties of Semiconductors* (Butterworths, London 1959)
2.113 R.A. Smith: *Semiconductors* (Cambridge University Press, Cambridge 1968)
2.114 K. Seeger: *Semiconductor Physics* (Springer, Wien–New York 1973)
2.115 D.E. Aspnes: Phys. Rev. B **14**, 5331 (1976)
2.116 C.H. Chen: Phys. Status Solidi b **83**, 347 (1977)
2.117 N.J. Shevchik, J. Tejeda, M. Cardona: Phys. Rev. B **8**, 2833 (1973)
2.118 E.E. Koch, P.Y. Yu, C.M. Penchina, M. Cardona: Phys. Status Solidi b **53**, 327 (1972)
2.119a S.G. Louie, J.R. Chelikovsky, M.L. Cohen: Phys. Rev. Lett. **34**, 155 (1975)
2.119b K. Sturm: Phys. Rev. Lett. **40**, 1599 (1978)
2.120 A. Faessler: In *Vacuum Ultraviolet Radiation Physics*, ed. by E.E. Koch, R. Haensel, C. Kunz (Pergamon/Vieweg, London–Braunschweig 1974) p. 801
2.121 H.W.B. Skinner: Philos. Trans. R. Soc. London A **239**, 95 (1940)
2.122 D.H. Tomboulian: In *Röntgenstrahlen, X-Rays*, ed. by S. Flügge, Handbuch der Physik, Bd. 30 (Springer, Berlin, Göttingen, Heidelberg 1957) p. 246
2.123 *Landolt :Börnstein Tables*, Vol. I, Pt. 4 (Springer, Berlin, Göttingen, Heidelberg 1955) p. 769
2.124 *Soft X-Ray Band Spectra*, ed. by D.J. Fabian (Academic Press, New York 1968)
2.125 *Band Structure Spectroscopy of Metals and Alloys*, ed. by D.J. Fabian, L.M. Watson (Academic Press, New York 1973)
2.126 *X-Ray Spectroscopy*, ed. by K. Azároff (McGraw-Hill, New York 1974)
2.127 G. Wiech: Z. Phys. **207**, 428 (1967)
2.128 J. Klima: J. Phys. C **3**, 70 (1970)
2.129 J. Drahokoupil, A. Simůnek: J. Phys. C **7**, 610 (1974)
2.130 T. Grandke, L. Ley, M. Cardona: Phys. Rev. Lett. **38**, 1033 (1977)
2.131 C.J. Powell: Surf. Sci. **44**, 29 (1974)
2.132 W.E. Spicer: In *Optical Properties of Solids* ed. by B.O. Seraphin (North-Holland, Amsterdam 1976) p. 631
2.133 W.D. Grobman, D.E. Eastman, J.L. Freeouf: Phys. Rev. B **12**, 4405 (1975)
2.134 M. Klasson, A. Berndtsson, J. Hedman, R. Nilsson, R. Nyholm, C. Nordling: J. Electron Spectros. **3**, 427 (1974)
2.135 E.O. Kane: Phys. Rev. **159**, 624 (1967)
2.136 J.E. Rowe, H. Ibach: Phys. Rev. Lett. **32**, 421 (1974)
2.137 V.G. Aleshin, Yu.N. Kucherenko, V.V. Nemoshkalenko: Solid State Commun. **20**, 913 (1976)
2.138 W. Braun, A. Goldmann, M. Cardona: Phys. Rev. B **10**, 5069 (1975)
2.139 W.E. Spicer: J. Phys. (Paris) **34**, C 6–19 (1973)
2.140 J. Matzusaki: Stanford Electronics Lab., Stanford University, Rpt-No 5220-3 (1975)
2.141 K.C. Pandey, J.C. Phillips: Phys. Rev. B **9**, 1552 (1973)
2.142 J.F. Janak, A.R. Williams, V.L. Moruzzi: Phys. Rev. B **11**, 1522 (1975)

2.142a G.G. MacFarlane, T.P. McLean, J.E. Quarrington, V. Roberts: Phys. Rev. **108**, 1377 (1957)
2.143 V.G. Aleshin, Yu.N. Kucherenko: J. Electron Spectros. **8**, 411 (1976)
2.144 V.V. Nemoshkalenko, V.G. Aleshin, Yu.N. Kucherenko: Solid State Commun. **20**, 1155 (1976)
2.145 R.G. Cavell, S.P. Kowalczyk, L. Ley, R.A. Pollak, B. Mills, D.A. Shirley, W. Perry: Phys. Rev. B **7**, 5313 (1973)
2.146 P. Nielssen: Phys. Rev. B **6**, 3739 (1972). The 4*s* levels of Se, which lie ~14 eV below the top of the valence band, were incorrectly interpreted in this paper
2.147 L. Ley, R.A. Pollak, F.R. McFeely, S.P. Kowalszyk, D.A. Shirley: Phys. Rev. B **9**, 600 (1974)
2.148 D.E. Eastman, W.D. Grobman, J.L. Freeouf, M. Erbudak: Phys. Rev. B **9**, 3473 (1974)
2.149 N.J. Shevchik, J. Tejeda, C.M. Penchina, M. Cardona: Solid State Commun. **11**, 1619 (1972)
2.150 M.L. Cohen, T.K. Bergstresser: Phys. Rev. **141**, 789 (1966)
2.151 N.J. Shevchik, J. Tejeda, M. Cardona, D.W. Langer: Phys. Status Solidi b **59**, 87 (1973); **60**, 345 (1973)
2.152 M.Y. Au-Yang, M.L. Cohen: Solid State Commun. **6**, 855 (1968)
2.153 R. Harrison: Philos. Mag. **22**, 131 (1970)
2.154 J.N. Zemel: J. Phys. (Paris) C **4**, 9 (1968)
2.155 A.L. Hagström, A. Fahlmann: Proc. of the 5th Int. Conf. on Vacuum UV Spectroscopy, Vol. 2, ed. by M.C. Castex, M. Pouey, N. Pouey, (C.N.R.S., Paris 1977) p. 232
2.156 W.E. Spicer, G.J. Lapeyre: Phys. Rev. A **139**, 565 (1965)
2.157 G.B. Fischer, W.E. Spicer: J. Non-Cryst. Solids **8**, 978 (1972)
2.158 I. Abbati, L. Braicovich, B. De Michelis: J. Phys. C **7**, 3661 (1974)
2.159 M. Cardona, D.W. Langer, N.J. Shevchik, J. Tejeda: Phys. Status, Solidi b **58**, 127 (1973)
2.160 P.C. Kemeny, M. Cardona: J. Phys. C **9**, 1361 (1976)
2.161 N.J. Shevchik, D.W. Langer, M. Cardona: Phys. Status Solidi b **57**, 245 (1973)
2.162 F.R. McFeely, S. Kowalczyk, L. Ley, R.A. Pollak, D.A. Shirly: Phys. Rev. B **7**, 5228 (1973)
2.163 T. Grandke, L. Ley, M. Cardona: Phys. Rev. Lett. **38**, 1033 (1977) and unpublished data
2.164 S.E. Kohn, P.Y. Yu, Y. Petroff, Y.R. Shen, Y. Tsang, M.L. Cohen: Phys. Rev. B **8**, 1477 (1973)
2.165a A. Baldereschi: Phys. Rev. B **7**, 5212 (1973)
2.165b P.C. Kemeny, J. Azoulay, M. Cardona, L. Ley: Nuovo Cimento B **39**, 709 (1977)
2.165c T. Grandke, L. Ley: Phys. Rev. B **16**, 832 (1977)
2.166 L. Ley, R.A. Pollak, S.P. Kowalczyk, F.R. McFeely, D.A. Shirley: Phys. Rev. B **8**, 641 (1973)
2.167 J. Tejeda, N.J. Shevchik, W. Braun, A. Goldmann, M. Cardona: Phys. Rev. B **12**, 1557 (1975)
2.168 S. Kono, T. Ishii, T. Sagawa, T. Kobayasi: Phys. Rev. B **8**, 795 (1973)
2.169 M.G. Mason: Phys. Rev. B **11**, 5094 (1975)
2.170a D.E. Eastman, M. Kuznietz: Phys. Rev. Lett. **26**, 846 (1971)
2.170b H.R. Phillip: J. Appl. Phys. **43**, 2835 (1972)
2.170c J.A.R. Samson: Adv. At. Mol. Phys. **2**, 178 (1976)
2.170d H.J. Hagenau, W. Gudat, C. Kunz: J. Opt. Soc. Am. **65**, 742 (1975)
2.171 D.E. Eastman, J.L. Freeouf: Phys. Rev. Lett. **34**, 395 (1975)
2.172 W. Braun, A. Goldmann, M. Cardona: Phys. Rev. B **10**, 5069 (1974)
2.173 R.S. Bauer, W.E. Spicer: Phys. Rev. B **14**, 4539 (1976)
2.174 S.F. Lin, W.E. Spicer, R.S. Bauer: Phys. Rev. B **14**, 4551 (1976)
2.175 R.S. Bauer, S.F. Lin, W.E. Spicer: Phys. Rev. B **14**, 4527 (1976)
2.176 I.T. McGovern, R.H. Williams: J. Phys. C **9**, L 337 (1976)
2.177 P.M. Williams, F.R. Shepherd: J. Phys. C **6**, L 36 (1973)
2.178 A.R. Beal, J.C. Knights, W.Y. Liang: J. Phys. C **5**, 3531 (1972)
2.179 R.H. Williams, P. Kemeny, L. Ley: Solid State Commun. **19**, 495 (1976)
2.180 A.J. Grant, T.M. Griffiths, G.D. Pitt, A.D. Yoffe: J. Phys. C **8**, L 17 (1975)
2.181 G.K. Wertheim, F.J. DiSalvo, D.N.E Buchanan: Solid State Commun. **13**, 1225 (1973)
2.182 D.W. Fischer: Phys. Rev. B **8**, 3576 (1973)
2.183 C. Sugiura, I. Suzuki, J. Kashiwakura, Y. Gohshi: J. Phys. Soc. Jpn. **40**, 1720 (1976)

2.184 C.Y. Fong, M.L. Cohen: Phys. Rev. B **5**, 3095 (1972)
2.185a R.B. Murray, R.H. Williams: J. Phys. C **6**, 3643 (1973)
2.185b G. Margaritondo, J.E. Rowe, M. Schlüter, H. Kasper: Solid State Commun. **22**, 753 (1977)
2.185c L. Ley, R.H. Williams, P.C. Kemeny: Nuovo Cimento B **39**, 715 (1977)
2.186 G. Harbeke, E. Tosatti: Phys. Rev. Lett. **28**, 1567 (1972)
2.187a J. Azoulay, L. Ley: Solid State Commun. **22**, 557 (1977)
2.187b J. Robertson: Solid State Commun. **26**, 791 (1978)
2.188 S.P. Kowalczyk, L. Ley, F.R. McFeely, D.A. Shirley: Solid State Commun. **17**, 463 (1975)
2.189 P. Thiry, R. Pincheaux, D. Dagneaux, Y. Petroff: In [Ref. 2.11 XII, p. 1324]
2.190 L.W. James, J.L. Moll: Phys. Rev. **183**, 740 (1969)
2.191 W. Gudat, E.E. Koch, P.Y. Yu, M. Cardona, C.M. Penchina: Phys. Status Solidi b **52**, 505 (1972)
2.192 D.E. Eastman, J.L. Freeouf: Phys. Rev. Lett. **33**, 1061 (1974)
2.193 D.E. Eastman, J.L. Freeouf: Phys. Rev. Lett. **34**, 1624 (1975)
2.194 J. Van Laar, A. Huijser, T.L. Van Rooy: Surf. Sci. **62**, 472 (1977)
2.195 M. Altarelli: In [Ref. 2.155, Vol. 2, p. 1]
2.196 N. Shevchik: Private communication
2.197 T. Grandke, L. Ley, M. Cardona: Phys. Rev. (in press)
2.198 B. Feuerbacher, N.E. Christensen: Phys. Rev. B **10**, 2373 (1974)
2.199 H.R. Roloff, H. Neddermeyer: Solid State Commun. **21**, 561 (1977)
2.200 J. Stöhr, P.S. Wehner, R.S. Williams, G. Apai, D.A. Shirley: Phys. Rev. B **17**, 587 (1978)
2.201 G.V. Hansson, S.A. Flodström: Phys. Rev. B **17**, 473 (1978)
2.202 D.L. Rogers, C.Y. Fong: Phys. Rev. Lett **34**, 660 (1975)
2.203 P.J. Feibelman, D.E. Eastman: Phys. Rev. B **10**, 4932 (1974)
2.204 K.L. Chopra: *Thin Film Phenomena* (McGraw-Hill, New York 1969)
2.205 R. Grigorovici: In *Amorphous and Liquid Semiconductors*, ed. by J. Tauc (Plenum Press, New York 1974) p. 45
2.206 D.E. Sayers, F.W. Lyttle, E.A. Stern: In [Ref. 2.13 IV, p. 401]
2.207 D.E. Sayers, F.W. Lyttle, E.A. Stern: In [Ref. 2.13 V, p. 403]
2.208 R. Koplow, T.A. Rowe, B.L. Averbach: Phys. Rev. **168**, 1068 (1968)
2.209 A. Seeger, K.P. Chik: *Abstracts of Intern. Conf. on Amorphous and Liquid Semiconductors, Cambridge, England* 1969, p. 96
2.210 D.E. Polk: J. Non-Cryst. Solids **5**, 365 (1971)
2.211 D. Turnbull, D.E. Polk: J. Non-Cryst. Solids **8–10**, 19 (1972)
2.212 G. Lucovsky: In [Ref. 2.13 V, p. 1099]
2.213 E. N. Economou, M.H. Cohen, K.F. Freed, E.S. Kirkpatrick: In *Amorphous and Liquid Semiconductors*, ed. by J. Tauc (Plenum Press, New York 1974) p. 108
2.214 M.H. Brodsky, S. Kirkpatrick, D. Weaire (eds.): *Tetrahedrally Bonded Amorphous Semiconductors* (The American Institute of Physics, New York 1974), AIP Conf. Proc. No. 20
2.215 G. Lucovsky, F.L. Galeener (eds.): *Structure and Excitations of Amorphous Solids* (The American Institute of Physics, New York 1976), AIP Conf. Proc. No. 31
2.216 P.G. LeComber, J. Mort (eds.): *Electronic and Structural Properties of Amorphous Semiconductors* (Academic Press, New York 1973)
2.217 J. Tauc (ed.): *Amorphous and Liquid Semiconductors* (Plenum Press, New York 1974)
2.218 L. Ley, S.P. Kowalczyk, R. Pollak, D.A. Shirley: Phys. Rev. Lett. **29**, 1088 (1972)
2.219 T.M. Donovan, W.E. Spicer: Phys. Rev. Lett. **21**, 1571 (1968)
2.220 T.M. Donovan, W.E. Spicer, J.M. Bennett: Phys. Rev. Lett. **22**, 1058 (1969)
2.221 D.T. Pierce, W.E. Spicer: Phys. Rev. B **5**, 3017 (1972)
2.222 D.E. Eastman, W.D. Grobman: In [Ref. 2.11 XI, p. 889]
2.223 G. Wiech, E. Zöpf, D.J. Fabian, L.M. Watson: In *Band Structure Spectroscopy of Metals and Alloys*, (Academic Press, New York 1973) p. 637
2.224 A.H. Clark: Phys. Rev. **154**, 750 (1970)
2.225 T.M. Donovan, W.E. Spicer, J.M. Bennett, E.J. Ashley: Phys. Rev. B **2**, 397 (1970)
2.226 T.B. Light: Phys. Rev. Lett. **22**, 1058 (1969)

2.227 F.W. Lyttle: In *Physics of Non-Crystalline Solids*, ed. by J.A. Prins (North-Holland, Amsterdam 1965) p. 12
2.228 T.M. Donovan, E.J. Ashley, W.E. Spicer: Phys. Lett. **32** A, 86 (1970)
2.229 S. Koc, O. Renner, M. Zavetova, J. Zemek: Czech, J. Phys. B **22**, 1296 (1972)
2.230 F. Herman, J.P. Van Dyke: Phys. Rev. Lett. **21**, 1575 (1968)
2.231 S.C. Moss, J.F. Graczyk: Phys. Rev. Lett. **23**, 1167 (1969)
2.232 J.F. Graczyk, S.C. Moss: In [Ref. 2.13 VIII, p. 658]
2.233 T.M. Donovan, K. Heinemann: Phys. Rev. Lett. **27**, 1794 (1971)
2.234 I. Ohdomari, M. Ikeda, H. Yoshimoto: Phys. Lett. A **64**, 253 (1977)
2.235 W.A. Harrison: *Solid State Theory* (McGraw-Hill, New York 1970) p. 212ff.
2.236 D. Brust: Phys. Rev. **186**, 768 (1969)
2.237 D. Brust: Phys. Rev. Lett. **23**, 1232 (1969)
2.238 K. Maschke, P. Thomas: Phys. Status Solidi **39**, 453 (1970)
2.239 B. Kramer: Phys. Status Solidi **41**, 649 (1970)
2.240 B. Kramer: Phys. Status Solidi b **47**, 501 (1971)
2.241 K. Maschke, P. Thomas: Phys. Status Solidi **41**, 743 (1970)
2.242 B. Kramer, K. Maschke, P. Thomas: Phys. Status Solidi b **48**, 635 (1971)
2.243 B. Kramer, K. Maschke, P. Thomas: Phys. Status Solidi b **49**, 525 (1972)
2.244 R.W. Shaw, Jr., N.V. Smith: Phys. Rev. **178**, 985 (1969)
2.245 F.C. Brown, O.M.P. Rustgi: Phys. Rev. Lett **28**, 497 (1972)
2.246 D.E. Eastman, J.L. Freeouf, M. Erbudak: In [Ref. 2.214, p. 95]
2.247 W.E. Spicer, T.M. Donovan: Phys. Rev. Lett. **24**, 595 (1970)
2.248 B. Kramer: In *Advances in Solid State Physics*, ed. by O. Madelung (Vieweg, Braunschweig 1972) p. 133ff.
2.249 J.D. Joannopoulos, M.L. Cohen: In *Solid State Physics*, Vol. 31, ed. by H. Ehrenreich, F. Seitz, D. Turnbull (Academic Press, New York 1976) p. 71ff.
2.250 D. Weaire, M.F. Thorpe: Phys. Rev. Lett. **27**, 1581 (1971)
2.251 D. Weaire, M.F. Thorpe: Phys. Rev. B **4**, 2508 (1971)
2.252 M.F. Thorpe, D. Weaire: Phys. Rev. B **4**, 3518 (1971)
2.253 M.F. Thorpe, D. Weaire, R. Alben: Phys. Rev. B **7**, 3777 (1973)
2.254 A.F. Ioffe, A.R. Regel: Prog. Semicond. **4**, 239 (1960)
2.255 C. Domb: Adv. Phys. **9**, 145 (1960)
2.256 K. Husumi: J. Chem. Phys. **18**, 682 (1950)
2.257 H. Richter, G. Breitling: Z. Naturforsch. **13** a, 988 (1958)
2.258 M.V. Coleman, D.J.D. Thomas: Phys. Status Solidi **22**, 593 (1967); Phys. Status Solidi **24**, K 111 (1967)
2.259 R. Grigorovici, R. Mănăilă: Thin Solid Films **1**, 343 (1968)
2.260 S. Mader: J. Vac. Sci. Tech. **8**, 247 (1971)
2.261 I.B. Ortenburger, W.E. Rudge, F. Herman: J. Non-Cryst. Solids **8–10**, 653 (1972)
2.262 D. Weaire, A.R. Williams: Phys. Status Solidi b **49**, 619 (1972)
2.263 R. Alben, S. Goldstein, M.F. Thorpe, D. Weaire: Phys. Status Solidi b **53**, 545 (1972)
2.264 D. Henderson, F. Herman, I.B. Ortenburger: In *Amorphous and Liquid Semiconductors*, see [Ref. 2.13 V]
2.265 D. Henderson, I.B. Ortenburger: J. Phys. C **6**, 631 (1973)
2.266 J.D. Joannopoulos, M.L. Cohen: Solid State Commun. **11**, 549 (1972)
2.267 J.D. Joannopoulos, M.L. Cohen: Phys. Rev. B **7**, 2644 (1973)
2.268 M.L. Cohen: In [Ref. 2.11 XI, p. 731]
2.269 J.D. Joannopoulos, M.L. Cohen: Phys. Rev. B **8**, 2733 (1973)
2.270 J.D. Joannopoulos, M.L. Cohen: Solid State Commun. **13**, 1115 (1973)
2.271 R. Alben, D. Weaire, P. Steinhardt: J. Phys. C **6**, L384 (1973)
2.272 F.C. Choo, B.Y. Tong, J.R. Swenson: Phys. Lett. A **50**, 255 (1974)
2.273 B.Y. Tong, J.R. Swenson, F.C. Choo: Phys. Rev. B **10**, 3338 (1974)
2.274 W.Y. Ching, C.C. Ling: Phys. Rev. Lett. **34**, 1223 (1975)
2.275 W.Y. Ching, C.C. Ling, D.L. Huber: Phys. Rev. B **14**, 620 (1976)
2.276 W.Y. Ching, C.C. Ling, L. Guttmann: Phys. Rev. B **16**, 5488 (1977)

2.277 M.J. Kelly, D.W. Bullett: J. Non-Cryst. Solids **21**, 155 (1976)
2.278 F. Yndurain, J.D. Joannopoulos, M.L. Cohen, L. Falicov: Solid State Commun. **15**, 617 (1974)
2.279 J.D. Joannopoulos, F. Yndurain: Phys. Rev. B **10**, 5164 (1974)
2.280 J.P. Gaspard, F. Cyrot-Lackmann: J. Phys. C. **6**, 3077 (1973)
2.281 P.E. Meek: J. Phys. C. **10**, L59 (1977)
2.282 D.E. Polk, D.S. Boudreaux: Phys. Rev. Lett. **31**, 92 (1973)
2.283 P. Steinhardt, R. Alben, M.G. Duffy, D.E. Polk: Phys. Rev. B **8**, 6021 (1973)
2.284 P. Steinhardt, R. Alben, D. Weaire: J. Non-Cryst. Solids **15**, 199 (1974)
2.285 D. Henderson, F. Herman: J. Non-Cryst. Solids **8–10**, 359 (1972)
2.286 D. Henderson: J. Non-Cryst. Solids **16**, 317 (1974)
2.287 N.J. Shevchik, W. Paul: J. Non-Cryst. Solids **8–10**, 381 (1973)
2.288 G.A.N. Connell, R.J. Temkin: Phys. Rev. B **9**, 5323 (1974)
2.289 R.J. Temkin, W. Paul, G.A.N. Connell: Adv. Phys. **22**, 581 (1973)
2.290 J.P. Gaspard: In [Ref. 2.214 p. 170]
2.291 I.B. Ortenburger, D. Henderson: In [Ref. 2.214, p. 1511]
2.292 J. Treusch, B. Kramer: Solid State Commun. **14**, 169 (1974)
2.293 P.N. Sen, F. Yndurain: Phys. Rev. B **15**, 5076 (1977)
2.294 G.A.N. Connell, W.Paul: J. Non-Cryst. Solids **8–10**, 215 (1972)
2.295 G.A.N. Connell: Phys. Status Solidi b **53**, 213 (1972)
2.296 J.E. Smith, Jr., M.H. Brodsky, B.L. Crowder, M.I. Nathan: In *Proc. Int. Conf. on Light Scattering in Solids*, ed. by M. Balkanski (Flammarion, Paris 1972) p. 330
2.297 M. Wihl, M. Cardona, J. Tauc: J. Non-Cryst. Solids **8–10**, 172 (1972)
2.298 J.S. Lannin: Solid State Commun. **11**, 1523 (1972)
2.299 J.S. Lannin: In [Ref. 2.214, p. 260]
2.300 J.D. Joannopoulos, M.L. Cohen: Phys. Rev. B **10**, 1545 (1974)
2.301 F. Yndurain, J.D. Joannopoulos: Phys. Rev. B **11**, 2957 (1975)
2.302 B. Kramer, K. Maschke, P. Thomas: Phys. Status Solidi b **48**, 635 (1971)
2.303 L. Ley, F.R. McFeely, S.P. Kowalczyk, D.A. Shirley: Unpublished
2.304 R.A. Pollak: In [Ref. 2.214, p. 90]
2.305 M.K. Bahl, R.O. Woodall, R.L. Watson, K.J. Irgolic: J. Chem. Phys. **64**, 1210 (1976)
2.306 J. Stuke, G. Zimmerer: Phys. Status Solidi b **49**, 513 (1972)
2.307 Y.L. Yarnell, J.L. Warren, R.G. Wenzel, S.H. Koenig: IBM J. Res. Dev. **8**, 234 (1964)
2.308 D.M. Adams: *Inorganic Solids* (Wiley and Sons, New York 1974) p. 225ff.
2.309 J.C. Phillips: *Bonds and Bands in Semiconductors* (Academic Press, New York 1973)
2.310 R.H. Vallance: *Textbook of Inorganic Chemistry*, Vol. VI (C. Griffin, London 1938) p. 39
2.311 H. Stöhr: Z. Anorg. Allg. Chem. **242**, 138 (1939)
2.312 G. Breitling, H. Richter: Mater. Res. Bull. **4**, 19 (1969)
2.313 T.S. Moss: *Photoconductivity in the Elements* (Butterworth, Washington, D.C. 1952)
2.314 L. Holland: *Vacuum Deposition of Thin Films* (Chapman and Hall, London 1963)
2.315 G.N. Greaves, J.C. Knights, E.A. Davis: In [Ref. 2.13V, p. 369]
2.316 J.C. Knights, J.E. Mahan: Solid State Commun. **21**, 983 (1977)
2.317 J.C. Knights: In *Structure and Excitation of Amorphous Solids*, AIP Conf. Proc. No. **31** (New York 1976) p. 296
2.318 H. Krebs, R. Steffen: Z. Anorg. Allg. Chem. **327**, 224 (1964)
2.319 G.N. Greaves, E.A. Davis: Philos. Mag. **29**, 1201 (1974)
2.320 J. Robertson: J. Phys. C. **8**, 3131 (1975)
2.321 M.J. Kelly, D.W. Bullett: Solid State Commun. **18**, 593 (1976)
2.322 J.C. Knights: Solid State Commun. **16**, 515 (1975)
2.323 C. Rainsin, G. Leveque, S. Robin-Kandare: J. Phys. C **9**, 2887 (1976)
2.324 G.N. Greaves, E.A. Davis, J. Bordas: Philos. Mag. **34**, 265 (1976)
2.325 C. Raisin, G. Leveque, J. Robin, S. Robin-Kandare: In *Vacuum Ultraviolet Radiation Physics*, ed. by E.E. Koch, R. Haensel, C. Kunz (Pergamon/Vieweg, London-Braunschweig 1974) p. 502
2.326 C. Raisin, G. Leveque, J. Robin: Solid State Commun. **14**, 723 (1974)

2.327 J. Bordas, J. West: Philos. Mag. **34**, 501 (1976)
2.328 W.Y. Liang, A.R. Beal: J. Phys. C **9**, 2823 (1976)
2.329 J.S. Lannin, J.M. Calleja, M. Cardona: Phys. Rev. B **12**, 585 (1975)
2.330 J.S. Lannin: Phys. Rev. B **15**, 3863 (1977)
2.331 L. Ley, R.A. Pollak, S.P. Kowalczyk, F.R. McFeely, D.A. Shirley: Phys. Rev. B **8**, 641 (1973)
2.332 N.J. Shevchik: Philos. Mag. **35**, 261 (1977)
2.333 L. Ley, R.A. Pollak, S.P. Kowalczyk, D.A. Shirley: Phys. Lett. A **41**, 455 (1972)
2.334 F.R. McFeely, L. Ley, S.P. Kowalczyk, D.A. Shirley: Solid State Commun. **17**, 1415 (1975)
2.335 L.M. Falicov, S. Golin: Phys. Rev. **137**, A871 (1965)
2.336 S. Golin: Phys. Rev. **140**, A993 (1965)
2.337 L.M. Falicov, P.J. Lin: Phys. Rev. **141**, 562 (1966)
2.338 L.G. Ferreira: J. Chem. Phys. Solids **28**, 1891 (1967)
2.339 S. Golin: Phys. Rev. **166**, 643 (1968)
2.340 S. Mase: J. Phys. Soc. Jpn. **14**, 584 (1959)
2.341 D.W. Bullett: Solid State Commun. **17**, 695 (1975)
2.342 D.W. Bullett: Philos. Mag. **36**, 1529 (1977)
2.343 H. Krebs, W. Holz, K.H. Worm: Chem. Ber. **90**, 1031 (1957)
2.344 P.M. Smith, A.J. Leadbetter, A.J. Apling: Philos. Mag. **31**, 57 (1975)
2.345 A.J. Leadbetter, P.M. Smith, P. Seyfert: Philos. Mag. **33**, 441 (1976)
2.346 J.P. Van Dyke: J. Non-Cryst. Solids **12**, 263 (1973)
2.347 A. Bienenstock, F. Betts, S.R. Ovshinsky: J. Non-Cryst. Solids **2**, 347 (1970)
2.348 D.B. Dove, M.B. Heritage, K.L. Chopra, S.K. Bahl: Appl. Phys. Lett. **16**, 138 (1970)
2.349 A. Bienenstock: J. Non-Cryst. Solids **11**, 447 (1973)
2.350 R. Tsu, W.E. Howard, L. Esaki: J. Non-Cryst. Solids **8–10**, 364 (1972)
2.351 G.B. Fisher, W.E. Spicer: J. Non-Cryst. Solids **8–10**, 978 (1972)
2.352 G.B. Fisher, I. Lindau, B.A. Orlowski, W.E. Spicer, Y. Verhelle, H.E. Weaver: In [Ref. 2.13V, p. 621]
2.353 N.J. Shevchik, J. Tejeda, D.W. Langer, M. Cardona: Phys. Status Solidi b **57**, 245 (1973)
2.354 F. Betts, A. Bienenstock, S.R. Ovshinsky: J. Non-Cryst. Solids **4**, 554 (1970)
2.355 F. Betts, A. Bienenstock, D.T. Keating, J.P. de Neufville: J. Non-Cryst. Solids **7**, 417 (1972)
2.356 Y. Verhelle, A. Bienenstock: (Unpublished) cited in [Ref. 2.349]
2.357 A. Feiltz, J.J. Buettner, F.J. Lippmann, W. Maul: J. Non-Cryst. Solids **8–10**, 64 (1972)
2.358 G. Lucovsky, J.P. de Neufville, F.L. Galeener: Phys. Rev. B **9**, 1591 (1974)
2.359 G. Lucovsky, F.L. Galeener, R.C. Keezer, R.H. Geils, H.A. Six: Phys. Rev. B **10**, 5134 (1974)
2.360 G.B. Fisher, J. Tauc, Y. Verhelle: In [Ref. 2.13V, p. 1259]
2.361 N.F. Mott: Adv. Phys. **16**, 49 (1967)
2.362 G. Lucovsky, R.M. White: Phys. Rev. B **8**, 660 (1973)
2.363 G.B. Fisher: In *Physics of Structurally Disordered Solids*, ed. by S.S. Mitra (Plenum Press, London 1975) p. 703
2.364 W.R. Salaneck, K.S. Liang, A. Paton, N.O. Lipari: Phys. Rev. B **12**, 725 (1975)
2.365 S.G. Bishop, N.J. Shevchik: Phys. Rev. B **12**, 1567 (1975)
2.366 Z. Hurych, D. Davis, D. Buczek, C. Wood, G.J. Lapeyre, A.D. Baer: Phys. Rev. B **9**, 4392 (1974)
2.367 N.J. Shevchik, S. Bishop: Solid State Commun. **17**, 269 (1975)
2.368 L. Laude, B. Fitton, M. Anderegg: Phys. Rev. Lett. **26**, 637 (1971)
2.369 L. Laude, B. Fitton: J. Non-Cryst. Solids **8–10**, 971 (1972)
2.370 L.D. Laude, B. Fitton, B. Kramer, K. Maschke: Phys. Rev. Lett. **27**, 1053 (1971)
2.371 M. Schlüter, J.D. Joannopoulos, M.L. Cohen, L. Ley, S.P. Kowalczyk, R.A. Pollak, D.A. Shirley: Solid State Commun. **15**, 1007 (1974)
2.372 R.A. Powell, W.E. Spicer: Phys. Rev. B **10**, 1603 (1974)
2.373 R.H. Williams, J.I. Polanco: J. Phys. C **7**, 2748 (1974)
2.374 *Elemental Sulfur*, ed. by D. Mayer (J. Wiley and Sons, New York 1965)
2.375 P. Nielsen: Phys. Rev. B **10**, 1673 (1974)

2.376 N.V. Richardson, P. Weinberger: J. Electron Spectrosc. **6**, 109 (1975)
2.377 W.R. Salaneck, N.O. Lipari, A. Paton, R. Zallen, K.S. Liang: Phys. Rev. B **12**, 1493 (1975)
2.378 W.R. Salaneck, C.B. Duke, A. Paton, C. Griffiths, R.C. Keezer: Phys. Rev. B **15**, 1100 (1977)
2.379 G.B. Fisher, R.B. Shalvoy: Cited in [Ref. 2.363]
2.380 A.F. Holleman, E. Wiberg: *Lehrbuch der anorganischen Chemie* (de Gruyter, Berlin 1960) p. 218
2.381 G. Lucovsky, A. Mooradian, W. Taylor, G.B. Wright, K.C. Keezer: Solid State Commun. **5**, 113 (1967)
2.382 R.M. Martin, G. Lucovsky, K. Helliwell: Phys. Rev. B **13**, 1370 (1976)
2.383 B. Kramer: Phys. Status Solidi **41**, 725, 649 (1970)
2.384 I. Chen: Phys. Rev. B **2**, 1053, 1060 (1970)
2.385 N.F. Mott, E.A. Davies: *Electronic Processes in Non-Crystalline Materials* (Clarendon Press, Oxford 1971)
2.386 M.H. Cohen, H. Fritzsche, S.R. Ovshinsky: Phys. Rev. Lett. **22**, 1065 (1969)
2.387 N.F. Mott: Philos. Mag. **24**, 935 (1971)
2.388 M.H. Cohen: J. Non-Cryst. Solids **4**, 391 (1970)
2.389 N.F. Mott, R.A. Street: Philos. Mag. **36**, 33 (1977)
2.390 W.E. Spear: In [Ref. 2.13V, p. 1]
2.391a A. Madan, P.G. LeComber, W.E. Spear: J. Non-Cryst. Solids **20**, 239 (1976)
2.391b S.R. Ovshinsky, A. Madan: Nature **276**, 482 (1978)
2.392 M. Pollak, M.L. Knotek, H. Kurtzman, H. Glick: Phys. Rev. Lett. **30**, 856 (1973)
2.393 W.E. Spear: Adv. Phys. **26**, 312 (1977)
2.394 J. Tauc: In *Amorphous and Liquid Semiconductors*, ed. by J. Tauc (Plenum Press, New York 1974) p. 159
2.395 T.M. Donovan, W.E. Spicer, J.M. Bennet, E.J. Ashley: Phys. Rev. B **2**, 397 (1970)
2.396 M.L. Theye: In [Ref. 2.13V, p. 479]
2.397 W.I. Kinney, G.S. Cargill, III: Phys. Status Solidi A **40**, 37 (1977)
2.398a M.H. Brodsky, R.S. Title, K. Weiser, G.D. Pettit: Phys. Rev. B **1**, 2632 (1970)
2.298b M.H. Brodsky, D.M. Kaplan, J.F. Ziegler: In [Ref. 2.11XI, p. 529]
2.399 J.E. Fischer, T.M. Donovan: J. Non-Cryst. Solids **8–10**, 202 (1972)
2.400 R.F. Willis, L.D. Laude, B. Fitton: Surf. Sci. **37**, 395 (1973)
2.401 L.D. Laude, R.F. Willis, B. Fitton: In [Ref. 2.13V, p. 278]
2.402 L.D. Laude, R.F. Willis, B. Fitton: Solid State Commun. **12**, 1007 (1973)
2.403 C.W. Peterson, J.H. Dinan, T.E. Fischer: Phys. Rev. Lett. **25**, 861 (1970)
2.404 D.T. Pierce, W.E. Spicer: Phys. Rev. Lett. **27**, 1217 (1971)
2.405 T.E. Fischer, M. Erbudak: Phys. Rev. Lett. **18**, 1220 (1971)
2.406 M. Erbudak, T.E. Fischer: J. Non-Cryst. Solids **8–10**, 965 (1972)
2.407 L.D. Laude, R.F. Willis, B. Fitton: Phys. Rev. Lett. **29**, 472 (1972)
2.408a W.E. Spear, P.G. LeComber: J. Non-Cryst. Solids **8–10**, 727 (1972)
2.408b B. von Roedern, L. Ley, M. Cardona: Solid State Commun. (in press)
2.409 J.P. Walter, M.L. Cohen: Phys. Rev. B **4**, 1877 (1971)
2.410 J.C. Phillips: *Bonds and Bands in Semiconductors* (Academic Press, New York 1973)
2.411a L. Pauling: *The Nature of the Chemical Bond* (Cornell University Press, Ithaca, New York 1960)
2.411b L. Pauling: *The Nature of the Chemical Bond*, 1st ed. (Cornell University Press, Ithaca, New York 1939) p. 72ff.
2.412 R.S. Mullikan: J. Chem. Phys. **46**, 497 (1949)
2.413 C.A. Coulson, L.B. Redei, D. Stocker: Proc. R. Soc. London **270**, 352 (1962)
2.414 J.A. Van Vechten: Phys. Rev. **187**, 1007 (1969)
2.415 J.C. Phillips, J.A. Van Vechten: Phys. Rev. B **2**, 2147 (1970)
2.416 J.C. Phillips: Rev. Mod. Phys. **42**, 317 (1970)
2.417 W.A. Harrison: Phys. Rev. B **10**, 767 (1974)
2.418 W.A. Harrison, S. Ciraci: Phys. Rev. B **10**, 1516 (1974)
2.419 H. Schulz, K.H. Thiemann: Solid State Commun. **23**, 815 (1977)
2.420 G. Lucovsky, R.M. Martin, E. Burstein: Phys. Rev. B **4**, 1367 (1971)

2.421 R.M. Martin: Phys. Rev. B **5**, 1607 (1972)
2.422 P. Vogl: J. Phys. C **11**, 251 (1978)
2.423 S.P. Kowalczyk, L. Ley, F.R. McFeely, D.A. Shirley: J. Chem. Phys. **61**, 2850 (1974)
2.424 D.R. Penn: Phys. Rev. **128**, 2093 (1962)
2.425 P.J. Stiles: Solid State Commun. **11**, 1063 (1972)
2.426 D. Shiferl: Phys. Rev. B **10**, 3316 (1974)
2.427 E. Mooser, W.B. Pearson: Acta Crystallogr. **12**, 1015 (1959)
2.428 T.H. DiStefano, W.E. Spicer: Phys. Rev. B **7**, 1554 (1973)
2.429 S.P. Kowalczyk, F.R. McFeely, L. Ley, R.A. Pollak, D.A. Shirley: Phys. Rev. B **9**, 3573 (1974)
2.430 R.T. Poole, J.G. Jenkin, J. Liesegang, R.C.G. Leckey: Phys. Rev. B **11**, 5179, 5190 (1975)
2.431 P.C. Kemeny, J. Azoulay, M. Cardona, L. Ley: Nuovo Cimento B **39**, 709 (1977)
2.432 R.B. Shalvoy, G.B. Fisher, P.J. Stiles: Phys. Rev. B **15**, 1680 (1977)
2.433 R.B. Shalvoy, G.B. Fisher, P.J. Stiles: Phys. Rev. B **15**, 2021 (1977)
2.434 W.D. Grobman, D.E. Eastman, M.L. Cohen: Phys. Lett. A **43**, 49 (1973)
2.435 D.J. Chadi, M.L. Cohen, W.D. Grobman: Phys. Rev. B **8**, 5587 (1973)
2.436 K. Unger, H. Neumann: Phys. Status Solidi b **64**, 117 (1974)
2.437 K. Siegbahn, C. Nordling, A. Fahlman, R. Nordberg, K. Hamrin, J. Hedman, G. Johansson, T. Bergmark, S.-E. Karlsson, I. Lindgren: *ESCA Atomic Molecular and Solid State Structure Studied by Means of Electron Spectroscopy* (Almqvist and Wiksells, Uppsala 1967)
2.438 C.S. Fadley, S.B.M. Hagstrom, M.P. Klein, D.A. Shirley: J. Chem. Phys. **48**, 3779 (1968)
2.439 C.J. Vesely, D.W. Langer: Phys. Rev. B **4**, 451 (1971)
2.440 G. Hollinger, P. Kumurdjian, J.M. Mackowski, P. Pertosa, L. Porte, Tran Minh Duc: J. Electron Spectrosc. **5**, 237 (1974)
2.441 M.P. Tosi: Solid State Phys. **16**, 1 (1964)
2.442 C. Kittel: *Introduction to Solid State Physics* (J. Wiley and Sons, New York 1971) p. 118
2.443 N.F. Mott, R.W. Gurney: *Electronic Processes in Ionic Crystals* (Dover, New York 1964) p. 66
2.444 P.H. Citrin, R.W. Shaw, Jr., A. Packer, T.D. Thomas: In Proc. Int. Conf. Electron Spectroscopy, Asilomar USA 1971, ed. by D.A. Shirley (North-Holland, Amsterdam 1972) p. 691
2.445 R.T. Poole, J.G. Jenkin, J. Liesegang, R.C.G. Leckey: Phys. Rev. B **11**, 5179 (1975)
2.446 R.T. Poole, J. Szajman, R.C.G. Leckey, J.G. Jenkin, J. Liesegang: Phys. Rev. B **12**, 5872 (1975)
2.447 S.P. Kowalczyk, L. Ley, R.F. McFeely, R.A. Pollak, D.A. Shirley: Phys. Rev. B **9**, 381 (1974)
2.448 C.D. Wagner, P. Biloen: Surf. Sci. **35**, 82 (1973)
2.449 W. Eberhardt, G. Kalkoffen, C. Kunz, D.E. Aspnes, M. Cardona: Phys. Status Solidi b **88**, 135 (1978)
2.450 D.J. Hnatowich, J. Hudis, M.L. Perlman, R.C. Ragaini: J. Appl. Phys. **42**, 4883 (1971)
2.451 D.S. Urch, M. Webber: J. Electron Spectrose. **5**, 791 (1974)
2.452 S.T. Pantelides, W.A. Harrison: Phys. Rev. B **11**, 4049 (1975)
2.453 K. Hübner: Phys. Status Solidi b **68**, 223 (1975)
2.454 K. Hübner, M. Schäfer: Phys. Status Solidi b **76**, K63 (1976)
2.455 B. Feuerbacher, B. Fitton, R.F. Willis: *Photoemission from Surfaces* (J. Wiley and Sons, New York 1977)
2.456 I. Tamm: Physikalische Zeitschrift der Soviet Union **1**, 733 (1932)
2.457 W. Shockley: Phys. Rev. **56**, 317 (1939)
2.458 S.G. Davisson, J.D. Levine: Solid State Phys. **25**, 1 (1970)
2.459 M. Henzler: Surf. Sci. **25**, 650 (1971)
2.460 F. Forstmann: *Theory of Imperfect Crystalline Solids*, Trieste Lectures 1970 (Internat. Atomic Energy Agency, Vienna 1971) p. 511
2.461 R.O. Jones: In *Surface Physics of Phosphors and Semiconductors*, ed. by C.A. Scott, C.E. Reed (Academic Press, New York 1973) p. 95
2.462 W. Mönch: In *Advances in Solid State Physics*, Vol. 13 (Pergamon Press, New York 1973)

2.463 C.B. Duke, A.R. Lubinsky, B.W. Lee, D. Mark: J. Vac. Sci. Technol. **13**, 761 (1976)
2.464 C.B. Duke: *Proc. of NATO Advanced Studies Institute on Surface Plasticity*, Hohegeiss, Germany (1975), ed. by R.M. Latanision, J.T. Fourie (Noordhoff, Leyden 1977) p. 165
2.465 J. Bardeen: Phys. Rev. **71**, 717 (1947)
2.466 A. Many, Y. Goldstein, W.B. Grover: *Semiconductor Surfaces* (North-Holland, Amsterdam 1965)
2.467 D.R. Frankl: *Electrical Properties of Semiconductor Surfaces* (Pergamon, Oxford 1967)
2.468 E.W. Plummer: "Photoemission and Field Emission Spectroscopy" in *Interactions on Metal Surfaces*, Topics in Appl. Phys., Vol. **4**, ed. by R. Gomer (Springer, Berlin, Heidelberg, New York 1975) p. 144
2.469 B.F. Lewis, T.E. Fischer: Surf. Sci. **40**, 371 (1974)
2.470 H.D. Hagstrum, T. Sakurai: Phys. Rev. Lett. **37**, 615 (1976)
2.471 G. Chiarotti, P. Chiaradia, S. Nannarone: Surf. Sci. **49**, 315 (1975)
2.472 R. Ludeke, L. Esaki: Phys. Rev. Lett. **33**, 653 (1974)
2.473 J.E. Rowe: Solid State Commun. **15**, 1505 (1974)
2.474 F. Forstmann: In [Ref. 2.455, Chap. 8]
2.475 F.G. Allen, G.W. Gobeli: Phys. Rev. **127**, 150 (1962)
2.476 A. Thanailakis: J. Phys. C **8**, 655 (1975)
2.477 C. Sebenne, D. Bolmont, G. Guichar, M. Balkanski: Phys. Rev. B **12**, 3280 (1975)
C.A. Sebenne, G.M. Guichar, G. Garry: *Proc. Int. Symp. on Photoemission*, Noordwijk, The Netherlands, 1976, ed. by R.F. Willis, B. Feuerbacher, B. Fitton, C. Backx (European Space Agency, Paris 1976) p. 115
2.478 J.E. Rowe, H. Ibach, H. Froitzheim: Surf. Sci. **48**, 44 (1975)
2.479 G.W. Gobeli, F.G. Allen: Phys. Rev. **127**, 141 (1962)
2.480 F.G. Allen, G.W. Gobeli: J. Appl. Phys. **35**, 597 (1964)
2.481 B. Feuerbacher, R.F. Willis: J. Phys. C **9**, 169 (1976)
2.482 W. Gudat, D.E. Eastman: In [Ref. 2.455, Chap. 11]
2.483 J.A. Appelbaum, D.R. Hamann: Rev. Mod. Phys. **48**, 479 (1976)
2.484 D.E. Eastman, W.D. Grobman: Phys. Rev. Lett. **28**, 1378 (1972)
2.485 L.F. Wagner, W.E. Spicer: Phys. Rev. Lett. **28**, 1381 (1972)
2.486 L.F. Wagner, W.E. Spicer: Phys. Rev. B **9**, 1512 (1974)
2.487 M. Henzler: Surf. Sci. **36**, 109 (1973)
2.488 J.E. Rowe, S.B. Christman, H. Ibach: Phys. Rev. Lett. **34**, 874 (1975)
2.489 M. Schlüter, K.M. Ho, M.L. Cohen: Phys. Rev. B **14**, 550 (1976)
2.490 M.M. Traum, J.E. Rowe, N.E. Smith: J. Vac. Sci. Technol. **12**, 298 (1975)
2.491 J.E. Rowe, H. Ibach: Phys. Rev. Lett. **32**, 421 (1974)
2.492 J.A. Appelbaum, D.R. Hamann: Phys. Rev. B **12**, 1410 (1975)
2.493 J.A. Appelbaum, D.R. Hamann: Phys. Rev. Lett. **31**, 106 (1973)
2.494 J.A. Appelbaum, D.R. Hamann: Phys. Rev. Lett. **32**, 225 (1974)
2.495 K.C. Pandey, J.C. Phillips: Phys. Rev. Lett. **34**, 1450 (1975)
2.496 K.C. Pandey, J.C. Phillips: Phys. Rev. B **13**, 750 (1976)
2.497 K.C. Pandey, J.C. Phillips: Phys. Rev. Lett. **32**, 1433 (1974)
2.498 M. Schlüter, J.R. Chelikowsky, S.G. Louie, M.L. Cohen: Phys. Rev. Lett. **34**, 1385 (1975)
2.499 M. Schlüter, J.R. Chelikowsky, S.G. Louie, M.L. Cohen: Phys. Rev. B **12**, 4200 (1975)
2.500 M. Schlüter, J.R. Chelikowsky, M.L. Cohen: Phys. Lett. A **53**, 217 (1975)
2.501 I.P. Batra, S. Ciraci: Phys. Rev. Lett. **34**, 1337 (1975)
2.502 S. Ciraci, I.P. Batra: Solid State Commun. **18**, 1149 (1976)
2.503 D. Haneman: In *Surface Physics of Phosphors and Semiconductors*, ed. by C.G. Scott, C.E. Reed (Academic Press, London 1975) p. 2
2.504 K. Ueda, F. Forstmann: 7th Int. Vac. Cong. 3rd Int. Conf. on Solid Surf. (Dobrozemski et al., Vienna 1977)
2.505 J.A. Appelbaum, D.R. Hamann: Phys. Rev. B **6**, 3166 (1972)
2.506 K.C. Pandey: J. Vac. Sci. Technol. **15**, 440 (1978)
2.507 J.J. Lander, J. Morrison: J. Chem. Phys. **37**, 729 (1962)
2.508 J. Van Laar, A. Huijser, T.L. Van Rooy: J. Vac. Sci. Technol. **14**, 894 (1977)

2.509 A. Huijser, J. Van Laar: Surf. Sci. **8**, 342 (1967)
2.510 A.R. Lubinsky, C.B. Duke, B.W. Lee, P. Mark: Phys. Rev. Lett. **36**, 1058 (1976)
2.511 J.E. Rowe: J. Vac. Sci. Technol. **13**, 798 (1976)
2.512 W. Gudat, D.E. Eastman: J. Vac. Sci. Technol. **13**, 831 (1976)
2.513 J.A. Knapp, G.J. Lapeyre: J. Vac. Sci. Technol. **13**, 757 (1976)
2.514 J.E. Rowe: J. Vac. Sci. Technol. **13**, 248 (1976)
2.515 K.C. Pandey, T. Sakurai, H.D. Hagstrum: Phys. Rev. Lett. **35**, 1728 (1975)
2.516 T. Sakurai, H.D. Hagstrum: Phys. Rev. B **12**, 5349 (1975)
2.517 T. Sakurai, K.C. Pandey, H.D. Hagstrum: Phys. Lett. A **56**, 204 (1976)
2.518 T. Sakurai, H.D. Hagstrum: J. Vac. Sci. Technol. **13**, 807 (1976)
2.519 T. Sakurai, H.D. Hagstrum: Phys. Rev. B **14**, 1593 (1976)
2.520 B. Von Roedern, L. Ley, M. Cardona: Phys. Rev. Lett. **39**, 1576 (1977)
2.521 H.D. Hagstrum, G.E. Becker: Phys. Rev. B **8**, 1580 (1973)
2.522 T.E. Fischer: Crit. Rev. Solid State Sci. **6**, 401 (1976)
2.523 J.E. Rowe, G. Margaritondo, H. Ibach, H. Froitzheim: Solid State Commun. **20**, 277 (1976)
2.524 C.M. Garner, I. Lindau, J.N. Miller, P. Pianetta, W.E. Spicer: J. Vac. Sci. Technol. **14**, 372 (1977)
2.525 T.C. McGill: J. Vac. Sci. Technol. **11**, 935 (1974)
2.526 S. Kurtin, T.C. McGill, C.A. Mead: Phys. Rev. Lett. **22**, 1433 (1969)
2.527 C.A. Mead, W.G. Spitzer: Phys. Rev. A **134**, 713 (1964)
2.528 V. Heine: Phys. Rev. A **138**, 1689 (1965)
2.529 S.G. Louie, J.R. Chelikowsky, M.L. Cohen: Phys. Rev. B **15**, 2154 (1977)
2.530 C.M. Varma, K.C. Pandey: Private communication
2.531 G. Margaritondo, J.E. Rowe, S.B. Christman: Phys. Rev. B **14**, 5396 (1976)
2.532 J.C. Inkson: J. Vac. Sci. Technol. **11**, 943 (1974)
2.533 W.E. Spicer: Crit. Rev. Solid State Sci. **6**, 317 (1976)
2.534 J.L. Freeouf, D.E. Eastman: Crit. Rev. Solid State Sci. **5**, 245 (1975)
2.535 J.E. Rowe: J. Vac. Sci. Technol. **13**, 798 (1976)
2.536 M. Erbudak, T.E. Fischer: Phys. Rev. Lett. **29**, 732 (1972)
2.537 J.E. Rowe: Phys. Lett. A **46**, 400 (1974)
2.538 T. Murotani, K. Fujiwara, M. Nishijima: Jpn. J. Appl. Phys., Suppl. **22**, 409 (1974)
2.539 J.E. Rowe, M.M. Traum, N.V. Smith: Phys. Rev. Lett. **33**, 1333 (1973)
2.540 G. Margaritondo, J.E. Rowe: Phys. Lett. A **59**, 464 (1977)
2.541 J.E. Rowe, H. Ibach: Surf. Sci. **43**, 481 (1974)
2.542 A. Koma, R. Ludeke: Phys. Rev. Lett. **35**, 107 (1975); Surf. Sci. **55**, 735 (1976)
2.543 G.W. Gobeli, F.G. Allen: Surf. Sci. **2**, 402 (1964)
2.544 T. Murotani, K. Fujiwara, M. Nishijima: Phys. Rev. B **12**, 2424 (1975)
2.545 J.E. Rowe: Solid State Commun. **17**, 673 (1975)
2.546 C.M. Garner, I. Lindau, J.N. Miller, P. Pianetta, W.E. Spicer: J. Vac. Sci. Technol. **14**, 372 (1977)
2.547 D.E. Eastman, J.L. Freeouf: Phys. Rev. Lett. **33**, 1601 (1974)
2.548 R. Ludeke, A. Koma: Phys. Rev. Lett. **34**, 817 (1975)
2.549 R. Ludeke, A. Koma: Crit. Rev. Solid State Sci. **5**, 259 (1975)
2.550 K. Jacobi: Surf. Sci. **51**, 29 (1975)
2.551 W. Ranke, K. Jacobi: Proc. 4th Int. Symp. Surf. Phys., Eindhoven, Holland (1976)
2.552 W. Ranke, K. Jacobi: Solid State Commun. **13**, 705 (1973)
2.553 W.E. Spicer, I. Lindau, P.E. Gregori, C.M. Garner, P. Pianetta, P.W. Chye: J. Vac. Sci. Technol. **13**, 780 (1976)
2.554 J.A. Knapp: Ph.D. Thesis, Montana State Univ. (1976)
2.555 W.E. Spicer, P.W. Chye, P.E. Gregory, T. Sukegawa, A. Babalola: J. Vac. Sci. Technol. **13**, 233 (1976)
2.556 G.M. Guichar, C.A. Sebenne, G.A. Garry: Phys. Rev. Lett. **37**, 1158 (1976)
2.557 W. Gudat, D.E. Eastman, J.L. Freeouf: J. Vac. Sci. Technol. **13**, 250 (1976)
2.558 J.L. Freeouf, D.E. Eastman: Phys. Rev. Lett. **34**, 1624 (1975)
2.559 D.E. Eastman, J.L. Freeouf: Phys. Rev. Lett. **33**, 1601 (1974)

2.560 D.E. Eastman, J.L. Freeouf: Solid State Commun. **13**, 1815 (1973)
2.561 R.S. Bauer, R.Z. Bachrach, S.A. Flodstrom, J.C. McMenamin: J. Vac. Sci. Technol. **14**, 378 (1977); Proc. Int. Symp. Photoemission, Noordwijk, Netherlands 1976, ed. by R.F. Willis, B. Feuerbacher, B. Fitton, C. Backx (European Space Agency, Paris 1976) p. 103
2.562 F. Garcia-Moliner, F. Flores: J. Phys. C **9**, 1609 (1976)
2.563 W.D. Grobman, D.E. Eastman, J.L. Freeouf: Phys. Rev. B **12**, 4405 (1975)
2.564 J.A. Appelbaum, D.R. Hamann: Crit. Rev. Solid State Sci. **6**, 357 (1976)
2.565 L.M. Falicov, F. Yndurain: J. Phys. C **8**, 147, 1563 (1975)
2.566 K. Hirabayashi: J. Phys. Soc. Jpn. **27**, 1475 (1969)
2.567 J.A. Appelbaum, G.A. Baraff, D.R. Hamann: Phys. Rev. B **11**, 3822 (1975); Phys. Rev. Lett. **35**, 729 (1975); Phys. Rev. B **12**, 5749 (1975)
2.568 J.A. Appelbaum, D.R. Hamann, K.H. Tasso: Phys. Rev. Lett. **39**, 1487 (1977)
2.569 J.C. Phillips: Surf. Sci. **40**, 459 (1973)
2.570 K. Hirabayashi: J. Phys. Soc. Jpn. **27**, 1475 (1969)
2.571 D.J. Chadi, M.L. Cohen: Solid State Commun. **16**, 691 (1975)
2.572 F. Yndurain, L.M. Falicov: J. Phys. C **8**, 1571 (1975)
2.573 J.R. Chelikowski, M.L. Cohen: Phys. Rev. B **13**, 826 (1976)
2.574 J.D. Joannopoulos, M.L. Cohen: Phys. Rev. B **10**, 5075 (1974)
2.575 P.E. Gregory, W.E. Spicer, S. Ciraci, W.A. Harrison: Appl. Phys. Lett. **25**, 511 (1974)
2.576 J.A. Appelbaum, G.A. Baraff, D.R. Hamann: Phys. Rev. B **14**, 1623 (1976)
2.577 C. Calandra, G. Santoro: J. Vac. Sci. Technol. **13**, 773 (1976)
2.578 A.N. Mariano, K.L. Chopra: Appl. Phys. Lett. **10**, 282 (1967)

3. Unfilled Inner Shells: Transition Metals and Compounds

S. Hüfner

With 25 Figures

The knowledge of the electronic structure of a compound as given by its band structure is a basic requirement for the understanding of its macroscopic properties. In ions with unfilled inner shells, such as the 3*d*, 4*d*, and 5*d* transition-metal ions, many of their properties (e.g., magnetism, catalytic activity and metal-nonmetal transitions) are governed by the *d*-electron states; the determination of the structure of these electrons is therefore one of the clues for the understanding of many of the unique properties of this class of materials.

In this chapter we shall be concerned with a review of photoelectron energy distribution curves (EDC), of transition metals and their compounds, and of the relation of these EDC's to band structure. We will discuss mainly the EDC's obtained for valence bands.

3.1 Overview

There is a large literature on the core level spectra of transition metals and their compounds. But the study of these, has, so far, not contributed very much to the understanding of the band structure. We shall therefore not deal with these spectra here other than as an aid in the analysis of valence band spectra (where necessary). In addition, there is an excellent recent review on core level shifts in transition metals and their compounds, and we do not want to duplicate this material [3.1].

It is only fair to point to a number of reviews and conference proceedings from recent years that contain a great number of excellent articles on material covered in this article. In these proceedings one can also find many original contributions on the subject of this review [3.2–8]. The available work can be divided into two classes of substances, namely transition metal compounds (TMC) (insulating) and transition metals (including alloys and intermetallic compounds).

It seems appropriate first to discuss briefly the relation of an experimental EDC to the band structure in order to make the presentation of the experimental material more readable. Photoemission experiments can be described by a three-step model ([3.9] see also [Ref. 1.1, Chap. 2]): photo-excitation of an electron, travelling of that electron in the solid to the surface, and escape of the electron through the surface into the vacuum. In this model the energy distribution of electrons with initial energy E_i (relative to the Fermi energy) as excited by photons of energy $\hbar\omega$ is given by [3.10]

$$N(E_i, \hbar\omega) \sim \sum_{n,n^*} \int dk^3\, T_{nn^*}(\boldsymbol{k}) D_{nn^*}(\boldsymbol{k}) \{|P_{nn^*}(\boldsymbol{k})|^2 \delta[E_{n^*}(\boldsymbol{k}) - E_n(\boldsymbol{k}) - \hbar\omega]\, \delta[E_i - E_n(\boldsymbol{k})]\}, \quad (3.1)$$

where $T_{nn^*}(\boldsymbol{k})$ represents the surface transition term, $D_{nn^*}(\boldsymbol{k})$ is the electron transport term, and the term in braces represents the optical excitation. In this term $P_{nn^*}(\boldsymbol{k})$ is the electric dipole matrix element between the initial and the final state; n and n^* number the occupied and empty bands between which the transitions take place.

In order to determine an experimental $E(\boldsymbol{k})$ relation, one has to measure the kinetic energy of an electron photoexcited in a single crystal and its direction in space. If the relation between the momentum of the electron in the solid and outside is known, a measurement of $E_n(\boldsymbol{k})$ is possible. Because of the existence of a translation lattice parallel to the crystal surface one may assume that $k_{||}$ is conserved, to a reciprocal lattice vector, in passing the crystal surface, see [Ref. 1.1, Chap. 6]. $k_\perp$ is modified by the crystal potential due to the removal of translational symmetry along this direction by the presence of the surface. Therefore, at present only two-dimensional band structures (and, e.g., overlayer band structures) can be determined experimentally in that fashion with sufficient accuracy. However, such experiments are dealt with in detail in [Ref. 1.1, Chap. 6], so we shall abandon this interesting field [3.10–17] (see also Sect. 2.6). Most of the photoemission experiments described here are performed on polycrystalline samples with angle-integrated detection such that no $\boldsymbol{k}$ dependences can be measured directly any way; that means that one is always measuring a quantity related in some way to the density of states (DOS) of the sample under investigation. We shall therefore always discuss the experimental EDC's in terms of DOS. This implies that the comparison between theory and experiment must be quite indirect at times.

It must be mentioned now in what way an EDC is related to a DOS. It is useful to distinguish three regions of $\hbar\omega$, the exciting energy: $\hbar\omega < 50$ eV, the UPS (ultraviolet photoemission spectroscopy) regime; $50\,\text{eV} < \hbar\omega < 200$, the SXPS (soft X-rays photoelectron spectroscopy) regime, the domain of most synchrotron radiation experiments performed so far see Chap. 6; and $\hbar\omega > 200$ eV, the XPS (X-rays photoelectron spectroscopy) regime.

In all interpretations of EDC's the factors $T_{nn^*}(\boldsymbol{k})$ and $D_{nn^*}(\boldsymbol{k})$ are considered constant. In the UPS regime transitions occur between occupied and empty bands and the structure of the EDC is interpreted in terms of structure in the so-called energy distribution of the joint density of states (EDJDOS). Initial state features can then be recovered from an UPS spectrum as a structure that shows stationary initial energy as a function of $\hbar\omega$. In comparisons between UPS spectra and EDJDOS the matrix elements $P_{nn^*}(\boldsymbol{k})$ are often assumed to be constant.

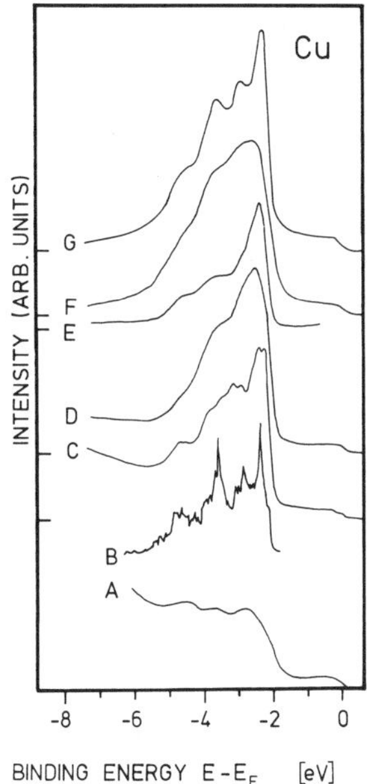

Fig. 3.1. Comparison of Cu photoelectron spectra obtained at various photon energies $\hbar\omega$.
A $\hbar\omega = 11.4$ eV [3.68].
C, *D* and *E* $\hbar\omega = 21.2$, 40.8, 48.4 eV [3.86].
F, *G* $\hbar\omega = 1487$ eV [3.70], measured spectrum (*F*) and result of deconvolution (*G*).
B Theoretical DOS [3.137]

In the SXPS regime the structures in the EDC's stay constant with varying photon energy $\hbar\omega$; however, the relative intensities can and do vary considerably with $\hbar\omega$. In the summation in (3.1) the index n^* can be dropped, because the final states (f) are now a quasicontinuum, but the matrix element $P_{nf}(\boldsymbol{k})$ can still vary with $\hbar\omega$.

Finally, in the XPS regime one may, with reservations, assume a near constancy of $P_{nf}(\boldsymbol{k})$ and in this approximation an XPS spectrum represents a DOS function. We add, however, that the $P_{nf}(\boldsymbol{k})$ matrix elements vary with atomic number Z and are a function of the nl quantum numbers, sometimes making a direct comparison between a DOS and an XPS spectrum difficult.

Some of the points mentioned can be demonstrated in Fig. 3.1 and also in Figs. 3.16–18. Figure 3.1 shows EDC's of Cu taken at energies of $\hbar\omega = 11$ (*A*), $\hbar\omega = 21$ (*C*), $\hbar\omega = 41$ (*D*), $\hbar\omega = 48$ (*E*) and $\hbar\omega = 1487$ eV (*F*). In addition, a DOS histogram (*B*) and a deconvoluted version of the 1487 eV spectrum (*G*) is shown. It can be seen that the $\hbar\omega = 11$ eV spectrum shows peaks in the EDC at 2.1 (weak) 2.8, 3.6, and 4.5 [eV], which reappear in all other spectra – and the theoretical DOS. The intensities in the theoretical DOS, however, are best matched by the 1487 eV (XPS) spectrum, especially in its deconvoluted version (*G*).

Spectrum C has been obtained with an instrumental resolution of 60 meV – its structures, although the most detailed reported so far, are at least 0.1–0.2 eV wide. This shows that the resolution obtainable in EDC's is now no longer limited by the resolution of the technique, but by "intrinsic" reasons, most likely the lifetime of the states.

The discussion has not mentioned such effects as final-state effects, energy-loss mechanisms and surface-states phenomena which also affect the EDC's. They will be mentioned only where their consideration is important.

3.2 Transition Metal Compounds

3.2.1 The Hubbard Model

Transition Metal Compounds (TMC) can be insulators (e.g., NiO), show a metal-nonmetal transition (e.g., VO_2) or can be high-conductivity metals (e.g., ReO_3). It is not a trivial question why so seemingly similar compounds show such different properties, and the theory that accounts for this startling fact has been given by *Hubbard* [3.18] and elaborated upon by numerous other workers [3.19–24].

In the most simple formulation of band theory, TMC's (except those with a nd^0 or a nd^{10} configuration) are solids with a nonfilled band, and thus should all be metals. *Mott* [3.22] was the first to realize the importance of the correlation of the d-electrons for the actual band structure of these materials. *Hubbard* [3.18] later put this idea on a more solid mathematical footing. If conductivity in a TMC is to be achieved by d-electrons, e.g., in a $3d$ TMC, a transition of the following form is required

$$3d^n + 3d^n \rightarrow 3d^{n-1} + 3n^{n+1} \,.$$

An energy U is needed for this process. In a conductor this correlation energy U has to be provided by the energy available within the bandwidth in order to make a free movement of the d-electrons possible. If U is larger than the bandwidth this movement is no longer possible and the material becomes a semiconductor or an insulator (Mott insulator).

The important parameters that determine the metal-nonmetal transition in the framework of the Hubbard model are thus the correlation energy U and the bandwidth W, which is mainly given by the overlap of the d-orbitals with the orbitals of the ligands. The theoretical or experimental determination of these two parameters presents problems. The bandwidth can be estimated from EDC's. The parameter U is, however, much harder to come by. Let us take as an example NiO. Atomic data [3.25] yield an energy $U = 18$ eV for the transition $2d^8 \rightarrow d^7 + d^9$; since the bandwidth is around 1 eV in this compound (as can be seen from the UPS and XPS spectra) within the Hubbard argument NiO should be an insulator, which it is. On the other hand, NiS, with a metal-nonmetal transition, has about the same bandwidth; by the above reasoning U in NiS must therefore be an order of magnitude reduced from the atomic estimate of 18 eV. Then, of course, it is hard to believe why it should not also be of the order of a few eV in NiO. Indeed various estimates give $U \approx 3$ eV in NiO, the reduction from the free-ion value obviously produced by screening. It is worth adding that *Herring* [3.26] has estimated $U \approx 1$ eV for Ni metal. (The bandwidth here is about 3 eV.) (The interpretation of the metal-nonmetal transition in NiS in terms of the Hubbard model is, however, not unchallenged [3.27].) So far Hubbard's ideas have not been incorporated into band-structure calculations.

There are, however, many band-structure calculations for TMC, many of them by *Mattheiss* [3.27–31], and their results will be discussed together with the experimental results for the various compounds.

3.2.2 Final-State Effects in Photoemission Spectra

The photoemission process as such presents a major disturbance of the sample on an atomic scale. It is therefore to be expected that the EDC's represent this "disturbed" state rather than the ground state in which one is actually interested. These final-state effects will be treated at various places in this volume, and we shall therefore restrict the material to what is relevant for an interpretation of the valence band spectra of TMC's.

a) Satellites

In the one-electron interpretation of photoemission experiments each photon $\hbar\omega$ can only excite one electron, and the energy of this electron reflects its initial-state energy. The real states of a solid, however, are not one-electron states, and therefore a one-electron interpretation of EDC's is not always appropriate: there is always a finite probability that two electrons are ejected by one photon. This process gives rise to the so-called satellite lines which are a major problem in the interpretation of EDC's of TMC's. To illustrate this fact, we compare in Fig. 3.2 the $2p$ spectra of Cu_2O and of CuO, which have d^{10} and d^9 configurations, respectively [3.32].

The spectrum of Cu_2O shows the $2p_{1/2}-2p_{3/2}$ spin-orbit split doublet. The spectrum of CuO is markedly different: both of the $2p$ lines are accompanied by so-called satellite lines, which are 10 eV below the main lines. From the intensity of these lines one realizes that they constitute a major portion of the XPS spectrum. The interpretation of these satellite lines is still somewhat controversial, but in the case of the TMC most authors agree [3.33–35] that they are caused by charge transfer transitions from the ligands (O^{2-} ions for the oxides) into the empty d-states of the central ion. Thus in the present case in CuO, with a d^9 configuration, there is one empty d-state which can get populated by a charge transfer transition from a neighboring oxygen ligand; in Cu_2O, with a d^{10} ground state configuration such a transition cannot occur, which explains the absence of the strong satellites in this material. One of the

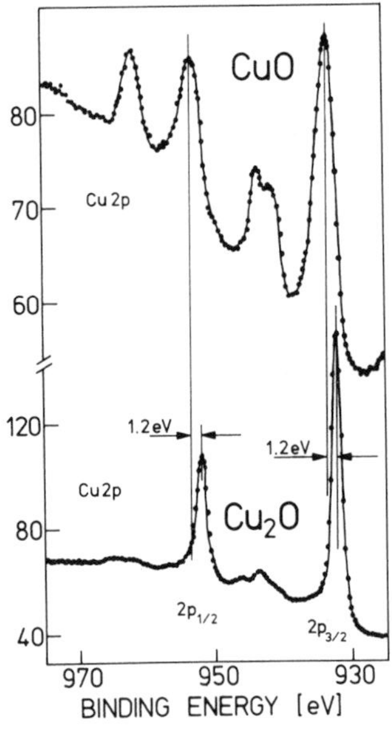

Fig. 3.2. XPS spectrum of the Cu $2p_{1/2}$–$2p_{3/2}$ core levels in Cu_2O (bottom) with a d^{10} configuration and CuO (top) with a d^9 configuration [3.32]. The CuO spectrum shows strong satellites

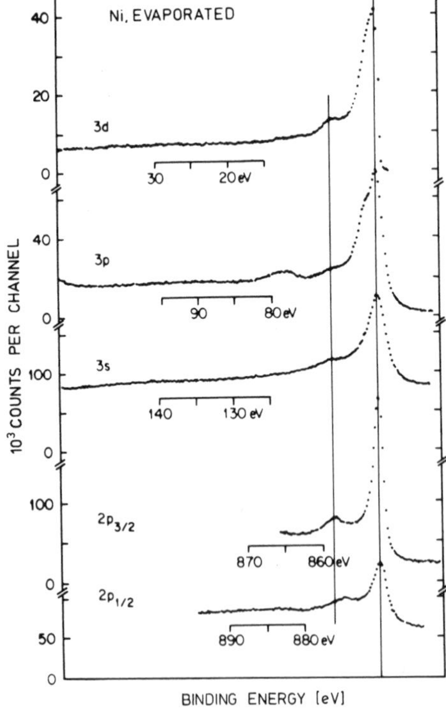

Fig. 3.3. XPS spectra of evaporated Ni showing the 6 eV satellite. The spectra have been lined up so as to bring the peaks of the main lines into coincidence. The $3p$ spectrum shows additional structure presumably due to multiplet effects [3.99]

best arguments in favor of this kind of interpretation is the nonoccurrence of these satellites in d^{10} configuration (as in Cu_2O) whereas they can be seen in d^0 compounds. The selection rules governing these satellites are still a matter of debate.

Satellites may also occur in the spectra of metals; they are, however, less common and also the mechanism is slightly different. To illustrate this fact, Fig. 3.3 shows the XPS spectra of all the levels with binding energies below 1000 eV for Ni metal; they all show a satellite, 6 eV below the main line. This applies even to the valence band. The most likely interpretation here is that this satellite is caused by a two-hole state on a Ni ion. *Kotani* and *Toyozawa* [3.36] have worked out the theory of satellites in metals in detail and give the following conceptually simple interpretation. Upon photoionization, the effective charge acting on the valence electrons is increased by one. Thus empty *d*-states that were before photoionization above the Fermi energy are now pulled down below it. These *d*-states can be either full or empty, the empty case corresponding to the satellite described above.

One should point out that the satellites seen in compounds like NiO lend themselves to the same kind of interpretation. Here the filling of the empty *d*-state, pulled down by photoionization, occurs via charge transfer from the anion states.

Put in more general terms, satellites occur if the photoionization is accompanied by a transition which promotes an electron from a filled valence band state into an empty conduction band state. In that sense, the asymmetric lines observed in the core level spectra of metallic systems [3.37–39], can be interpreted as being satellite lines. Here the potential, which the photohole represents, scatters the conduction electrons from filled into empty conduction states, i.e., across the Fermi boundary. Since there are an "infinite" number of states above and below the Fermi energy, one does not observe a discrete satellite but a tail to the high-binding energy side of a core line; for "zero" energy transfer at exactly the Fermi energy the transition probability becomes infinite (infrared catastrophe).

In summary we note that satellites (discrete and continuous ones) are a property inherent in the photoemission spectra (of core levels *and* of valence bands) and they have to be properly recognized before a spectrum can be designated as a genuine valence band, or core level spectrum.

b) Multiplet and Crystal-Field Splitting

Photoemission experiments measure by their very virtue not the initial-state energy, but the difference in energy between the ground state and the singly ionized state. If an ion has an unfilled outer shell, the hole created through photoionization can couple to the spin and orbital angular momentum of the unfilled shell and thereby produce a structure not necessarily present in the initial state [3.40]. These effects have been discussed by D. A. Shirley in [Ref. 1.1, Chap. 4] for core states, but they can also play an important part in valence bands. In an insulator a core state is well localized and so is the photohole. Thus it can safely be assumed that the state produced by the coupling of the hole to the open shell configuration lives long enough to be detected as a final state. The argument becomes slightly more involved for a valence band of an insulator. If we look at a perfect insulator, with a large gap, the above reasoning certainly also holds. If, however, the gap is small, and we are perhaps dealing with a defect structure, such that many carriers are around, one might envision that the hole in the valence band is readily screened, so that practically no final-state interaction can be observed. This argument holds all the more for metals. Here one may assume safely that the screening charge moves in readily such that, e.g., in a valence band one observes essentially an initial-state picture with only minor modification (see the *Kotani-Toyozawa* satellites of Fig. 3.3). As far as core levels in metals go, the argument is similar to that for insulators; hence one can observe multiplet splitting. One has to pay attention, however, to the fact that then the valence band may have one additional d-electron, namely the screening charge. This is important for the discussion of the multiplet structure in the core levels of Ni; here the screening leads to a closed d^{10} configuration for the photoionized ion, giving no multiplet coupling.

We shall now discuss the multiplet coupling for valence bands in insulating TMC's. Data will be presented for $3d$ bands only because they are the only ones that have been thoroughly investigated so far. There are two energies of similar magnitude that determine the energies of TMC valence bands, namely, the crystal-field energy and the Hund's rule energy. Let us assume for the moment for the sake of simplicity that the crystal field has octahedral symmetry. Then a d-orbital is split into a lower, triply degenerate t_{2g} state and a doubly degenerate e_g state. If the Hund's rule coupling is larger than the crystal-field energy, one will first fill the t_{2g} orbitals, then the e_g orbitals with electrons of parallel spin. After 5 electrons have been accommodated in this way in the same order, electrons of opposite spin will be accommodated starting again with the t_{2g} levels. We can illustrate this for three typical cases, d^3, d^5, and d^6 by the diagrams in Fig. 3.4. (For a detailed discussion of this point see [3.41].)

Let us now discuss photoionization for these three cases separately. For a d^3 configuration photoionization will always result in a d^2 configuration with a $^3T_1$ ground state [3.42]. This state is seen in the XPS spectrum of Cr_2O_3 (in Cr_2O_3 the Cr ion has the d^3 configuration) (Fig. 3.5) which shows in the valence band region the O$2p$ band and on top of it a single $3d$ level ($3T_1$). The position of the O$2p$ spectrum can be deduced from a spectrum of $K_2Cr_2O_7$ where Cr is in its hexavalent state, and has a d^0 configuration; it is also realized that the O$2p$ band changes its position little in different compounds [3.43].

Figure 3.6 shows UPS spectra of a number of $3d$ transition metal oxides and we want to focus for the moment on those of Cr_2O_3 and MnO. Whereas for Cr_2O_3 the UPS spectrum is very similar to the XPS spectrum of this compound with a one-hump d-band structure, the MnO spectrum shows a two-hump structure; this is interpreted in terms of final states as photoionization of an e_g electron (leading to a ^5E final state) and photoionization of a t_{2g} electron

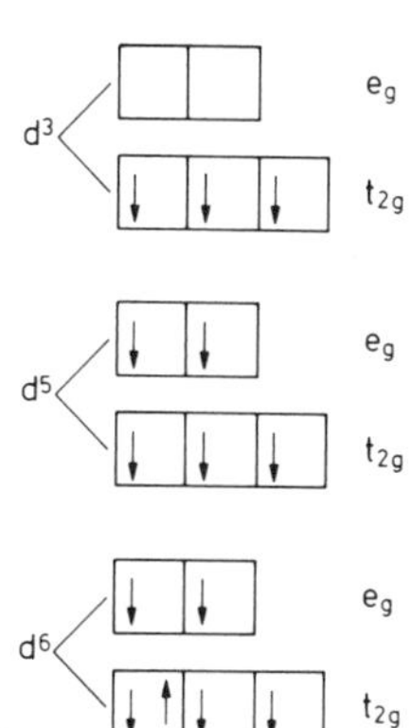

Fig. 3.4. Construction of the ground states of the d^3, d^5, and d^6 configuration in an octahedral field using the strong crystal-field approach, by assuming the spin coupling (Hund's rule energy) to be stronger than the crystal-field energy

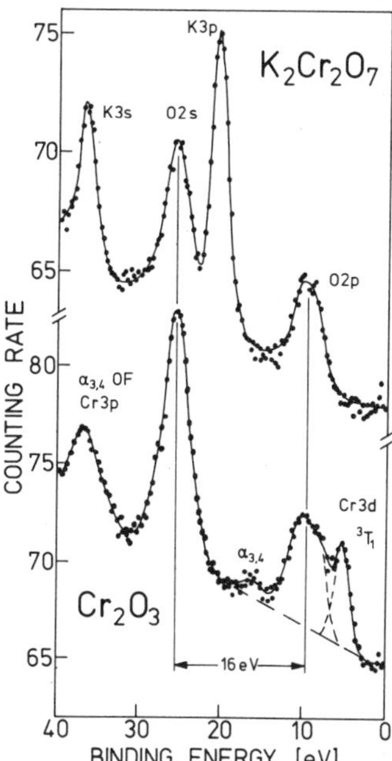

Fig. 3.5. The valence band structure of Cr_2O_3 compared with that of $K_2Cr_2O_7$. The O 2*s* lines have been brought into coincidence [3.43]

(leading to a ^5T final state): Note that the d^4 configuration of the final state represents one hole in the filled orbital shell (d^5). Hence for the d^4 Hund's rule ($S=2$) multiplet the orbital crystal field one-electron splitting (with reversed spin) is expected. The splitting of the two *d*-hump is thus a direct measure of the $e_g - t_{2g}$ crystal-field splitting [3.44].

Finally we have a look at the XPS spectrum of FeF_2 where the ground state is d^6 (see Fig. 3.7). Photoionization leads either to an excited $^6A_1$ state or to a number of (Hund's rule ground states) quartets. The lines resulting from the two states are readily seen in the spectrum. Here the separation of these two lines is a rough measure of the Hund's rule coupling energy, because the multiplicity of the two final states is different [3.43].

In order to be reasonably complete, we shall also briefly outline the determination of the exact intensities of the various final-state components of the *d*-electrons [3.41] [3.140]. The state of *n*-electrons can be constructed from those with $n-1$ electrons by adding one electron, where the coefficients that determine the strength of the various components of the $n-1$ state wave functions in a *n*-state wave function are the so-called coefficients of fractional parentage. These coefficients of fractional parentage can be obtained by

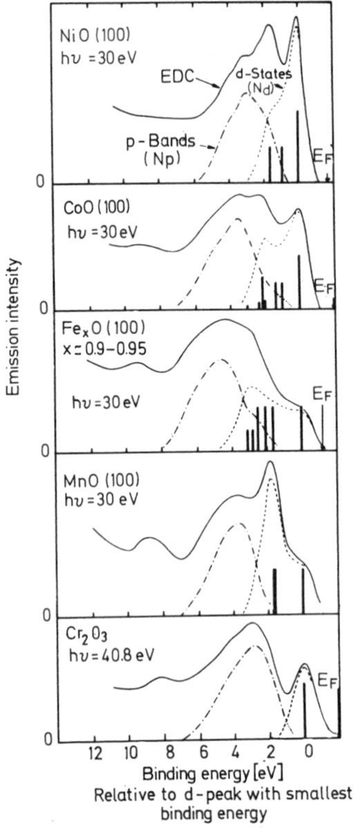

Fig. 3.6. UPS spectra, decomposed into partial d- and p-components N_d and N_p for NiO, CoO, Fe_xO, MnO, and Cr_2O_3. The vertical lines denote calculated $3d^{n-1}$ final-state ionization potentials (see text). The weak structures near 8–10 eV for CoO, Fe_xO, MnO, and Cr_2O_3 are attributed to multielectron satellite peaks [3.44]

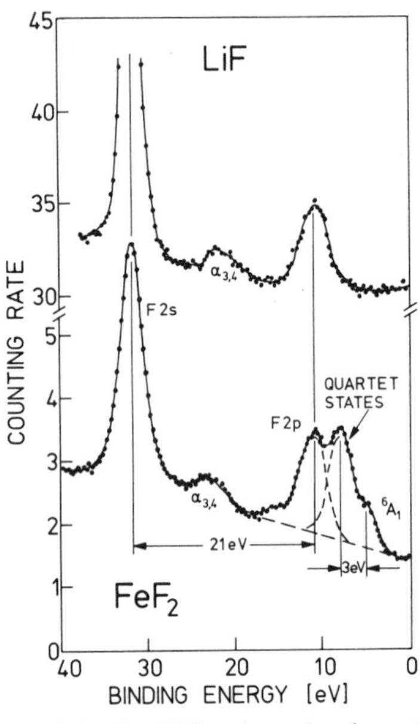

Fig. 3.7. The XPS valence band spectrum of FeF_2 compared with that of LiF. The latter spectrum has been shifted so as to bring the F 2s lines into coincidence. States are labeled in terms of the final states [3.43]

standard procedures in atomic theory [3.141]. Photoemission is now just the reverse process, namely from a state with n-electrons that with $n-1$ electrons is produced, by the emission of one electron. It is apparent that the state $n-1$, which is the one after the emission of the electron, contains those states from which the state n can be constructed with the addition of an electron, and that the intensities with which the $n-1$ final states appear, are then essentially given by the square of the corresponding coefficients of fractional parentage multiplied by an orbital and spin momentum coupling coefficient and coefficients for degeneracies [3.41].

From the examples it is evident that the photoemission valence bands of TMC's have to be interpreted in terms of final-state energies; crystal-field and exchange energy are responsible for the actual splitting patterns. The diagrams for these energies have been calculated by *Tanabe* et al. [3.42] and they give the correct positions of the final states, as can be seen in the comparison of Fig. 3.6, where the bars indicate the energies expected as final states from these diagrams. In these estimates the crystal-field parameters Dq were used for t_{2g} orbitals as for divalent ions, for e_g orbitals as for 2.5 valent ions, and the Racah parameter B was reduced by 0.7 from its divalent value [3.44]. These "adjustments" indicate that the described crude model can only approximately account for the observed spectra.

3.2.3 Transition Metal Oxides

The transition metal oxides are perhaps that single class of TMC's for which there exists the largest body of experimental and theoretical information [3.19–20, 23, 30, 31]. They were therefore also chosen in the present context to serve as a guide through the various phenomena related to the band structure. The discussion will thus be split up into a section on (Mott) insulators, one on the metal-nonmetal transition, and one on a typical metal. As a general policy we shall deal with some rather detailed and well-documented examples instead of reproducing all the available data.

a) MnO, CoO, NiO: Mott Insulators

In many respects the $3d$ transition metal monoxides, with a more than half-filled $3d$ shell, can be regarded as typical examples of Mott insulators, i.e., compounds for which the Bloch-Wilson theory does not make the correct prediction with respect to their electronic properties (electrical conductivity). An approach that starts out with the atomic energy levels adds the Madelung potential and then broadens these energy levels somewhat, in order to account for the overlap in the crystal leads to the following "band structure" [3.23] (compare Fig. 3.11). The $O2s^2$ band lies about 20 eV below the Fermi energy, 4 eV below the Fermi energies lies the $O2p^6$ band and on top of this, one finds the $3d^n$ "band"; here we have quite arbitrarily positioned the Fermi energy at the top of the $3d^n$ band. The (empty) $4s$ band lies 2 to 4 eV above the top of the $3d^n$ band, depending on the ion. If in such a band structure conductivity cannot be obtained (although the $3d^n$ shell is not full) it is due to the impossibility of transferring a $3d$ electron to an adjacent site, meaning that the $3d$–$3d$ correlation energy U is responsible for the insulating nature. Among the transition metal oxides NiO has always received special attention and a great deal of experimental evidence on its electronic structure has been accumulated. We shall therefore start the discussion with this compound [3.24].

NiO is an insulating antiferromagnet ($T_N = 523$ K) with a room temperature conductivity typical for a good insulator. Optical data show a "gap" of approximately 4 eV and a great similarity of the spectra of NiO and Ni impurities in the isomorphic insulator MgO (Ni:MgO) stressing again the

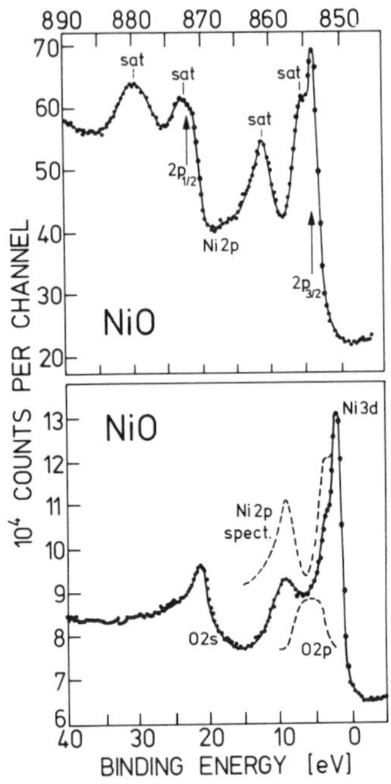

Fig. 3.8. XPS spectrum of the Ni-$2p$ region of NiO (upper part). In addition to the $2p$ lines separated by the spin-orbit splitting, a number of extra satellite lines ("sat") are observed. These are most likely due to multielectron excitations. Valence band region of NiO, with the Ni-$2p$ region superimposed (lower part). The Ni-$2p$ line has been made to coincide with the Ni-$3d$ line [3.138]

localized nature of the $3d$ orbitals [3.23, 24]. It is therefore tempting to try to construct an energy level diagram for NiO by starting from the atomic energy levels slightly. Although there is currently general agreement that this gives the Madelung potential. Overlap effects are then introduced by broadening the levels slightly. Although there is current general agreement that this gives the correct ordering of the orbitals, there is still considerable disagreement about the bandwidth of the $3d$ electrons and about the $p-d$ hybridization.

Photoemission experiments have been used in determining at least the position and approximate width of the various bands. The interpretation of the spectra was, however, plagued by the occurrence of very strong satellites [3.45]. Figure 3.8 shows an XPS spectrum of NiO covering the $2p$ and the valence band region. Instead of just the expected $2p_{1/2}$–$2p_{3/2}$ doublet, one sees in this spectrum strong satellites. The satellite 6 eV below the main line in the $2p$ spectrum is most probably due to a O$2p$→Ni $3d$ charge transfer transition, whereas the nearby satellites cannot be identified with confidence at this point. If one overlays the $2p_{3/2}$ spectrum and the valence band spectrum, one sees that the structure in the valence band at 9 eV is most likely also due to a satellite. The position of the O$2p$ band can then only be obtained by comparing the XPS spectra of a number of different oxides, and as an example Fig. 3.9 gives those of TiO_2 (no $3d$ electrons), NiO ($3d^8$ configuration) and Cu_2O (d^{10} configuration). The position of the O$2p$ band is well recognized in TiO_2 (insulator) and Cu_2O (semiconductor): since it is nearly the same in these two compounds it was

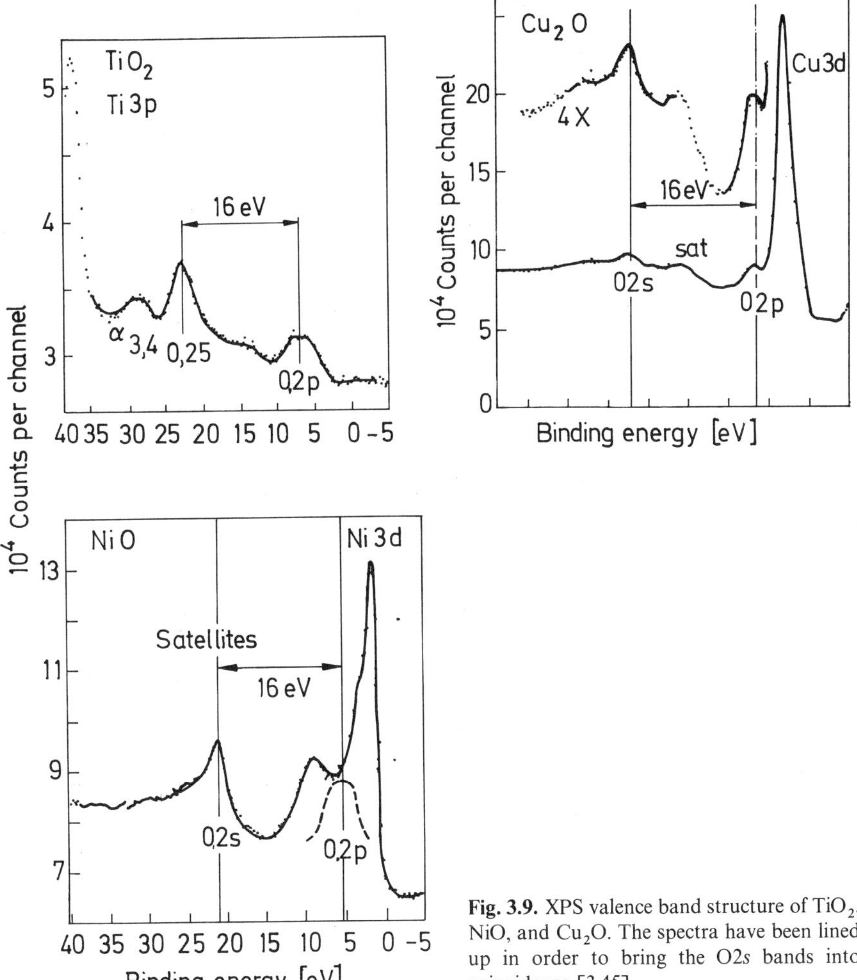

Fig. 3.9. XPS valence band structure of TiO_2, NiO, and Cu_2O. The spectra have been lined up in order to bring the O2*s* bands into coincidence [3.45]

placed also at the same energy in NiO although it is not directly visible in its XPS spectrum. The structure in the 3*d* band of NiO is the same as seen, e.g., in $NiCl_2$ or NiF_2 and is therefore interpreted as $3d^7$ final-state structure, using the *Tanabe-Sugano* [3.42] diagrams.

This interpretation of the XPS valence band structure is supported by UPS measurements of *Eastman* and *Freeouf* [3.44]; their data, obtained for different uv photon energies (from 20 to 78 eV), is shown together with the XPS spectrum of *Wertheim* and *Hüfner* [3.45] in Fig. 3.10. The UPS experiments with varying photon energies are well suited to detect the orbital nature of a specific peak, because the photoionization cross sections for different orbitals

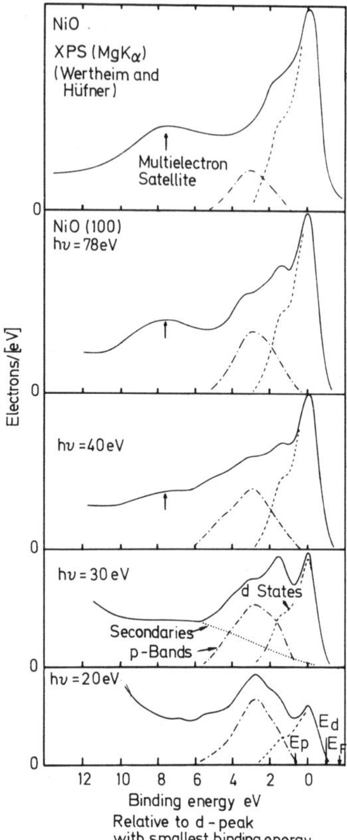

Fig. 3.10. Photoelectron spectra for NiO taken with various photon energies between 20 and 78 eV and with Mg K_α radiation. The zero of energy is placed at the d-state peak and E_F, E_d, and E_p are the Fermi energy, d-state edge, and p-band edge. Partial d-state emission intensities (dashed lines) and p-band intensities (broken lines) are shown (see text) [3.44]

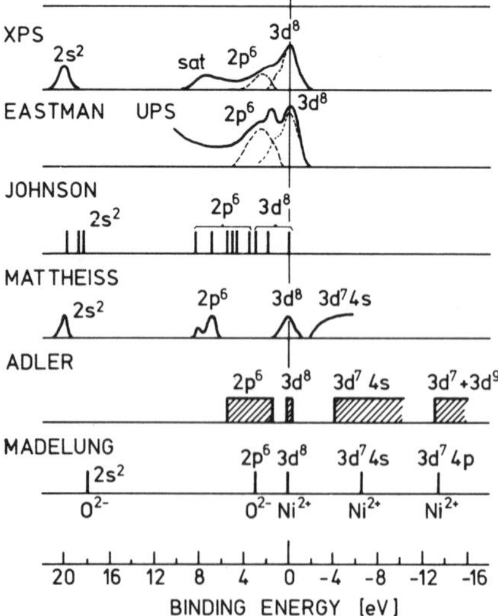

Fig. 3.11. Experimental and theoretical results for the valence band structure of NiO. In every case the zero of energy has been made to coincide with the center of the 3d band

vary differently with energy; the d cross section increases with increasing energy relative to the p cross section, allowing discrimination between the two [3.46] (see Sect. 2.4.4). The $h\nu = 20$ eV UPS spectrum shows the O2p band and the 3d final-state structure, whereas at this energy the intensity of the satellite is negligible. With increasing uv energy the emergence of the satellite is seen and also the relative intensity of the O2p and the Ni 3d bands converges towards the X-ray limit, thus supporting the placement of the O2p level by *Wertheim* and *Hüfner* [3.35, 45].

Figure 3.11 summarizes the available DOS information on NiO. The different data have been lined up with the Ni $3d^8$ state. It does look as if the Madelung approach represents the position of the energy levels fairly well; the *Adler* and *Feinleib* [3.24] diagram, which in essence relies on it, gives therefore the likely position of most of the bands.

The two available band-structure calculations, the augmented plane wave (APW) calculation of *Mattheiss* [3.30] and the molecular cluster calculation by *Johnson* et al. [3.47], both seem to overestimate the $p-d$ hybridization. In both of these calculations the O2p level appears at a position where the experiments place the multielectron satellite. In addition, in Johnson's calculation the width of the 3d state seems to be slightly larger than observed experimentally (the observed width is $\lesssim 1.5$ eV).

Very recently *Kunz* and *Surrat* [3.142] have reported results of a self-consistent Hartree-Fock energy-band calculation including correlation effects for FeO, CoO, and NiO. The results of these calculations are in much better agreement with the data from the photoemission experiments than those of the previous calculations, giving rather narrow metal d-bands on top of the oxygen p-bands with a small overlap between the two. Self-consistency and the inclusion of correlation effects seem therefore to be necessary to come to a correct band structure of a TMC.

The structure of the unoccupied states can be obtained from optical experiments. There are two such investigations available in the literature that are not totally consistent in their interpretation [3.48, 49]. Both groups agree that the gap of 4 eV is due to the $3d^8 \rightarrow 3d^7\,4s$ transition and that the transitions observed in the gap are due to excited states of the d^8 configuration. The interpretations differ, however, with respect to the structure in the optical spectrum between 4 and 10 eV. *Powell* and *Spicer* [3.49] gave evidence that the structure in that region is due to $d^7 4s$ final-state structure. On the other hand, *Messick* et al. [3.48] favor an interpretation that places $2d^8 \rightarrow d^9 + d^7$ transitions in this energy interval. We know of no way to distinguish between these two alternatives at this point (see, however, the recent work in [3.183]).

We have dealt with NiO in some detail because is shows all the features also observed for other TMC's. The main conclusion is that photoemission experiments can locate quite accurately the position of the anion and cation valence bands. Final-state structure determines the shape of the d-band. An interpretation in terms of a simple Madelung picture seems to be superior to more sophisticated molecular cluster or band-structure calculations.

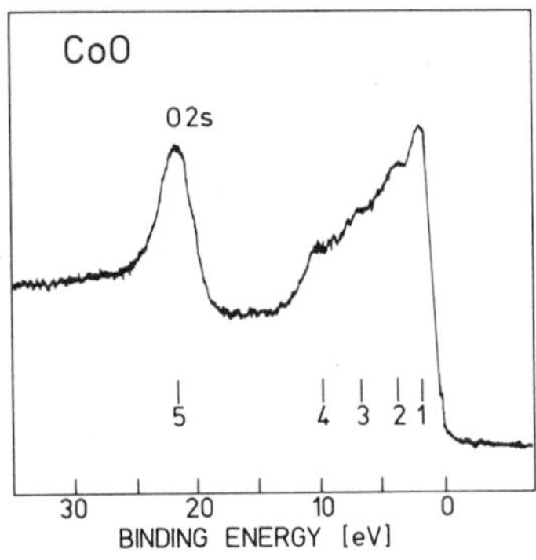

Fig. 3.12. XPS valence band structure of CoO; the peaks are *1, 2* $3d^6$ final-state structure; *3* O2*p* band; *4* charge transfer satellite; *5* O2*s* band [3.50]

MnO, FeO, CoO. The situation in these two oxides is very similar to NiO, the only difference in the EDC's being that the satellites get increasingly weaker as one proceeds from NiO to CoO and to MnO. Both CoO and MnO are very good insulators, indicating a sizeable Hubbard gap [3.23].

An XPS spectrum of CoO [3.50] is shown in Fig. 3.12. The peaks which are numbered 1, 2, 3, 4, 5 have the following interpretation: 1, 2) *d*-final-state structure; 3) O2*p* band; 4) charge transfer (O2$p \rightarrow$Co3*d*) satellite; 5) O2*s* band. The analysis of the O2*s* band and O2*p* band comes from a comparison with the spectra of other oxides. The identification of peak 4 as a charge transfer satellite rests on a comparison with the satellites found with other core levels in CoO, mainly the 2*p* levels. The 3*d* final-state structure can be identified with the help of the *Tanabe* and *Sugano* [3.42] diagrams.

In Fig. 3.6 we find a comparison of the UPS of valence bands of Cr_2O_3, MnO, Fe_xO, CoO, and NiO taken with $\hbar\omega = 30$ eV. At this photon energy the transition probability for the *d* states and the charge transfer satellites is low with respect to the O2*p* cross section and therefore the position of the latter is clearly visible. One can see that the O2*p* band lies roughly at the same position for all the oxides, whereas the 3*d* band exhibits the final-state structure.

The main findings from the EDC's of insulating transition metal oxides are that the O2*s*–O2*p* separation is roughly the same irrespective of the transition-metal ion and that the *d*-bands have to be interpreted in terms of final-state structure. These findings have been substantiated by investigations of other transition metal series, namely $TMCl_2$ [3.143], TMF_2 [3.144, 145], and $KTMF_3$ [3.146], thus their general applicability can be assumed.

b) VO_2: A Nonmetal-Metal Transition

In compounds where the Hubbard gap (see Sect. 3.2.1) is small, the electrons can by thermal excitation populate the conduction band. The gap can then collapse by screening and the material can go from a nonmetallic state into a metallic one. This transition ("Mott transition") can be caused also by doping or by applying hydrostatic pressure. It is tempting to try to detect the

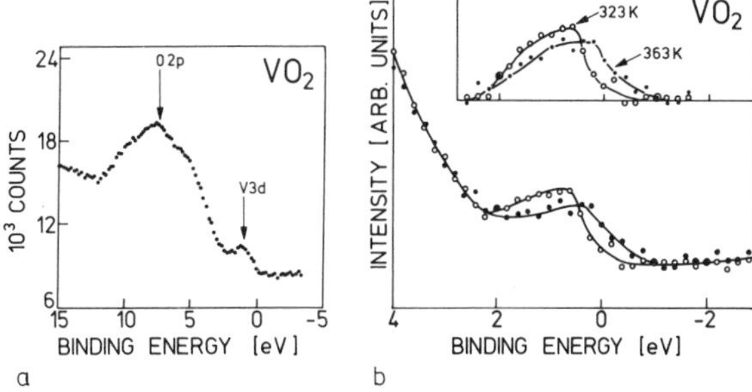

Fig. 3.13. a An XPS spectrum of VO_2 in the region of low binding energy collected at room temperature. The position of the V 3*d* level, located at the tail of the broad O 2*p* band, is indicated [3.12]. **b** An XPS spectrum of the vanadium 3*d* band above and below the transition temperature. The insert shows these spectra after correction for the background including the tail of the O 2*p* line. Full circles, $T = 363$ K; open circles, $T = 323$ K [3.51]

associated shift in the valence band by photoelectron spectroscopy. Such an experiment has been performed, e.g., for VO_2 [3.51–53]. This oxide has a nonmetal-to-metal transition at 340 K, at which temperature the electrical conductivity changes by several orders of magnitude. The transition is of first order. The foregoing discussion makes it evident that the valence band must be the 3*d* band, which is above the O2*p* band; this ordering is shown by the XPS spectrum in Fig. 3.13a. Figure 3.13b gives a more detailed view of the 3*d* band below ($T = 323$ K) and above ($T = 363$ K) the transition temperature. The position of the *d*-band is shifted toward the Fermi energy as the transition temperature is raised. In order to estimate a lower limit for the Hubbard gap from the experimental data it was assumed that in the nonmetallic state the Fermi level is pinned to the bottom of the conduction band. Thus the spectra can be analyzed quantitatively leading to a Hubbard gap of ~ 0.3 eV at $T = 0$ K, in agreement with the results obtained from other data (e.g., resistivity measurements) [3.51]. If the above assumption does not hold, the Hubbard gap would be larger than 0.3 eV.

The XPS spectra are in reasonable agreement with the DOS deduced from a band-structure calculation of *Gupta* et al. [3.164]. Very similar effects have been seen in an XPS study of NbO_2 which undergoes a metal-nonmetal transition at 1100 K [3.147].

c) ReO_3: A Typical Metal

Conductivity measurements show that ReO_3 is a metal with a resistivity of the same order of magnitude as that of Cu (although with a stronger temperature dependence) [3.23]. Since the isostructural oxide WO_3, which has no 5*d* electron, is a very good insulator, it is appealing to suggest that conduction in ReO_3 takes place in the 5*d* band. This band, which is situated on top of the O2*p* band, contains one 5*d* electron. An APW band-structure calculation of ReO_3

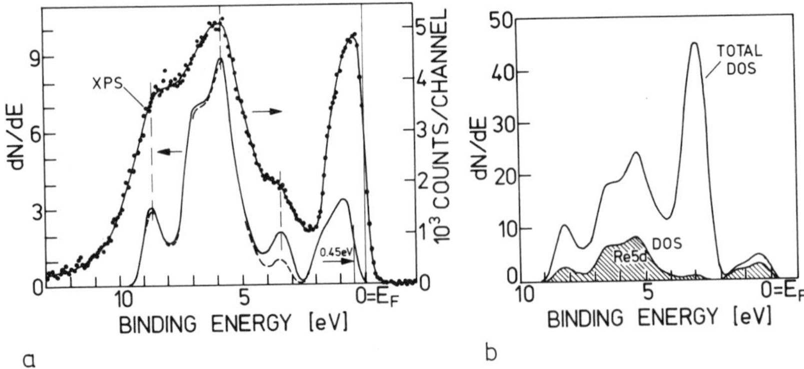

Fig. 3.14. a Comparison of XPS data on vacuum cleaved ReO_3 with the Re 5*d* component of the density of states (dashed line), and a composite density of states containing the 5*d* and 2.5% of the oxygen 2*p* component (solid line) [3.54]. **b** Calculated density of states of ReO_3 [3.31] with a Gaussian smoothing of 0.55 eV. The shaded area indicates the Re 5*d* contribution [3.54]

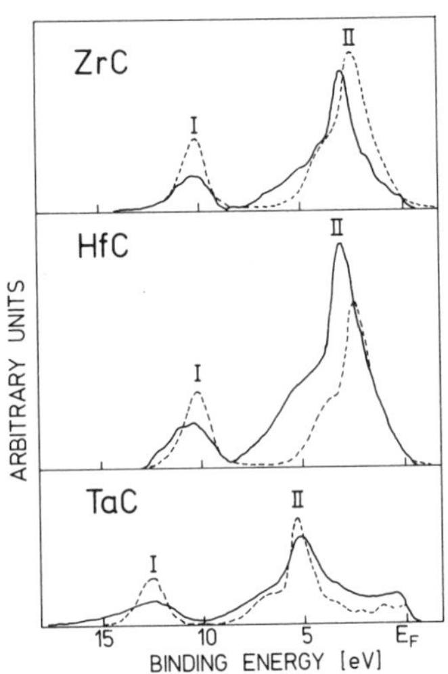

Fig. 3.15. Comparison between the experimental energy distribution curves (XPS, solid lines) and the theoretical ones (dashed lines) of ZrC, HfC, and TaC [3.56]

by *Mattheiss* [3.31] was able to reproduce the optical data and the Fermi surface topology well. Problems arise, however, in the interpretation of the XPS spectrum, which one would assume to give a reasonable representation of the DOS. The clue to this apparent puzzle was given by *Wertheim* et al. [3.54]. Since the facts connected with the interpretation of the ReO_3 XPS spectrum may have wider applicability we shall reproduce the reasoning here.

The total DOS and the partial 5*d* DOS of ReO_3 are reproduced in Fig. 3.14b, whereas Fig. 3.14a contains a high resolution XPS spectrum. The most dramatic difference between the XPS spectrum and the total DOS is a strong peak around 3 eV in the latter that is only visible as a shoulder in the XPS spectrum. The Re $5d$ part in the DOS is, however, quite similar to the XPS spectrum. This suggests that the XPS spectrum reflects mostly a $5d$ DOS with only minor contributions from the O$2p$ levels. To verify this, Fig. 3.14a contains two DOS curves: a solid curve representing the $5d$ DOS with 1/40 of the O$2p$ DOS added to it, and a broken curve that gives the Re $5d$ DOS only. The calculated DOS has been shifted by 0.45 eV to lower energies in order to get the best agreement with the measured XPS spectrum. One can see that the agreement between the calculated DOS and the XPS spectrum is quite good if the mentioned adjustment is made; this gives one confidence in the procedure. The decomposition into partial densities of states (discussed above) emphasizes again the fact that in order to compare DOS calculations and XPS spectra in multicomponent systems, careful consideration of the photoionization cross sections has to be made [3.46]. Data like those just presented can in turn be used to investigate hybridization.

3.2.4 Miscellaneous Compounds

The oxides of d-ion compounds have been dealt with in detail because they illustrate the problems and successes encountered in measuring and analyzing EDC's of TMC's. A great number of compounds have been investigated and we shall not list all the results here. As before, we shall rather present typical results – namely some transition metal carbides. These compounds are quite interesting for technological applications: they are metallic, have very high melting points, have very high hardness and are superconductors. Their simple rocksalt structure makes them particularly suited for theoretical investigations. Photoemission spectra of some transition metal carbides have been obtained by a number of groups [3.55–57]. The X-ray induced EDC's of these compounds are dominated by two peaks: one, at least 10 eV below E_F, consists mainly of the C$2s$ orbitals, while the other, nearer to E_F, is a combination of the C$2p$ and the d orbitals. By measuring soft X-ray EDC's [3.57] one can discriminate between the d- and p-orbitals. A comparison of XPS spectra of ZrC, HfC, and TaC with DOS curves obtained from APW calculations is shown in Fig. 3.15 [3.56]; one can note a very good agreement between experiment and theory. One should, however, mention that the calculations and the experiments of Fig. 3.15 have been performed by the same group and that comparison with theoretical curves from other authors is not as good. This may indicate the usefulness of having experimental data in adjusting certain parameters of the band-structure calculations.

3.2.5 The Correlation Energy U

In the Hubbard model the relation between the $3d$ bandwidth W and the Coulomb correlation energy U determine whether a particular compound is an insulator or a metal. Photoemission data can be used to estimate both of these quantities. Since all these estimates are still quite uncertain we shall only give rough numbers. The bandwidth of a typical TMC is of the order of 1–2 eV. The correlation energy can be estimated from the chemical shift. From the equation $2d^n \rightarrow d^{n-1} + d^{n+1}$, which schematically describes the transfer of an electron from an atom to another, we see that the correlation energy is the ionization energy of the d^n configuration minus the electron affinity of the d^n configuration; the latter number is, however, the ionization energy of the d^{n+1} configuration. Therefore the difference between the d^n and the d^{n+1} ionization energies give an estimate of U (see also Sect. 4.1.3 for the correlation energy of f-electrons in rare earths). These ionization energies are given directly in the EDC's; the chemical shift between a d^n and a d^{n+1} compound of a particular line gives directly the correlation energy U. These shifts have to be taken between compounds with the same ligands, and they should of course have the same crystal structure, which is rarely the case. However, the differences in crystal structures will introduce uncertainties of less than 1 eV, which will not invalidate the rough estimates intended here. The chemical shift data for compounds containing $3d$ elements give values of U between 1 and 3 eV (see e.g., the shift between CuO and Cu_2O in Fig. 3.1), much reduced from the atomic estimates, which give 10–20 eV [3.26]. There is still another way to estimate U. If, as stated, the satellites correspond to an excitation of electrons from the anion to the empty d-states the difference in energy between the occupied $3d$ states and the satellites minus that of the occupied $3d$ states and the anion band is also a rough measure of U. Again using actual data one obtains values for U between 1 and 3 eV.

These considerations show that photoemission data can be used to determine parameters for the Hubbard model [3.35].

3.3 *d*-Band Metals: Introduction

The analysis of the electronic structure of the d-band metals has now begun to give a consistent picture if the results from different experimental techniques are combined. Whereas a metal like Na or Mg can approximately be described by a free-electron model, d-band metals suffer from the problem that the d-states have partially localized and partially itinerant character, a fact which makes their description difficult. Also, data which probe the band structure far enough below the Fermi energy to allow a test of the calculations over a wide energy range have only recently become available. The analysis of EDC's in terms of theoretical DOS has been done extensively for d-band metals [3.58–60], notably for noble metals.

The usual pattern is to apply to a calculated EDJDOS or DOS the necessary corrections in order to compare it with a measured photoemission spectrum. The opposite procedure, to apply corrections to a measured spectrum in order to recover an experimental DOS is rarely used. The following corrections have to be considered:

1) *Experimental resolution.* It can be measured for instance by the means of the width of Fermi edge of a noble metal and is at best 0.01 eV for an UPS and SXPS experiment and 0.25 eV for an XPS experiment. It is, however, close to impossible to deconvolute the instrumental resolution from an experimental spectrum completely. Therefore some broadening is usually applied to a calculated EDJDOS or DOS curve before comparing it with an experimental curve.

2) *Matrix element modulation.* This effect is implicit in the description of UPS spectra in terms of EDJDOS. The angular and radial dependence of $P_{n^*n}(\boldsymbol{k})$ can give rise to an intensity variation with energy and, in principle, the integral in (3.1) has to be performed explicity before a comparison with a DOS can be made [3.61]. In the XPS regime it is generally assumed that variations in the matrix elements are only due to changes in the atomic character of the initial states. On the other hand, it is an empirical fact that if an XPS spectrum is compared with a DOS, the lower part of the d-bands is always less intense in the measured spectra than we would expect from the calculations. One can try to explain this intensity mismatch by calculating the actual matrix elements from the e_g and the t_{2g} contributions to the wave functions; attempts along this line have been tried [3.62], although they were not completely successful (this approach has been disputed by *Stöhr* et al. [3.61]). Having realized the experimental fact, one can assume that any theory that will explain it, will, to first order, produce a linear variation of the intensity across the band. Such an intensity variation can then be applied empirically to the calculated DOS's in order to obtain a better agreement with the measured XPS spectrum. This intensity variation has the form

$$I(E_i)=\frac{1}{1+\lambda E_i}D(E_i), \tag{3.2}$$

where E_i is the energy measured with respect to the center of the d-band, $D(E_i)$ is the theoretical DOS and λ is an empirical parameter ($\lambda<0$) for which there is at present no convincing explanation (but see [3.63]).

3) *Inelastic "tail"*. On its way to the sample surface the photoexcited electron can scatter inelastically by interacting one or several times with other electrons, phonons, etc., thereby losing energy of any given amount up to its kinetic energy. In the same manner it can also excite one or multiple plasmons. The plasmon excitation is weak in transition metals and does not produce a measurable disturbance of the valence bands. The nonplasmon inelastic tail is, however, seen in all the data. Its magnitude can be measured at the core levels in metals that have no tail (or only a weak one) due to the hole conduction electron coupling (see [Ref. 1.1, Chap. 5]) like, e.g., Au.

4) *Hole conduction electron coupling.* If in a metal a photohole is produced it acts as a scatterer for the conduction electrons producing electron-hole pairs [3.37, 38]. This effect gives a line-shape function calculated by *Doniach* and

Šunjić [3.39] and verified for core levels in many metals and alloys [3.64]. This shape function distorts also the DOS, and has to be taken care of by a corresponding convolution. The effect is discussed in detail in [Ref. 1.1, Chap. 5].

5) *Surface effects.* So far we have not mentioned the most conspicuous correction: that for surface states. The mean escape depth for electrons of 10 to 1500 eV is not more than 30 Å, probably considerably less than that [3.65]. It is therefore reasonable to suspect that an appreciable amount of intensity in the EDC's comes from areas of the sample under investigation that reflect properties characteristic of the surface rather than of its bulk. The experimental evidence accumulated in the last few years has, however, demonstrated that this is (fortunately?) not so. So far in no XPS spectrum has a measurable contribution from surface states been detected. In UPS and SXPS spectra with the class of substances considered in this article, only the single-crystal spectra of W [3.11] and Ni [3.66] have shown an appreciable intensity from surface *d*-states. In addition, very recent UPS work on Cu, Ag, and Au single crystals has also shown strong surface-state signals originating from the *L* points in the Brillouin zone. Their intensity is mostly found near the Fermi energy. We add that special care with the sample preparation has to be taken in order to make the surface state EDC's visible. Already quite small adsorbate layers make it disappear. Therefore we shall not discuss in detail the surface-state properties (see, e.g., [3.67]) and we shall only briefly mention the problems related with them in our discussion of Ni metal.

6) *Momentum broadening.* If $2\pi/\lambda_e$ (λ_e being the electron mean free path in the solid) is not negligible compared to the final electron momentum, this momentum gets smeared [3.10]. This effect can also broaden the structure in EDC's and modulate their intensity. So far this effect has only been observed in Cu for photon energies around 100 eV, where it is also most likely to occur due to the short mean free path [3.61].

All the mentioned corrections have been applied at various places in the analysis of experimental data. We shall mention them as we discuss specific cases.

3.3.1 The Noble Metals Cu, Ag, Au

From all the *d*-band metals it is the noble metals and the ferromagnetic metals which have received the most attention, and to both of these groups we shall therefore devote a separate section. The noble metals are easy to handle experimentally (due to their relative chemical inertness) and relatively easy to handle theoretically (due to the filled *d*-shell); therefore, our present understanding of their photoelectric properties is quite good. We shall deal here with Cu in some detail and then give only the corresponding results for Ag and Au.

All photoemission spectra of Cu (we mention only a few representative references, because the literature on this material is too extensive to be reproduced completely) [3.3, 6, 8, 61, 68–71] show a strong peak beginning at 2 eV below the Fermi energy. This is due to the emission from the 3*d* states, as

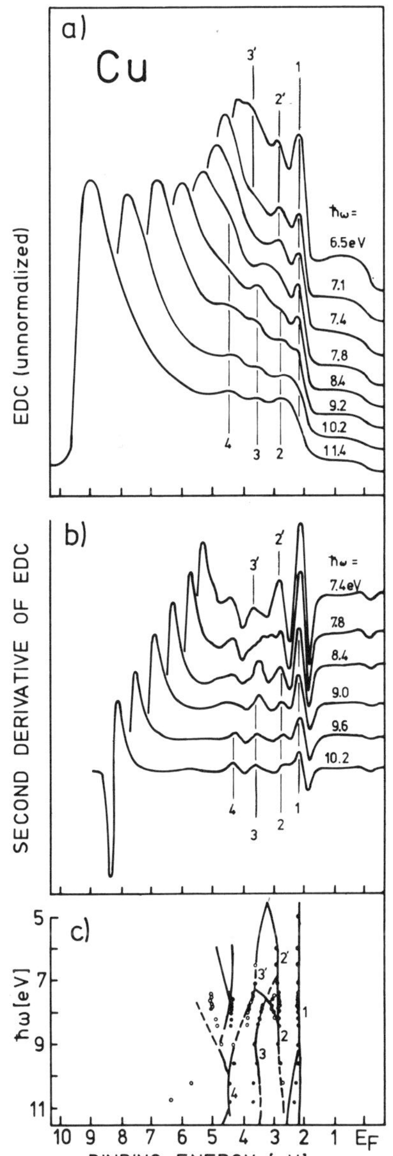

Fig. 3.16. **a** Photoelectron energy distributions of cesiated Cu referred to initial-state energy [3.68] as obtained for several photon energies between 6 and 12 eV.
b Second derivative of the photoelectron energy distribution measured on cesiated Cu [3.68].
c Peak positions in the experimental UPS spectra (full and open circles) of Cu compared with peak loci in the theoretical EDJDOS (solid curves) [3.68]

already identified in early experiments. The structure of the UPS spectra was analyzed a number of times in terms of known band structures, the definitive work having been done by *Smith* [3.68]. In order to bring out more clearly the structure from the UPS spectra, he measured second derivatives instead of the direct photocurrent. The advantage of second derivative vs. direct photocurrent spectra is shown in Fig. 3.16b. These spectra were then analyzed in terms of

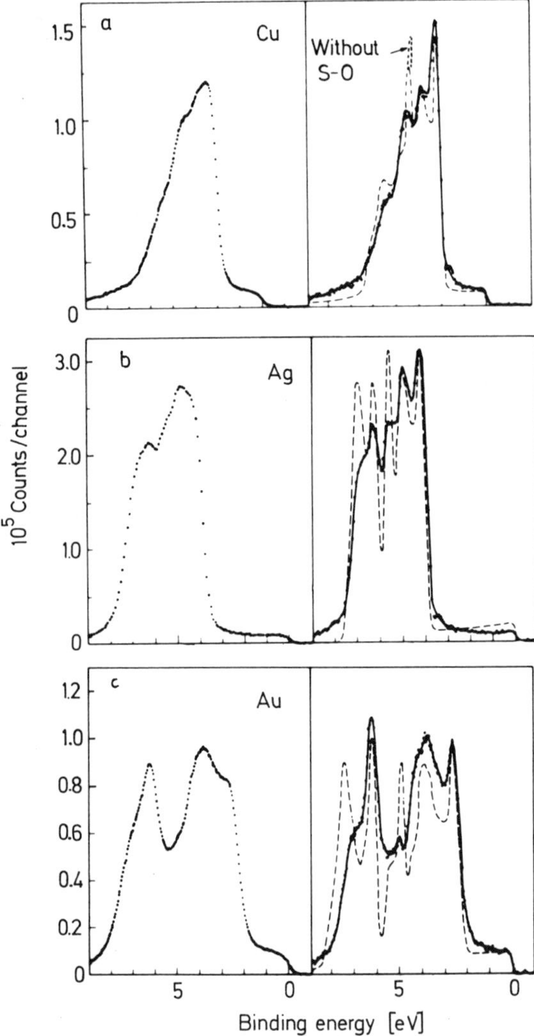

Fig. 3.17. a–c. X-ray photoemission data for **a** clean evaporated copper, **b** silver cleaned by ion sputtering, and **c** clean evaporated Au. The data, after subtraction of background, are shown on the left, the result of a deconvolution [3.75] on the right. A smoothed density of states is shown dashed for comparison [3.70]

EDJDOS. For the calculations the APW band structure of *Burdick* [3.72] (which was obtained by using *Chodorow*'s [3.73] potential) was adopted and numerical results were produced with the interpolation scheme of *Hodges* and *Ehrenreich* [3.74]. The results of this analysis are reproduced in Fig. 3.16c which shows as solid lines the calculated positions of structures in the EDJDOS as a function of exciting uv energy and the data points. Given the remarkably good agreement between theory and experiment one can be hopeful that the calculations produce also a high accuracy DOS. The analysis of that DOS in terms of an XPS spectrum as performed by *Wertheim* et al. [3.70] is illustrated in Fig. 3.17a. The raw XPS spectrum shows some structure although it is not

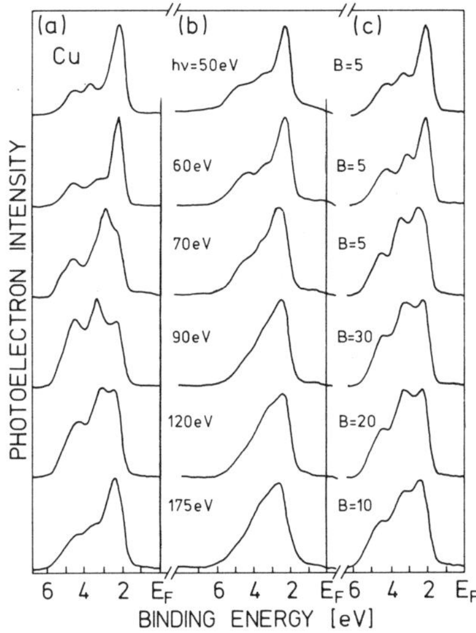

Fig. 3.18. **a** SXPS spectra calculated for Cu assuming k-conservation. **b** Experimental results for Cu; the data have been corrected for inelastic background. **c** SXPS spectra calculated for Cu assuming k-broadening in the final state. The broadening factor B is given in percent [3.61]

very pronounced. *Wertheim* et al. [3.70] applied a deconvolution procedure to this spectrum [3.75], the results of which, shown also in Fig. 3.17a, display quite spectacularly structure that is barely visible in the original spectrum. Superimposed on the deconvoluted spectrum is a DOS with the small spin-orbit interaction included in addition to *Burdick*'s [3.72] original calculation; this DOS curve has been smoothed in order to allow for the residual broadening in the deconvoluted spectrum. The agreement between the theoretical DOS and the measured XPS spectrum is very good, the only deviation being a smaller intensity at the bottom of the d-band for the measured spectrum. No corrections for final-state effects (which are known to be small in this case) have been applied.

Figure 3.18 shows EDC's for evaporated Cu taken with photon energies from 50 to 175 eV [3.61]. These data indicate that although the peaks do not change their position with exciting energy, the relative intensities do. A calculation using *Smith*'s [3.68] band structure represents the data reasonably well; since it can be shown that the radial dependence of $P_{nf}(\boldsymbol{k})$ is small over the width of the d-band, one is obviously observing a variation of the angular part of $P_{nn*}(\boldsymbol{k})$ [3.76]. In addition *Stöhr* et al. [3.61] have tried to correct for the residual discrepancy by taking momentum broadening [3.10] into account in an approximate way. The broadening factor B of the total momentum (given in percent) is the amount by which the momentum in the final state is smeared, thus making more transitions available. It has been applied to all states simultaneously.

Figure 3.17 also shows a comparison of a deconvoluted EDC for Ag and Au with a smoothed DOS. The agreement between theory and experiment becomes progressively worse as one moves on from Cu to Ag and Au. We think this is due to the fact that spin-orbit coupling gets increasingly important in the mentioned order; spin-orbit coupling was, however, inserted only in a one-parameter way into the interpolation scheme. Since it is well known that it varies across the d-band, this rough procedure may be responsible for the observed deviations.

A detailed XPS investigation of the valence band of Ag was also performed by *Barrie* and *Christensen* [3.148] with results similar to those obtained by *Wertheim* et al. [3.70].

The data shown so far in this section indicate that the photoemission process and the band structure of the noble metals is reasonably well understood. One would, however, wish to determine directly the $E(\boldsymbol{k})$ curve experimentally. Steps in this direction have been taken recently by performing single-crystal measurements on the noble metals [3.12–14, 77–86, 139, 149–152, 165, 180–182] in the UPS, SXPS, and XPS regime. A demonstration of the power of this method is indicated in Fig. 3.19 where UPS spectra ($\hbar\omega = 21$ eV) from the (100), (110), and (111) face of Ag together with a spectrum from a polycrystalline sample are shown; the single-crystal spectra correspond to a mapping of the band structure in the ΓX, ΓKX and ΓL direction. These spectra show considerable detail and, as expected, a strong variation for the different crystal faces. The interpretation of these data is, however, still under question; at this point it looks as if the (110) EDC corresponds roughly to a one-dimensional DOS in the ΓKX direction (implying small effects of the final states), whereas the two other faces seem to yield EDC's which correspond to EDJDOS [3.81]. This is admittedly a very puzzling result and further work is needed to clarify the situation. Perhaps the most impressive set of data in this field for noble metals has been presented by *Ilver* and *Nilsson* [3.83]. They measured the UPS spectra (16.8 eV) for two faces of Cu (Cu [111], Cu [100]) with a fixed polar angle ($\theta = 45°$) for the detection of the photoelectrons (normal incidence for the light) and a variation of the azimuthal angle ϕ between 0° and 360°. Figure 3.19b taken from this work shows a few representative spectra and a polar plot of the peak positions (full curves). The dots are the transition energies calculated for direct transitions. There is very good agreement between theory and experiment showing that this is an excellent way to check calculated $E(\boldsymbol{k})$ curves. We add that also the interpretation of the XPS data on single crystals is still a matter of debate [3.77, 78, 151, 152].

Single crystal XPS spectra naturally also show a variation of the spectra as a function of the electron detection angle relative to the crystal axes. Since in this energy regime there is less structure in the final states than in the UPS regime, the changes in the XPS spectra are not as dramatic as those in the UPS spectra. *Baird* et al. [3.77] in their interpretation of the single-crystal spectra came to the conclusion that they represent essentially a one-dimensional DOS; this results from their assumption of direct transitions with constant matrix elements to plane wave final states. In contrast, *Wehner* et al. [3.151] assume that the angular variation of the spectra is produced essentially by a transition matrix element variation between the initial-state tight-binding Bloch functions

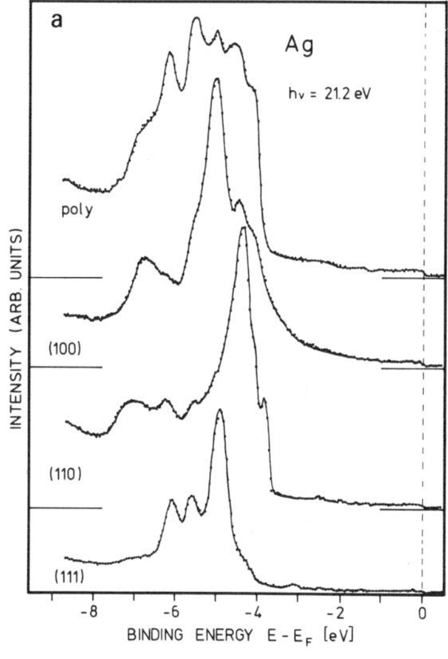

Fig. 3.19. **a** UPS spectra of polycrystalline Ag [3.86] compared to spectra taken of single crystals of Ag along the (100), (110), and (111) direction [3.150]. **b** Angular resolved photoemission of Cu in the (100) plane [3.83]; the light of 16.8 eV has normal incidence, the detection angle with respect to the nornal, i.e., the polar angle is 45°. The azimuthal angle ϕ has been varied between 0° and 360°. The left part of the figure shows representative spectra, whereas the right part shows a polar plot of the peak positions. The dots are the results of calculation, the full lines correspond with the experimental data. The fourfold symmetry expected for the (100) face is clearly visible

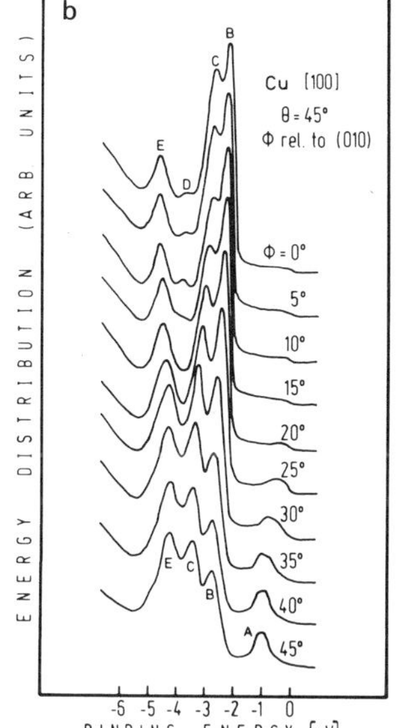

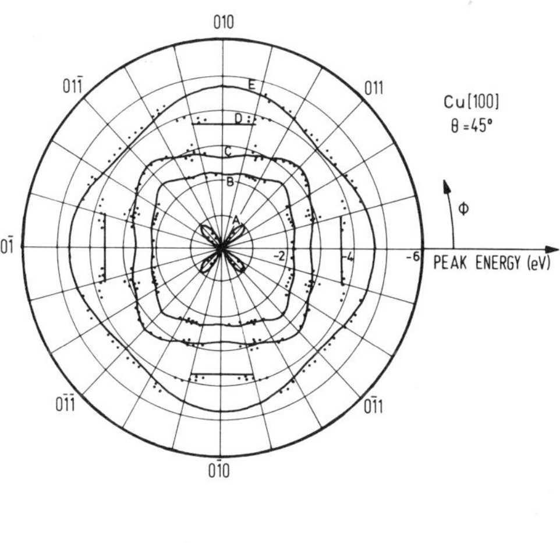

and the final-state plane-wave functions, where the total DOS is weighed in each direction by this transition matrix element. At this point the experimental data are not accurate enough to arrive at a clear decision which of the two models is the appropriate one for the interpretation of the single-crystal XPS spectra.

In looking at the available data some general statements can, however, be made. Single-crystal spectra should eventually allow us to map the $E(\boldsymbol{k})$

relation – the final object of photoemission experiments. At the present exploratory stage, however, a thorough understanding of the EDC's is still needed. By inspecting the available material, it looks as if UPS data were more useful than SXPS and XPS data for two reasons: they contain more structure, obviously because of final-state effects and they can be obtained at considerably higher resolution (50 meV against ~ 0.5 eV or, at most 0.25 eV).

We finally want to mention that the noble metals show a surface photoemission peak close to Fermi energy [3.79, 150]. Its origin is not yet completely understood.

3.3.2 The Ferromagnets: Fe, Co, Ni

Much of the work on the band structures of metallic systems has centered around the ferromagnetic *d* metals Fe, Co, and Ni. Although it looked for some time as if one could account for the bulk of their properties by the *Stoner–Wohlfarth* [3.87, 88] band model, experiments in the last decade have cast some doubt on that simple model. In brief, that band model states that the band structure of these ferromagnetic metals can be made up by shifting the spin-up and spin-down bands with respect to each other as a result of exchange interaction. We do not want to discuss the problems in this area because an excellent and exhaustive review is available [3.26]. Instead we review the photoemission experiments and see what information they yield on the band structure. Since Fe is best documented, we shall deal with this material in detail and then add the available information on Co and Ni. Figure 3.20 shows an XPS spectrum of an evaporated Fe sample [3.89]. This spectrum is similar in shape to that of *Ley* et al. [3.90] and *Hüfner* and *Wertheim* [3.91]. However, *Ley* et al. [3.90] gave a different placement of the Fermi energy. For the spectrum shown in the figure various checks as to the position of the Fermi energy were performed: they all agreed with the results given. Figure 3.21 shows a comparison of the measured XPS with a calculated DOS curve [3.91]. The latter was folded with the instrumental resolution (0.55 eV), with a linewidth function (to account for the lifetime of the band states), with a *Doniach-Šunjić* [3.39] line shape (to take into consideration the core hole-conduction electron coupling), with the function (3.2) (in order to correct for the variation of the matrix element over the bandwidth) and finally the inelastic tail was added. The procedure for application of the lifetime broadening deserves a comment. At the Fermi energy the lifetime is relatively long, and it gets shorter for states deeper in the band. We found the best agreement between the theoretical and experimental XPS spectrum if the lifetime width was assumed to be energy dependent in the form of a Fermi distribution function with width V where the half height point of the distribution function was positioned at V below E_F. The maximum lifetime width was Γ_h. Table 3.1 contains the optimal parameters used for the convolution of the DOS curves of Fig. 3.21. The agreement with the theoretical DOS of *Callaway* et al. [3.92, 93] with that of *Cornwell* et al. [3.94] and that of *Wakoh* and *Yamashita* [3.95] is as

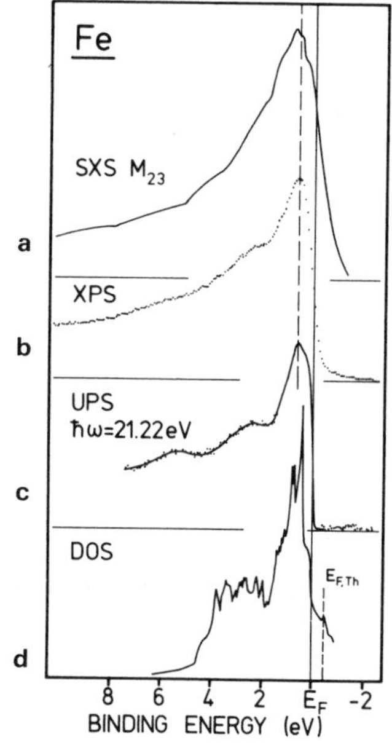

Fig. 3.20. a–d DOS information on Fe metal. **a** Corrected (for lifetime of the $M_{2,3}$ state and instrumental resolution) $M_{2,3}$ emission spectrum [3.98]. **b** XPS spectrum ($\hbar\omega = 1487$ eV [3.89]. **c** UPS spectra ($\hbar\omega = 21$ eV) [3.97]. **d** DOS function [3.92]

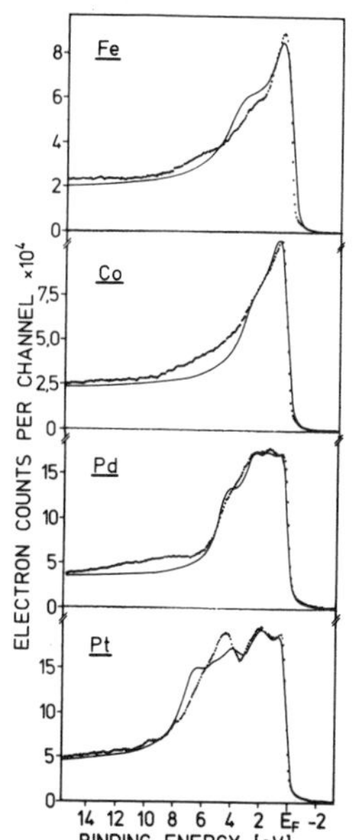

Fig. 3.21. XPS spectra for Fe, Co, Pd, and Pt (solid line) and the convoluted DOS histogram for these metals [3.91]

Table 3.1. Correction parameters used to calculate the EDC of Fe, Co, Pd and Pt from the theoretical DOS (see also [Ref. 1.1, Chap. 5]

	Fe	Co	Ni	Pd	Pt
α_{VB} [a]	0.2	0.18	0.10	0.1	0.15
α_{CORE} [b]	0.4	0.32	0.24	0.19	0.19
$\lambda[\mathrm{eV}]^{-1}$ [c]	−0.13	−0.13	−0.32	−0.23	−0.13
$\Gamma_h[\mathrm{eV}]$ [d]	0.78	0.78	0.62	0.47	0.47
$V[\mathrm{eV}]$ [e]	1.44	0.94	0.62	0.31	0.31

[a] α_{VB}: is the asymmetry parameter as defined by the line-shape function in [3.39].
[b] α_{CORE}: Same as α_{VB} but for core lines, taken from [3.64] and this work.
[c] λ: Parameter in the intensity modulation function $I(E) \propto (1+\lambda E)^{-1}$ [3.63].
[d] Γ_h: Lorentzian lifetime width (FWHM).
[e] V: Width of the Fermi distribution that governs the variation of Γ_h near the Fermi energy.

good as that of Fig. 3.21. A major discrepancy exists, however, for the band structure of *Duff* and *Das* [3.96].

Figure 3.20 summarizes the available information for experimental DOS on Fe metal. It is encouraging to see that the XPS, the UPS spectrum [3.97, 153] and the theoretical DOS of *Callaway* et al. [3.92, 93] are in such good

agreement, thus supporting each other. The DOS curve obtained from the M_{23} soft X-ray emission spectrum (SXS) [3.98] is, however, different and shows more structure although it has inherently less resolution than the UPS (due to the 1 eV linewidth of the M_{23} core levels). The correspondence between SXS, theoretical DOS and DOS from other experimental techniques has to be further investigated in order to resolve these differences that are also found for other materials.

It is worth pointing out an interesting result which arises from the data of Fig. 3.20. The resolution of the UPS spectrum is 0.06 eV and that of the XPS 0.55 eV. Nevertheless the two spectra show the same amount of details (except for the structure around 6 eV which is presumably a two-hole state [3.99] and has nothing to do with the DOS; this state is more clearly visible in the UPS data). This shows that lifetime effects, which according to our analysis (Table 3.1) are of the order of the resolution of the XPS data, play an important role. This point is substantiated if the same sort of comparison between UPS and XPS spectra is made for a noble metal like Au: for this material the UPS [3.79, 154] shows considerably more details than the XPS spectrum, indicating a larger lifetime of the *d*-band states in Au than in Fe. This demonstrates that in open *d*-shell metals (like Fe) the lifetime of the deeper lying bands is shortened considerably, presumably due to the high mobility of the *d*-electrons. This is another manifestation of the fact that in the open shell metals the *d*-electrons account for much of the screening. In Au the *d*-electrons are well below the Fermi energy: they hardly contribute to the conductivity or to the screening.

We can summarize that for the itinerant ferromagnet Fe there is reasonable agreement between the XPS spectrum and the DOS derived on the basis of the Stoner-Wohlfarth band model. The results for Co are very similar to those for Fe, as can be seen from the comparison of the EDC and the corrected DOS [3.100] in Fig. 3.21. Again a sizeable lifetime broadening (Table 3.1) is required in order to produce the very structureless experimental *d*-band.

Ni metal is perhaps the material on which the most photoemission experiments have been performed. Although recently a large number of investigations on this metal have appeared, a number of puzzling questions still remain. Early UPS [3.3] and XPS [3.35] experiments seemed to indicate, that the bandwidth of Ni, measured at the bottom of the band, is smaller than anticipated from band-structure calculations. This is not easy to understand, because the band-structure calculations seem to fit the EDC's of Cu, the neighboring element with the same crystal structure, quite nicely. Early spin polarization experiments [3.107, 108] showed a positive (majority) spin polarization, in contrast to the expectations from the Stoner-Wohlfarth band model of ferromagnetism. More recent spin polarized photoemission experiments [3.166] do however show a negative (minority) spin polarization very close to the photoemission threshold in apparent agreement with the Stoner-Wohlfarth model [3.167]. A negative spin polarization is also found in field emission experiments [3.106, 168]; here the polarization depends on the crystal face one is looking at and is very sensitive to small amounts of adsorbates. The different sizes of the electron polarization found in photoemission and field emission experiments is a consequence of the fact, that the former experiment samples predominantly the bulk electronic states whereas the latter samples the surface electronic states. Early temperature-dependent photoemission experiments [3.109] did not show any variation in the EDC's as the sample temperature was raised through the Curie temperature; this was interpreted as a breakdown of the Stoner-Wohlfarth model. On the other hand, *Rowe* and *Tracy* [3.110] could explain the absence of a magnetic shift in the EDC's of Ni by a cancellation of the magnetic and the thermal expansion shift.

Very recently a great number of photoemission experiments on polycrystalline [3.104, 180] and single crystal [3.66, 102, 103, 106, 169–175] samples of Ni have appeared with the objective to come to a better understanding of its electronic properties. These investigations have added much information, although some of the problems remain.

Figures 3.22a, b show a comparison of XPS spectra of Ni with two theoretical spectra [3.104]; in Fig. 3.22a the DOS of *Wang* and *Callaway* [3.105] has been used to construct the curve termed Ni–DOS. The bare theoretical DOS was folded with the instrumental resolution function, a many-body correction, a finite lifetime, a matrix element modulation factor (for the parameters used see Table 3.1), and a background function. The agreement between the experimental and the theoretical curve is reasonable, although not good. The peak at 6 eV in the measured spectrum is due to a satellite [3.99] and should therefore not be reproduced in the theoretical spectrum. On the other hand, it seems to be impossible to define with any reasonable accuracy a bandwidth from the experimental spectrum, although it looks as if it is slightly narrower than the theoretical one. It is disturbing that the very high "background" of the experimental spectrum cannot be accounted for in the theoretical spectrum. This indicates that probably many-body effects are more important than estimated by *Höchst* et al. [3.104]. In a further attempt to understand the XPS spectrum of Ni it is compared in Fig. 3.22b with a synthetic spectrum that has been produced by adding up two XPS spectra of Cu, shifted by 0.3 eV (the experimentally observed exchange splitting [3.66, 172]). This synthetic spectrum reproduces the main features of the Ni *d*-band quite well; since the bandwidth of Cu is smaller than that of Ni, this is another indication that the *d*-band width, as measured by a photoemission experiment, is smaller than that calculated [3.105, 176].

One might hope, that the single-crystal spectra give more convincing results. However, so far they seem to contradict each other. *Heimann* and *Neddermeyer* [3.66], *Dietz* et al. [3.171], and *Sagurton* et al. [3.174] indicate that their data are in agreement with band-structure calculations [3.105, 176], whereas *Williams* et al. [3.102], and *Eastman* et al. [3.172] find a band narrowing if they compare their data with the band-structure calculations. Judging from the data, those of *Eastman* et al. [3.172] seem to be the most reliable ones and their summary is therefore presented in Figs. 3.22c, d. Figure 3.22c shows the dispersion relations in the $\Gamma\Lambda L$ direction; the experiments have

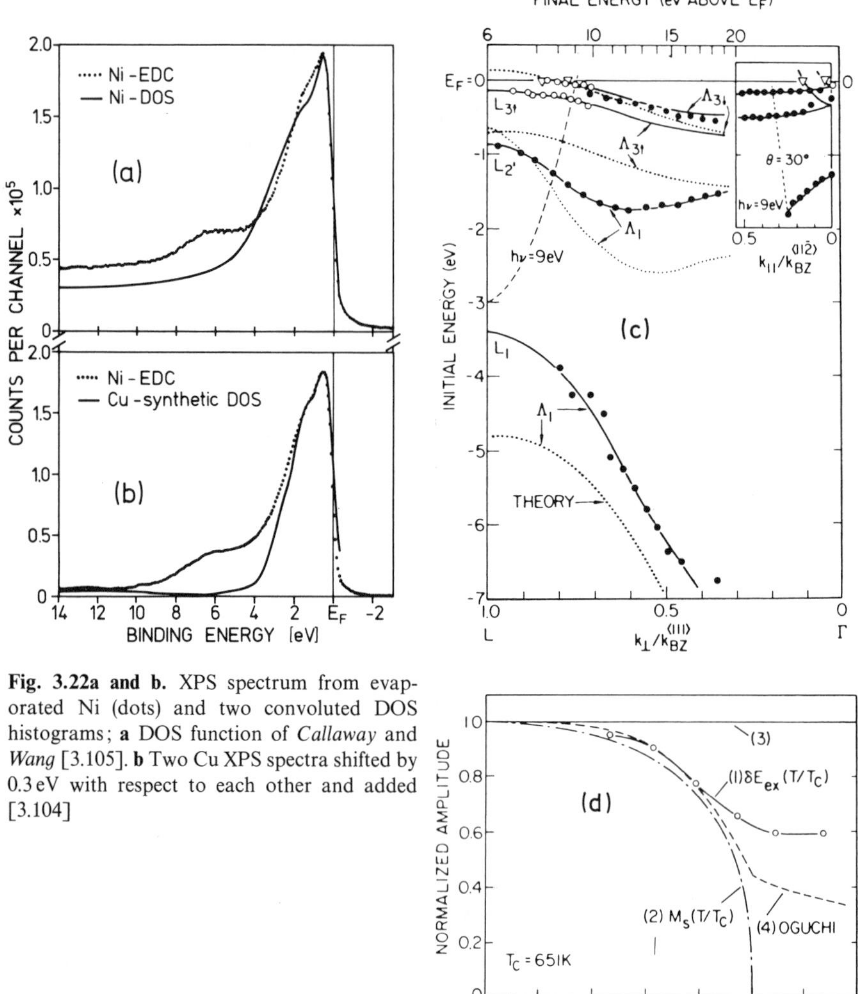

Fig. 3.22a and b. XPS spectrum from evaporated Ni (dots) and two convoluted DOS histograms; **a** DOS function of *Callaway* and *Wang* [3.105]. **b** Two Cu XPS spectra shifted by 0.3 eV with respect to each other and added [3.104]

Fig. 3.22. c Experimental E-vs-$k_{\perp}^{\langle 111\rangle}$ energy-band dispersions for Ni (111) along Λ. A single conduction band is involved for $\hbar\omega < 20$ eV and is shown for $\hbar\omega = 9$ eV (shifted down by $\hbar\omega$). The hollow circles are data taken for Ni (111) + ~0.1 Torr-s O_2, which removes a surface state at −0.25 eV, but does not significantly affect bulk emission. The triangles denote bands crossing the Fermi surface as given by de *Haas* van *Alphen* data [3.105, 176]. The light dotted curves are calculated bands [3.176]. The inset shows E-vs-k dispersion for $k_{\parallel}$ in the $(\bar{1}\bar{1}2)$ direction for $h\nu = 9$ eV. The zone-boundary momenta are $k_{BZ}^{\langle 111\rangle} = 1.55\ \text{Å}^{-1}$ and $k_{BZ}^{\langle\bar{1}\bar{1}2\rangle} = 1.46\ \text{Å}^{-1}$ [3.172]. The dotted line shows the position at which the data for Fig. 3.22d have been obtained. **d** Experimental temperature-dependent d-band splitting, compared with several theoretical models (see text) [3.172]

been performed with synchrotron radiation at normal emission ($k_\perp$). For normal emission only one free-electronlike conduction band is involved for the direct transitions. The experimental results (points and triangles and full curves) are contrasted with the theoretical bands [3.176]. One realizes that the experimental band energies at critical points [$E_F - L_1$, $E_F - (\Lambda_1)_{min}$] are noticeably smaller than the calculated ones. In addition, the measured exchange splitting is also much smaller (~ 0.3 eV) than the calculated one. The insert in Fig. 22c shows the dispersion relation along the [$\bar{1}\bar{1}2$] direction and also indicates where for the azimutal angle $\theta = 30°$ and $h\nu = 9$ eV the temperature dependence of the exchange splitting has been measured. These data are presented in Fig. 3.22d and show a splitting at 293 K of 0.31 ± 0.03 eV and at 693 K of 0.19 eV. Thus for the first time definitely a temperature effect in the EDC of Ni has been observed. Figure 3.22d shows also three possible temperature-dependent splittings: (2) the measured saturation magnetization, which is a measure of the long-range order, (3) a temperature-independent curve, which would correspond to a localized intra-atomic interaction, and (4) the short-range order parameter curve due to *Oguchi* [3.177]. The experiments on single crystals [3.66, 173] also show that there is a peak near the Fermi energy that is due to a surface state. Among transition metals W and Ni represent the cases where surface states have been quite definitely observed in UPS experiments; interestingly these are also the metals which seem to have lower mean free paths than other materials [3.65]. There remain some unsolved problems with respect to the interpretation of the photoemission experiments on Ni. The strong peak observed by *Smith* et al. [3.103] 4.5 eV below E_F is still not explained and seems hard to reconcile with a very much narrower band structure. The most important question is however whether the apparent band narrowing observed in the photoemission experiments is an initial- or a final-state effect. Optical experiments [3.178] show a bandwidth of 5 eV in agreement with the theoretical predictions; these measurements should not suffer very much from many-body effects. Photoemission experiments in Ni show strong many-body effects, a large asymmetry in the core lines, and the 6.5 eV satellite accompanying all lines, also the valence band. It remains to be worked out whether such effects can indeed lead to an apparent narrowing of the *d*-bands in the photoemission spectra [3.179].

3.3.3 Nonmagnetic *d*-Band Metals

For other *d*-band metals, all of which have by now been investigated more or less extensively by photoemission methods, the agreement between band-structure derived DOS and EDC's is generally not as good as for the noble metals. As an example we have included in Fig. 3.21 a comparison of the XPS spectra of Pd and Pt with the suitably corrected DOS [3.91, 111, 112]. One may conclude this section by stating that the description of EDC's by first principle band structures is well advanced and that the remaining discrepancies do not seem to be fundamental. A direct mapping of $E(\boldsymbol{k})$ curves by angular resolved photoemission experiments on single crystals is, however, still desirable.

A question that deserves further study concerns the importance of surface photoemission in the UPS experiments. Surface photoemission has been found in the noble metals where it does not interfere with the d-band structure. The most detailed study of this effect has been performed for single crystals of W [3.155] where the "surface peak" coincides with the top of the d-band.

3.4 Alloys

Alloys have considerable technological importance and therefore the understanding of their electronic structure deserves attention. The first model to describe the band structure of transition metal alloys (such as CuNi alloys) was the *Mott* and *Jones* [3.113] rigid-band model. It assumed that in an alloy the two constituents have a common d-band and that the effect of adding one to the other is merely to shift the Fermi energy in such a way as to accommodate the extra (or missing) electrons. This model has had remarkable successes considering its simplicity; e.g., it correctly predicts the disappearance of magnetism in the CuNi alloys at 45% Cu.

Other properties were, however, less well described by the rigid-band model, and thus models were developed that start out from the local properties of the ions in an alloy and then introduce the d-orbital conduction electron interaction as a perturbation. This resulted in the virtual bound state model [3.114, 115] for the description of dilute impurities and for the minimum polarity model [3.116] in alloys of arbitrary concentration. Photoemission has played an important part in revealing the DOS curves in the two cases and we shall therefore handle them separately.

3.4.1 Dilute Alloys: The Friedel-Anderson Model

The properties of dilute alloys cannot be described in the rigid-band model. For instance it is not understandable in this model that FeCu is magnetic: the d-states in Cu are two eV below the Fermi energy and the addition of some Fe would not change this situation appreciably, leaving that alloy diamagnetic. The solution to this problem was provided by *Friedel* [3.114] and put on a rigorous mathematical footing by *Anderson* [3.115]. The basic concept of the Friedel-Anderson model is the screening of the impurity charge by the conduction electrons. It is assumed that the d-levels of a transition metal like Fe, Ni, or Pd that is brought into an s-band metal like Cu or Ag retain their identity as d-orbitals. Their position is determined by Coulomb and exchange interactions, and their width by the $s-d$, crystal-field, and spin-orbit interactions. In order to extract the $s-d$ interaction from a measured linewidth, one has to know the latter two interactions. For the crystal-field interaction one

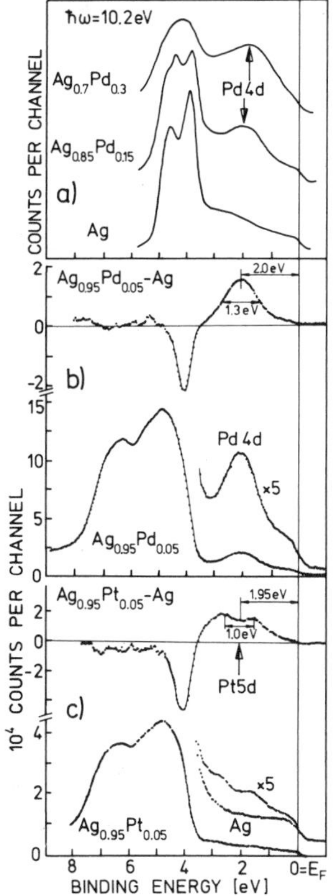

Fig. 3.23a–c. Photoemission spectra of virtual bound states: **a** UPS spectra of Ag and AgPd alloys [3.118]. **b** XPS spectrum of $Ag_{0.95}Pd_{0.05}$ and difference spectrum with respect to Ag metal [3.64]. **c** XPS spectrum of $Ag_{0.95}Pt_{0.05}$ and difference spectrum with respect to Ag metal [3.64]

can obtain an order of magnitude estimate from the band-structure calculations for the transition metals; if these data provide any guidance, one can safely neglect it, because it should be about 0.05 eV for Cu and for Au [3.58, 59, 117]. (This is the parameter Δ in the paper by *Smith* [3.58, 59] and should not be confused with the Γ_{12}–$\Gamma_{25'}$ splitting which is of course much larger.) The spin-orbit splitting cannot be dismissed so easily. If again we assume that it has the same magnitude as in the transition metals themselves, it can be of the order of 1 eV for 4*d* and 5*d* virtual bound states and thus be in the detectable limit of present photoemission experiments [3.59]. We now want to discuss experiments on nonmagnetic virtual bound states for which the situation is most clearly established. The classical system for this class is Pd$\underline{\text{Ag}}$. Silver has a 4 eV wide $s-p$ band below the Fermi energy before the onset of its own 4*d* band. The UPS spectra of *Norris* and *Myers* [3.118] taken on $Pd_{0.15}Ag_{0.85}$ shows the Pd state clearly sitting on the flat $s-p$ band, 2 eV below the Fermi energy with a width of about 1 eV (Fig. 3.23). This experiment confirms the picture of *Friedel* and *Anderson*. We shall defer discussion of these data for a moment.

Table 3.2. Virtual bound state parameters for CuPd, AgPd, AuPd, and AgPt

System	Γ_d [eV]	Γ_{ex} [eV][a]	Γ_{sd} [eV][b]	Method
$Cu_{0.80}Pd_{0.20}$	1.6	1.0	0.70	XPS[c]
$Cu_{0.90}Pd_{0.10}$	1.75	0.83	0.45	XPS[c]
$Cu_{0.95}Pd_{0.05}$	1.75	0.90	0.55	XPS[c]
$Cu_{0.86}Pd_{0.14}$		No numerical results		Opt[d]
$Cu_{0.90}Pd_{0.14}$		No numerical results		UPS[e]
$Ag_{0.90}Pd_{0.10}$	2.0	1.3	1.05	XPS[c]
$Ag_{0.95}Pd_{0.05}$	2.0	1.3	1.05	XPS[c]
$Ag_{0.99}Pd_{0.01}$	1.95	1.0	0.70	XPS[c]
$Ag_{0.90}Pd_{0.10}$	2.6	1.8		Opt[d]
$Ag_{0.85}Pd_{0.15}$	2.2	1.0	1.0	UPS[f]
$Ag_{0.97}Pd_{0.03}$	1.91	0.46	0.46	Opt[g]
$Au_{0.90}Pd_{0.10}$	1.4	1.2	0.95	XPS[c]
$Au_{0.95}Pd_{0.05}$	1.55	0.75	0.35	XPS[c]
$Au_{0.90}Pd_{0.10}$	2.0	1.15	1.15	Opt[d]
$Au_{0.905}Pd_{0.095}$	1.65	0.36	0.36	Opt[h]
$Ag_{0.90}Pt_{0.10}$	2.1[i]	~1.2	~0.95	XPS[c]
$Ag_{0.95}Pt_{0.05}$	1.95[j]	~1.2	~0.95	XPS[c]

[a] Experimental full width at half maximum.
[b] Width corrected for instrumental resolution.
[c] [3.134].
[d] [3.122].
[e] [3.123].
[f] [3.118], 0.5 eV was subtracted but for spin-orbit splitting.
[g] [3.121].
[h] [3.135–3.136].
[i] Spin-orbit splitting: $\frac{5}{2}\cdot\xi = 1.2$ eV.
[j] Spin-orbit splitting: $\frac{5}{2}\cdot\xi = 1.0$ eV.

Figure 3.23 shows also another example, namely an XPS spectrum of $Pt_{0.05}Ag_{0.95}$. Here the situation is very similar to that in AgPd, the only difference being that the virtual bound state displays a double hump. This splitting of the state can be interpreted as arising from spin-orbit interaction; if the splitting is extrapolated to zero concentration one obtains a value of (0.80 ± 0.20) eV, yielding a spin-orbit coupling constant ξ of $\xi = 0.32$ eV. This value is somewhat lower than the value $\xi = 0.42$ eV determined for the d^9 configuration from an analysis of the atomic spectra of Pt [3.25], but one has to keep in mind that the experimental value obtained from photoemission has an uncertainty of about 25%. Other measurements of the spin-orbit splittings of virtual bound states have given values closer to the atomic limit, yet since these data have been obtained on cosputtered films of otherwise not miscible constituents, the meaning of these results may be open to questions [3.119, 120]. We conclude from these data that spin-orbit splitting of about the atomic size (probably somewhat smaller) is exhibited by the virtual bound states and has to be taken into account in the analysis of the linewidth.

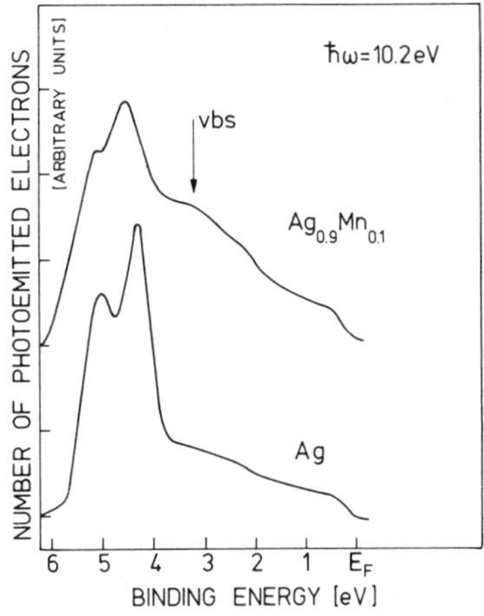

Fig. 3.24. UPS spectra of Ag and $Ag_{0.9}Mn_{0.1}$ showing virtual bound states (vbs) at about 3 eV below E_F [3.123]

In analyzing the $\underline{Pd}Ag$ data spin-orbit interaction should thus not be neglected. The smallest width measured in a photoemission experiment is 1.0 eV for Pd_1Ag_{99} [3.64]; after correction for instrumental resolution this yields an intrinsic width of the bound state of 0.7 eV. If the atomic spin-orbit interaction of 0.5 eV is assumed in this case, this leaves a width due to $s-d$ scattering of ~(0.3–0.4) eV for the $Pd\underline{Ag}$ state. A little bit disturbing is the fact that optical experiments yield an experimental width of 0.46 eV for $Pd_{0.03}Ag_{0.97}$, smaller than seen in the photoemission experiments [3.121]. If this number is corrected for spin-orbit splitting the intrinsic width could not be more than 0.2 eV. Table 3.2 lists the widths and positions (with respect to the Fermi energy) of some nonmagnetic bound states, uncorrected for crystal-field interaction and spin-orbit interaction. These data lead to the conclusion that the width is usually of the order of 0.5 eV, about an order of magnitude smaller than the accepted value [3.115]. The data on nonmagnetic bound states of d-metal impurities may thus be summarized by saying that the Friedel-Anderson model describes the data well, and that the width of the states is surprisingly narrow.

In the theoretical model the magnetic case is distinguished from the nonmagnetic by a splitting of U of the five spin-up and the five spin-down states. Although this picture is the backbone of the magnetism of dilute alloys, it is disquieting to realize that there is at best one system, $Mn\underline{Ag}$, where the spin-up and spin-down states may possibly have been observed. In this system optical experiments [3.122] show two humps, whereas the UPS experiments (Fig. 3.24) show only one [3.123, 124]; these data were interpreted as evidence for one bound state 3 eV below and another one 2 eV above the Fermi energy. Recent optical measurements on the $Mn\underline{Ag}$ system [3.156] show, however, only one peak at the position where it had also been seen in the UPS data. *Miller* argues that the second peak in the earlier optical data [3.122] was probably produced by impurities; thus, the data on this system have to be taken with caution. Attempts to obtain the bound state parameters in MnCu and FeAu and especially in FeCu have so far produced inconclusive results. This is a very unsatisfactory situation, especially in view of the fact that much of the magnetic data on these systems has been analyzed in terms of the bound-state model.

Recently, attempts have been successful to detect the virtual bound states of Mn, Fe, Co, Ni, and Cu impurities in aluminium [3.157]. Except for MnAl which is a spin-fluctuating system, these systems are unfortunately all non-magnetic. They all show a Lorentzian-type state with widths slightly larger than found for *d*-metal impurities in noble metals.

Hirst [3.125] has expressed doubts about the bound state model in its presently accepted form. He argues that rather than starting from an orbital and filling it up with electrons, one has to account first for the interconfigurational coupling of these electrons (Hund's rule energy) before dealing with the $s-d$ interaction. Although it is generally accepted that his point of view is valid for rare earth impurities, one wonders whether it does not apply to *d*-metal impurities in view of the narrowness of the nonmagnetic states which indicates a small $s-d$ hybridization. The experimental data on the nonmagnetic bound state, however, cannot as such distinguish between the two approaches.

3.4.2 Concentrated Alloys: The Coherent Potential Approximation

The now accepted approach to describe the electronic structure of transition metal alloys is the coherent potential approximation, as developed by *Soven* [3.116] and by *Velicky* et al. [3.126] and applied successfully to a number of binary alloy systems. Physically this model starts from a position opposite to that of the rigid band model. It assumes that the charge exchange between the two constituents of a binary alloy is very small. This results from the realization that a tight-binding model accounts well for the wave functions of the *d*-electrons in metals, meaning that they have a high degree of localization. Therefore one expects that, to first order, the electronic states of an ion are not changed in going from the metal to an alloy. Thus one can get a rough idea about an actual DOS of an alloy by adding up the DOS of the constituents multiplied with their respective concentrations. In an actual case two further effects do, of course, appear. The alloy lattice loses its periodicity compared to that of the pure metals, resulting in a smearing out of the sharp DOS features. Secondly, the bandwidth of the *d*-bands of the constituents will be, in some way, proportional to the concentration, since it is mainly caused by the overlap with the like neighbors whose number is a function of the concentration. The coherent potential approximation is a scattering theory. It is based on the notion that the alloy can be replaced by a single scatterer and a potential which is determined such that it produces no further scattering of the electrons. Calculations with this approach have been performed among other systems for the AuCu and the AuAg [3.127], the CuNi [3.128, 129], and the AgPd [3.130]

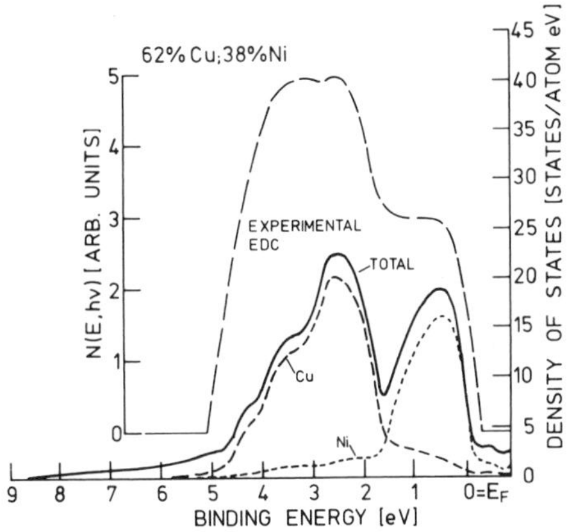

Fig. 3.25. Densities of states for a 62% Cu, 38% Ni alloy obtained with the CPA. Shown is the total DOS function and the Cu and Ni partial DOS function. Also shown is an experimental electron distribution curve taken from the photoemission work of *Seib* and *Spicer*. The EDC is for an incident photon energy of 10.2 eV [3.131]

system, all well-behaved alloy systems which are miscible over the whole concentration range. In general, the calculations reproduce the experimental results quite well. As an example we show in Fig. 3.25 the UPS spectrum [3.131] for a sample of $Cu_{0.62}Ni_{0.38}$ and the corresponding DOS as calculated by *Stocks* et al. [3.132] in the coherent potential approximation (CPA). One can clearly see in the alloy DOS the two sub-bands which are approximately at the position of the bands of the pure metals, thus confirming the underlying ideas.

The two most thoroughly investigated systems are the CuNi system [3.35, 131, 158] and the AgPd system [3.35, 118, 159], where the work of *Hüfner* et al. [3.35] was done by XPS and the other data were obtained by UPS. All these measurements confirm the CPA predictions in considerable detail. They show that the position of the bands of the two constituents stays roughly constant as a function of concentration. The shape of the bands is given by a Lorentzian in the low concentration limit and widens with increasing concentration attaining structure gradually as the concentration gets larger.

By varying the photon energy, *Yu* et al. [3.158] could investigate the bulk and surface DOS of the CuNi system independently. They found for the surface DOS that it consists of two peaks 2 eV apart. The magnitudes of these peaks are correlated with the surface composition, but neither the shape nor the energy position of these peaks depends much on changes in bulk composition, surface crystallinity, and surface local environment. This seems to indicate that the surface atoms retain much of their atomic character. For adsorption processes this means that they are governed by local molecular orbitals, whereas band-structure effects play only a minor role.

In one case, namely $Pd_{0.775}Cu_{0.06}Si_{0.165}$, the crystalline sample and the metallic glass form of the alloy have been investigated [3.160]. The authors found only very small changes in the XPS spectra of the two forms, indicating that there is little difference in the electronic structure of the glass and the crystal alloy.

3.5 Intermetallic Compounds

Intermetallic compounds do in a sense represent special cases of alloys, at a fixed concentration with chemically ordered structures. This may make the covalent bonding in them stronger than in metals and alloys but leave their properties otherwise very similar to those of the alloys. Thus as far as the DOS goes, a great similarity is expected and found with the situation in alloys. The subbands of the constituents stay roughly intact, having less structure and less width than in the constituents. An interesting experimental approach for the preparation of intermetallic compounds in XPS investigations has been employed by *Fuggle* et al. [3.133]. For AuAl intermetallic compound they evaporated a layer of Au on one of Al and then heated this layer sample in the XPS instrument in order to interdiffuse the two ion species. In this way the compounds Au_2Al, AuAl, and $AuAl_2$ could be prepared and their spectra investigated.

Useful examples for photoelectron spectra of intermetallic compounds are, e.g., those of the A 15 superconductors V_3Si [3.161, 162] and Nb_3Sn [3.163]. They show the same quality of agreement between theory and experiment as has been previously found for the TMC. We note that the sharp singularity in the DOS at the Fermi energy predicted by some theoretical models for these compounds could so far not be observed due to a lack of experimental resolution.

3.6 Summary, Outlook

The many experiments performed in recent years have shown that photoemission can locate the various bands in compounds and alloys of transition metals as well as in transition metals themselves. Care must, however, be exercised if structures in the EDC's are to be compared with theoretical DOS. The contribution of final-state effects has to be investigated in each case before a correlation between theory and experiment can be performed. In metals with an open *d*-shell the lifetime of the states seems to be in some cases sufficiently short to smear out structure in the spectra: the same effect can be produced by momentum broadening due to the short escape length at energies around 100 eV.

The agreement between theoretical DOS curves and the DOS information extracted from photoemission experiments is surprisingly good, better for 3*d* elements than for 5*d* elements, most likely because of the problem of inserting the spin-orbit coupling into the calculations. For dilute and concentrated alloys the models (virtual bound state – coherent potential) are quite well confirmed by the experimental data. In insulating compounds, band-structure calculations seem to overestimate the hybridization and, as such, the bandwidth of the 3*d* bands. An approach that starts out from the free ions and then solely corrects for the Madelung potential seems to describe the experimental data better than the APW band-structure calculations.

Future experiments will certainly be dominated by two developments: the use of single-crystal specimens and the use of synchrotron radiation. This should allow one to directly map the $E(\boldsymbol{k})$ relations; in addition the intensities in the spectra, and especially their variation with energy, will allow for a determination of the eigenfunctions.

Acknowledgement. I want to thank G.K. Wertheim, J.H. Wernick, H. Höchst, A. Goldmann, and P. Steiner, with whom I have had the pleasure to work at various stages, and who have contributed to my understanding of the material presented in this summary. In addition, it is gratefully acknowledged that A. Goldmann has carefully checked the manuscript.

This work was supported in part by the "Deutsche Forschungsgemeinschaft".

References

3.1 R.E. Watson, M.L. Perlman: Structure and Bonding **24**, 83 (1975)
3.2 D.J. Fabian (ed.): *Soft X-ray Band Spectra* (Academic Press, New York 1968)
3.3 D.E. Eastman: In [Ref. 3.4, p. 487]
3.4 D.A. Shirley (ed.): *Electron Spectroscopy* (North-Holland, Amsterdam 1972)
3.5 D.J. Fabian, L.M. Watson (eds.): *Band Structure Spectroscopy of Metals and Alloys* (Academic Press, New York 1973)
3.6 N.V. Smith: Crit. Rev. Solid State Sci. **2**, 45 (1972)
3.7 L.H. Bennett (ed.): Electronic Density of States, N.B.S. Special Publication No. 323 (U.S. GPO, Washington, D.C. 1971)
3.8 D.E. Eastman: In *Techniques of Metals Research VI*, ed. by E. Passaglia (Interscience, New York 1972) 413
3.9 C.N. Berglund, W.E. Spicer: Phys. Rev. **136**, A1030, A1044 (1964)
3.10 P.J. Feibelman, D.E. Eastman: Phys. Rev. B **10**, 4932 (1974)
3.11 N.E. Christensen, B. Feuerbacher: Phys. Rev. B **10**, 2349, 2373 (1974)
3.12 H. Becker, E. Dietz, U. Gerhardt, H. Angermüller: Phys. Rev. B **12**, 2084 (1975)
3.13 P.O. Nilsson, L. Ilver: Solid State Commun. **17**, 667 (1975)
3.14 D.R. Lloyd, C.M. Quinn, N.V. Richardson: J. Phys. C **8**, L371 (1975)
3.15 C. Caroli, D. Lederer-Rozenblatt, B. Roulet, D. Saint James: Phys. Rev. B **8**, 4552 (1973)
3.16 W.L. Schaich, N.W. Ashcroft: Phys. Rev. B **3**, 2451 (1971)
3.17 N.V. Smith, M.M. Traum: Phys. Rev. B **11**, 2087 (1975)
3.18 J. Hubbard: Proc. R. Soc. A **176**, 328 (1963)
3.19 J.A. Wilson: Adv. Phys. **26**, 143 (1972)
3.20 J.B. Goodenough: In *Progress in Solid State Chemistry*, Vol. 5, ed. by H. Reiss (Pergamon, New York 1971) p. 145
3.21 N.F. Mott, Z. Zinamon: Rep. Prog. Phys. **33**, 881 (1970)
3.22 N.F. Mott: Proc. Phys. Soc. London Sect. A **62**, 416 (1949)
3.23 D. Adler: *Solid State Physics*, Vol. 21, ed. by F. Seitz, D. Turnbull, H. Ehrenreich (Academic Press, New York 1968)
3.24 D. Adler, J. Feinleib: Phys. Rev. B **2**, 3112 (1970)
3.25 C.E. Moore: *Atomic Energy Levels*, NBS Circ. No. 467 (U.S. GPO, Washington, D.C. 1949)
3.26 C. Herring: In *Magnetism*, Vol. IV, ed. by G.T. Rado, H. Suhl (Academic Press, New York 1966)
3.27 L.F. Mattheiss: Phys. Rev. B **10**, 995 (1974)
3.28 L.F. Mattheiss: Phys. Rev. **181**, 987 (1969)
3.29 L.F. Mattheiss: Phys. Rev. B **2**, 3918 (1970)
3.30 L.F. Mattheiss: Phys. Rev. B **5**, 290, 306 (1972)

3.31 L.F. Mattheiss: Phys. Rev. B **6**, 4718 (1972)
3.32 A. Rosencwaig, G.K. Wertheim: J. Electron Spectrosc. **1**, 493 (1973)
3.33 K.S. Kim: J. Electron Spectrosc. **3**, 217 (1974)
3.34 K.S. Kim: Chem. Phys. Lett. **26**, 234 (1974)
3.35 S. Hüfner, G.K. Wertheim, J.H. Wernick: Phys. Rev. B **8**, 4511 (1973)
3.36 A. Kotani, Y. Toyozawa: J. Phys. Soc. Jpn. **35**, 1073, 1082 (1973); **37**, 912 (1974)
3.37 G.D. Mahan: Phys. Rev. **163**, 612 (1967)
3.38 P. Nozières, C.T. de Dominicis: Phys. Rev. **178**, 1097 (1969)
3.39 S. Donaich, M. Šunjić: J. Phys. C **3**, 285 (1970)
3.40 G.K. Wertheim: *Electronic States of Inorganic Compounds*, ed. by P. Day (Reidel, Dordrecht 1975) p. 393
3.41 P.A. Cox: Structure and Bonding **24**, 59 (1975)
3.42 S. Sugano, Y. Tanabe, H. Kamimura: *Multiplets of Transition Metal Ions in Crystals* (Academic Press, New York 1970)
3.43 G.K. Wertheim, H.J. Guggenheim, S. Hüfner: Phys. Rev. Lett. **30**, 1050 (1973)
3.44 D.E. Eastman, J.L. Freeouf: Phys. Rev. Lett. **34**, 395 (1975)
3.45 G.K. Wertheim, S. Hüfner: Phys. Rev. Lett. **28**, 1028 (1972)
3.46 A. Goldmann, J. Tejeda, N.J. Shevchik, M. Cardona: Phys. Rev. B **10**, 4388 (1974)
3.47 K.H. Johnson, R.P. Messmer, J.W.D. Connolly: Solid State Commun. **12**, 313 (1973)
R.P. Messmer, C.W. Tucker, Jr., K.H. Johnson: Surf. Sci. **42**, 341 (1974)
3.48 L. Messick, W.C. Walker, R. Glosser: Phys. Rev. B **6**, 3941 (1972)
3.49 R.J. Powell, W.E. Spicer: Phys. Rev. B **2**, 2182 (1970); see also
R.J. Powell: Stanford Electronic Laboratory Tech. Rpt. **52**, (1967) unpublished
3.50 K.S. Kim: Phys. Rev. B **11**, 2177 (1975)
3.51 C. Blaauw, F. Leenhouts, F. van der Woude, G.A. Sawatzky: J. Phys. C **8**, 459 (1975)
3.52 G.K. Wertheim: J. Franklin Inst. **298**, 289 (1974)
3.53 G.K. Wertheim, M. Campagna, H.J. Guggenheim, J.P. Remeika, D.N.E. Buchanan: AIP Conf. Proc. **24**, 235 (1975)
3.54 G.K. Wertheim, L.F. Mattheiss, M. Campagna, T.P. Pearsall: Phys. Rev. Lett. **32**, 997 (1974)
3.55 H. Ihara, Y. Kumashiro, A. Itoh: Phys. Rev. B **12**, 5465 (1975)
3.56 H. Ihara, M. Mirahayashi, H. Nakagawa: Phys. Rev. B **14**, 1707 (1976)
3.57 A.L. Hagström, L.I. Johansson, B.E. Jacobsson, S.B.M. Hagström: Solid State Commun. **19**, 647 (1976)
3.58 N.V. Smith, L.F. Mattheiss: Phys. Rev. B **9**, 1341 (1974)
3.59 N.V. Smith: Phys. Rev. B **9**, 1365 (1974)
3.60 M.M. Traum, N.V. Smith: Phys. Rev. B **9**, 1353 (1974)
3.61 J. Stöhr, F.R. McFeely, G. Apai, P.S. Wehner, D.A. Shirley: Phys. Rev. B **14**, 4431 (1976)
3.62 V.V. Nemoshkalenko, B.G. Aleshin, Yu.N. Kucherenko, L.M. Sheludchenko: Solid State Commun. **15**, 1745 (1974)
3.63 N.J. Shevchik: Phys. Rev. B **13**, 4217 (1976)
3.64 S. Hüfner, G.K. Wertheim, J.H. Wernick: Solid State Commun. **17**, 1585 (1975)
3.65 C.J. Powell: Surf. Sci. **44**, 29 (1974)
3.66 P. Heimann, H. Neddermeyer: J. Phys. F **6**, L257 (1976)
3.67 B. Feuerbacher, R.F. Willis: J. Phys. C **9**, 169 (1976)
3.68 N.V. Smith: Phys. Rev. B **3**, 1862 (1971)
3.69 W.E. Spicer: In [Ref. 3.7, p. 139], and references therein
3.70 G.K. Wertheim, D.N.E. Buchanan, N.V. Smith, M.M. Traum: Phys. Lett. **49**A, 191 (1974)
3.71 L. Lindau, L. Wallden: Solid State Commun. **9**, 209 (1971)
3.72 G.D. Burdick: Phys. Rev. **129**, 138 (1963)
3.73 M.I. Chodorow: Thesis, Mass. Inst. of Technology (1939) unpublished
3.74 L. Hodges, H. Ehrenreich: In *Methods in Computational Physics*, Vol. 8, ed. by B. Alder, S. Fernbach, M. Rotenberg (Academic Press, New York 1968)
3.75 G.K. Wertheim: J. Electron. Spectrosc. **6**, 239 (1975)
3.76 G.D. Mahan: Phys. Rev. B **2**, 4334 (1970)
3.77 R.J. Baird, L.F. Wagner, C.S. Fadley: Phys. Rev. Lett. **37**, 111 (1976)
3.78 F.R. McFeely, J. Stöhr, G. Apai, P.S. Wehner, D.A. Shirley: Phys. Rev. B **14**, 3273 (1976)
3.79 P. Heimann, H. Neddermeyer, H.F. Roloff: Phys. Rev. Lett. **37**, 775 (1976)

3.80 R.S. Williams, P.S. Wehner, J. Stöhr, D.A. Shirley: Surf. Sci. **75**, 215 (1978)
3.81 D. Liebowitz, N.J. Shevchik: Phys. Rev. B**16**, 2395 (1977); **17**, 3825 (1978)
3.82 L.F. Wagner, Z. Hussain, C.S. Fadley: Solid State Commun. **21**, 257 (1977)
3.83 L. Ilver, P.O. Nilsson: Solid State Commun. **18**, 677 (1976)
3.84 E. Dietz, H. Becker, U. Gerhardt: Phys. Rev. Lett. **36**, 1397 (1976)
3.85 S.P. Weeks, J.E. Rowe: Solid State Commun. **27**, 885 (1978)
3.86 A. Goldmann, F. Battye, L. Kasper, S. Hüfner: Unpublished results (1977)
3.87 E.C. Stoner: Proc. R. Soc. London Ser. A**154**, 656 (1936); A**165**, 372 (1938)
3.88 E.P. Wohlfarth: Rev. Mod. Phys. **25**, 211 (1953)
3.89 H. Höchst, A. Goldmann, S. Hüfner: Z. Phys. **24**B, 245 (1976)
3.90 L. Ley, O.B. Deboussi, S.P. Kowalczyk, F.R.McFeely, D.A. Shirley: Phys. Rev. B**16**, 5372 (1977)
3.91 S. Hüfner, G.K. Wertheim: Phys. Lett. **47**A, 349 (1974)
3.91 H. Höchst, S. Hüfner, A. Goldmann: Phys. Lett. **57**A, 265 (1976)
3.92 R.A. Tawil, J. Callaway: Phys. Rev. B**7**, 4242 (1973)
3.93 M. Singh, C.S. Wang, J. Callaway: Phys. Rev. B**11**, 287 (1975)
3.94 J. Cornwell, D.M. Hum, K.C. Wong: Phys. Lett. A**26**, 365 (1968)
3.95 S. Wakoh, J. Yamashita: J. Phys. Soc. Jpn. **21**, 1712 (1966)
3.96 K.J. Duff, T.P. Das: Phys. Rev. B**3**, 192 (1971)
3.97 M. Pessa, P. Heimann, H. Neddermeyer: Phys. Rev. B**14**, 3488 (1976)
3.98 A.J. McAlister, J.R. Cuthill, R.C. Dobbyn, M.L. Williams, R.E. Watson: Phys. Rev. B**12**, 2973 (1975)
3.99 S. Hüfner, G.K. Wertheim: Phys. Lett. **51**A, 299 (1975)
3.100 E.P. Wohlfarth: J. Appl. Phys. **41**, 1205 (1970)
3.101 T.T.A. Nguyen, R.C. Cinti, S.S. Choi: J. Phys. (Paris) Lett. **37**, L-111 (1976)
3.102 P.M. Williams, P. Butcher, J. Wood, K. Jacobi: Phys. Rev. B**14**, 3215 (1976)
3.103 R.J. Smith, J. Anderson, J. Heimann, G.J. Lapeyre: Solid State Commun. **21**, 459 (1977)
3.104 H. Höchst, S. Hüfner, A. Goldmann: Z. Phys. B**26**, 133 (1977)
3.105 C.S. Wang, J. Callaway: Phys. Rev. B**9**, 4897 (1974)
3.106 N. Müller: Phys. Lett. **54**A, 415 (1975)
3.107 G. Busch, M. Campagna, H.C. Siegmann: Phys. Rev. B**4**, 746 (1971)
3.108 P.M. Tedrow, R. Meservey: Phys. Rev. Lett. **26**, 192 (1971)
3.109 D.T. Pierce, W.E. Spicer: Phys. Rev. B**6**, 1787 (1972)
3.110 J.E. Rowe, J.C. Tracy: Phys. Rev. Lett. **27**, 799 (1971)
3.111 F.M. Mueller, A.J. Freeman, J.O. Dimmock, A.M. Furdyna: Phys. Rev. B**1**, 4617 (1970)
3.112 F.M. Mueller, J.W. Garland, M.H. Cohen, K.H. Bennemann: Ann. Phys. N.Y. **67**, 19 (1971)
3.113 N.F. Mott, J. Jones: *Theory of the Properties of Metals and Alloys* (Dover, New York 1958)
3.114 J. Friedel: Nuovo Cimento (Suppl.) **7**, 287 (1958)
3.115 P.W. Anderson: Phys. Rev. **124**, 41 (1961)
3.116 P. Soven: Phys. Rev. **156**, 809 (1967); **178**, 1136 (1969)
3.117 Y. Yafet: Phys. Lett. **26**A, 481 (1968)
3.118 C. Norris, H.P. Myers: J. Phys. F**1**, 62 (1971)
3.119 N.J. Shevchik, A. Goldmann: J. Electron Spectrosc. **5**, 631 (1974)
3.120 N.J. Shevchik: J. Phys. F**5**, 1860 (1975)
3.121 A.B. Callender: Dissertation, Princeton University (1970), quoted in [Ref. 3.136]
3.122 H.P. Myers, L. Walldén, A. Karlsson: Philos. Mag. **18**, 725 (1968)
3.123 L.E. Walldén: Solid State Commun. **7**, 593 (1969)
3.124 L.E. Walldén: Philos. Mag. **21**, 571 (1970)
3.125 L.L. Hirst: Phys. Kondens. Mater. **11**, 255 (1970)
3.126 B. Velicky, S. Kirkpatrick, H. Ehrenreich: Phys. Rev. **175**, 747 (1968)
3.127 P.O. Nilsson: Phys. Kondens. Mater. **11**, 1 (1970)
3.128 S. Kirkpatrick, B. Velicky, H. Ehrenreich: Phys. Rev. B**1**, 3250 (1970)
3.129 J. Yamashita, S. Wakoh, S. Asano: J. Phys. Soc. Jpn. **31**, 1620 (1971)
3.130 G.M. Stocks, R.W. Williams, J.S. Faulkner: J. Phys. F**3**, 1688 (1973)
3.131 D.H. Seib, W.E. Spicer: Phys. Rev. B**2**, 1676 (1970)
3.132 G.M. Stocks, R.W. Williams, J.S. Faulkner: Phys. Rev. B**4**, 4390 (1971)
3.133 J.C. Fuggle, L.M. Watson, D.J. Fabian, P.R. Norris: Solid State Commun. **13**, 507 (1973)

3.134 S. Hüfner, G.K. Wertheim, J.H. Wernick: Solid State Commun. **17**, 419 (1975)
3.135 F. Abelès (ed.): *Colloquium on the Optical Properties and Electonic Structure of Metals and Alloys*, Paris (1965) (North-Holland, Amsterdam 1966)
3.136 F. Abelès (ed.): *Optical Properties of Solids* (North-Holland, Amsterdam 1972) p. 93
3.137 F.M. Mueller: Phys. Rev. **153**, 659 (1967)
3.138 S. Hüfner, G.K. Wertheim: Phys. Rev. B**7**, 5086 (1973)
3.139 P. Heimann, H. Neddermeyer: Solid State Commun. **26**, 279 (1978)
3.140 P.S. Bagus, J.L. Freeouf, D.E. Eastman: Phys. Rev. B**15**, 3661 (1977)
3.141 J.S. Griffith: *Theory of Transition Metal Ions* (Cambridge University, Cambridge, England 1964)
3.142 A.B. Kunz, G.T. Surrat: Solid State Commun. **25**, 9 (1978)
3.143 T. Ishii, S. Kono, S. Suzuki, I. Nagakura, T. Sagawa, R. Kato, M. Watanake, S. Sato: Phys. Rev. B**12**, 4320 (1975)
3.144 S.P. Kowalczyk, L. Ley, F.R. McFeely, D.A. Shirley: Phys. Rev. B**15**, 4997 (1977)
3.145 R.T. Poole, J.D. Riley, J.G. Jenkin, J. Lieegang, R.C.G. Leckey: Phys. Rev. B**13**, 2620 (1976)
3.146 H. Onuki, F. Sugawara, M. Hirano, Y. Yamaguchi: J. Phys. Soc. Jpn. **41**, 1807 (1976)
3.147 G. Thornton, A.F. Orchard, C.N.R. Rao: Phys. Lett. **54**A, 235 (1975)
3.148 A. Barrie, N.E. Christensen: Phys. Rev. B**14**, 2442 (1976)
3.149 J. Stöhr, G. Apai, P.S. Wehner, F.R. McFeely, R.S. Williams, D.A. Shirley: Phys. Rev. B**14**, 5144 (1976)
3.150 H.F. Roloff, H. Neddermeyer: Solid State Commun. **21**, 561 (1977)
3.151 P.S. Wehner, J. Stöhr, G. Apai, F.R. McFeely, D.A. Shirley: Phys. Rev. Lett. **38**, 169 (1977)
3.152 L.F. Wagner, Z. Hussain, C.S. Fadley, R.J. Baird: Solid State Commun. **21**, 453 (1977)
3.153 L.G. Petersson, R. Melander, D.P. Spears, S.B.M. Hagström: Phys. Rev. B**14**, 4177 (1976)
3.154 P. Heimann, H. Neddermeyer, H.F. Roloff: J. Phys. C**10**, L17 (1977)
3.155 B. Feuerbacher, N.E. Christensen: Phys. Rev. B**10**, 2373 (1974)
3.156 M.D. Miller: Ph. D. Thesis, Princeton University (1974); Xerox University Microfilms, Ann. Arbor, Michigan
3.157 P. Steiner, H. Höchst, S. Hüfner: J. Phys. F**7**, L105 (1977)
3.158 K.Y. Yu, C.R. Helms, W.E. Spicer, P.W. Chye: Phys. Rev. B**15**, 1629 (1977)
3.159 A.D. McLachlan, J.G. Jenkin, R.C.G. Leckey, J. Liesegang: J. Phys. F**5**, 2415 (1975)
3.160 S.R. Nagel, G.B. Fisher, J. Tauc, B.G. Bagley: Phys. Rev. B**13**, 3284 (1976)
3.161 J. Riley, J. Azoulay, L. Ley: Solid State Commun. **19**, 993 (1976)
3.162 P.O. Nilsson, I. Curelaru, T. Jarlborg: Phys. Status Solidi B**79**, 277 (1977)
3.163 H. Höchst, S. Hüfner, A. Goldmann: Solid State Commun. **19**, 899 (1976)
3.164 M. Gupta, A.J. Freeman, D.E. Ellis: Phys. Rev. B**16**, 3338 (1977)
3.165 G. Apai, J. Stöhr, R.S. Williams, P.S. Wehner, S.P. Kowalczyk, D.A. Shirley: Phys. Rev. B**15**, 584 (1977)
3.166 W. Eib, S.F. Alvarado: Phys. Rev. Lett. **37**, 444 (1976)
3.167 E.P. Wohlfarth: Phys. Rev. Lett. **38**, 524 (1977)
3.168 M. Landolt, M. Campagna: Phys. Rev. Lett. **38**, 663 (1977)
3.169 S.P. Weeks: Phys. Rev. B**17**, 1738 (1978)
3.170 L.G. Peterson, R. Erlandsson: Phys. Rev. B**17**, 3006 (1978)
3.171 E. Dietz, U. Gerhardt, C.J. Maetz: Phys. Rev. Lett. **40**, 892 (1978)
3.172 D.E. Eastman, F.J. Himpsel, J.A. Knapp: Phys. Rev. Lett. **40**, 1514 (1978)
3.173 F.J. Himpsel, D.E. Eastman: Phys. Rev. Lett. **41**, 507 (1978)
3.174 M. Sagurton, D. Liebowitz, N.J. Shevchick: Solid State Commun. **25**, 955 (1978)
3.175 J.B. Pendry, J.F.L. Hopkinson: J. Phys. F**8**, 1009 (1978)
3.176 C.S. Wang, J. Callaway: Phys. Rev. B**15**, 298 (1977)
3.177 T. Oguchi: Prog. Theor. Phys. (Kyoto) **13**, 148 (1955)
3.178 J.L. Erskine, E.A. Stern: Phys. Rev. Lett. **30**, 1329 (1973)
3.179 S. Doniach: AIP Conf. Proc. **5**, 549 (1972)
3.180 D.E. Eastman, J.A. Knapp, F.J. Himpsel: Phys. Rev. Lett. **41**, 825 (1978)
3.181 R.J. Baird, C.S. Fadley, L.F. Wagner: Phys. Rev. B**15**, 666 (1977)
3.182 R.S. Williams, P.S. Wehner, J. Stöhr, D.A. Shirley: Phys. Rev. Lett. **39**, 302 (1977)
3.183 R. Merlin, T.P. Martin, A. Polian, M. Cardona, B. Andlauer, D. Tannhauser: J. Magnetism and Mag. Materials **9**, 83 (1978)

4. Unfilled Inner Shells: Rare Earths and Their Compounds

M. Campagna, G. K. Wertheim, and Y. Baer

With 35 Figures

In this chapter we review the present status of our understanding of the photoemission process from rare-earth (RE) metals, their compounds and alloys, and discuss the impact of this technique on various aspects of the electronic structure of RE solids. As is the case in other areas of electron spectroscopy (e.g., Auger, Leed, Electron Energy Loss) and, especially in those areas related to surfaces, the volume of experimental information is not matched by our theoretical understanding. A few of these aspects will be discussed in the following paragraphs. This problem results from the inherent many-electron nature of the $4f$ states and from our inability to treat correlated many-electron shells within the available schemes for calculating the electronic structure of solids. This aspect of the $4f$ electrons, together with the fact that they often have binding energies smaller than uncorrelated valence electrons, is responsible for the variety of electronic properties displayed by RE materials. The desire to elucidate the peculiar behavior of some RE solids has presented a challenging task to the electron-spectroscopists. The steady improvements of the experimental techniques in this field, especially in photoemission spectroscopy will, in turn, provide challenging information to test theories, not necessarily specific to RE solids.

4.1 Background

4.1.1 Where Are the 4*f* Levels Located?

Although optical spectroscopy provided us with a quite detailed understanding of the absorption and emission spectra of the rare earths during the 1960's [4.1,2], it was toward the end of that period that questions like the one above were asked and answered. During that period the Eu chalcogenides and some other RE sulfides (especially GdS) received greater attention than even the metals themselves. The first photoemission studies, which showed unambiguously that the $4f$ levels could indeed have lower photoionization energy than the valence electrons, were performed on such materials. They involved energy distribution curves (EDC) [4.3] and spin polarization [4.4] measurements. Although the high spin polarization ($\sim 30\%$) near photothreshold in EuS films was a clear indication of the low $4f$-binding energy, E_B^{4f}, in the Eu chalcogenides, the EDC measurement using photon energies up to 10.2 eV on GdS gave ambiguous results regarding the position of the $4f$ levels in this metallic sulfide. However, in that work it was nevertheless established that in

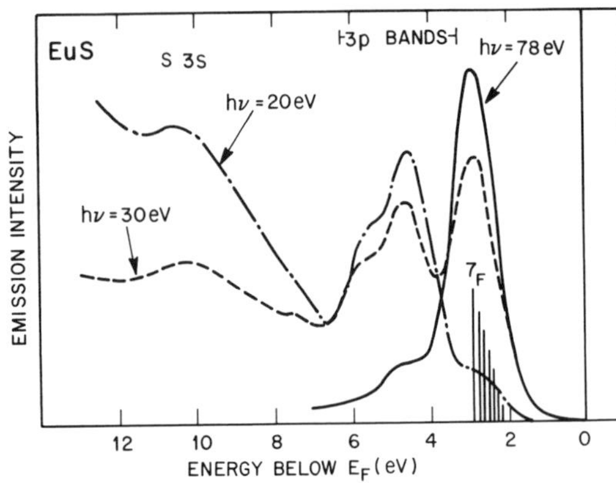

Fig. 4.1. Ultraviolet Photoemission Spectra (UPS) of EuS measured with synchrotron radiation, showing mainly *s*-, *p*-, and *d*-band emission at lower photon energies $h\nu$. 4f multiplet emission dominates at higher energies, e.g., $h\nu = 78$ eV. [4.5]

most cases 4f photoemission would be best studied using high-energy photons e.g., soft X-rays, in part because E_B^{4f} just exceeded the LiF cutoff, but especially because the photoionization cross sections were small. We provide an a posteriori demonstration of this fact in Figs. 4.1, 2. The emission intensity from the $4f^7$ levels relative to the S 3p band in EuS (Fig. 4.1) increases dramatically from 20 to 78 eV photon energy. These data were taken using the synchrotron radiation of the Cambridge Electron Accelerator [4.5], with an energy resolution of about 0.7 at 78 eV. The EuO data, which we shall discuss in greater detail in Sect. 4.3.2, were obtained using AlK_α radiation, $h\nu = 1486$ eV, and an energy resolution of about 0.55 FWHM. It can be seen from Fig. 4.2 that at X-ray energies the 4f emission intensity is perhaps 50 times greater than that of the O(2p) valence band.

4.1.2 Multiplet Intensities Versus Total Photoelectric Cross Sections at 1.5 keV

Having convinced ourselves that 4f states are indeed best studied using photon energies greater than 80 eV, we now address the question of the detailed understanding of the 4f photoemission spectrum.

The following questions are pertinent:

I) How large is the *total* emission intensity (photoionization cross section) of the $4f^n$ configuration relative to that of other outer levels (usually valence or conduction band states)? The answer to this question, was computed by *Scofield* [4.6] and is shown graphically for $h\nu = 1.5$ keV in Fig. 4.3. This calculation indicates that for most RE's the $4f^n$ configuration in alloys or compounds should be readily distinguished from emission occurring at the same energy due to other electron states [4.7] by virtue of the larger cross section[1]. The XPS data which we present in this chapter largely confirm the

1 Tables like those of *Scofield* [4.6] are not available for low photon energies. A simple semiempirical diagram based on calculations and experimental data was given by *Eastman* and *Kutznietz* [4.7]. The applications of photon-energy dependent photoemission have been discussed by D.E. Eastman, in Proc. VUV Meeting, Hamburg (1974), Ed. E.E. Koch, C. Kunz, R. Hänsel, Pergamon-Vieweg (1974). The dependence of *s*-, *p*-, *d*-, and *f*-photo cross section on photo energy allows one to separate out the various *l* component of degenerate states. See Sect. 2.4.4 and Fig. 2.39.

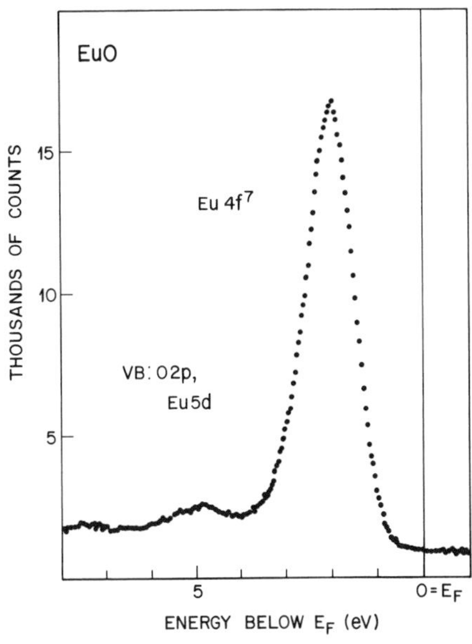

Fig. 4.2. X-ray photoemission spectrum (XPS) of valence band and $4f$ multiplets of EuO, confirming clearly that $4f$ electrons in RE solids are advantageously studied using soft X-ray photons of about 1.5 keV

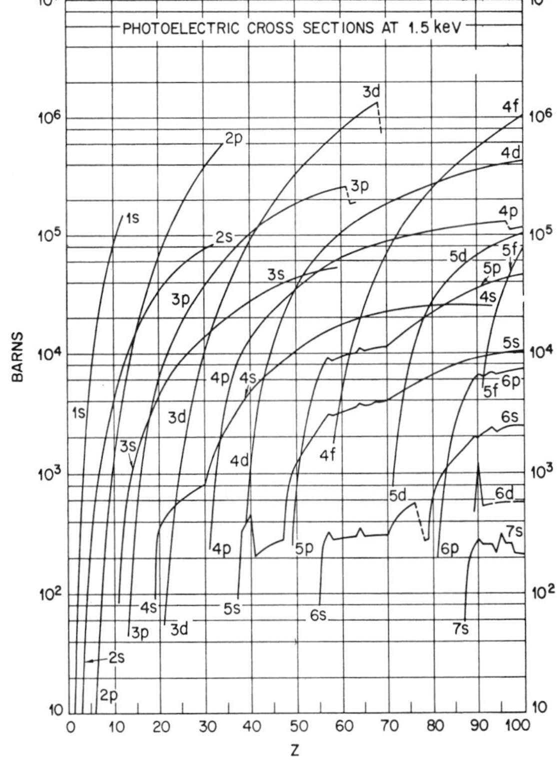

Fig. 4.3. Calculated photoelectric cross sections of atomic levels at 1.5 keV as a function of atomic number (plotted from data of [4.6]). See also [Ref. 2.2, Table 1.10]

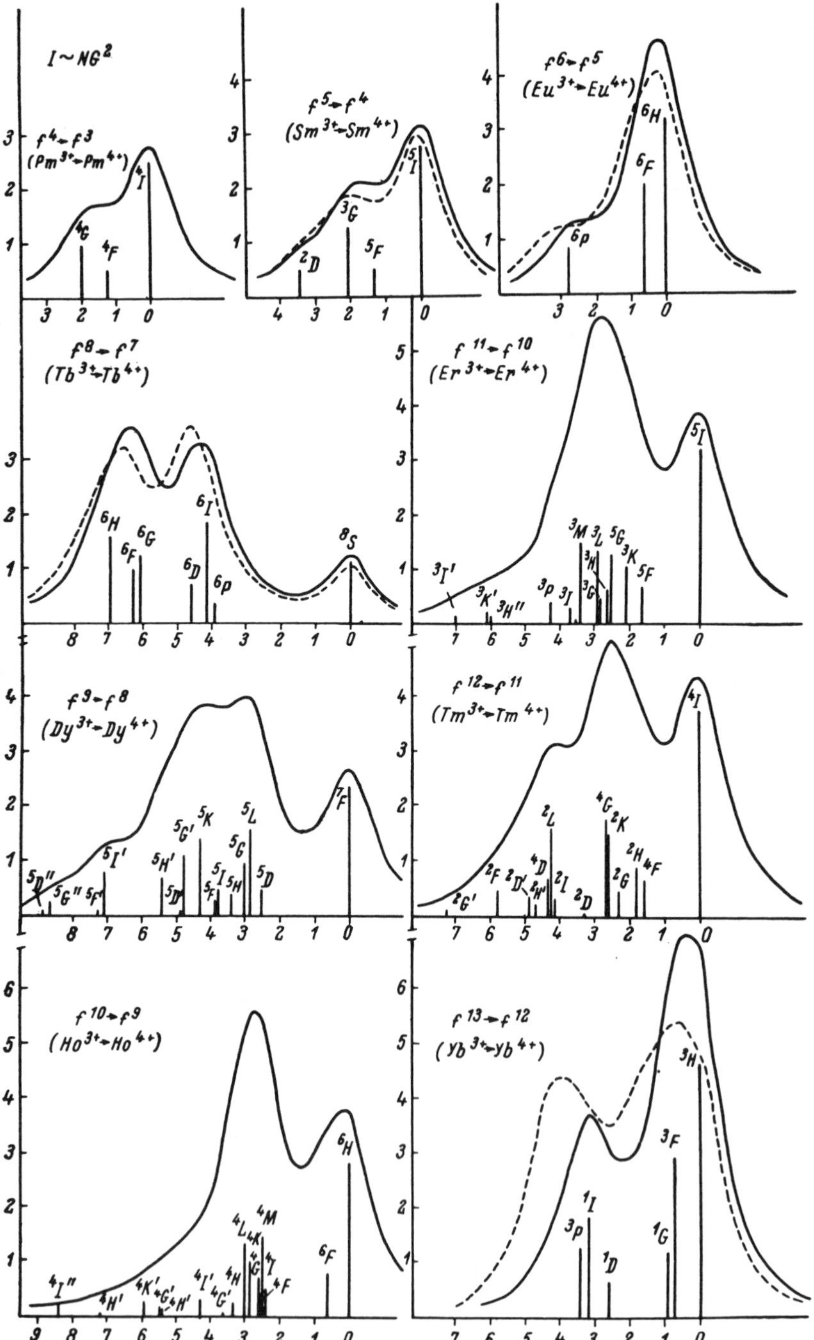

Fig. 4.4. Calculated spectra of $4f$ electrons (bars) with a broadening of 1.6 eV (continuous line); the dashed curves incorporate the spin-orbit interaction [4.9]

results of these calculations. It should be noted, regarding the emission intensity from extended states, that XPS does not reproduce the "total density of states" but rather a "cross section-weighted" density of states. The weighting factor is related to the matrix element corresponding to the orbital quantum numbers of the states making up the extended electron state. If the $4f$ signal can be unambiguously separated in a given spectrum, the next questions are:

II) What is the distribution in energy of the various final states, and how large are the corresponding intensities? Here we take advantage of the well-established fact that $4f$ electrons in solids are, under normal conditions, strongly atomic-like and do not participate appreciably in the bonding. Their photoemission spectrum will, therefore, coincide with the spectrum of the hole-state left behind by the photoemitted electron. Extensive many-electron calculations of the hole-state spectra for all the $4f^n \rightarrow 4f^{n-1} + e^-$ transitions have been done independently by *Cox* et al. [4.8] and *Zabolotskii* et al. [4.9]. In these calculations the outgoing photoelectron is approximated by a plane wave and the amplitude of the wave function of a given resulting state of the $n-1$ remaining electrons is evaluated. This amplitude is given by the coefficient of fractional parentage [4.10], which was calculated using the method developed by *Racah* for treating excitations of n-electron systems. Spherical symmetry and Russell-Saunders coupling are also assumptions underlying these calculations. The squared coefficients have been normalized to the number, n, of electrons in a given initial configuration $4f^n$. Since the available energy resolution at 1.5 keV is of the order of 0.25 to 0.5 eV, a summation over the J-levels belonging to the same term can often be made to facilitate comparison with experiment. For the sake of completeness we present in Table 4.1 and Fig. 4.4 the results of the calculations of [4.8, 9], respectively.

The energies of the final-state configurations can be taken from optical studies of trivalent RE ions in insulating solids or aqueous solutions [4.1, 2, 11]. For trivalent initial states we must take the optical spectra of the preceding rare-earth ion. The energies of the corresponding configurations have to be expanded from 7 to 13% above those of the optical spectra, because the $4f^{n-1}$ final state is on an ion with nuclear charge $Z+1$ compared to Z in the isoelectronic optical case. For divalent initial-state configurations such a correction is obviously not necessary since one takes the trivalent spectrum of the same ion.

4.1.3 Renormalized Atom Scheme and Thermodynamics

Having obtained an answer to the questions of relative multiplet energies and intensities, there remains the question regarding the absolute position of the $4f$ levels in the photoemission spectrum or, more conveniently, their position relative to the Fermi level in metallic systems. This problem has been attacked theoretically in two different ways. *Herbst* et al. [4.12] performed a detailed systematic investigation of the electronic properties of the RE metals within the

Table 4.1. States arising in ionisation from f^n free-ion configurations with first-order spin-orbit coupling. (Only multiplet components with normalized intensity $\geqq 0.1$ are listed)

Initial state		Final (LS) state and intensity		Multiplet components (J) and intensities	
f^2	$^3H_4$	^2F	2.000	$^5/_2$	1.714
				$^7/_2$	0.286
f^3	$^4I_{9/2}$	^3H	2.333	4	1.890
				5	0.424
		^3F	0.667	2	0.563
f^4	$^5I_4$	^4I	2.545	$^9/_2$	1.903
				$^{11}/_2$	0.599
		^4G	0.955	$^5/_2$	0.658
				$^7/_2$	0.263
		^4F	0.500	$^3/_2$	0.371
				$^5/_2$	0.114
f^5	$^6H_{5/2}$	^5I	2.758	4	1.755
				5	0.919
		^5G	1.266	2	0.513
				3	0.575
				4	0.165
		^5F	0.500	1	0.168
				2	0.234
		^5D	0.476	0	0.149
				1	0.224
f^6	$^7F_0$	^6H	3.143	$^5/_2$	0.898
				$^7/_2$	2.245
		^6F	2.000	$^5/_2$	1.428
				$^7/_2$	0.571
		^6P	0.857	$^5/_2$	0.816
f^9	$^6H_{15/2}$	^5H	1.000	6	0.233
				7	0.739
		^5G	0.817	5	0.195
				6	0.605
		^5F	0.310	5	0.232
		^5D	0.454	4	0.454
f^{10}	$^5I_8$	^6H	2.800	$^{11}/_2$	0.138
				$^{13}/_2$	0.646
				$^{15}/_2$	2.004
		^6F	0.800	$^{11}/_2$	0.727
		^4M	1.462	$^{21}/_2$	1.375
		^4L	1.307	$^{17}/_2$	0.134
				$^{19}/_2$	1.167
		^4K	1.189	$^{15}/_2$	0.161
				$^{17}/_2$	1.018
		^4I	1.000	$^{13}/_2$	0.154
				$^{15}/_2$	0.837
		^4H	0.379	$^{13}/_2$	0.318
		^4G	0.692	$^{11}/_2$	0.611
		^4F	0.405	$^9/_2$	0.405
f^{11}	$^4I_{15/2}$	^5I	3.182	6	0.169
				7	0.736
				8	2.263
		^5G	1.193	5	0.241
				6	0.913
		^5F	0.625	5	0.550
		^3M	1.462	10	1.400
		^3L	1.307	9	1.206

f^7	${}^8S_{7/2}$	${}^7F$	7.000	0	0.143
				1	0.429
				2	0.714
				3	1.000
				4	1.286
				5	1.571
				6	1.857
f^8	${}^7F_6$	${}^8S$	1.143	$7/2$	1.143
		${}^6I$	1.857	$15/2$	0.303
				$17/2$	1.500
		${}^6H$	1.571	$11/2$	0.101
				$13/2$	0.424
				$15/2$	1.030
		${}^6G$	1.286	$9/2$	0.134
				$11/2$	0.434
				$13/2$	0.694
		${}^6F$	1.000	$7/2$	0.138
				$9/2$	0.383
				$11/2$	0.460
		${}^6D$	0.714	$5/2$	0.107
				$7/2$	0.307
				$9/2$	0.301
		${}^6P$	0.429	$5/2$	0.230
				$7/2$	0.199
f^9	${}^6H_{15/2}$	${}^7F$	2.333	5	0.583
				6	1.674
		${}^5L$	1.545	9	0.136
				10	1.400
		${}^5K$	1.364	8	0.211
				9	1.131
		${}^5I$	1.182	7	0.240
				8	0.913
		${}^3K$	1.154	7	0.117
				8	1.033
		${}^3I$	0.364	7	0.320
		${}^3H$	0.846	6	0.747
		${}^3G$	0.454	5	0.415
		${}^3F$	0.382	4	0.382
f^{12}	${}^3H_6$	${}^4I$	3.677	$11/2$	0.212
				$13/2$	0.914
				$15/2$	2.536
		${}^4G$	1.688	$7/2$	0.106
				$9/2$	0.419
				$11/2$	1.157
		${}^4F$	0.667	$7/2$	0.151
				$9/2$	0.486
		${}^2L$	1.545	$17/2$	1.500
		${}^2K$	1.364	$15/2$	1.288
		${}^2I$	0.263	$13/2$	0.242
		${}^2H$	1.000	$11/2$	0.909
		${}^2G$	0.396	$9/2$	0.359
		${}^2F$	0.470	$7/2$	0.436
		${}^2D$	0.296	$5/2$	0.926
f^{13}	${}^2F_{7/2}$	${}^3H$	4.714	4	0.321
				5	1.375
				6	3.018
		${}^3F$	3.000	2	0.357
				3	0.875
				4	1.768
		${}^3P$	1.286	0	0.107
				1	0.375
				2	0.804
		${}^1I$	1.857		
		${}^1G$	1.286		
		${}^1D$	0.714		
		${}^1S$	0.143		

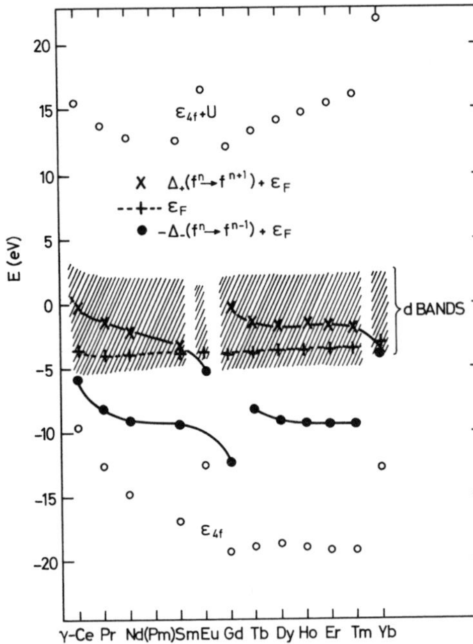

Fig. 4.5a. One-electron ε_{4f} and multielectron Δ_- predictions of promotion energy of $4f$ electrons plotted with respect to the d-bands and Fermi levels of RE metals. Multielectron prediction of the correlation energy: $U_{\text{eff}}=(\Delta_-)+(\Delta_+)\sim 7\,\text{eV}$; one-electron prediction: $U=F^\circ(4f,4f)$ Slater integral $\sim 27\,\text{eV}$. [4.12]

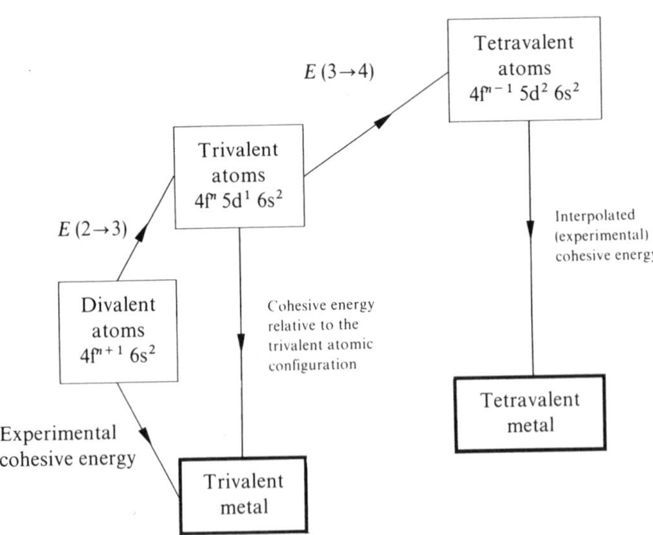

Fig. 4.5b. Estimate of $4f$ promotion energy based on simple thermodynamical arguments. $E(2\to3)$ and $E(3\to4)$ denote the atomic excitation energies between the configurations indicated [4.16]

framework of the renormalized atom scheme of *Watson* et al. [4.13]. More recently these calculations have been extended to include relativistic effects [4.14]. One of the major results of these very interesting calculations is presented in Fig. 4.5a. The simplest computed quantity, which is directly comparable to the experimental binding energy, is the position of the $4f$ states relative to the Fermi level. The single-particle prediction, ε_{4f} neglecting screening and relaxation in the photoemission final state is in glaring disagreement with the experimental data. It is now recognized that the position of the photoemission peaks of metals corresponds to adiabatic final states. To account for this situation, *Herbst* et al. [4.12, 14] have repeated their calculations for the configurations $4f^{n-1}5d^{m+1}6s^1$ and $4f^{n+1}5d^{m-1}6s^1$ (the metallic configuration is $4f^n5d^m6s^1$). The neutrality is conserved within the Wigner-Seitz cell by a corresponding change of the d-population at the Fermi level. The multielectron prediction for the $4f$ binding energy relative to E_F is now the energy difference, Δ_-, between the renormalized atom with the configuration of the initial state and with the configuration of the final state. Put another way, Δ_- is the energy required to promote a $4f$ electron into a d-state at E_F. Similarly, Δ_+ is the energy necessary to promote a $5d$ electron at E_F into a $4f$ state. It is shown below (see Sect. 4.3.1 and Fig. 4.13) that XPS binding energies [4.15] for Hund's rule ground states are in excellent agreement with the theoretical estimate, Δ_-. The agreement demonstrates the power of the renormalized atom approximation as applied to rare-earth metals. The fact that Eu and Yb are divalent is unambiguously predicted. Another interesting result is that in Sm Δ_+ is only ~ 0.5 eV. This value, which represents the energy required to form a divalent site is smaller than the probable error of the calculation. This provides an initial indication that Sm metal could contain a small divalent component as discussed in Sect. 4.3.1.

Johansson [4.16] has found a very simple way of exploiting the atomic nature of the $4f$ electrons in metals to estimate their promotion energy. He takes advantage of the complete relaxation of the final states by expressing the binding energy of $4f$ electrons relative to E_F as the energy difference between a trivalent and a tetravalent site in the metal. The cohesive energy of an atom in a trivalent metal, $E_c(2\to3)$, is known from experiments, but the free atom has a divalent configuration. By adding the energy $E_a(2\to3)$ required to excite the trivalent atomic configuration, one can calculate $E_c^3 = E_c(2\to3) + E_a(2\to3)$, the cohesive energy without change of configuration, which is not directly available. One can proceed now to a tetravalent atom and add the corresponding free atom excitation energy $E_a(3\to4)$. Finally, to come back to a tetravalent metal the corresponding cohesive energy E_c^4 must be subtracted. This cycle, which we reproduce in Fig. 4.5b, makes it possible to write the $4f$ promotion energy

$$\Delta_- = E_c^3 + E_a(3\to4) - E_c^4 = E_a(3\to4) - (E_c^4 - E_c^3)\,.$$

Johansson observed then that for neighboring elements of the periodic table, the binding energy difference between trivalent and tetravalent metals is systematically of the order of 2 eV/atom. Using this value for $(E_c^4 - E_c^3)$ and collecting atomic spectroscopic data for $E(3 \rightarrow 4)$, the calculation of the $4f$ promotion energy, based exclusively on experimental data, can be performed very easily. These results have also been plotted in Fig. 4.13. In view of the simplicity of this approach the agreement with the experimental values, as we shall see in Sect. 4.3.1, is surprisingly good.

Before closing this section we note that the experimental screening and relaxation energies can be estimated by measuring the $4f$ level position in an RE atom in the gas form and in the solid form. We shall briefly touch on this point in Sect. 4.3.2.

4.1.4 Multiplet and Satellite Structure in Photoemission from Core Levels Other than 4*f*

Final-state multiplet structure of the kind discussed above is not confined to the $4f$ shell but has also been identified [4.17, 18] in some d-group transition-metal ions where the final states are crystal-field states. This makes for much greater variety in the XPS data for a given ion. The concept of multiplet structure, furthermore, arises also in the photoionization of inner complete shells of ions with outer open shells [4.19]. Under those circumstances the final states of the ion are obtained by multiplet coupling of the two incomplete shells and allowing all electronic transitions which conserve the symmetry of the Koopmans' states. Historically the first multiplet effects to be elucidated in solids were those of core s-levels in S-state transition-metal ions, e.g., Mn^{2+} and Fe^{3+} [4.20, 21]. The multiplet splittings are largest when all of the interacting states are within a shell with a given principal quantum number, because the overlap integrals are then strongest. However, this condition also leads to the most complex spectra, because all of the different configurations with the same symmetry, which can be obtained by rearranging the available number of electrons within the shell, must be considered. When the core hole is in a different shell, much of this complexity vanishes, because the energy for excitation between shells is large, and the exchange coupling is much weaker [4.22]. Nevertheless, the full complexity of a core hole multiplet spectrum often remains unresolved in experimental data.

The spectra of inner core levels, e.g., $2p$ in $3d$ group transition metals or $3d$ in the rare earths, are further complicated by so-called shake-up satellites [4.23, 24]. In essence, the relaxation energy of the outer shells following core ionization is sufficient to cause electronic transitions in the solid. A single-ion point of view has generally been found to be inadequate. Instead, one must consider transitions in the band structure of the solid with the core-ionized atom. One likely consequence is a change-transfer excitation to the ionized atom, especially to a metal atom from the anion valence band. Suffice it to say that the possibility of simultaneous complications from multiplet coupling and outer shell shake-up have made the interpretation of inner-shell spectra of rare-earth and transition-metal ions very difficult.

4.2 Techniques

Cardona and *Ley* have reviewed the general techniques used in photoelectron spectroscopy in [Ref. 1.1, Chap. 1]. Here we concentrate on some further technical aspects characteristic of RE research.

4.2.1 The Need of High Resolution in Rare-Earth Studies

From the discussion in the preceding section, we expect the XPS spectra of $4f$ and deeper core levels in RE solids to be particularly complex, so that the highest available resolution is required in order to separate all the components which may fall into a narrow energy range. The intensities of adjacent components can vary greatly (e.g., $4f$ and valence states), in which case not only the instrumental line*width* but also the line *shape* becomes of major importance. The tail of a very intense signal can hinder the observation of weak structures which are in fact well separated from the dominant signal. The total instrumental line shape, including X-ray source and electron spectrometer contributions, should have Gaussian rather than Lorentzian character. Most of the RE studies available today have been performed with commercial instruments equipped with multidetector system and X-ray monochromator and using the principle of dispersion compensation [4.25]. These instruments combine a reasonably good resolution (FWHM of 0.55 to 0.66 eV) with a high sensitivity. The sample surface must be flat and precisely positioned within the instrument. These conditions may be difficult to satisfy, depending on the procedure for surface preparation. Simpler instruments with monochromatized X-ray sources have also been developed [4.25–27]. They yield a better linewidth but this improvement is achieved at the cost of a major reduction of the sensitivity. This is not a serious drawback in the study of the intense levels of the rare earths. Data accumulation time can be increased if surfaces can be kept contamination-free for long time and greater X-ray intensities can be obtained with rotating anodes. An instrument of this kind [4.28] has been used for studying the metallic rare earths. The total instrumental width of the symmetrical line is slightly less than 0.3 eV. Typical background counting rates of 3 counts/min at 0.5 eV above E_F makes it easy to detect very weak signals in this energy range. A comparison of spectra of the outermost levels of metallic dysprosium recorded by 3 different instruments is shown in Fig. 4.6. Spectrum (*A*) was obtained with a standard instrument using the natural Mg K_α line. The intensity observed around E_F is mainly due to excitations of $4f$ lines by the $K_{\alpha 3,4}$ satellites. Spectrum (*B*) was recorded with a commercial instrument equipped with an X-ray monochromator [4.29]. An improvement of the resolution is observed and the valence band can be discerned since the $K_{\alpha 3,4}$ satellites are absent. The spectrum (*C*), measured with the instrument of highest resolution [4.28], shows much better resolved structure and the valence band emerges clearly from the tail of the intense $4f$ peaks. It is not obvious that

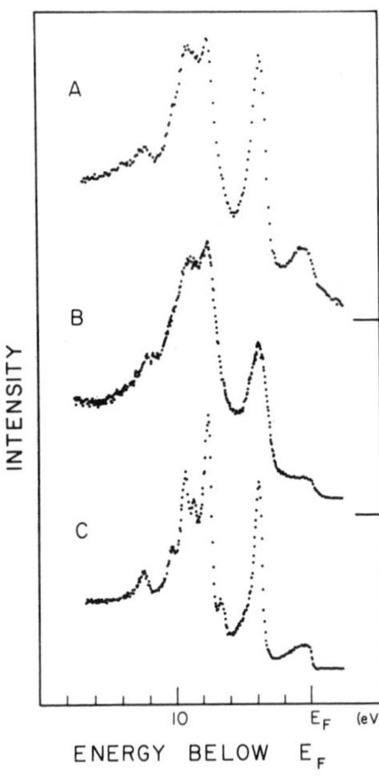

Fig. 4.6. Comparison of the spectra of the outermost levels of metallic dysprosium recorded by different instruments: *A* Commercial instrument without monochromator (Mg K_α, resolution ~1.2 eV). *B* Commercial instrument with monochromatized Al K_α source ([4.29], resolution ~0.6 eV). *C* Modified commercial instrument with monochromatized Al K_α source ([4.28], resolution ~0.3 eV)

further improvement in resolution would be rewarding in every case. The lifetime broadening of the 4*f* final states has been found in certain cases to exceed the available resolution of commercial instruments, as we shall see below. In fact, these instruments have been quite successful in view of the amount of fundamental information on rare earths already obtained.

4.2.2 Sample Preparation

a) Pure Metals

The easiest way to prepare a clean surface of a metallic RE is by in situ evaporation of the pure element from a tungsten filament basket. After a few preliminary evaporations, the material is well outgassed and only negligible O 1*s* signal is usually detected in the spectrum of freshly evaporated film. In the early XPS studies, which were performed in a vacuum of 5×10^{-7} Torr [4.30], the evaporation had to be repeated every 5 min. Most current instruments provide UHV conditions. Evaporation can be performed in a residual atmosphere as low as 10^{-10} Torr, and measurement below 10^{-11} Torr. Under these conditions surface contamination remains negligible for at least 10 h, which is sufficient for data accumulation. Thin films produced by evaporation tend to be polycrystalline with a texture which may depend critically on the substrate and the evaporation conditions. This should not have important effects on XPS measurements of localized levels like the 4*f* states. Bulk samples can also be used, but they require cleaning in situ. Cycles of argon-ion bombardment and annealing might be used. If the surface structure is not likely to be important, a simple and effective in situ cleaning is obtained by scraping the sample with a sharp tool made of a hard material like tungsten carbide. This procedure has been used successfully to clean Ce and a number of other RE metals and alloys.

b) Chalcogenides, Borides, and Alloys

As previously mentioned, certain commercial instruments can attain high resolution only if the surface of the sample is perfectly flat and accurately positioned in the instrument.

Optically flat surfaces of a large number of RE compounds have been obtained by vacuum cleaving single crystals. A piece, typically $2 \times 3 \times 4\ mm^3$, is cut from a larger single crystal and mounted in such a way as to allow cleaving along a low-index plane in vacuum. RE chalcogenides are very suitable for this process, because most of them crystallize in the NaCl structure, and cleave on [100] planes. For materials that cannot be readily cleaved in that way, e.g., the hard RE borides, filing with a diamond file in vacuum has been found to be a crude, but effective surface preparation procedure. RE intermetallics are generally brittle and can be cleaned by removing macroscopic amounts of surface material by vacuum scraping with a tungsten carbide blade. The progress of cleaning can be monitored in terms of the relative intensity of the O 1*s* signal.

4.3 Results

4.3.1 Metals

a) Identification of the Outermost Levels

In the pioneer XPS studies of rare-earth metals [4.30, 31], resolution and statistics were rather poor, but extended structures in an energy range 15 eV below E_F could be identified. They were already interpreted as being due mainly to multiplet splitting of 4*f* final states. The presence of the first peak near E_F in the light RE was, however, not understood. It was shown in a subsequent study of these metals [4.32] that, except for Ce, valence and 4*f* states appear as separate structures. This quite unusual situation had not been encountered in other elements. The metallic rare earths show two distinct partially occupied states: a normal valence band cut by the Fermi level and the 4*f* states. The large change in Coulomb energy resulting from a unit decrease in the number of 4*f* electrons on a single site explains why these levels appear as partially occupied levels well below E_F in most XPS spectra. The 4*f* states play an important role in determining the valence, because they can function as a reservoir, able to accept or donate electrons. The energy required to promote a

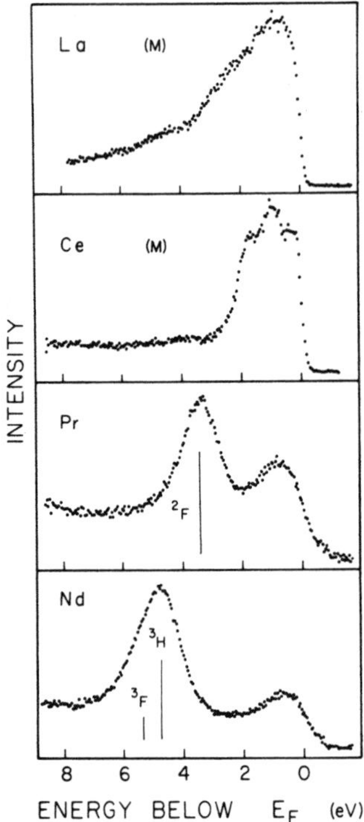

Fig. 4.7. XPS spectra of some light rare-earth metals recorded with MgK_α or monochromatized AlK_α radiation (*M*) [4.50]. The vertical scales are different for the various spectra. The corresponding $4f$ final-state multiplets and their strenghts, as indicated by vertical bars, were obtained as described in [4.32]

$4f$ electron into a valence state can be overcome by the corresponding gain of cohesive energy in forming the solid. This happens in the condensation of most rare-earth metals which have the configuration $4f^{n+1}6s^2$ as free atoms and $4f^n(spd)^3$ in the metallic state.

b) The Light Rare Earths

The XPS spectra of the outer levels of the light RE metals [4.32, 50] are shown in Fig. 4.7. Lanthanum, which is trivalent with the configuration $4f^0\,(spd)^3$ is included for reference. The emission intensity is high at the Fermi level of La and then decreases, forming a tail with weak structures. The bottom of the band is probably located around 4 eV. The bandwidth of 3.5 eV found in the relativistic augmented plane wave (RAPW) calculation of *Fleming* et al. [4.33] for *dhcp* La is in agreement with the experimental value. This calculation yields a set of flat bands near E_F, responsible for a high density of states with *d*-character, followed by broader *sp* bands. This band structure, which is rather similar to those of 3*d* metals, satisfactorily explains the measured spectrum. The two following metals Pr and Nd are also trivalent and are expected to have band structures similar to that of La. This seems to be the case in spite of their overlap with the $4f$ states. The final states after $4f$ photoemission in Pr are

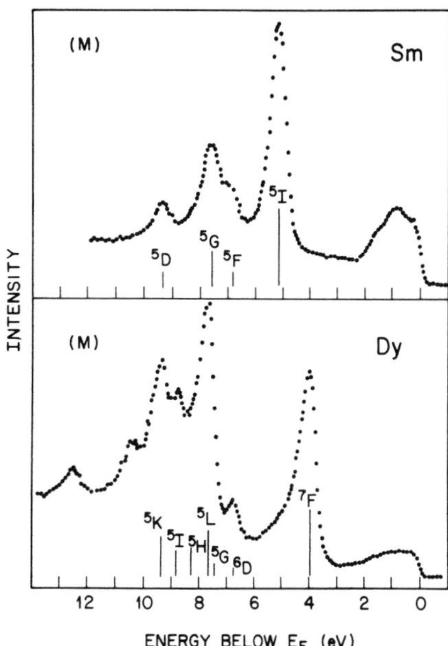

Fig. 4.8. High resolution spectrum of metallic samarium and dysprosium. The corresponding $4f$ final-state multiplets and their strenghts, as indicated by vertical bars, were obtained as described in [4.15]

$^2F_{5/2}$ and $^2F_{7/2}$. Their intensity ratio is about 6 and their separation small, $\sim 1/4$ eV. Therefore only one $4f$ peak appears in the Pr spectrum. (The J-splitting will be disregarded, except when it could influence the shape of a well-resolved line or when the components can be separated.) The $4f$ spectrum of Nd contains a large peak (^3H) with only a weak indication of the less intense ^3F final state. There is a strong indication that both the Pr and Nd $4f$ spectra have lifetime broadening of ~ 0.8 eV, exceeding the instrumental width. The increase in the $4f$ binding energies with atomic number is clearly illustrated by the simple spectra of Pr and Nd. The core potential becomes stronger as a consequence of the imperfect mutual screening of the nuclear charge by electrons belonging to the same shell.

Metallic samarium is usually considered to be trivalent, having five $4f$ electrons with parallel spins. In this case the initial ground state is $^6H_{5/2}$, and the final states are ^5I, ^5F, ^5G, and ^5D. The corresponding lines are clearly observed in the spectrum of Fig. 4.8, upper part. The valence band is well separated from the $4f$ structures. The UPS spectra of Sm recorded at different photon energies [4.34] show a shoulder which might be identified with the maximum of the XPS valence band spectrum. Recently a careful examination of the $4d$ spectrum of Sm metal (see Sect. 4.3.2) gave indication that either the surface or the bulk of Sm is in an intermediate valence state or else that Sm is divalent at the metal-vacuum interface [4.35]. Experiments have been suggested which may distinguish between these alternatives [4.35]. The maximum at 0.9 eV and the weak shoulder between 1 and 2 eV would then correspond to the multiplet ^6H and ^6F originating from Sm^{2+}. In support of this hypothesis it is

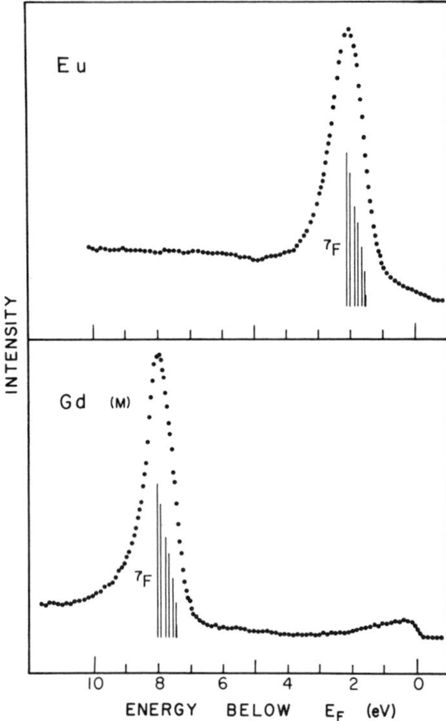

Fig. 4.9. XPS spectra of Eu and Gd

seen from a comparison with Dy (see Fig. 4.8, lower part) that the valence band in Sm is abnormally high.

In divalent Eu metal, the intense $4f$ level (shown in Fig. 4.9) is located near the Fermi level, leaving one no possibility of studying the valence states by XPS. The UPS spectrum of Eu shows a single peak below E_F [4.36], but no further details can be extracted from these data. The half-filled $4f$ shell in Eu yields a very simple spectrum and has a very low energy because of the large gain of exchange energy. For this reason Eu is divalent, the energy gained by forming a $4f^7$ configuration being larger than the loss of cohesive energy. This situation is clearly recognized in Fig. 4.9, where the single ^7F peak is less tightly bound by 6 eV in Eu than in Gd, which is trivalent with the same $4f^7$ configuration. The Gd spectrum, measured with high resolution, yields an experimental $4f$ linewidth of 1.0 eV. The different J-levels have been drawn in the spectrum as taken from optical data of Eu^{3+}. No correction has been made in this case for the larger nuclear charge in Gd. The separation of these levels is not large enough to be reflected in the experimental spectra. But it can be shown that a superposition of lines of *Doniach–Šunjić* type [4.37][2] with a proper choice of the parameters can accurately reproduce the experimental line

2 For a discussion of the status of theory and experiment of this many-body aspect in photoemission, see [Ref. 1.1, Chap. 5].

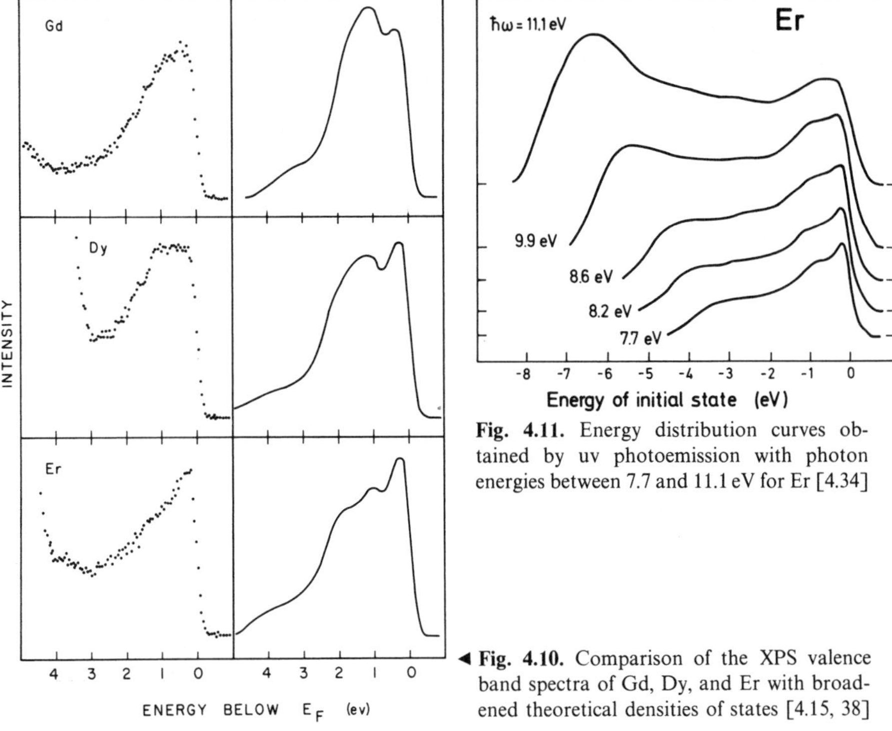

Fig. 4.11. Energy distribution curves obtained by uv photoemission with photon energies between 7.7 and 11.1 eV for Er [4.34]

◀ **Fig. 4.10.** Comparison of the XPS valence band spectra of Gd, Dy, and Er with broadened theoretical densities of states [4.15, 38]

shape. The hump observed near E_F in the Eu spectrum is an indication of the valence states.

c) The Heavy Rare Earths

Except for Yb, which is divalent, the heavy rare-earth metals are all trivalent with *hcp* structure. It can be anticipated that they will all have similar band structures. As in the light elements, the region near E_F has *d*-character, changing to *sp*-symmetry toward the bottom of the band. In Fig. 4.10 those parts of the XPS valence band spectra of Gd, Dy, and Er lying at lower energies than the 4*f* structures, are compared with a broadened RAPW densities of states [4.15] computed by *Keeton* and *Loucks* [4.38]. Rather satisfactory agreement is found if one takes into account the strong overlap with the tails of the 4*f* peaks. UPS should allow a study of the whole valence band of these metals. The spectrum of Er [4.34] shown in Fig. 4.11 is in good agreement with XPS data; however, the secondary electrons obscure the bottom of the band at higher photon energy. Photoemission therefore yields confirmation that all heavy trivalent rare-earth metals gave similar valence bands with dominantly *d*-character near E_F.

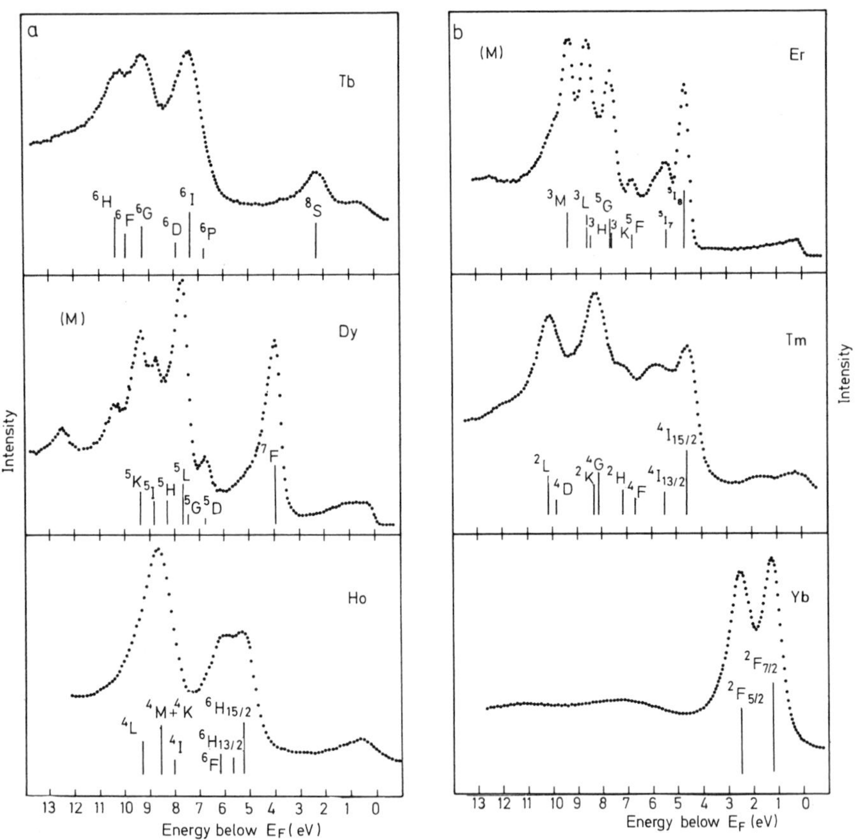

Fig. 4.12a and b. XPS spectra of the heavy rare earth recorded with Mg K_α or monochromatized Al K_α radiation (*M*). The corresponding 4*f* final-state multiplets and their strenghts, as indicated by vertical bars, were obtained as described in [4.39]

The 4*f* spectra of the heavy rare earths shown in Fig. 4.12 are very rich in structures. Their interpretation is very simple, however, and is again a strong support for the validity of the theory outlined in Sect. 4.1.2. In the spectrum of Tb, the exchange splitting, first identified in the REF_3 [4.39], is very large. Photoemission of the minority-spin electron gives rise to the ^8S state, which is well separated from the other multiplet lines produced by the emission of electrons belonging to the half-filled spin-up shell. In the next two elements, Dy and Ho, these groups of final-state multiplets remain distinct, but, as expected, their separation and intensity difference decrease. The spin-orbit splitting also becomes important and in Ho *J*-multiplets must already be taken into account. The situation becomes more complicated in the spectra of Er and Tm where the structures corresponding to the two different *S*-values overlap. The final state of lowest energy is, however, still produced by emission of a minority-spin electron. The behavior of Yb and Lu with filled $4f$ shell is quite similar to that of Eu and Gd, with half-filled $4f$ shell, except that the final state is split by the spin-orbit interaction. The two states $^2F_{5/2}$ and $^2F_{7/2}$ have an energy separation of 1.25 eV for Yb and 1.4 eV for Lu (the spectrum of Lu is not shown in Fig. 4.12). The binding energy shift between the divalent and the trivalent metal is here again almost exactly 6.0 eV.

The interesting aspect of these results is not the study of the excited states of a $4f$ shell; here XPS cannot compete with uv absorption in insulating solids, which offers much higher resolution and is far more accurate. The identification of the lines is, however, important because the spectra are quite distinctive and can be used as a "fingerprint" of the number of $4f$ electrons present in the *initial* state. This offers a convenient way to determine the valence of rare-earth atoms in compounds. This method of determination will be illustrated later in this chapter.

The discussion of experimental $4f$ spectra in metals has so far been limited to the relative positions and intensities of the lines. A comparison of the absolute binding energy relative to the Fermi level with theory will be given in Sect. 4.3.1 (see also Sect. 4.1.3).

d) Cerium

The fascinating properties of metallic Ce are evident in its rich phase diagram. No attempt will be made here to review this field exhaustively. We shall consider only some aspects of the $\gamma \rightarrow \alpha$ phase transition. The interpretation of magnetic and lattice constants measurement [4.40] indicates a change of valence from 3.06 to 3.67 at the transition. The original explanation was that the $4f$ electron in the γ phase is promoted to a conduction state in the α phase. Different versions of this promotion model have subsequently been proposed [4.41, 42]. *Hirst* [4.43], in connection with the development of the theory of configuration crossover, presents the $\gamma \rightarrow \alpha$ transition as a typical example for a transition between a stable configuration and a state characterized by rapid interconfiguration fluctuations (ICF). In this model, the fractional valence of Ce is considered to be a time average over the fluctuations between $4f^1$ (spd)^3 and $4f^0$ (spd)^4. None of these models ever question the localisation of the $4f$ states in both γ and α Ce. *Johansson* [4.44] pointed out, however, that in the rare-earth series, the $4f$ orbitals of Ce have the largest spatial extension. He also presents arguments supporting a relatively small value of the effective Coulomb correlation energy, U_{eff}, in Ce. Considering this transition in the Hubbard model, the condition $U_{\text{eff}} > W$ ($W = 4f$ bandwidth) for a localized state is fulfilled in the γ phase, but the situation might be reversed in the α phase. This would mean that the γ phase is the low carrier density side of a Mott transition. In the α phase the $4f$ electrons would then be expected to form a narrow metallic band. Finally, an estimate of the cohesive energy of tetravalent Ce yields an unrealistically high value which militates against any electron

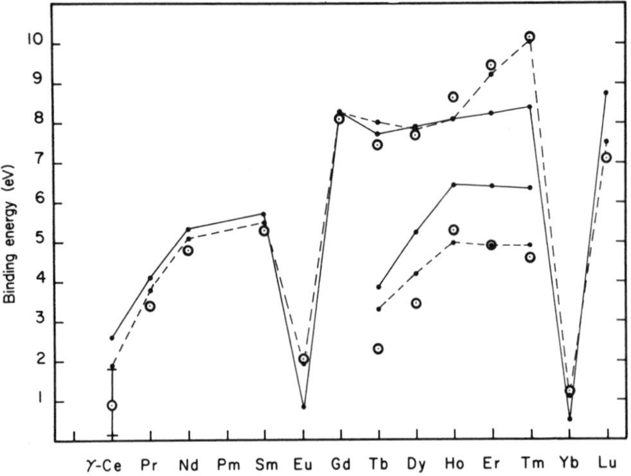

Fig. 4.13. Comparison of the $4f$ promotion energies: ⊙ Experimental XPS data [4.15]. —●— Renormalized atom approach [4.12, 14]. (See Sect. 4.1.3 and Fig. 4.5a). —●— Thermodynamical approach [4.16]. (See Sect. 4.1.3 and Fig. 4.5b)

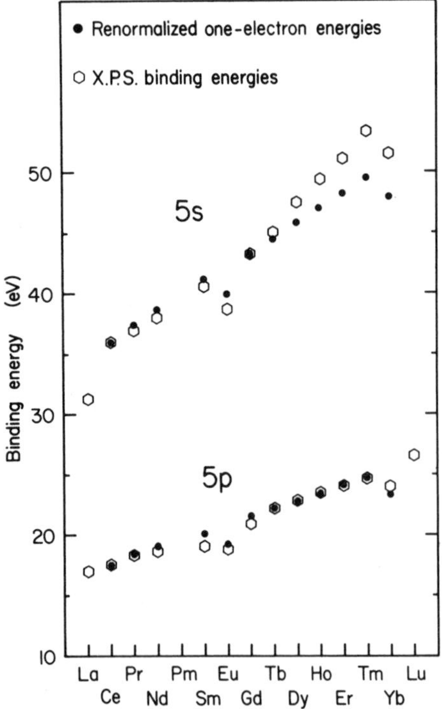

Fig. 4.14. Comparison of the $5s$ and $5p$ binding energies in rare-earth metals. (Experimental XPS data from [4.50]; renormalized atom approach from [4.49]

promotion model. Whether the γ phase of Ce is an f^1 system is also not yet well established.

The XPS valence band spectrum of γ Ce is reproduced in Fig. 4.7. It is tempting to consider the leading peak at 0.9 eV as due to the localized $4f$ state. A comparison with a uv spectrum [4.45] seems to confirm this interpretation. The XPS measurements have been repeated on bulk samples and films evaporated on different substrates at room temperature and liquid N_2 temperature. The main structure was discernible in all spectra, but with variable intensity. Attempts have been made to induce the $\gamma \to \alpha$ transition in situ. It has been shown [4.46] that by heating Ce to 400 °C and cooling it fast to liquid N_2 temperature, 10 to 20% of α phase is obtained; this is sufficient to detect a change in the valence band and in core-line multiplets. This partial change of phase could easily be measured by dilatometric techniques on free samples, but any clamping of the sample to the holder hinders the transition. The best way to produce pure α Ce is to apply hydrostatic pressure during the cooling. Since this cannot be realized in the present XPS apparatus, no spectrum of α Ce is so far available. Further efforts will have to be devoted to this intriguing problem. The chemical shift of X-ray emission lines in this Ce phase transition has been recently studied [4.47] and a substantial effect found. The similarity of the shifts observed in the phase transitions of Ce and SmS [4.48] is considered by the authors as strong evidence favoring the $4f$ promotion mechanism. It remains to be shown, however, that a Mott transition [4.44] of the $4f$ states cannot account equally well for the observed shift.

e) The 4*f* Promotion Energy

The energy separation of the $4f$ level from the Fermi level Δ_- can be directly deduced from Figs. 4.7, 9, 12. The theoretical estimates of Δ_- discussed in Sect. 4.1.3 have been plotted in Fig. 4.13, along with the experimental points. We conclude from this comparison that the renormalized atom scheme [4.12, 14] and the simple scheme of *Johansson* [4.16] give results in gratifying agreement with experiment. The renormalized atom scheme has also been used to calculate the binding energies of the $5s$ and $5p$ shells in rare-earth metals [4.49]. These values, corresponding to the lowest energy final state, are compared in Fig. 4.14 to XPS data [4.50]. The agreement is again very good.

4.3.2 Compounds and Alloys: Stable $4f^n$ Configurations

a) Rare-Earth Halides

XPS data for La and Ce halides [4.51] are shown in Fig. 4.15. Data for the REF_3 series are also available [4.39]. Due to their chemical stability, these are among the few RE compounds which do not need special attention with regard to surface preparation and vacuum conditions. Comparison of the spectra of CeF_3 and $CeCl_3$, with those of LaF_3 and $LaCl_3$ allows unambiguous identifi-

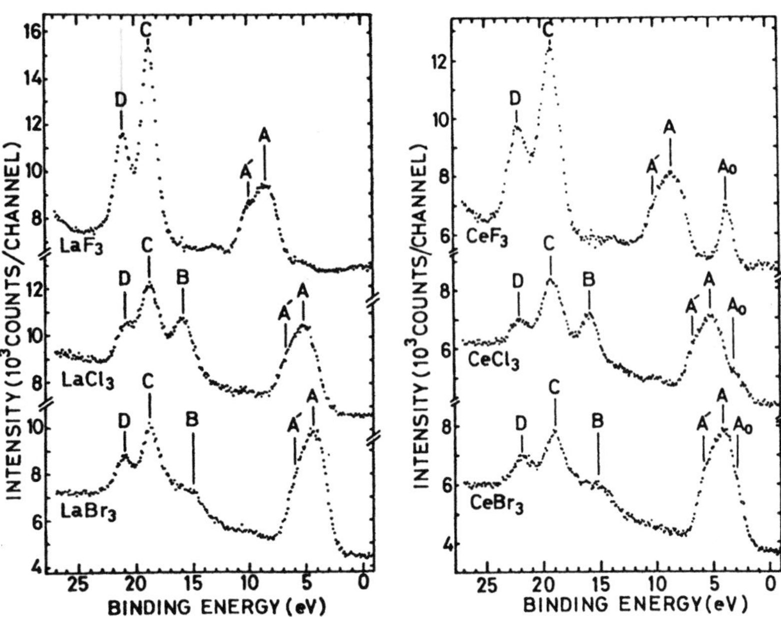

Fig. 4.15. XPS spectra of valence band region of La and Ce halides [4.51] obtained with AlK_α radiation

cation of the Ce $4f^1$ electron peak, labelled A_0 in Fig. 4.15. This peak is the only qualitative difference between the corresponding Ce and La spectra. Peaks A and A′, C and D, and B are well-understood features of the spectra, corresponding to valence bands, $5p$ spin-orbit doublet, and Cl $3s$ and Br $4s$ electrons, respectively. In the course of our investigation we have found that the $4f$ spectra of the REF_3 exhibit large nonlifetime broadening, so that improved instrumental resolution does not result in significant improvement over spectra taken at lower (~ 1 eV) resolution. The data in Fig. 4.15 are therefore the only ones discussed here.

b) Chalcogenides and Pnictides

Some of these materials have received widespread attention in the last 10 to 15 years because their physical properties closely approach those of ideal models, e.g., EuO is a good approximation to a Heisenberg ferromagnet. A comprehensive collection of references can be found in a recent conference proceeding [4.52]. The chalcogenides are metallic when the RE ion is formally trivalent like in GdS, and semiconducting if the RE is divalent like in EuS. The pnictides are also believed to be intrinsically metallic, the only exception being found among the nitrides (GdN is semiconducting, but CeN metallic). In all cases, at least at room temperature where most of the data have been taken, the conductivity of the sample is so large that charging effects do not present a

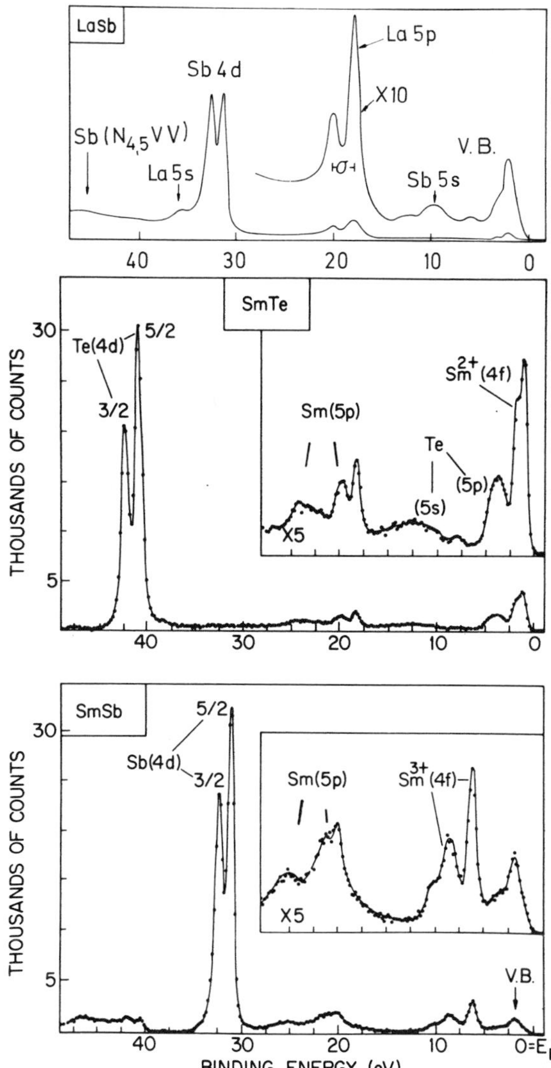

Fig. 4.16. Overview of valence band and low binding energy core levels in LaSb, SmTe, and SmSb

problem and good resolution and accurate binding energies can be obtained. In many cases the spectrometer resolution exceeds the lifetime width of the $4f$ hole.

The low binding energy part (i.e., $E_B \leqq 50$ eV) of the XPS spectrum of RE solids is very rich in structure. We show three representative examples of this region in Fig. 4.16. These are for metallic-like LaSb (f^0), SmSb (f^5) and semiconducting SmTe (f^6). The RE $5s$, $5p$, and $4f$ electrons all have binding energies smaller than 50 eV (see also Fig. 4.14). The calculated cross sections of the $5s$ and $5p$ electrons are of the order of $3-4\times10^3$ and $1-1.5\times10^4$ barns

[4.6], e.g., fairly constant over the RE series. In contrast that of the $4f$ electrons increases from 1.8×10^3 (Ce) to 7×10^4 barns (Yb), largely due to the increase in the number of $4f$ electrons. This is well confirmed by the experiment (see Fig. 4.16). Due to the $5p-4f$ multiplet coupling, the $5p$ spectrum of the ions between Pr and Tm is significantly different from that of a simple spin-orbit doublet, examplified by the case of La. A detailed, theoretical description of the $5p$ spectra has not yet been given, although a successful attempt has been made in the case of Sm [4.53]. From Fig. 4.16 it is also evident that the binding energy of the $5p$ electrons of Sm decreases by about 2 eV due to the screening of one additional $4f$ electron. Despite its lower binding energy, most of the charge of the $4f$ shell resides inside the $5p$ shell [4.54].

In Figs. 4.17–19 we show XPS spectra of RE–Sb and CeAs. In the Ce compounds (Fig. 4.17) the $4f^1$ electron is not unambiguously resolved. This is a results of the relative photoionization cross sections:

$$\sigma(\mathrm{Ce}4f^1)/\sigma(\mathrm{As}4p) \sim 1.1, \quad \sigma(\mathrm{Ce}4f^1)/\sigma(\mathrm{Sb}5p) \sim 1.3 .$$

Furthermore, as will become clear in the discussion of the RE borides, we expect the $4f^1$ level of Ce to be strongly lifetime broadened. In both CeSb and CeAs the $4f$ level is located near E_F (within 1 to 3 eV) overlapping the valence band. This fact must account for the peculiar magnetic behavior of these compounds and for the anomalously small crystal-field splitting [4.55]. The spectra of Figs. 4.18, 19 also fully confirm the finding in the case of the metals regarding the validity of the method of fractional parentage (Sect. 4.3.1). In the case of the antimonides of the heavy RE, the 0.55 eV instrumental width exceeds the lifetime width of the $4f$ hole. It can, however, be shown that the $4f$ spectra of the antimonides closely resemble those of the metals. This is achieved [4.56] by deconvolution of the spectra of Figs. 4.18, 19 so as to simulate the energy resolution actually attained in the case of the metals ($\sim$250 meV).

This comparison demonstrates that in the RESb and in the metals from Pr to Tm the $4f$ electrons are indeed negligibly influenced by the chemical environment. The success of the method of fractional parentage is more clearly demonstrated by the analysis shown in Fig. 4.20. Here the experimental XPS spectra of SmTe and SmSb are compared with theoretical spectra, obtained by folding the *Doniach-Šunjić* line shape (see [Ref. 1.1, Chap. 5]) and the instrumental resolution function into the theoretical intensities. Best agreement is found if a lifetime broadening of 150 and 300 meV, and asymmetry parameters of 0.05 and 0.1 are used for SmTe and SmSb, respectively. Some of the discrepancy in the fit for SmSb around 9 to 10 eV is most likely ascribable to emission from the Sb $5s$ level located in that energy region according to the LaSb data. This rather rough analysis is in agreement with two expectations: first, that the asymmetry parameter α will increase in going from semiconducting SmTe to metallic SmSb (according to the point of view of Chap. 5, α should be zero in a semiconductor), and second, that the lifetime width will be proportional to the number of $4f$ holes, $q=15-n$, available for recombination

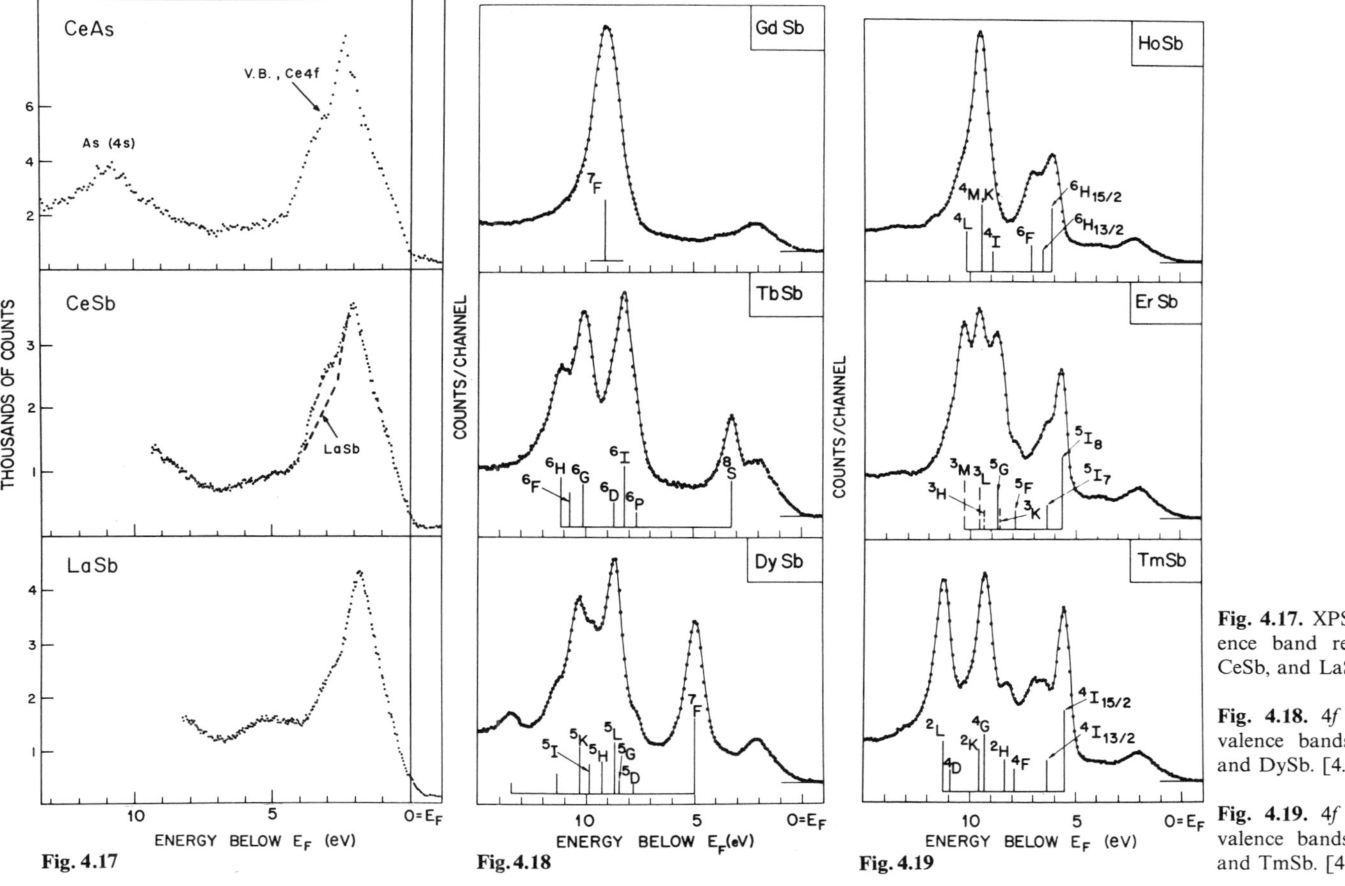

Fig. 4.17. XPS spectra of valence band region of CeAs, CeSb, and LaSb

Fig. 4.18. 4*f* multiplets and valence bands of Gd-, Tb-, and DySb. [4.56]

Fig. 4.19. 4*f* multiplets and valence bands for Ho-, Er-, and TmSb. [4.56]

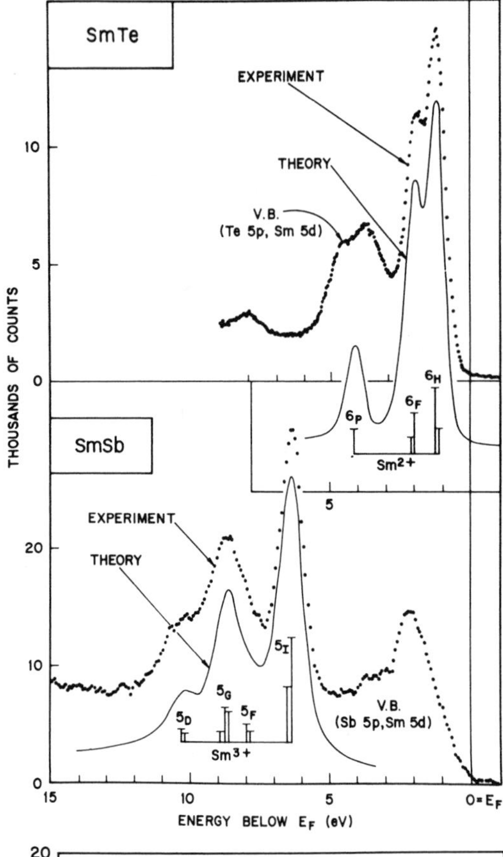

Fig. 4.20. Experimental and theoretical 4*f* multiplets of SmTe and SmSb. (See text)

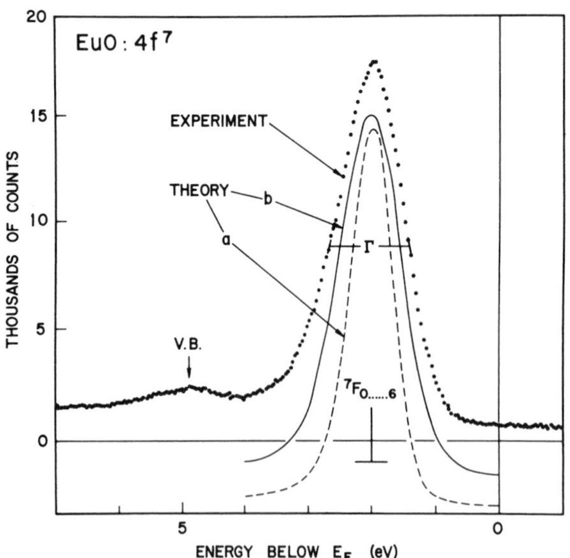

Fig. 4.21. Experimental and theoretical 4*f* multiplets of EuO. (See text)

after photoemission. It may also increase with increasing binding energy, E_B. These two are competing effects, since E_B increases and q decreases with increasing number of $4f$ electrons. It will be seen later that the variation of q is more important. The analysis of Fig. 4.20 can in principle be extended to all $4f^n$ configurations.

An accurate analysis of the $4f$ spectrum sometimes requires an independent determination of the emission from states other than $4f$ but degenerate with them. This has been done in the case of the borides and is reported below. It is of course also possible that the $4f$ emission is not superimposed on any other or else that its intensity is so much larger that no correction is required. These cases are typified by EuO and YbTe, respectively. We treat here only that of EuO [4.57].

c) Phonon Broadening in EuO

The XPS $4f$ spectrum of EuO is shown in Fig. 4.21, along with a convolution of the 7 J-multiplets (using intensities from Table 4.1) with the experimental resolution function. The result, curve a, is much too narrow. Curve b, which fits the data well, is the result of additional Gaussian broadening of curve a by 780 meV. The origin of this nonlifetime broadening lies in phonon excitations. The effect is large in EuO due to the small mass of oxygen. This is the first time that such large effects have been detected in $4f$ spectra, although their importance in X-ray spectroscopy, in general, has been stressed by a number of authors [4.58]. The present analysis is not influenced by the presence of other emission, because the $4f^7$ levels in EuO are known to be the highest occupied levels lying in the band gap, and to have a calculated photo cross section 78 times larger than the O(2p) states, at 1.5 keV. The data for EuO have been shown here, in part, in order to give an indication of the kind of $4f$ photoemission line-shape analysis we should expect to be able to perform quite generally in the future. Another example of line-shape analysis is given in the next section. It was generated by the observation that the $4f$ linewidth depends both on relative position of extended and localized $4f$ states and on the number of holes q available for recombination after photoemission.

The XPS spectra of YS [4.59], without $4f$ electrons, and of NdS, NdSb, and NdBi, f^3 systems, are shown in Fig. 4.22. HoSb and HoS are compared in Fig. 4.23. From the data of YS we can identify the S 3p-derived valence bands in the range from 4 to 8 eV below E_F, and the $4d^1$ conduction band of about 1 to 1.5 eV width. The relative strength of the $4f$ emission, estimated from calculated cross sections is $\sigma(\mathrm{Nd}\,4f^3)/\sigma(\mathrm{S}\,3p) \simeq 5.5$. For NdS we, therefore, expect the ^3H and ^3F structure of the $4f$ emission to be dominant. Comparison with the spectra of NdSb and NdBi indicates that, for reasons not yet clear, the lifetime of the $4f$ hole in the sulfide is much smaller than in NdSb and NdBi. The same conclusion follows from Fig. 4.23 in the case of HoS and HoSb, where it is made evident by the long wings attached to the main $4f$ emission peaks. A detailed analysis of $4f$ hole lifetime in the sulfides is not yet available.

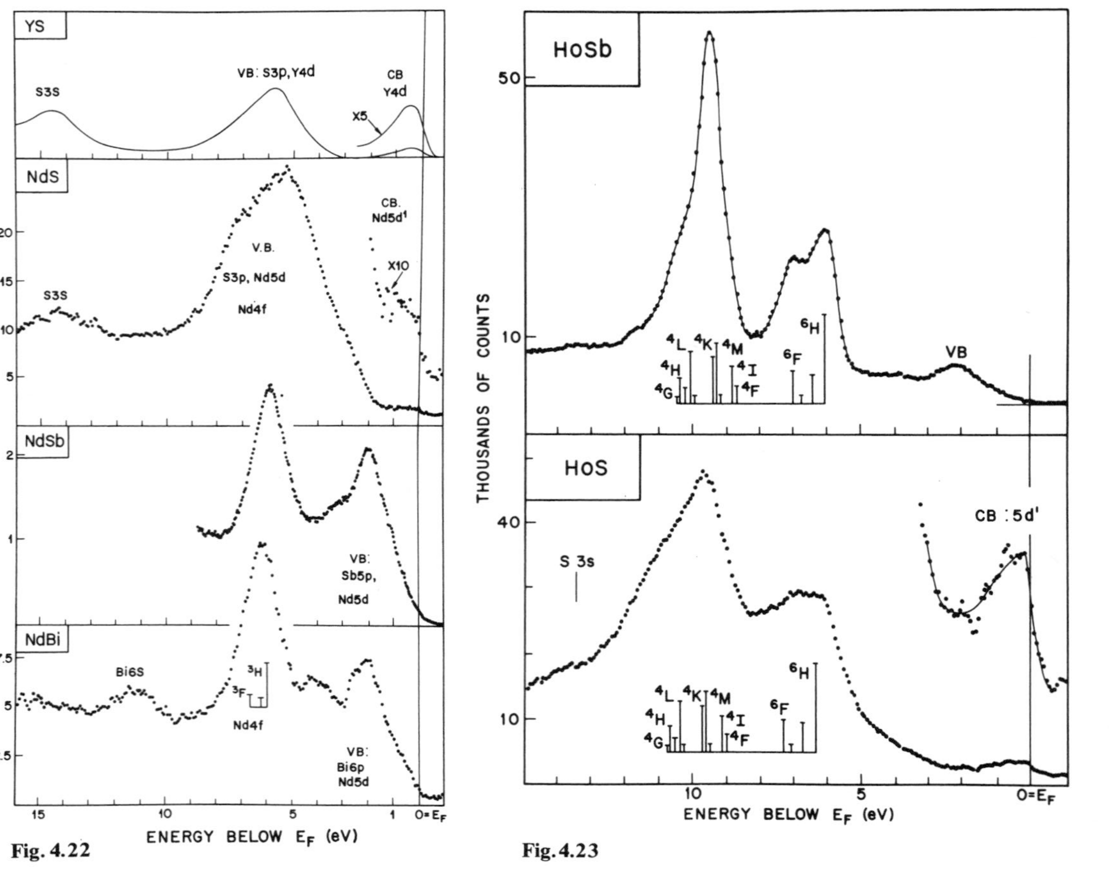

Fig. 4.22. XPS valence band spectra of YS [4.59], NdS, NdSb, and NdBi

Fig. 4.23. XPS valence band spectra of HoSb and HoS

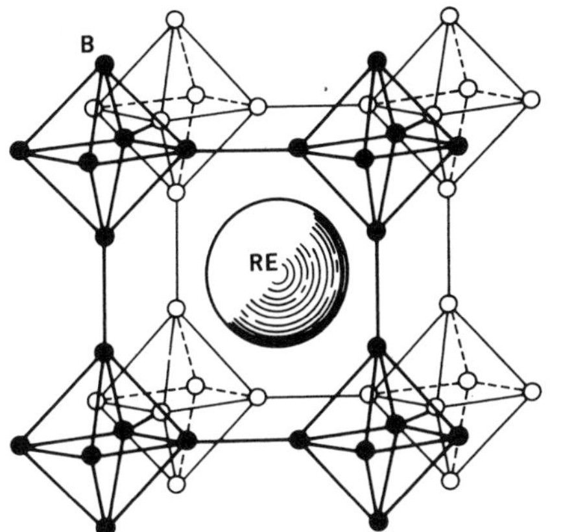

Fig. 4.24. Structure of rare-earth hexaborides, REB_6

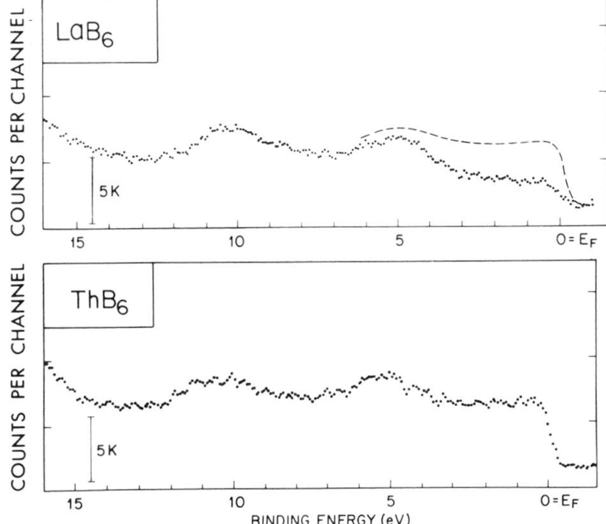

Fig. 4.25. XPS spectra of valence band region of metallic LaB_6 and ThB_6. The dashed line reproduces the ThB_6 spectrum. [4.93]

d) Interatomic Auger Transitions in Rare-Earth Borides

The borides are especially attractive from the point of view of XPS simply because $\sigma(4f^n)$ is so much larger than σ(B2*s*, 2*p*). Furthermore they are very hard and stable compounds [4.60]. In the case of the hexaborides this is due to the very strong covalent bonding between the boron octahedra (see Fig. 4.24). The electronic structure of the REB_6 has been the subject of a number of investigations, starting in 1954 [4.61]. In general there is agreement among the various band-structure calculations with regard to the bandwidth and band position. More recent investigations of the electronic structure of CaB_6 and LaB_6 [4.62] show good agreement with the earliest calculation and confirm the semiconducting and metallic behavior of these two compounds, respectively. XPS valence band spectra of La- and ThB_6 are shown in Fig. 4.25 [4.63]. The very wide, mostly boron derived valence bands, extending from about 3 to 12 eV below E_F are rather featureless. In going from LaB_6 to ThB_6 an extra electron is added to the conduction band, which is responsible for the increased emission between E_F and 2 or 3 eV. It locates the position in energy of the states having metallic *d*-character. The favorable cross section ratio $\sigma(\mathrm{RE}4f)/\sigma(\mathrm{B}sp)$ is apparent from the data of NdB_6 in Fig. 4.26, upper part, when compared to the spectrum of LaB_6, dashed line. The 4*f* photoemission line shape was investigated for the whole REB_6 series and also for some other borides. The results are shown in Figs. 4.26, 27. These data, which contain a superposition of the 4*f* and valence band emission, can, however, be analyzed quantitatively because the valence band contribution is relatively featureless. It was subtracted under the assumption that it is like the one of LaB_6 throughout the series. Theoretical 4*f* spectra were then generated numerically for comparison (see Fig. 4.28). The individual line shapes were of the *Doniach–Šunjić* type [4.37] (with singularity index $\alpha \approx 0.15$) convoluted by the experimental resolution function. The lifetime widths were found to be large (1 eV for SmB_6, as compared to 0.1 to 0.15 eV for Sm metal) and were measured with good accuracy. The Lorentzian shape of the lines, most clearly recognized from the wings of the line of the GdB_6 spectrum (Figs. 4.26, 28), clearly indicates that these widths are due to a lifetime effect.

The lifetime widths for the whole series of hexaborides (Fig. 4.29) exhibit a rather simple behavior: the recombination rate is proportional to the number of holes in the 4*f* level in the final state, as suggested by *Hirst* [4.64]. For an RE ion with initial configuration $4f^n$, the final-state configuration is $4f^{n-1}$, and the number of holes for recombination is $15-n$. The linear relationship $\hbar/\tau \propto 15-n$ is obeyed with surprising accuracy. The magnitude of these linewidths can be accounted for on the basis of an interatomic Auger effect [4.63], whose high efficiency stems from the large number of bands in the hexaborides. We refer the reader to the original literature [4.63] for a detailed discussion of this point. Further support for the importance of the Auger process was obtained by investigating another RE boride, GdB_N. The density of states of the valence bands decreases with *N*, and the lifetime width should do the same, this has also been observed [4.63]. The measured linewidths for $N=2$, 4, and 6 are 0.50, 0.85, and 1.00 ± 0.04 eV, respectively. Although a quantitative discussion would require detailed knowledge of the band structures of these materials, the monotonic behavior of $\hbar/\tau$ with *N* was clearly demonstrated. Equivalently, one can relate the decreasing lifetime to the increase of the number of first neighbor anions (12, 18, and 24, respectively). This shows the connection with the previously invoked interatomic Auger process [4.65a, b][3]. The interatomic Auger processes therefore seem to have general validity as the mechanism which limits the lifetime of 4*f* holes, not only in borides and sulfides (see Figs. 4.22, 23) but in all RE systems.

---

3 Interatomic Auger processes were previously invoked to explain linewidth effects in Auger and XPS spectra of insulating compounds, see [4.65].

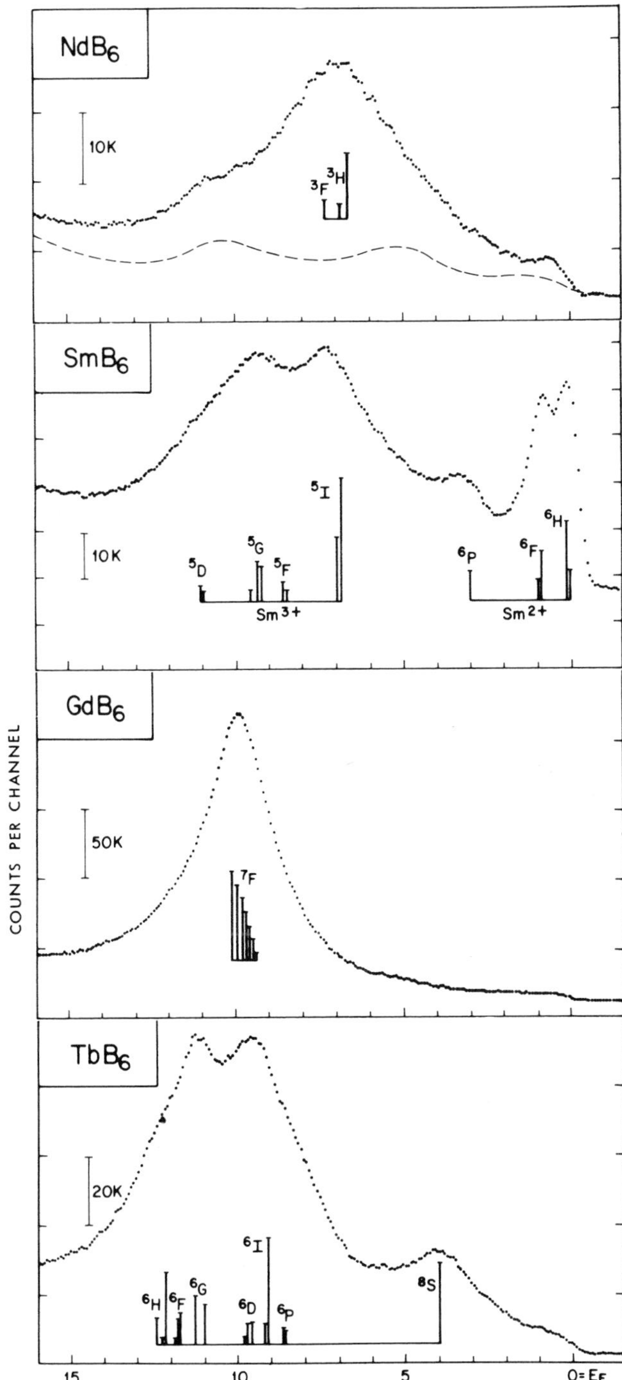

Fig. 4.26. XPS spectra of $4f$ multiplets in Nd-, Sm-, Gd-, and TbB_6. [4.93]

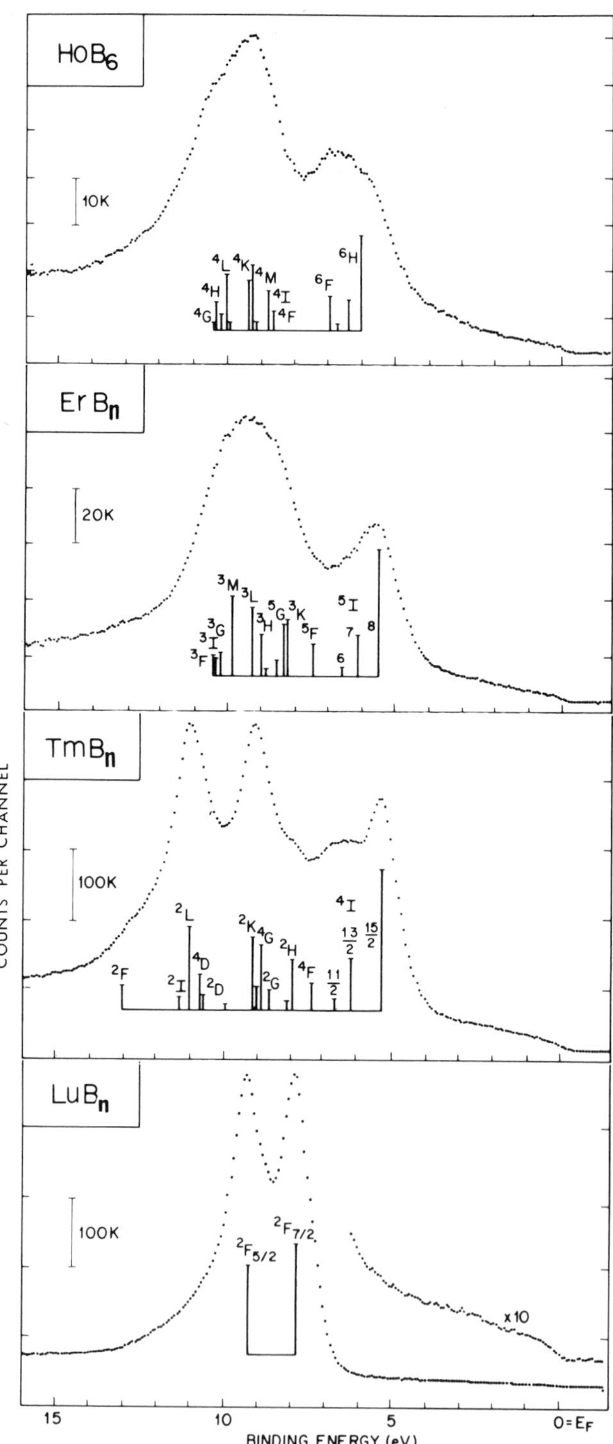

Fig. 4.27. XPS spectra of HoB_6 and other borides; B_n means that these samples were not single phases. X-ray analysis showed in the cases of Er, Tm, and Lu the presence of $-B_{12}$ phases (where the important quantity for our analysis, the number N of nearest-neighbor B atoms to the RE ion is 24, the same as in the REB_6) and some $-B_4$ phases (where $N=18$). [4.93]

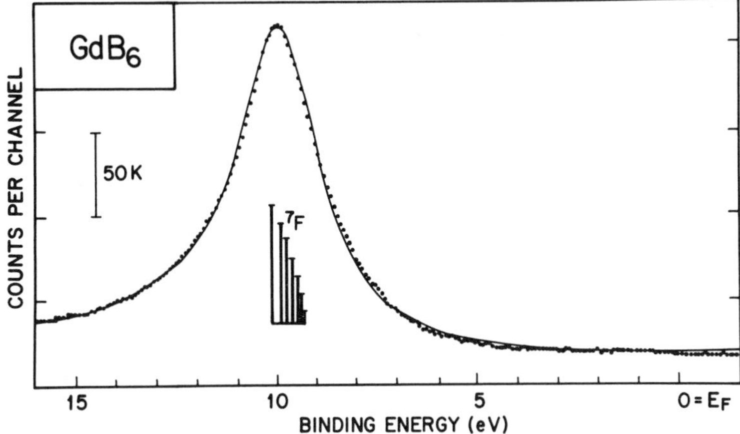

Fig. 4.28. Example of a fit to the $4f$ spectrum of GdB_6. The points are experimental after valence band subtraction (using the valence band of LaB_6). The full line is theoretical [4.63]

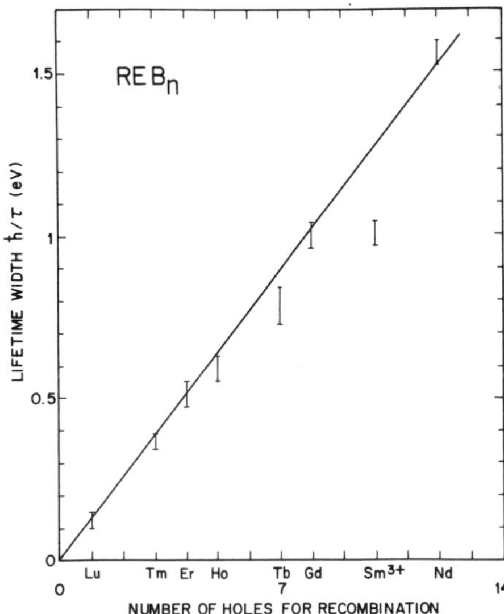

Fig. 4.29. $4f$-lifetime width (half width at half heights for a Lorentzian), plotted versus the number of available holes for recombination of an electron into the $4f$ level RE borides. Straight line: $\hbar/\tau = (15 - n) \times 0.125$. [4.63]

e) Rare-Earth Intermetallics

XPS research in this field has so far been quite limited. Unless a careful analysis of the various contributions from other degenerate states is made, it is difficult to draw conclusions regarding the detailed behavior of $4f$ levels either in the

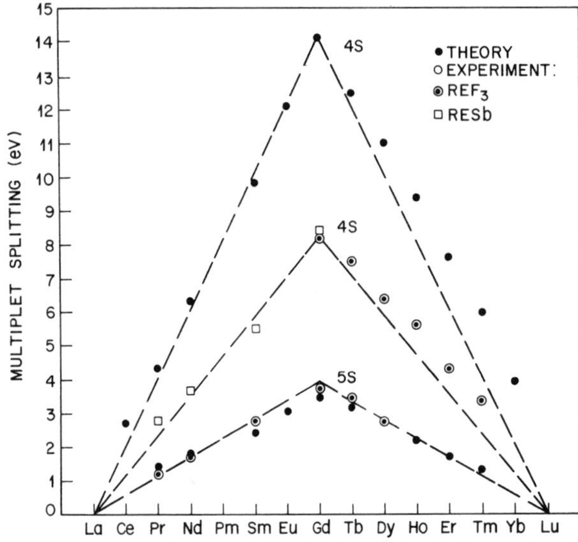

Fig. 4.30. Systematics of the multiplet splittings of 4*s* and 5*s* levels for a number of compounds compared with theory [4.12]

initial or in the final state, or even to gain information on charge transfer in the alloys. The investigations of *Güntherodt* and *Shevchik* [4.66], and *Cuthill* et al. [4.67] on the GdFe and $RECo_5$ systems, respectively, confirm this point.

f) 4*s* and 5*s* Multiplet Splittings

The multiplet splitting of core *s*-shells in the RE's indirectly provides information regarding the population of the 4*f* shell. The splitting is proportional to the atomic Slater integral G^3 and to $2S+1$, where S is the initial-state spin of the *f*-shell [4.68]. These multiplet splittings are quite insensitive to the chemical state of the RE, having been found to be the same in the trifluoride [4.69] and in the metal [4.70] with the same valence, i.e., 4*f* occupancy. We show in Fig. 4.30 that the RESb conform to this pattern. This provides a further indication of the core-like nature of the 4*f* electrons.

Comparison with theoretical calculations yields very good agreement for the 5*s* shell because inter-shell correlation effects are generally small. A surprising result is that intra-shell correlations reduce the 4*s* shell splitting uniformly across the whole RE series by a factor of 2. These correlations presumably involve mixing with configurations like $4s^2\,4p^5\,4d^9\,4f^{n+1}$ [4.22].

These multiplet splittings can, in principle, be used to study Intermediate Valence (IV) materials, but the rather large lifetime width of the 4*s* hole state makes them relatively unattractive for this purpose. The 5*s* shell is more difficult to use because both the photoelectric cross section and the multiplet splitting are smaller. Nevertheless discrete 5*s* final states have been seen in IV compounds of Yb, but this feature has not so far been exploited.

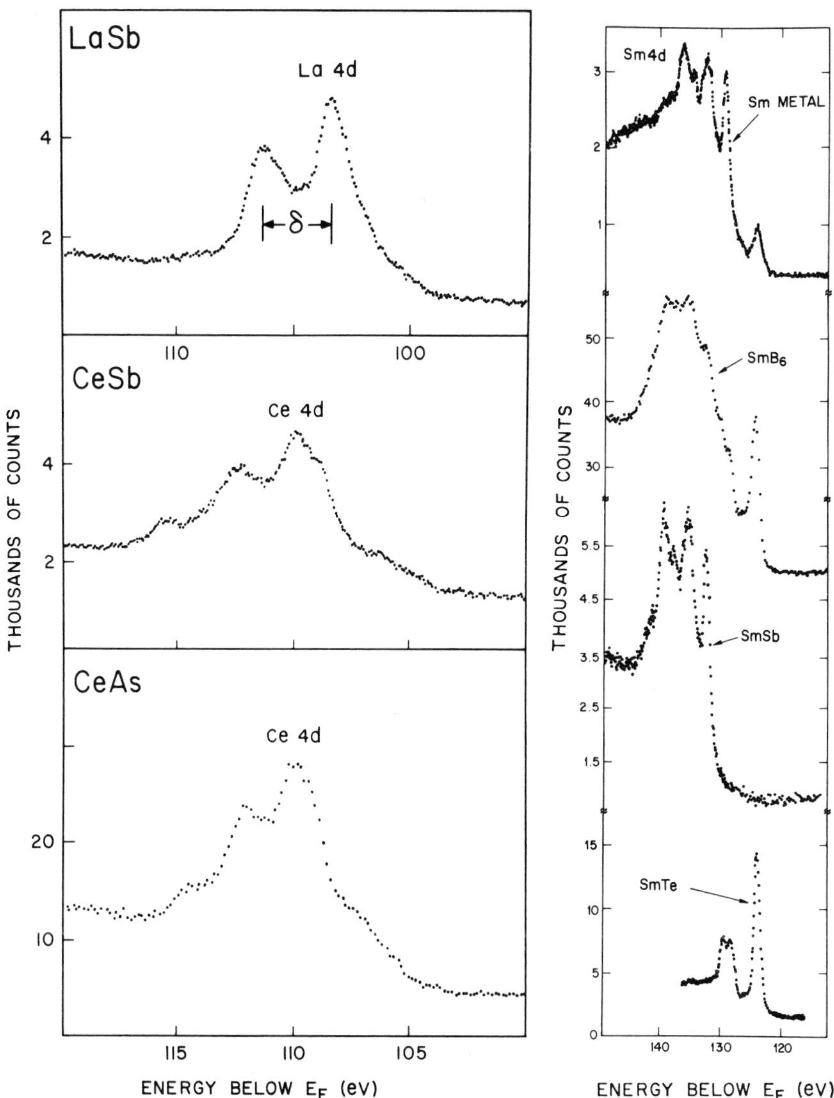

Fig. 4.31. XPS spectra of 4*d* electrons in LaSb, CeSb, and CeAs, Sm metal, intermediate valence SmB_6, divalent SmTe and trivalent SmSb. δ represents the spin-orbit splitting

g) Spectra of 3*d* and 4*d* Electrons of Rare-Earth Solids

On the basis of the photoelectric cross sections [4.6] alone, the 3*d* and 4*d* electron should be the most advantageous one for the study of the RE (see Fig. 4.3). The presence of strong shake-up satellites on 3*d* spectra and complex multiplet structure on 4*d* spectra has generally made them less useful for purposes other than chemical analysis.

Spectra of the 4*d* shell in the RE metals Ce, Sm, Eu, Gd, Tb, Dy, Ho Er, Tm, and Yb have been reported [4.71, 72], but only the relatively simple case of Eu with $^8S_{7/2}$ initial state has been analyzed [4.71]. The spectra of $4f^0$ La and $4f^{14}$ Yb and Lu, of course, exhibit only spin-orbit splitting (see Fig. 4.31 for LaSb). The complexity of 4*d* spectrum even in a 4*f* or $4f^{13}$ system can readily be appreciated by reference to the calculation of *Signorelli* and *Hayes* [4.73] for Ce^{3+} and Yb^{3+}. Comparison of the data for CeF_3 with those for CeSb and CeAs and even with those for Ce metal [4.71] shows that the structures are all very similar, and in reasonable agreement with theory [4.73]. This is an indication of the atomic nature of the 4*f* shell even in Ce^{3+}.

In the case of Sm the 4*d* spectra of Sm^{2+}, $4f^6$, and Sm^{3+}, $4f^5$ are well separated in energy and sufficiently distinct, that it has been possible to use them to reconstruct the spectrum of the IV compound SmB_6 (see Fig. 4.31) thereby obtaining an independent measure of the average Sm valence [4.74]. Comparison of these spectra with those of Sm metal [4.71] clearly shows the presence of a Sm^{2+} signal in the metal. This raises the intriguing possibility that Sm metal is itself in an IV state [4.35]. However, it is also possible that the divalent state is confined to the surface layer [4.75].

In the case of Eu successful use of the 4*d* spectrum has been made in an analysis of the IV compound $EuRh_2$ [4.76]. Data for trivalent $EuPt_2$ were used to identify the divalent component.

Spectra of the deeper 3*d* shell are generally available only in the first half of the RE series because the binding energy approaches the Al K_α energy at Tm. An interesting early finding [4.77, 78] here were strong satellites in insulating La, Ce, and Pr compounds. The proposal [4.77, 78] that they represent ligand-to-4*f* charge transfer excitation now seems to have been confirmed in further studies [4.73, 79]. The full complexity of the satellite structures associated with these spectra, however, still remains to be fathomed. As a simple example we show in Fig. 4.32, the 3*d* spectra of LaSb, $4f^0$, and CeSb, $4f^1$, two compounds with *metallic conductivity*. The main peak in LaSb at 834 eV presumably corresponds to screened final state in the absence of excitations; the broad structure at 2.5 eV greater BE should then represent a shake-up satellite state. That leaves a broad state at *smaller* BE which does not fit into the normal satellite concept. Its shift, S_1, is of the order of the Coulomb correlations energy, suggesting that it corresponds to a $4f^2$ final state, i.e., Ce^{2+}. This tentative and novel proposal appears to contradict the identification of a *higher* energy satellite with ligand to 4*f* charge transfer in *insulating* compounds. One can argue, however, that in metal, where final states are fully screened, the order of valence states should be as proposed here. Similar satellites appear in the 3*d* spectra of many RE metals. This clearly is a problem and its solution would greatly improve our understanding of satellites in general.

The main peak of the $3d_{5/2}$ spectrum of CeSb is clearly broadened compared to that of LaSb. This suggests that multiplet coupling cannot be

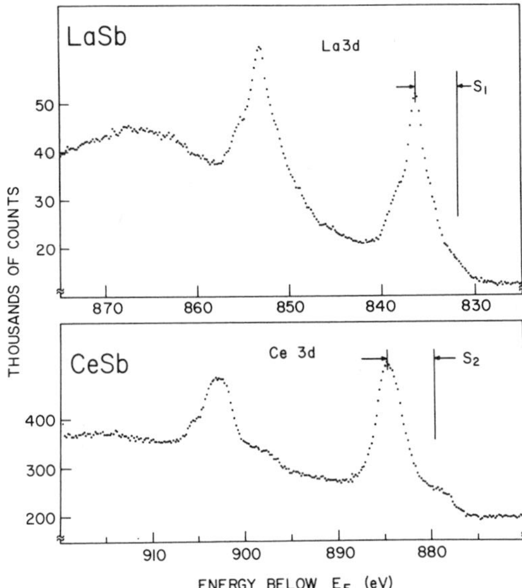

Fig. 4.32. XPS spectra of 3*d* levels of LaSb and CeSb

neglected in 3*d* spectra. The full complexity of such spectra has not been resolved experimentally, but the calculations of *Spector* et al. [4.80] for Pr^{3+} and Nd^{3+}, which result in 107 and 386 discrete final state *J*-levels, serve to exhibit it eloquently. For a discussion of the additional influence of the ligands on 3*d* spectra of lanthanide compounds see *Berthou* et al. [4.81].

h) 4*f* and 4*d* Binding Energy: Atom Versus Solid

Before treating the Intermediate Valence compounds, we briefly address the question of the meaning of binding energies measured by XPS. The question of the change in binding energy of a core level between the free atom and solid environment has in recent years received increasing attention[4]. This question is also of major relevance to theory because it is related to such problems as the screening of impurities in metals. So far only rough estimates based on *atomic* calculations are available [4.82, 83]. The uncertainty of the various approaches is related to the question of the various contributions to the so-called relaxation energy, e.g., intra- versus extra-atomic screening, and the importance of chemical effects. In Table 4.2 we compare data of 4*d* and 4*f* XPS binding energies, E_B referenced in each case to the vacuum level E_∞, for Eu atom [4.84], Eu metal and EuO [4.85]. From this comparison we deduce that

$$\Delta[E_B(4d,\text{ atom}) - E_B(4d,\text{ metal or EuO})] \sim 6\text{–}7\ \text{eV}$$

4 D.A. Shirley discussed this point at length [Ref. 1.1, Chap. 4].

Table 4.2. 4f and 4d XPS binding energies in atom and RE solid [eV]

Core level	E_B:gas phase[a]	E_B:metal[b]	E_B:EuO[c]
4d	137.5	131.2	130.8
4f	14	5	4

[a] Data for the gas phase, [4.85].
[b] ϕ_m= work function of Eu metal =3 eV, included.
[c] ϕ_{EuO}=2 eV, from [4.81], work function of EuO, included.

and

$$\Delta[E_B(4f, \text{atom}) - E_B(4f, \text{metal or EuO})] \sim 9\text{–}10\,\text{eV}\,.$$

In each case Δ is larger in EuO. The $4f$ levels, although the electron states with lowest binding energy in EuO, are found to have a Δ comparable to that of the much more tightly bound $4d$ states. We note also that, in the notation of *Herbst* et al. [4.37] (see Fig. 4.6)

$$\Delta = (-\Delta_- + \varepsilon_{4f}) \sim 1/2\,\text{Ry}$$

where ε_{4f} is the one-electron $4f$ energy. This theoretical estimate of Δ, which is not far from experimental values, identifies, as major contributions to Δ, the so-called screening and relaxation around the $4f$ hole. From experiment it appears that chemical effects (or shifts) on $4f$ electrons are not large; one would expect Δ to be different for Eu metal and EuO, an insulator with a 2 eV $4f-5d$ gap. The total effects of screening and relaxation in EuO and Eu metal are apparently of comparable magnitude despite the expected difference in screening and relaxation mechanism.

A major open question, in this context, concerns relaxation and screening in intermediate valence compounds, where we know that the $4f$ electrons in the ground state are pinned at the Fermi level.

4.3.3 Intermediate Valence (IV) Compounds

This is a rapidly expanding field of RE research, largely because many of the peculiar physical properties of IV RE compounds and alloys are not readily explained in the context of current solid-state theories. The application of photoelectron spectroscopy to the study of IV systems has been recently reviewed [4.86]. We shall therefore here add only a few recent examples, in order to demonstrate the power of photoelectron spectroscopy in the study of the electronic structure of IV compounds.

As was pointed out above, the multiplet structure can be used as a "fingerprint" to identify unambiguously the state of a RE ion in a solid in the

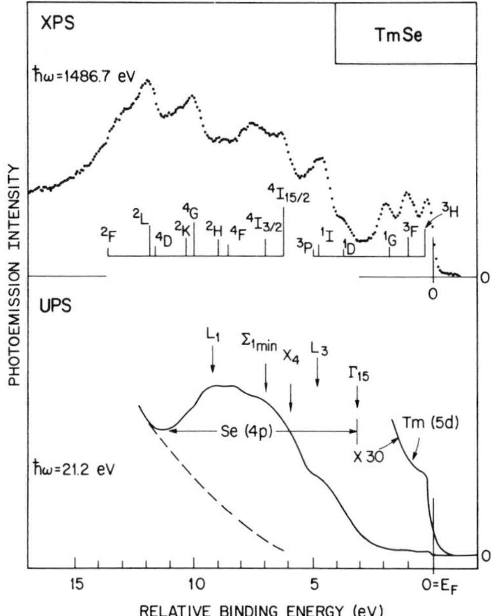

Fig. 4.33. UPS and XPS spectra of the valence band and the $4f$ region of the intermediate valence compound TmSe [4.87]

initial state prior to photoemission. From lattice constant measurements and other physical properties (susceptibility, resistivity, etc.) it has been concluded [4.86] that the Tm ions in TmSe are neither in the divalent $4f^{13}$ nor in the trivalent $4f^{12}\,5d^1$ state. The UPS spectrum of TmSe (see Fig. 4.33, lower part) confirms the metallic character of TmSe as resulting from Tm$5d$ electrons, while the XPS spectrum (Fig. 4.33, upper part) demonstrates unambiguously that TmSe is an IV material. Both valences of Tm can be identified in the final state through their corresponding multiplet structure. Within the language of the "measuring time of a technique", we can say that Tm ions in TmSe change valence, from 2+ to 3+ and vice versa, on a time scale much slower than the measuring time of the XPS technique.

Similar results have been obtained so far for many other compounds, confirming the picture deduced from the first measurement on TmSe [4.87]. Temperature-dependent studies on the system $Sm_{1-x}Y_xS$ [4.59] have further confirmed that the Sm $4f$ electrons are indeed responsible for the peculiar behavior of these alloys. This makes temperature-dependent XPS an ideal technique for studying temperature-dependent valence transitions in RE solids. The case of SmB_6 has been discussed at length in a recent publication [4.74]. Here too, both Sm valences can be easily detected by XPS (see Fig. 4.26). The ratio of the normalized areas under the two multiplets is in good agreement

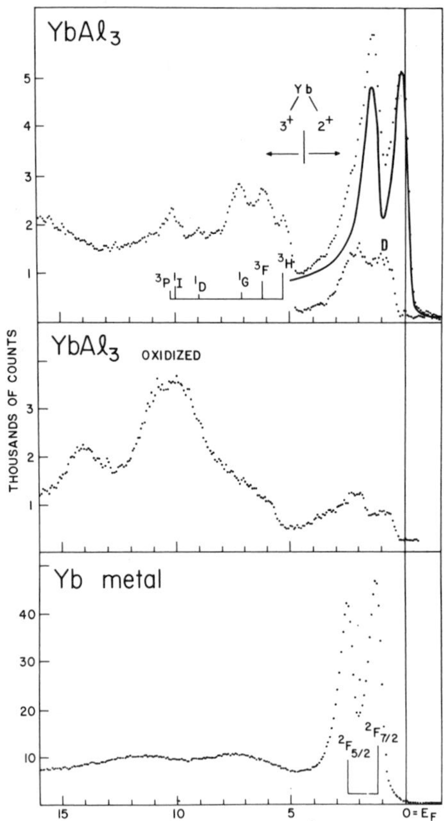

Fig. 4.34. XPS spectra of vacuum scraped and oxidized $YbAl_3$ and of Yb metal. The continuous line in the $YbAl_3$ case is obtained by convolution of the spin-orbit doublet ^2F with the instrumental resolution function and the many-body line shape [4.37]. *D* indicates the difference spectrum between the experimental points and the theoretical line

with the valence ratio, $Sm^{2+}/Sm^{3+} \sim 0.4$, deduced from other measurements. From the discussion in Sect. 4.3.2 we also know that the large $4f$ lifetime broadening is not related to the IV state of SmB_6 but is a result of the XPS process.

The IV phenomenon is expected to occur in compounds of Ce, Sm, Eu, Tm, and Yb. Examples for Ce (the metal), Sm and Tm have been given. We refer the reader to the literature for the case of Eu [4.76, 88]. We present in Fig. 4.34 a recent example concerning Yb [4.71]. Both $YbAl_3$ and $YbAl_2$ have been known for sometime as IV materials, as shown by their physical properties [4.89]. XPS on $YbAl_3$ confirms those findings. The cross section for emission from the Al conduction band and the Yb $5d$ states can, to a first approximation, be neglected when compared to that of the Yb $4f$, $\sigma(\text{Yb}4f^{14})/\sigma(\text{Al } 3s, p) \sim 33$. An estimate of the XPS ratio $Yb^{3+}/Yb^{2+} \sim 3$ is in reasonable agreement with the valence ratio inferred from lattice constant or high-temperature susceptibility measurements.

a) The Intra-Atomic Coulomb Correlation Energy U_{eff}

In Figs. 4.26 (SmB_6), 4.33, 34 the splitting between the signals of divalent and trivalent RE ions, of the order of 5 to 7 eV, can be identified as the experimental value of the intra-atomic Coulomb correlation U_{eff} [4.86]. It is the energy needed to make the transition $2 \times 4f^n \rightarrow 4f^{n+1} + 4f^{n-1}$, taking into account screening and relaxation. In the notation of the renormalized atom scheme $U_{\text{eff}} = (\Delta_-) + (\Delta_+)$, where Δ_+ is the energy necessary to convert one $5d$ electron from E_F into the $4f$ shell. U_{eff} has been estimated by *Herbst* et al. [4.12] and found to be of the order of 6.5 eV. This value is in good agreement with the experiment. The one-electron estimate, neglecting screening and relaxation, gives $U = 27$ eV at complete variance with experiment. It is this large value of U_{eff} that makes the $4f$ electron behave as if it were atomic-like in spectroscopy despite its apparent "itinerant" character evident in many other measurements. Clearly such a large correlation makes the IV phenomenon only a minor perturbation to the ionic structure of RE in IV materials.

4.4 Conclusions and Outlook

High-energy photoelectron spectroscopy of RE solids has now reached a degree of maturity well above the level of only 4 years ago. $4f$ binding energies, multiplet position and intensities are now well understood, both in metals and compounds. In contrast more work, both experimental and theoretical, is needed before such a statement can be made regarding $3d$-core levels. A full understanding of the $3d$ and $4d$ levels may be the only means by which ambiguous cases like those of Ce or Ce-alloys will be clarified. A field which will clearly receive increased attention in the near future is that of IV compounds. Especially from the theoretical point of view fundamental questions like the one concerning the screening of a "$4f$ hole at the Fermi level" still await answers. Temperature-dependent effects, like phase transitions, in which the $4f$ levels are known to play a major role, will also become the subject of more systematic investigations.

The combination of photoelectron spectroscopy with other electron spectroscopic techniques, a common feature in equipment available to physicists working in surface related problems, will certainly result in further progress of both RE research and photoelectron spectroscopy.

In closing we would like to point out that in the immediate future, the most interesting questions related to $4f$ electrons may well be those dealing with surface problems. We cite two examples: the existence of a magnetically nonsaturated layer at the surface of EuO, as detected by spin-polarized photoemission [4.90], see Fig. 4.35, and existence of surface phase transitions [4.91, 92]. The latter could be electronic in origin, e.g., of the type semiconductor-metal (or to an IV phase, as detected by electron energy-loss

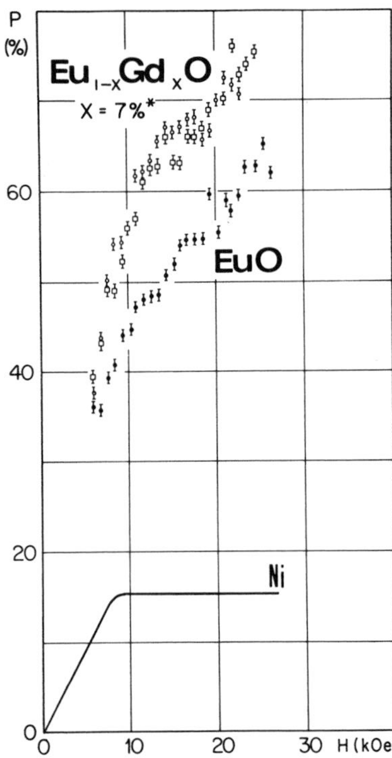

Fig. 4.35. Dependence of spin polarization P on magnetic field strength H for pure and Gd-doped single-crystal EuO and polycrystalline films of Ni at 10 and 4.2 K, respectively. For EuO and $Eu_{1-x}Gd_xO$ magnetic saturation is not observed, in contrast to Ni, indicating a nonsaturated surface layer well below $T_c \cong 70$ K for EuO and ~ 100 K for $Eu_{1-x}Gd_xS$

spectroscopy on SmS [4.92]) or magnetic in origin as proposed by *Fulde* [4.93] (and due to the change of the symmetry of the crystalline electric field acting on $4f$ levels of ions at the surface of RE compounds).

References

4.1 B.G. Wybourne: *Spectroscopic Properties of Rare Earths* (Wiley and Sons, New York 1965)
4.2 G.H. Dieke: *Spectra and Energy Levels of Rare Earth Ions in Cystals* (Wiley and Sons, New York 1969)
4.3 D.E. Eastman, F. Holtzberg, S. Methfessel: Phys. Rev. Lett. **23**, 226 (1969)
G. Busch, P. Cotti, P. Munz: Solid State Commun. **7**, 795 (1969)
4.4 G. Busch, M. Campagna, H.C. Siegmann: Solid State Commun. **7**, 775 (1969)
4.5 D.E. Eastman, F. Holtzberg, J.L. Freeouf, M. Erbudak: AIP Proc. **18**, 1030 (1973)
4.6 J.H. Scofield: J. Electron Spectrosc. **8**, 129 (1976)
4.7 D.E. Eastman, M. Kutznietz: J. Appl. Phys. **42**, 1396 (1971)
4.8 P.A. Cox, Y. Baer, C.K. Jørgensen: Chem. Phys. Lett. **22**, 433 (1973)
P.A. Cox: Structure and Bonding **24**, 59 (1975)
4.9 E.I. Zabolotskii, Yu.P. Irkhin, L.D. Finkel'shtein: Sov. Phys. Solid State **16**, 733 (1974)
4.10 G. Racah: Phys. Rev. **176**, 1352 (1949) and references cited therein
4.11 W.T. Carnall, P.R. Fields, K. Rajnak: J. Chem. Phys. **49**, 4412, 4443, 4447, 4450 (1968)
4.12 J.F. Herbst, N.D. Lowy, R.E. Watson: Phys. Rev. B **6**, 1913 (1972)
4.13 R.E. Watson, H. Ehrenreich, L. Hodges: Phys. Rev. Lett. **24**, 829 (1970)
L. Hodges, R.E. Watson, H. Ehrenreich: Phys. Rev. B **5**, 3953 (1972)

4.14 J.F. Herbst, R.E. Watson, J.W. Wilkins: Phys. Rev. B **13**, 1439 (1976)
4.15 Y. Baer, G. Busch: J. Electron Spectrosc. **5**, 611 (1974)
4.16 B. Johansson: J. Phys. F **4**, L169 (1974)
4.17 G.K. Wertheim, H.J. Guggenheim, S. Hüfner: Phys. Rev. Lett. **30**, 1050 (1973)
4.18 For a recent review of this subject see G.K. Wertheim: In *Electron Spectroscopy, Theory, Techniques, and Applications*, ed. by C.B. Brundle (Academic Press, New York, to be published)
4.19 J. Hedman, P.F. Heden, C. Nordling, K. Siegbahn: Phys. Letts. **29** A, 178 (1969)
4.20 C.S. Fadley, D.A. Shirley, A.J. Freeman, P.S. Bagus, J.V. Mallow: Phys. Rev. Lett. **23**, 1397 (1969)
C.S. Fadley, D.A. Shirley: Phys. Rev. A **2**, 1109 (1969)
4.21 A.J. Freeman, P.S. Bagus, J.V. Mallow: Int. J. Magn. **4**, 35 (1973)
4.22 See G.K. Wertheim: In *Electronic Structure of Inorganic Compounds*, ed. by P. Day (Reidel, Dordrecht-Holland 1975) p. 393 for a further discussion
4.23 D.P. Spears, M.J. Fischbeck, T.A. Carlson: Phys. Rev. A **9**, 1603 (1974)
4.24 T. Robert: Chem. Phys. **8**, 123 (1975)
M.A. Brisk, A.D. Baker: J. Electron Spectrosc. **7**, 197 (1975)
4.25 K. Siegbahn, D. Hammond, H. Fellner-Feldegg, E.F. Barnett: Science **176**, 245 (1972)
4.26 U. Gelius, E. Basilier, S. Svensson, T. Bergmark, K. Siegbahn: J. Electron Spectrosc. **2**, 405 (1973)
4.27 A. Barrie, I.W. Drummond, Q.C. Herd: J. Electron Spectrosc. **5**, 217 (1974)
4.28 Y. Baer, G. Busch, P. Cohn: Rev. Sci. Instrum. **46**, 466 (1975)
4.29 F.R. McFeely, S.P. Kowalczyk, L. Ley, D.A. Shirley: Phys. Lett. **45** A, 227 (1973)
4.30 P.O. Hedén, M. Löfgren, P.B.M. Hagström: Phys. Rev. **26**, 432 (1971)
4.31 S.B.M. Hagström, P.O. Hedén, H. Löfgren: Solid State Commun. **8**, 1245 (1970)
4.32 Y. Baer, G. Busch: Phys. Rev. Lett. **31**, 35 (1973)
4.33 G.S. Fleming, S.H. Liu, T.L. Loucks: Phys. Rev. Lett. **21**, 1524 (1968)
4.34 G. Brodén: Phys. Kondens. Mater. **15**, 171 (1972)
4.35 G.K. Wertheim, M. Campagna: Chem. Phys. Lett. **47**, 182 (1977)
4.36 G. Brodén, S.B.M. Hagström, P.O. Hedén, C. Norris: *Electronic Density of States*, ed. by L.H. Bennett, Natl. Bur. Stand. (US) Publ. **323**, 217 (1971)
4.37 S. Donaich, M. Šunjić: J. Phys. C **3**, 285 (1970)
4.38 S.C. Keeton, T.L. Loucks: Phys. Rev. **168**, 672 (1968)
4.39 G.K. Wertheim, A. Rosencwaig, R.L. Cohen, H.J. Guggenheim: Phys. Rev. Lett. **27**, 505 (1971)
4.40 K.A. Gschneidner, R. Smoluchowski: J. Less-Common Met. **5**, 374 (1963)
4.41 R. Ramirez, M. Falicov: Phys. Rev. B **3**, 2425 (1971)
4.42 B. Coqblin, A. Blandin: Adv. Phys. **17**, 281 (1968)
4.43 L.L. Hirst: J. Phys. Chem. Sol. **35**, 1285 (1974)
4.44 B. Johansson: Philos. Mag. **30**, 469 (1974)
4.45 C.R. Helms, W.E. Spicer: Appl. Phys. Lett. **21**, 237 (1972)
4.46 K.A. Gschneidner, R.O. Elliott, R.R. McDonald: J. Phys. Chem. Sol. **23**, 555 (1962)
4.47 V.A. Shaburov, I.M. Band, A.I. Grushko, T.B. Mezentseva, E.V. Petrovich, Yu.P. Smirnov, A.E. Sovestnov, O.I. Sumbaev, M.B. Trzhaskovskaya, I.A. Markova: Sov. Phys.-JETP **38**, 573 (1974)
4.48 V.A. Shaburov, A.E. Sovestnov, O.I. Sumbaev: Phys. Lett. **49** A, 83 (1973); Sov. Phys.-JETP **41**, 158 (1975)
4.49 J.F. Herbst, N.D. Lowy, R. Watson: unpublished. Improved calculations of the $5s$ binding energies (in analogy to the $\Delta_{-}(4f)$ of [Ref. 4.14]) have been recently performed by J.F. Herbst, R.E. Watson, Y. Baer: Phys. Rev. (to be published)
4.50 Y. Baer: Unpublished results
4.51 S. Sato: J. Phys. Soc. Jpn. **41**, 913 (1976)
4.52 *Proc. of Intern. Conf. on Magnetic Semiconductors*, Jülich, W. Germany (1975), ed. by W. Zinn (North-Holland, Amsterdam 1976)
4.53 J.F. Herbst, R.E. Watson, Y. Yafet, M. Campagna, G.K. Wertheim: Unpublished results
4.54 A.J. Freeman, R.E. Watson: Phys. Rev. **127**, 2058 (1962)

4.55 R.J. Birgenau, E. Buchner, J.P. Maita, L. Passel, K.C. Turberfield: Phys. Rev. B**8**, 5345 (1973)
4.56 M. Campagna, E. Bucher, G.K. Wertheim, D.N.E. Buchanan, L.D. Longinotti: Proc. 11th Rare Earth Res. Conf., Traverse City, Mich. 1974, (US AEC TIC, Oak Ridge, Tenn. 1974)
4.57 M. Campagna, G.K. Wertheim: Unpublished results
4.58 P.H. Citrin, P. Eisenberger, D.R. Hamann: Phys. Rev. Lett. **33**, 965 (1974)
4.59 R.A. Pollak, F. Holtzberg, J.H. Freeouf, D.E. Eastman: Phys. Rev. Lett. **33**, 820 (1974)
4.60 For a review of the properties of RE borides see K.E. Spear: *Material Science and Technology*, Vol. 4, (Academic Press, New York 1976) p. 91
4.61 H.C. Longuet-Higgins, M. de V. Roberts: Proc. R. Soc. London, Ser. A**224**, 336 (1954)
4.62 L.F. Mattheiss: To be published
A. Hasegawa, A. Yanase: To be published
4.63 J.-N. Chazalviel, M. Campagna, G.K. Wertheim, P.H. Schmidt, Y. Yafet: Phys. Rev. Lett. **37**, 919 (1976)
4.64 L.L. Hirst: Phys. Rev. Lett. **35**, 1394 (1975)
4.65a P.H. Citrin: Phys. Rev. Lett. **31**, 1164 (1973); J. Electron Spectrosc. **5**, 273 (1974)
4.65b R.M. Friedman, J. Hudis, M.L. Perlman: Phys. Rev. Lett. **29**, 630 (1972)
4.66 G. Güntherodt, N.J. Shevchik: AIP Conf. Proc. **29**, 174 (1975)
4.67 J.R. Cuthill, A.J. McAlister, N.E. Erickson: AIP Conf. Proc. **18**, 1039 (1973)
4.68 J.C. Slater: *Quantum Theory of Atomic Structure* (McGraw-Hill, New York 1960)
4.69 R.L. Cohen, G.K. Wertheim, A. Rosencwaig, H.J. Guggenheim: Phys. Rev. B**5**, 1037 (1972)
4.70 F.R. McFeely, S.P. Kowalczyk, L. Ley, D. A. Shirley: Phys. Letts. **49**A, 401 (1974)
4.71 S.P. Kowalczyk, N. Edelstein, F. R. McFeely, D.A. Shirley: Chem. Phys. Lett. **29**, 491 (1974)
4.72 W.C. Lang, B.D. Padalia, L.M. Watson, D.J. Fabian, P.R. Norris: Faraday Discuss. Chem. Soc. **60**, 37 (1975)
4.73 A.J. Signorelli, R.G. Hayes: Phys. Rev. B**8**, 81 (1973)
4.74 J.N. Chazalviel, M. Campagna, G.K. Wertheim, P.H. Schmidt: Solid State Commun. **19**, 725 (1976); Phys. Rev. B**14**, 4386 (1976)
4.75 G.K. Wertheim, G. Crecelius: Phys. Rev. Lett. **40**, 813 (1978)
4.76 I. Nowik, M. Campagna, G.K. Wertheim: Phys. Rev. Lett. **37**, 43 (1977)
4.77 G.K. Wertheim, R.L. Cohen, A. Rosencwaig, H.J. Guggenheim: In *Electron Spectroscopy* ed. by D.A. Shirley (North-Holland, Amsterdam 1972) p. 813
4.78 C.K. Jørgensen, H. Berthou: Chem. Phys. Lett. **13**, 186 (1972)
4.79 S. Suzuki, T. Ishii, T. Sagawa: J. Phys. Soc. Jpn. **37**, 1334 (1974)
4.80 N. Spector, C. Bonnelle, G. Dufour, C.K. Jørgensen, H. Berthou: Chem. Phys. Lett. **41**, 199 (1976)
4.81 H. Berthou, C.K. Jørgensen, C. Bonnelle: Chem. Phys. Lett. **38**, 199 (1976)
4.82 L. Ley, S.F. Kowalczyk, F.R. McFeely, R.A. Pollak, D.A. Shirley: Phys. Rev. B**8**, 2392 (1973)
4.83 R.E. Watson, M.L. Perlman, J.F. Herbst: Phys. Rev. B**13**, 2358 (1976)
4.84 C.S. Fadley, D.A. Shirley: Phys. Rev. A**2**, 1109 (1970)
4.85 P. Cotti, P. Munz: Phys. Kondens. Mater. **17**, 307 (1974)
4.86 M. Campagna, G.K. Wertheim, E. Bucher: Structure and Bonding **30**, 99 (1976)
4.87 M. Campagna, E. Bucher, G.K. Wertheim, D.N.E. Buchanan, L.D. Longinotti: Phys. Rev. Lett. **32**, 885 (1974)
4.88 K.H.J. Buschow, M. Campagna, G.K. Wertheim: Solid State Commun. **24**, 252 (1977)
4.89 E.E. Havinga, K.H.J. Buschow, H.F. van Daal: Solid State Commun. **13**, 621 (1973)
J.C.P. Klaase, W.C.M. Mattens, A.H. van Ommen, F.F. deBoer, P.F. deChatel: AIP Conf. Proc. **34**, 184 (1976)
4.90 M. Campagna, R. Sattler, H.C. Siegmann: AIP Conf. Proc. **18**, 1388 (1973)
F. Meier, H. Ruprecht: Comm. Phys. **1**, 137 (1976)
4.91 J.E. Rowe, M. Campagna, S.B. Christman, E. Bucher: Phys. Rev. Lett. **36**, 148 (1976)
4.92 P. Fulde: AIP Conf. Proc. **18**, 1 (1973)
4.93 Proc. 12th Rare-earth Research Conference, Vail, Colorado 1976 (unpublished)

5. Photoemission from Organic Molecular Crystals

W. D. Grobman and E. E. Koch

With 14 Figures

Molecular crystals are characterized by small lattice energies indicative of the weak bonding forces between the units constituting the crystal. The individual molecules, held together by these weak van der Waals forces, are only slightly affected by the solid environment. Thus the free molecule electronic structure largely determines the electronic structure of the crystal. It follows that in a first approximation (in the limit of vanishing intermolecular forces) the crystal may be thought of as an oriented gas. Interaction of the molecules in the crystalline state then perturbs the free molecule electronic structure producing new, solid-state effects in the electronic structure (e.g., crystal-field splitting) which differentiate it from an oriented gas.

Individual molecules are also the building blocks for donor-acceptor complexes. However, in this case a net charge transfer from one constituent to its partner takes place, adding ionic bonding to the van der Waals bonding and changing the zeroth-order electronic properties of the solid to those of molecular ions rather than those of neutral molecules.

For this reason, although the electronic properties of organic solids have been investigated for some time (see, e.g., [5.1–3]), interest in organic molecular crystals increased considerably with the recent discovery of unexpected metal-like behavior in some organic materials [tetrathiofulvalenium tetracyano-quinodimethane (TTF–TCNQ), for example]. The polymer $(SN)_x$ although basically inorganic, falls in the same category [5.4–10]. At that time it was realized that preparing and understanding even more interesting systems depends on gaining a detailed understanding of their solid-state properties, in particular their electronic structure. In addition, even classical systems such as alkanes and aromatic hydrocarbons have attracted new interest, since their simplicity lends them to more detailed studies, and a more extended basis for testing the validity of general conclusions can be provided. In fact, the absorption spectra of these systems in the gaseous and solid state [5.11, 12] as well as the photoelectron spectra in the gas phase [5.13–15] have been used as basic experimental tests of the predictive abilities of the various molecular orbital (MO) models for the electronic structure of organic π-electron systems [5.16–18]. Detailed photoemission experiments on organic solids, thin polycrystalline films or single crystals, are of more recent vintage although some early photoemission yield (e.g., [5.19–25]) and energy distribution curves (EDC) measurements (see, e.g., [5.21, 26–31]) exist. These older experiments were restricted to photon energies below the LiF cut-off for the exciting photons ($h\nu \lesssim 11.5$ eV), thus dealing mostly with the uppermost π-bonds. In the present review we have tried to emphasize more recent experimental progress. No attempt is made here to present an exhaustive treatment – we have rather tried to select examples which, in our opinion, are representative of some of the main directions of current research in this field. Thus much of the data presented consists of unpublished results obtained recently in our laboratory, rather than some of the older, more classical results in the literature.

We briefly mention also that the terms "photoemission", "photoyield", etc., are also often used in the literature of the photo-transport properties of organic materials, with a meaning different from that employed in this book. These photoconductivity and photoinjection studies form an extensive literature of their own – one which we do not attempt to discuss in this brief chapter. (A review is given in [5.32].) Our review is concerned with photoemission in which the excited electrons of interest leave the solid and are collected in vacuum without having undergone inelastic scattering in the solid. Thus the work presented in our review is complementary to the classical and very broad field of photoconductivity studies, and provides much new information not accessible to classical methods.

5.1 Some Experimental Aspects of Photoemission from Organic Molecular Crystals

5.1.1 Charging Effects

Probably the main reason why so little photoemission work has been performed on organic solids is the fact that they are usually very poor conductors. This fact led to a number of difficulties which are not present in photoemission studies of conducting materials – the main problems being charging effects and inadequate sample and surface characterization.

When electrons are emitted from the surface of an insulator a net positive charge remains, causing a decrease in the kinetic energies of the photoelectrons. In unfavorable cases (e.g., single crystals of $\gtrsim 0.1$ mm thickness) most of the structure in the energy distribution curves (EDC) will be washed out due to a spatially nonuniform charging of the sample surface. These charging effects have been discussed in connection with XPS studies in a number of papers [5.33–35]. Various methods, such as coating the samples with thin gold films, or smearing powdered solids onto metal substrates, or the use of an electron flood gun [5.36] have been tried. The use of a second, low photon energy light sources as a means for discharging the surface by the production and transport of carriers into the irradiated surface region by induced photoconduction has also been used advantageously [5.23, 27, 37]. Experiments with low intensities of the exciting light and with thin film samples (~ 100 Å thickness) gave the most satisfactory EDC's for insulators [5.37–39]. The main contribution to line broadening or smearing, nonuniform charging, could thus be almost completely eliminated.

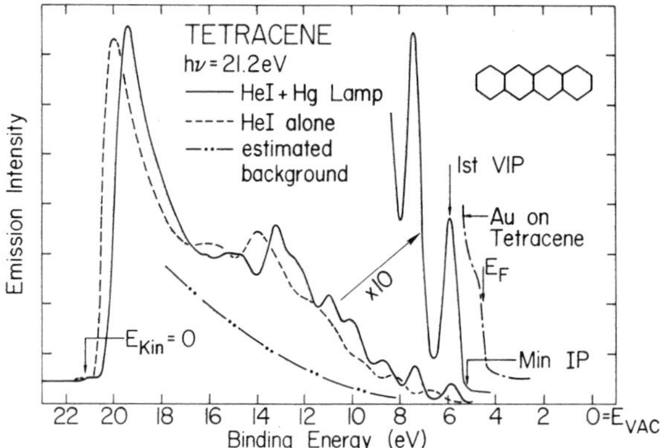

Fig. 5.1. He I (21.2 eV) photoelectron spectrum from polycrystalline tetracene films directly taken from a recording: the full curve is a spectrum where the sample is simultaneously illuminated with a low-energy $h\nu < 5.3$ eV Hg lamp; the dashed curve is the same spectrum without the additional lamp; the dashed double dotted curve indicates the estimated background; the dashed dotted curve is the photoemission from Au evaporated on top of the tetracene film. Min IP is the minimum ionization potential; 1st VIP is the first vertical ionization potential. All energies are referred to the vacuum level as the energy zero. [5.37]

In contrast, sample charging and sample preparation is usually not a problem in photoemission yield spectroscopy in which the number of photoemitted electrons per incident or absorbed photon is measured as a function of the energy of the exciting light, $h\nu$. These measurements can be performed with sufficiently high collecting voltages to avoid charging effects. The extrapolation of the yield curves according to $Y \propto (h\nu - I_c)^n$ where I_c is the threshold (Fowler's law for $n = 2$) is, however, not a straightforward procedure. Problems inherent in such an extrapolation have been discussed in a number of papers (e.g., [5.24, 25]).

The relatively high vapor pressure of most hydrocarbons (typical values range from $\sim 10^{-7}$ Torr for tetracene to $\sim 10^{-1}$ Torr for naphthalene at 300 K) implies that the surface is cleaned by steady sublimation from the sample. Thus procedures to clean the surface are generally not necessary. On the other hand, the slow evaporation changes the thickness of the sample, thereby changing the surface charge conditions so that in cases where the evaporation is too rapid, measurements on thin sublimed single crystals may not be possible.

As an illustration of some of these points, we show in Fig. 5.1 the experimentally determined photoemission spectrum for a thin polycrystalline film of tetracene [5.37]. The spectrum was obtained with He I (21.2 eV) radiation with a sample bias of -3 V on the Cu sample holder. The drastic influence of the secondary, Hg light source ($h\nu \lesssim 5$ eV) is obvious; the result of this influence is an enhancement of structure in the EDC and a shift of the spectral features towards higher kinetic energies. Thus this figure illustrates dramatically the fact that even though rather structureless spectra are observed when no special attention is given to charging effects, elimination of charging can permit sharp spectra to be revealed.

5.1.2 Secondary Electron Background

A feature common to most of the EDC's from organic materials is the *unusually* large background of inelastically scattered electrons, peaking near $E_{kin}=0$. In fact, this background makes it almost impossible to clearly distinguish features with binding energies larger than ~15 eV from the background. This background of secondary electrons is due to electron-electron and electron-phonon scattering processes. Electron-electron scattering has a threshold at a fairly large energy since the minimum energy for this process is equal to the excitation energy of the first singlet exciton E_S, of the order of 2–5 eV for these materials [5.12]. For example, for anthracene, electrons which are less than ~7 eV above the valence band edge have an unusually large yield and come from rather deep (~ several hundred Å) within the solid. Since this energy is greater than the minimum ionization energy of ~5 eV, one sees an unusually large secondary electron background in anthracene between ~0 and 2 eV of kinetic energy. Such an effect is common in many wide band gap organic insulators, and is one of the characteristic features of photoemission from organic solids. This effect is also seen in a few inorganic systems which have a large band gap and small electron affinity (see, e.g., [5.40]).

Finally, we note that in electron-phonon scattering one may differentiate further between an excitation of an intramolecular vibration (with a typical excitation energy of ~0.14 eV in the case of anthracene) and a loss due to scattering by a true crystal phonon with an average energy loss per collision in organic crystals of the order of 0.003 eV [5.21].

5.1.3 Electron Attenuation Length (Escape Depth) $\lambda_e(E)$

For the analysis of photoemission data the knowledge of the electron attenuation length $\lambda_e(E_{kin})$ is of fundamental importance (see Ref. 1.1, Chap. 1.]). Although for inorganic materials this quantity is well known and data have been compiled [5.41] and critically reviewed [5.42], the information available on the dependence of λ_e on E_{kin} for organic materials is meager. The few results reported in the literature do not lead to a consistent picture.

For instance, the electron escape length reported by *Pong* and *Smith* [5.43] for copper phthalocyanine (11 Å in thin films of thickness $d<40$ Å, and about 150 Å in thick films with $d>50$ Å) has been criticized by *Belkind* et al. [5.26] (in the case of the thin films) as being too small. These latter authors obtained 140 Å as the escape depth for tetracene films on a gold electron emitter when the electron energy was about 1 eV [5.26] above the vacuum level. Recently the electron attenuation length for electrons with kinetic energies ≦3 eV above the vacuum level has been determined by *Hino* et al. [5.44] for polycrystalline films of pentacene and perylene. In these experiments the hot electron current from a copper-iodide (CuI) substrate onto which the organic films were evaporated in situ was measured as a function of the thickness of the organic films. Thus values of 75 ± 10 Å for pentacene and 800 ± 80 Å for perylene films ($d \leqq 370$ Å) have been obtained. The large difference in these values obtained under very similar experimental conditions is quite surprising and has not yet been explained.

We mention also the measurements of the escape depth by *Berry* [5.45], *Chang* and *Berry* [5.46], and *Huang* and *Magee* [5.47] on a series of saturated hydrocarbons at 77 K, yielding values ranging from several Å to about 100 Å.

Recent studies of TTF–TCNQ and related compounds have yielded values of $\lambda_e(E_{kin})$ of $\lesssim 10$ Å [5.48] and ~ 4–5 Å [5.49] for $E_{kin} < 10$ eV above E_F. At much higher kinetic energies ($E_{kin} > 1000$ eV) X-ray photoemission spectroscopy results have suggested values greater than 50 Å for $\lambda_e(E_{kin})$ [5.50–52]. The short escape depths at low kinetic energy $E_{kin} \sim 5$ eV reported in these recent references are somewhat in contradiction with the small attenuation power of organic overlayers on ultrahigh vacuum-prepared substrates [5.53, 54], although these latter studies are for somewhat higher kinetic energies ($E_{kin} \sim 10$–15 eV). Further discussion of this matter is given in Sects. 5.6, 7.

Finally, we report on a very recent extensive series of measurements of $\lambda_e(E)$ using XPS for both organic solids (e.g., polyethylene, stearic acid, poly-vinyl chloride, etc.) as well as for the 3*d* transition metals and for the noble metals (5.47a). These measurements indicate that for $E \approx 1200$ eV, $\lambda_e(E)$ is about 14 Å for the transition and noble metals, but is about 60 to 70 Å for the organic solids, supporting the other literature which argues for long escape depths in organic films compared to metallic ones at high kinetic energy.

5.1.4 Vacuum Requirements

One of the most significant experimental differences in photoemission spectroscopy between organic solids and inorganic ones is the vacuum required. Due to the high vapor pressure and low chemical activity of many organic molecular solids, one can often obtain photoemission spectra intrinsic to the sample even under rather poor vacuum conditions ($\sim 10^{-5}$ Torr or better). For example, we can often prepare a sample in our laboratory and, within 5 min or less after introduction into the spectrometer, measure an intrinsic spectrum. However, one special warning obtained from our experience is that often this procedure does not give a reproducible result for the position of E_F with respect to the valence band edge as determined by subsequent metal overlayer evaporation. For this reason, the accuracy of work function measurements in a poor vacuum is questionable. This is probably due to the fact that we are dealing with variable amounts of band-bending under such conditions.

5.1.5 Effects of the Transmission Function of the Electron Energy Analyzer

Many of the figures in this chapter are based on our own not yet published spectra, obtained using a differentially pumped noble-gas discharge light source and a cylindrical-mirror electrostatic deflection energy analyzer. In this system, the analyzer pass energy is kept constant and spectra are scanned by retarding (or accelerating) electrons from the sample using a set of spherical grids centered on the position of electron emission. Spectra obtained in this manner are not subject to the severe transmission function distortion in intensity seen when the analyzer pass energy is scanned (as is often done in the 127° deflection analyzer commonly used for vapor phase spectra). This effect is illustrated by comparing the spectrum of TTF vapor in Fig. 5.10a, obtained with a 127° deflection analyzer, with that of TTF vapor in Fig. 5.11, obtained as described above. In the case of solids, all spectra which we have obtained using this method show the large secondary electron background discussed above, since we use an analyzer with a transmission function which does not vanish at zero kinetic energy.

5.2 Band Formation in Linear Alkanes

One striking feature of organic solids is the possibility of synthesizing an almost infinite number of slightly different molecules and of investigating systematic changes in the electronic structure of these molecules, and molecular crystals formed by these entities. An example of a physically realizable model of a quasi linear chain is the series of alkane molecules (formula C_nH_{2n+2}). Recently *Pireaux* et al. [5.55, 56] have studied the formation of a band structure by measuring the XPS valence band spectra of these molecules for $n=1$ to 13. By comparison of gas phase and solid phase spectra for $n=5$ to ∞, that is for n-pentane to polyethylene, they could show that $n=13$ (n-tridecane) already provides a convenient finite model for the band structure of an infinite linear polymer. This comparison also showed that the influence of intermolecular interactions on the band structure is rather small, the most significant difference between spectra from the two phases being a slight increase in the peak widths in the solid phase. As these saturated hydrocarbon molecules of the alkane series do not contain any polar group or dipole moment, it is quite plausible that the intermolecular interactions will not affect the solid-state spectra significantly.

The above observations are illustrated by the sequence of spectra displayed in Fig. 5.2. The XPS valence spectra consist of two parts – one originating from $C2p-H1s$ carbon-hydrogen bonding orbitals, and another in the region around 22 eV originating mostly from carbon $2s$ electrons of bonding character. The number of electronic levels in the latter region is equal to the number n of carbon atoms in the linear alkane chain. The $C2s$ molecular orbital splits with increasing number of carbon atoms, finally forming a band of overlapping levels of about 7.5 eV width and the typical intensity distribution characteristic of the density of states of a one-dimensional solid (see also Fig. 5.12). For $n=9$ (n-noneane), the envelope of the spectrum is already similar to that of the $n\rightarrow\infty$

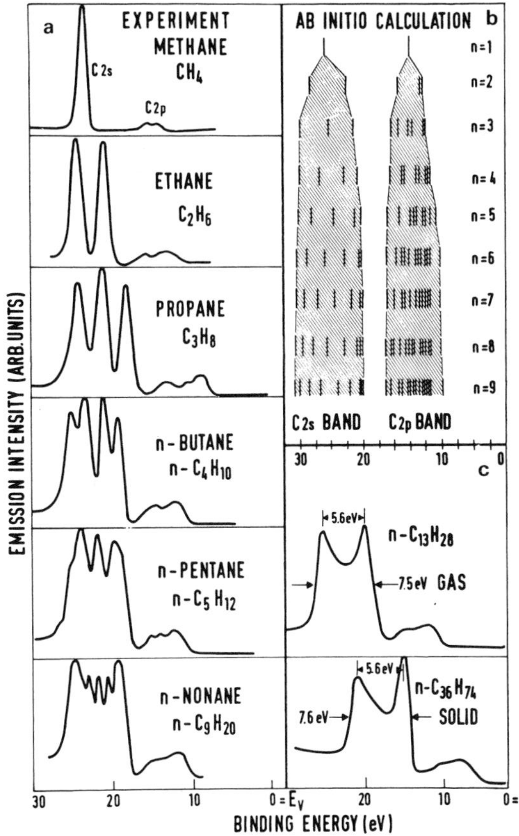

Fig. 5.2a–c. Valence band formation as studied by ESCA (electron spectroscopy for chemical analysis) for the linear alkane series. **a** Shows the valence electron spectra for methane, ethane, propane, n-Butane, n-Pentane and n-Nonane in the gas phase. **b** Gives the results of an ab initio analysis of the molecular orbitals of the alkanes. **c** Shows a comparison of the valence electron bands of gaseous n–$C_{13}H_{28}$ and solid n-$C_{36}H_{74}$. Note the close similarity of the two spectra. [5.55]

solid phase. For *n*-tridecane, the substructure in the C2*s* bands has entirely vanished. This spectrum already simulates the basic features of the electronic properties of polyethylene [5.57, 58].

We note in passing that the core levels have also been studied for these materials. The core level binding energies decrease with increasing number of carbon atoms in these molecules, due to relaxation effects [5.55, 59]. UPS studies of molecular crystals formed by the alkanes or by alkenes and alkines and their derivatives (all of which might be of considerable interest to the understanding and interpretation of UPS adsorption and surface chemistry studies [5.60]) have not yet been reported.

5.3 Aromatic Hydrocarbons

Molecular crystals formed by the aromatic hydrocarbons have been the subject of a number of photoemission studies (e.g., [5.1–3, 11, 12]). The electronic properties of the constituent molecules are largely determined by the sp^2-hybrid state leading to the formation of σ-bonds in the plane of the planar molecules and π-orbitals oriented perpendicular to the molecular planes. A large number of spectroscopic studies both in the gas and the solid phase [5.11, 12, 61] as well as theoretical calculations (see e.g., [5.16–18, 62–64]) have been performed in order for us to understand and describe their electronic structure. Because in the single crystals (for the structure of these molecular crystals see [5.65]) the units occupy definite and oriented positions, the optical spectral are anisotropic. It should thus be an interesting task to investigate the photoemission from single crystals as a function of both the polarization direction of the exciting light and the direction of emission of the outgoing electrons relative to the crystal axes. Such a program has, however, due to the experimental difficulties mentioned in Sect. 5.1, not yet been performed (below are some preliminary results).

In the following we shall discuss mainly photoemission results for molecular solids formed by members of the acene-series (benzene [5.53, 66–68]), naphthalene [5.69], anthracene (e.g., [5.19, 20, 23, 26]) tetracene (e.g., [5.21, 37, 70–73]) and pentacene [5.73, 74] which have been studied most thoroughly, with anthracene frequently being regarded as the prototype substance for organic molecular crystals. For photoemission work on other aromatic hydrocarbons, we refer to [5.21, 26, 69, 70, 73, 75]. We note that although the structure of these molecules is still fairly simple, the number of electrons is so large that ab initio calculations for the molecular orbitals (MO's) including *all* valence electrons have been reported only for benzene (e.g., [5.62]) and naphthalene [5.63]. Thus even for gas phase photoemission [5.13–15, 76–79] discussion has been mainly restricted to the π-electron bonds and only a few attempts have been made to partially disentangle the σ-bonds (e.g., [5.80]). Even less effort has been spent on calculating the band structure of the solids [5.81].

5.3.1 Acene

The He I, 21.2 eV spectra for solid polycrystalline films of benzene [5.53, 66], naphthalene [5.69], anthracene and tetracene [5.37, 69] are displayed in Fig. 5.3, where we compare them to the corresponding gas phase photoelectron spectra. Inspection of Fig. 5.3 reveals some general trends: while the EDC is still fairly simple for benzene, the spectra become more crowded with increasing number of aromatic rings; the lowest ionization potential (IP) decreases monotonically, and the Fermi level E_F moves closer to the top of the valence bands. Furthermore there is a remarkably good correspondence between the observed solid state EDC's and those of the gas phase. The rigid shift of $\Delta E_R \approx 1.15$ eV (by which one has to lower the binding energies of the gas phase in order to obtain coincidence for almost all pronounced structures in both phases) is nearly constant for all four molecules.

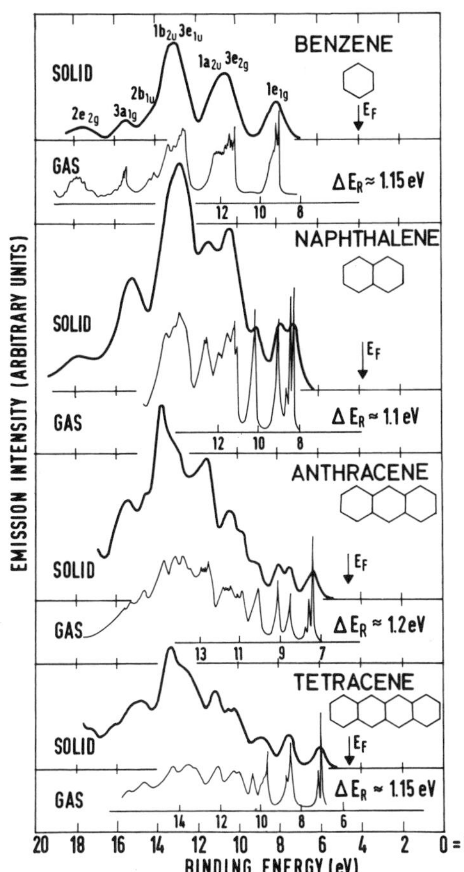

Fig. 5.3. Valence band spectra for solid benzene, naphthalene, anthracene and tetracene polycrystalline films, using He I radiation. The gas phase spectra. [5.13, 77] are shown for comparison. ΔE_R gives the rigid gas to solid relaxation shift by which the binding energy scales of the gas phase spectra have been shifted in order to obtain coincidence for most of the prominent bands. [5.69]

For a further discussion of the spectra we start from the assignment for the benzene spectrum. After a long debate [5.82, 83] about the relative ordering of the $2b_{1u}$ and $3a_{1g}$ orbitals (we use here the standard notation for the representations of the D_{6h} point group of benzene) and the $3e_{2g}$ and $1a_{2u}$ orbitals, the assignment given in Fig. 5.3 seems to be the generally accepted one. It is directly transferable to the EDC for solid benzene with some of the bands coalescing in the solid. We can then identify in the other compounds at least four distinct regions in the density of states: first the uppermost filled valence bands originating from pure π-orbitals; second, starting at about 10 eV binding energy, a region where bands from π- and σ-orbitals overlay; third, at even larger binding energies, from about 15 to 19 eV we have pure σ-bands; and fourth, at binding energies in excess of about 19 eV (not shown in Fig. 5.3), bands with essentially carbon 2s character are expected corresponding to the $2e_{2g}$ (19.2 eV), $2e_{1u}$ (23 eV) and $2a_{1g}$ (26 eV) bands in gaseous benzene (see Fig. 5.5 and 6).

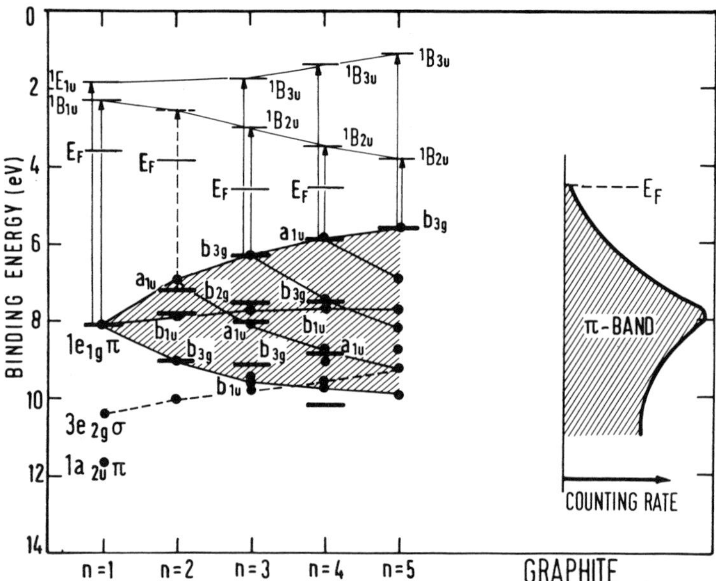

Fig. 5.4. Formation of the uppermost filled π-bands in solid acenes for $n=1$ (benzene) through $n=5$ (pentacene). The heavy solid bars give the experimentally determined binding energies of the filled bands, the dots represent the calculated binding energies [5.80] for the gas phase occupied orbitals adjusted at the $1e_{1g}$ π-band for solid benzene. The position of the Fermi level and the lowest $\pi-\pi^*$ transitions in the solid are also indicated. For comparison the EDC for graphite ([5.84]) is also shown. [5.69]

For the uppermost π-bands the results are displayed for the acenes together with calculated band positions [5.80] in Fig. 5.4. (For similar calculations see also e.g., [5.77].) For this comparison of the experimental results with the calculation actually performed for gaseous compounds, the calculated band positions have been shifted rigidly in order to obtain coincidence for the benzene $\pi-e_{1u}$ orbital. Again as in the case of the alkanes, the gradual formation of a homogeneous band can be observed with increasing size of the molecules. In this case it is a continuous band formed by the π-electrons leading to the limiting case of the graphite π-electron valence band [5.84].

Included in Fig. 5.4 are also the position of the Fermi level E_F and the lowest $\pi-\pi^*$ transitions observed for the solids [5.12]. The E_F values relative to the top of the valence band show a clear decreasing trend towards the situation encountered for graphite. We will return to a more detailed discussion of anthracene below.

5.3.2 Organometallic Phenyl Compounds

Because of their close similarity to the EDC's of solid benzene it seems appropriate at this point to briefly discuss recent results for the valence bands of a number of solid Sn, In, Sb, and Pb organometallic compounds containing (among others) phenyl (Ph) groups. These spectra together with data on the outermost *d*-core levels have been obtained recently by *Bancroft* et al. [5.85a] using synchrotron radiation as the photon source. Thus the dependence of the relative cross sections on the energy of the exciting light could also be studied.

The tetraphenyl tin (Ph_4Sn) spectrum is shown in Fig. 5.5 as an example. The resolution in the valence band region is good enough (0.3 eV at $h\nu = 57$ eV) to detect and assign a large number of valence band peaks. As can be seen from Fig. 5.5, the Ph_4Sn spectrum at 57 eV (curve *c*) is very similar qualitatively to the spectrum of gaseous benzene (curve *a*), with the shoulder at 5.7 eV binding energy being the only marked additional feature. This shoulder increases in intensity as the photon energy increases to 65 eV (curve *b*). *Bancroft* et al. [5.85a] conclude that the 5.7 eV shoulder is due to the t_2 Sn–C bonding molecular orbital, an interpretation consistent with the gas phase Me_4Sn (Me = Si, Ge, Pb) spectra [5.86a] in which the t_2 Me–C_1 peak lies at 9.7 eV (0.3 eV below the $1e_{1g}$ peak in benzene). The a_1 component of the Sn–C bonding orbital is expected at a binding energy of about 12.4 eV, and may be obscured by the $3a_{1g}$ benzene peak.

The variation of the areas of the major peaks with the energy of the exciting photons is shown in the insert of Fig. 5.5. It is apparent that band 4, which is mainly due to the benzene $3a_{1g}$ as well as the Sn–C a_1 orbital, remains at the same relative intensity as the first band, which is largely due to the $1e_{1g}$ benzene and the Sn–C t_2 orbital. This intensity information is consistent with the assignment of the 12.4 eV peak (16.9 eV) in the gas phase) to the $3a_{1g}$ orbital of pure *p*-character rather than the $2b_{1u}$ having *s*-character. Peaks 5 and 6 increase

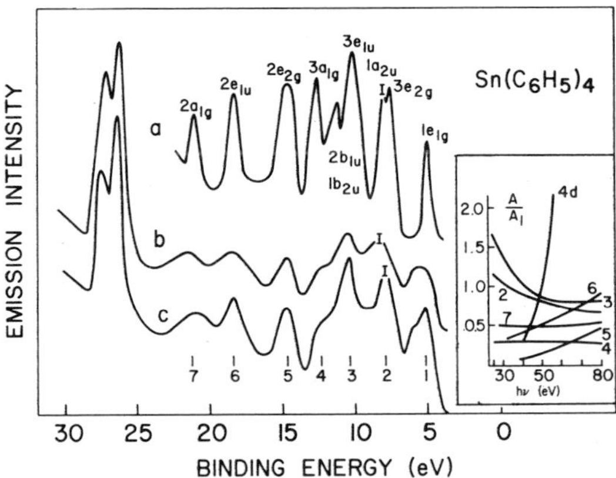

Fig. 5.5. Photoelectron spectra of benzene and tetraphenyl tin (Ph_4Sn). Curve *a* is the spectrum of gaseous C_6H_6 taken with 30 eV photons and 0.35 eV system resolution; Curve *b* is the spectrum for solid Ph_4Sn taken with 65 eV photons and 0.5 eV system resolution; Curve *c* is the spectrum for solid Ph_4Sn taken with 57 eV photons and 0.3 eV system resolution. The error bars indicate the standard deviation of the count. The *insert* shows the relative cross sections of major peaks in the Ph_4Sn spectra (normalized to the low-binding energy peak $1e_{1g}$) as a function of photon energy. [5.85a]

in intensity substantially, as expected for MO's having large C *s*-character (see e.g., 5.86b, c) and in agreement with the general discussion of the acene spectra given above. We note that the intensity of the 4*d*–Sn line increases with exciting photon energy in agreement with the cross section of Sn (see Fig. 5.5).

In these studies a marked broadening of the Sn 4*d* peaks at binding energies 25.69 and 26.73 eV was observed (5.85a) due to an unresolved ligand field splitting (compare Fig. 5.5 with [Ref. 5.86d, Fig. 5]). Electric field gradient splittings of Cd and Sn 4*d* energy levels in organometallic compounds have been studied and discussed in detail by *Bancroft* et al. [5.85b].

For further photoemission work on organometallic compounds including X-ray spectra see the references in [5.85] as well as [5.87, 88] and the review article by *Fenske* [5.89] on the molecular orbital theory, chemical bonding and photoemission spectroscopy of transition metal complexes.

5.3.3 Anthracene

Anthracene has been the most important material for the experimental and theoretical study of the electronic properties of molecular crystals. It forms a monoclinic crystal with two molecules per unit cell belonging to the space group C_{2h}^5, $P2_1/a$ (see [5.65]). The projection of the two molecules forming the unit cell onto the (001) face (*ab*-plane), which is a natural cleavage plane, is shown in the insert of Fig. 5.7.

Much detailed information about the excited states (Frenkel-type excitons) has been obtained from optical and electron energy loss studies [5.12, 61, 90]. Far less is known for anthracene about the deeper lying valence bands, where photoemission studies can be of paramount importance. A large amount of data is now available for anthracene, relevant for photocarrier generation (see e.g., [5.91, 92]), photoinjection studies from metal contacts (e.g., [5.92, 93]) and, to a lesser extent, for the exciton dynamics studied by photoemission (see Sect. 5.4). Here we confine ourselves to a discussion of the band structure, as revealed by photoemission experiments which have been reported over the years by a number of groups [5.19–26, 69].

EDC's of good quality from thin single crystals of anthracene have only recently become available [5.69]. In Figs. 5.6, 7, He I and He II spectra are shown, obtained from single crystals with the photons impinging perpendicularly onto the (001) *ab*-plane. In these experiments thin single crystal sublimation flakes have been mounted so that the electron energy analyzer, a cylindrical mirror type (CMA) device, accepts electrons from a 42° cone with the tip of the cone at the illuminated spot of the sample and the CMA axis (the axis of the cone) lying in the plane of the sample surface. By turning the sample around its normal, spectra could be obtained with *k* predominantly parallel, or *k* predominantly perpendicular, to the crystallographic *b*-axis, where *k* denotes the direction of the outgoing electrons. In Fig. 5.6 spectra averaged over the two directions are presented, while in Fig. 5.7 angle-resolved EDC's are shown.

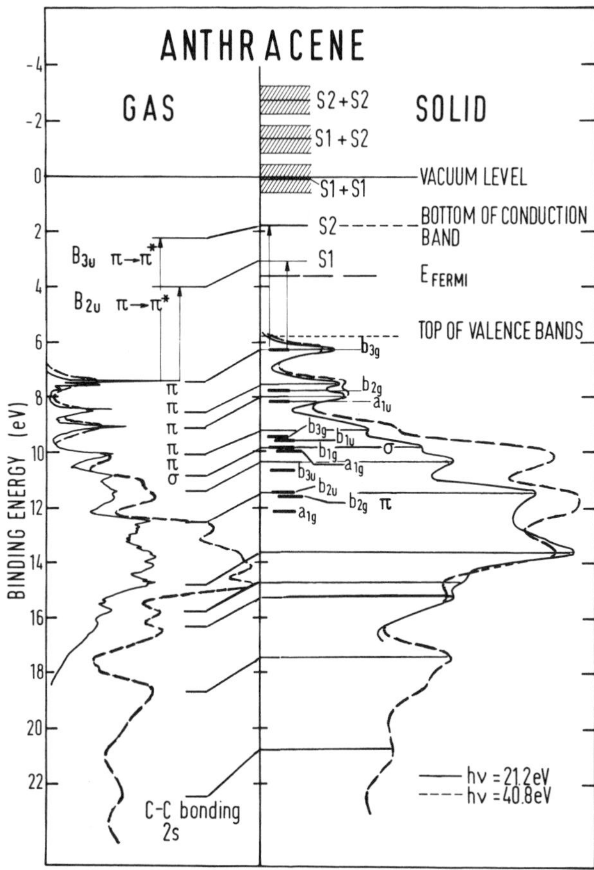

Fig. 5.6. Comparison of the He I (solid curves) and He II photoelectron spectra (dashed curves) for gaseous (left) and solid (right) anthracene. The lowest short axis polarized (B_{2u}) and long axis polarized (B_{3u}) transitions to singlet states and Frenkel excitons ($S1$ and $S2$) respectively are indicated. The shaded areas give the energies above the uppermost band available from exciton annihilation processes. For solid anthracene comparison is made with calculated band position according to [5.80] subjected to a rigid shift (see text). [5.69]

These data from single crystals clearly show a number of distinct individual sub-bands which in several cases are not broader, and sometimes are even more pronounced than in the gas phase. The change in relative intensities in going from He I to He II spectra should also be noted. Thus, for instance, a strong increase of the intensity at around 10 eV binding energy, the onset of the σ-bands is observed.

In Fig. 5.6 the filled bands are compared to the calculated binding energies for the highest 11 occupied electron orbitals as obtained using the CNDO (complete neglect of differential overlap) method by *Lipari* and *Duke* [5.80]. These uppermost 11 occupied bands include all 7 π-orbitals of anthracene except for the lowest π-band. For the comparison with experiment the calculated binding energies have been shifted rigidly by 1.96 eV to lower binding energies in order to obtain coincidence for the uppermost b_{3g} band. (Note that the relaxation shift (gas to solid) is only 1.2 eV (see Fig. 5.3). In this way an assignment of a large number of valence band peaks results. In

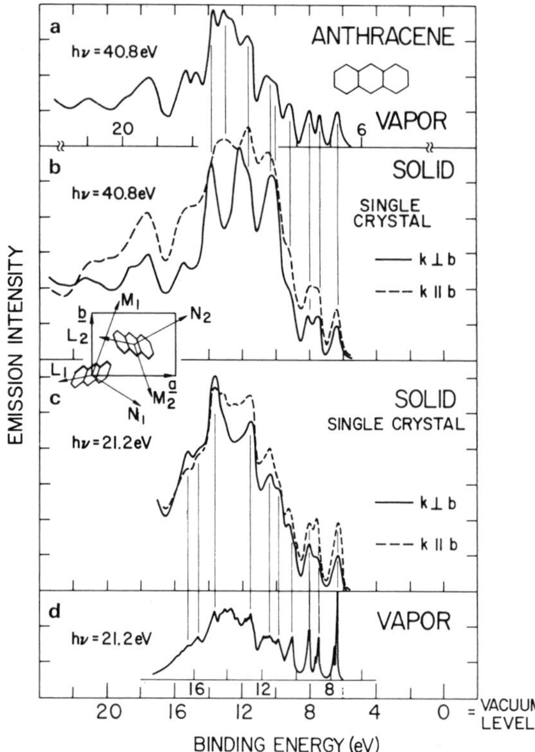

Fig. 5.7a–d. Photoelectron spectra from anthracene single crystals (background subtracted). The light [He I (Part c) and He II radiation (Part b)] was incident normal to the (001) crystal face. The insert shows the projection of the unit cell onto this face (*ab*-plane). Spectra have been recorded for two different directions *k* of the outgoing electrons with respect to the *b*-axis (see text). For comparison gas phase spectra for the two different photon energies (the He I spectrum is taken from [5.76]) are shown. [5.69]

particular the onset of the σ-bands is placed at a binding energy of about 10.0 eV. This is consistent with the strong enhancement of the band at around 10 eV when 40.8 eV radiation is used.

The implications of the band structure of anthracene for the analysis of the optical spectra have been discussed to some extent by *Cazaux* [5.94] and by *Cook* and *Lecomber* [5.95], although in both cases a knowledge of the deeper lying valence bands as provided by photoemission experiments was missing. There is general agreement as far as transitions in the energy range $h\nu \leqq 5$ eV are concerned that they originate from the uppermost b_{3g} band. In Fig. 5.6 we have indicated the two allowed $\pi-\pi^*$ transitions from the ground state [$A_{1g} \rightarrow B_{2u}$ (short axis polarized) and $A_{1g} \rightarrow B_{3u}$ (long axis polarized)] giving rise to the S1 and S2 singlet excitons in the solid phase at about 3.2 and 4.6 eV, respectively. Comparison with the position of the conduction band [5.32] shows that the latter is formed essentially by the S2 exciton states.

The assignment of higher lying excited states of energy $\geqq 5$ eV in a simple one-electron picture is not straightforward, since at high energy a number of transitions from deeper-lying valence bands become possible. It is anticipated that future angle-resolved photoemission experiments, including spectra from the (010) face (where the long axes of the molecules are almost parallel and within the surface), together with a careful analysis of such experiments will yield a complete and satisfactory picture of all states involved.

Finally we note that we have depicted in Fig. 5.6 final states denoted by $S1+S1$, $S1+S2$, and $S2+S2$ which may be reached in exciton–exciton annihilation processes, as discussed in the next section.

5.4 Photoemission Induced by Exciton Annihilation

Illumination of organic materials with strongly absorbed light of photon energies several electron volts smaller than the reported threshold for photoemission due to single photon absorption, can lead to electron emission. This phenomenon, involving the annihilation of two excited states, was first observed by *Pope* et al. [5.96, 97] on crystallites of anthracene, tetracene and perylene. In their experiments, the rate of emission was found to be proportional to the square of the light intensity, indicating an effect involving two excitations, which is analogous to an Auger process.

An interesting technique was employed in these studies: the discharging of small positively charged crystallites suspended between the condenser plates of a Millikan oil-drop apparatus was measured when the crystallites were exposed to photons with $h\nu \leqq E_{TH}$, the threshold energy for direct emission (5.8 eV). Analysis of the data for anthracene gave evidence that two singlet excitons (3.1 eV each) annihilate to yield one photoelectron.

Later *Pott* and *Williams* [5.98], *Haarer* and *Castro* [5.99], and *Wache* and *Karl* [5.100] studied the same phenomenon in conventional photoemission yield experiments on large anthracene single crystals in an attempt to identify the nature of the excited states involved in the annihilation process.

The analogous internal charge generation process via exciton–exciton interactions has been studied far more extensively [5.32]: The possibility that exciton–exciton interactions could produce internal ionization in organic crystals wass first suggested by *Northrop* and *Simpson* [5.101] and was given theoretical and experimental support by *Choi* and *Rice* [5.102], *Silver* et al. [5.103], and *Braun* [5.104]. In addition to singlet–singlet exciton annihilation a number of other processes involving two excitations have been observed to lead to charge carrier generation: direct two-photon absorption from the ground state [5.105]; internal ionization of the singlet–exciton state [5.106]; internal ionization of the triplet–exciton state [5.107]; and singlet–triplet annihilation [5.108]. All of these processes, as well as interaction with charge-transfer excitons (nearest-neighbor electron-hole pairs), or with conduction-band electrons, are in principle capable of yielding external photoemission, provided the resulting final energy is large enough to promote an electron above the threshold energy. Thus, for instance, in tetracene and pyrene, two of the lower singlet excitons $S1$ do not have sufficient energy to cause photoemission [5.96].

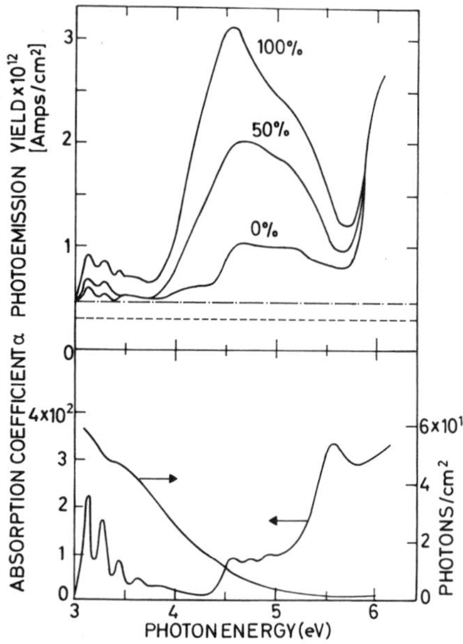

Fig. 5.8. *Upper part*: Anthracene photoemission yield spectrum taken for a single light source (dashed dotted curve) and in the presence of a second light source with $h\nu = 3.15$ eV at full (solid curve) and half (dashed curve) intensity. The shift in baseline below the dash-dotted line represents the current due to the 3.15 eV light by itself.
Lower part: Spectral output of the xenon lamp–monochromator combination (structureless curve) and the unpolarized absorption of crystalline anthracene (structured curve). [5.99]

Nevertheless double quantum photoemission has been observed for tetracene at around $h\nu \geqq 3.0$ eV [5.109] and the measured kinetic energy of the photoelectrons was such that two 2.9 eV entities must have been annihilated in the process. The first singlet–exciton energy for tetracene is 2.4 eV and the threshold energy for photoemission is ~5.4 eV. *Pope* et al. [5.109] therefore suggested that the 2.9 eV states found in the analysis of their data are charge transfer excitons.

Such charge transfer excitons had also been postulated as kinetic intermediates in the electron emission process by *Pott* and *Williams* [5.98] in order to explain the double quantum process occurring in anthracene when illuminated with light of energy 4 to 5 eV. However, *Haarer* and *Castro* [5.99] have shown by a double light source experiment that there is no need to postulate the participation of charge transfer excitons to explain the double quantum photoemission of anthracene for excitation energies larger than 4 eV. In this experiment a 600 W Xenon lamp served as the light source for two monochromators whose output could simultaneously illuminate the crystal surface. The photoelectron yield was measured at fields high enough (+300 V at a grid separated 0.5 cm from the emitting surface) to ensure a current independent of the voltage over the whole spectral region.

In Fig. 5.8 we reproduce the experimental results [5.99]. Photoemission yield curves from anthracene are shown for a single light source and in the presence of a second light source with $h\nu = 3.15$ eV at full and half intensity.

Three different processes manifest themselves at distinct energies in the photoyield:

I) for energies below $h\nu = 3.9\,\text{eV}$ the yield reproduces the absorption spectrum. The light intensity dependence varies as $I^{1.9\pm0.1}$. The presence of the second light source at 3.15 eV increases the yield by an amount proportional to the product of the intensity of the two light sources.

II) In the range above $h\nu \approx 3.9\,\text{eV}$ the yield has a complex spectral dependence. Here the light intensity dependence varies as $I^{1.5\pm0.1}$. There is a large increase of the photoemission current due to presence of the 3.15 eV light. The additional photoemission current depends linearly on the intensity of the 3.15 eV source.

III) In the energy range above 5.9 eV the strong photoemission due to the direct single photon process is unaffected by the presence of the second light.

For region I) these results are in agreement with the earlier conclusions [5.96–98] that two singlet Frenkel excitons ($S1$ at 3.15 eV energy) annihilate to produce an emitted electron. The large increase in photoemission current in the region above 3.9 eV, and its linear dependence on the intensity of the 3.15 eV light, shows that the enhanced photoemission process is due to a first-order interaction of Frenkel $S1$ excitons produced by the 3.15 eV light, with some particles produced by the primary excitation. *Haarer* and *Castro* [5.99] suggested as the most likely intermediates the bulk photogenerated electron produced by light of energy $h\nu \geqq 3.9\,\text{eV}$. We wish to point out that at about this energy level the low-energy onset of the absorption band due to the $S2$ exciton, which originates from a very intense, long axis polarised, molecular transition (see also the lower part of Fig. 5.8) is located [6.90]. Thus the enhanced photoemission can be attributed equally well to the annihilation of an $S1$ and $S2$ exciton (3.15 and 4.56 eV, respectively).

Since the higher exciton states are now fairly well known for a number of systems [5.12] it would be interesting to study the annihilation of these states in more detail. Use of polarised light to selectively excite different exciton states, and the possibility of energy analyzing the photoelectrons produced in the annihilation process, should yield new detailed information on the exciton dynamics. Preliminary EDC measurements on anthracene single crystals with exciting light below the threshold energy indicate [5.110] that there is at least more than one group of electrons (distinguished by their kinetic energies above threshold) originating from exciton–exciton annihilation. This shows that some of the higher kinetic energy intermediates have a lifetime long enough to take part effectively in the annihilation process prior to radiative decay or a radiationless transition to the $S1$ exciton state. Thus it seems that the fast photoemission process (fast compared to luminescence) can be exploited to study the exciton dynamics of higher exciton states in organic materials [5.111].[1] Further information may be obtained by time-resolved photoemission experiments, as carried out, e.g., by *Wache* and *Karl* [5.100] on anthracene

1 See *Schwentner* and *Koch* [5.111] for similar experiments on solid rare-gas molecular crystals.

with nanosecond pulses from an N_2 Laser in order to study the dynamics of the $S1$–$S1$ annihilation process.

It is our belief that studies in photoemission spectroscopy along these lines are particularly rewarding. The exciton annihilation process as described above is a very efficient way by which, subsequent to the absorption of a low-energy photon in the visible, higher energy states can be reached internally in an organic system. These processes take place on a time scale $\tau \leqq 10^{-12}$ s, and are of paramount importance in the interpretation and prediction of channels for energy dissipation in radiation chemistry and biology (see, e.g., [5.112, 113]).

5.5 Photoemission from Biological Materials

No survey of photoemission from organic solids would be complete without reference to work on biological materials. Although preparation and sample characterization is an even bigger problem than for simpler organic systems, some biologically important molecules such as phthalocyanines (PC), a number of dyestuffs, and nucleic acid bases can be prepared in thin film form by vacuum sublimation.

5.5.1 Phthalocyanines

The molecules of the metal PC's are derived from PC by replacing the two central hydrogen atoms with a metal (see inset, Fig. 5.9). The molecular structure of all the PC's is similar to that of chlorophyll and to that of blood pigments. Thus the electronic structure of these materials has been the subject of a number of investigations (see, e.g., references in [5.2] and [5.114]). In addition, molecular crystals of this class promise to have important technological applications as dyestuffs and in organic semiconductor devices [5.115, 116]. In Fig. 5.9 are shown recent photoemission data [5.69] for metal-free PC as well as for magnesium and lead PC. Also plotted for the sake of comparison are the results of earlier studies by *Vilesov* et al. [5.23], *Schechtman* and *Spicer* [5.27], and *Pong* and *Smith* [5.43], for copper PC. Even though the PC's are fairly large molecules, one can still observe a set of clearly distinct bands in the EDC's. Perhaps the most significant result of all the data is the fact that the EDC's are, within experimental error, the same for all the materials studied. From this, one may conclude that the density of the valence bands sampled in UPS studies is largely determined by the conjugated ring structure, since this is the part common to all molecules. Apparently, different substituents perturb the electronic structure only slightly.

Comparison to the EDC's of condensed benzene films (see Fig. 5.9) shows even further that, except for the highest band at 5.4 to 5.6 eV, the bands from 7 to ~14 eV have a close similarity to the benzene bands. In fact it had already

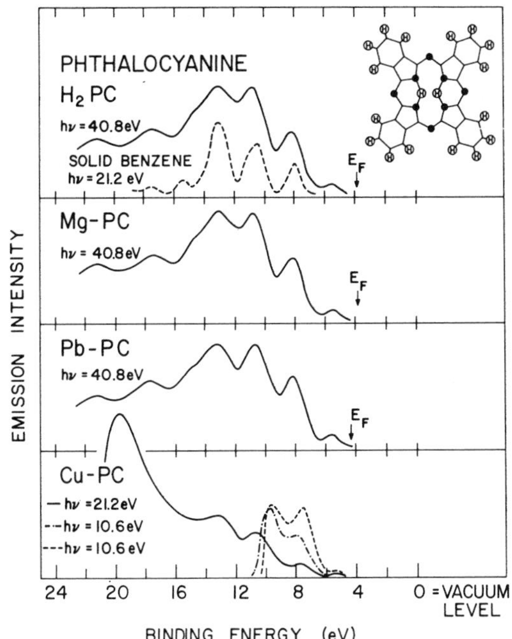

Fig. 5.9. Photoelectron spectra for H_2, Mg, Pb, and Cu-phthalocyanines. The insert shows the structure of the molecules (full circles indicate N-atoms; all positions not specifically indicated are occupied by C-atoms). The spectrum for solid benzene is given for comparison (dashed curve). Spectra for H_2, Mg, and Pb-phthalocyanine with the background of scattered electrons subtracted are from [5.69]; for the spectra for Cu-phthalocyanine the background has not been subtracted (full curve from [5.43], dashed curve from [5.27] and dash-dotted curve from [5.23])

been observed in uv optical studies [5.117][2] that some of the optical features of PC correspond to features in the benzene spectrum. Furthermore, neutron diffraction data [5.118] show that the molecule can be considered as almost completely planar and that the C–C distance within the phenyl rings is smaller (1.39 Å) than the C–C distance to the remaining molecular framework (1.49 Å). The uppermost band at around 5.5 eV binding energy is due to a π-molecular orbital ($4a_u$ in D_{2h} symmetry) with strong contributions from the carbon $2p$ orbitals situated on the C–N ring [5.114]. The binding energy of this orbital in the free molecule has been calculated as 5.71 eV [5.114], which is not far from the observed values in the solid films.

X-ray photoelectron spectroscopic studies of the electronic structures of a series of porphyrin and phthalocyanine compounds have been reported by *Zeller* and *Hayes* [5.119] and *Signorelli* and *Hayes* [5.120]. *Bettridge* et al. [5.121] have reported some difficulties with the gold decoration technique as a reference for binding energies when applied to phthalocyanine.

We mention also recent XPS and UPS studies by *Kawai* et al. [5.122], in which the chemisorption of various molecules (formic acid, HCOOH; pyridine C_5H_5N and water H_2O) over metal-phthalocyanines has been investigated.

2 For a detailed study of the near-infrared to vacuum ultraviolet absorption spectra and the optical constants, see *Schechtman* and *Spicer* [5.117].

5.5.2 Nucleic Acid Bases

In a recent study *Pong* and *Inouye* [5.123] measured the quantum yields of adenine, thymine, cytosine and uracil films for photon energies 7–23 eV thus extending previous work by *Subertova* et al. [5.124] on adenine. In addition, EDC's obtained using He I radiation have been reported in [5.123]. The EDC's exhibit up to four distinct features which can be associated with density-of-states structures below the vacuum level. Compared to the calculated energy of the highest occupied orbital for the isolated molecule, the photoemission thresholds were shifted by 1.5 to 2.0 eV towards smaller binding energies.

5.6 Valence Orbital Spectroscopy of Molecular Organic Conductors

The study of organic "metals" is a rather new field for physicists. Its interest was greatly stimulated by reports of unusually high dc conductivities for the salt of TCNQ (tetracyanoquinodimethane) with TTF (tetrathiofulvalene) for $T \leqq 300\,K$ [5.4, 5, 6]. Further interest was aroused when a different type of "organic" material, the long chain polymer $(SN)_x$ was found to have high room-temperature conductivity [5.7] and, later, to be superconducting at sufficiently low temperatures [5.8].

In the brief space available for discussion of molecular organic conductors, we will concentrate rather narrowly on the still small literature on photoemission (both UPS and XPS) from the valence levels of these two materials. Surveys of other types of work on these compounds are in [5.8–10]. Core level spectroscopy of these and related compounds are discussed in the next section.

We will start with a discussion of the charge transfer salt TTF-TCNQ, covering the nature of its electronic structure in relation to that of its constituents, the amount of charge transfer in this material, surface structure, and photoemission studies of the Se–substituted analogue of TTF. Next we discuss recent calculations of the electronic band structure of $(SN)_x$ and their relation to UPS and XPS valence level measurements. Finally, we discuss an important feature common to both TTF-TCNQ and $(SN)_x$ – namely the absence of a Fermi edge in photoemission. This result is the one experimental feature which is very different for these "organic metals" compared with *all* inorganic metals.

5.6.1 Valence Bands of TTF-TCNQ and Related Compounds

One of the central questions concerning the electronic structure of TTF-TCNQ is the degree of charge transfer – i.e., the electron charge per molecule transferred from the donor (TTF) stack to the acceptor (TCNQ) stack. Several photoemission studies of both neutral molecules (in the solid or vapor) and charge transfer salts have addressed this problem.

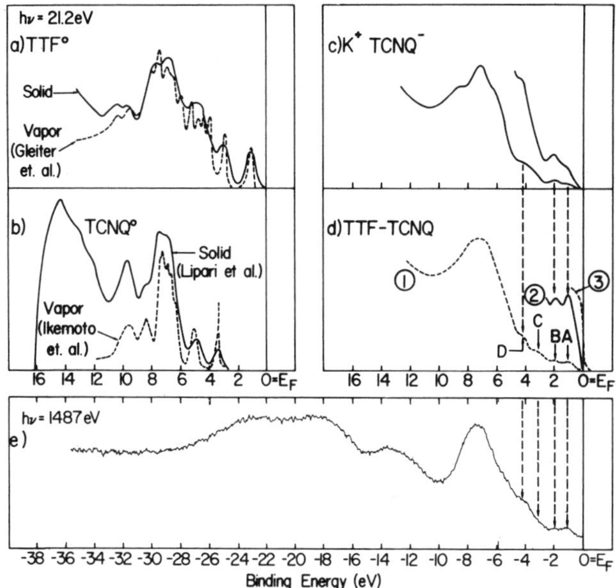

Fig. 5.10. **a** Comparison of UPS spectra ($h\nu = 21.2$ eV) of neutral TTF in the vapor and solid phase. Both spectra are obtained with a deflection-type energy analyzer operating in the mode in which pass energy is swept, thereby decreasing the amplitude of structure at lower kinetic (greater binding) energy. (Compare with Fig. 5.11, bottom curve.) This figure is taken from [5.126], in which the dashed curve was obtained from [5.134]. **b** Spectra of solid [5.127] and vapor phase [5.129] neutral TCNQ. Again the two spectra indicate very different energy analyzer transmission functions, as is evidenced by the dramatic difference in peak intensities at low kinetic energy. **c** and **d** These figures compare the high-lying orbitals of K^+ $TCNQ^-$ and of TTF–TCNQ, showing the proposed correspondence between orbitals of the two substances as discussed in [5.126–128, 48]. **e** This XPS spectrum of TTF–TCNQ shows the very close correspondence between the UPS and XPS spectra, even though the escape depth is probably quite different for the two materials. Spectra in **c**, **d**, and **e** are from [5.126]

We start by reviewing photoemission measurements on neutral TCNQ and TTF. One of the first valence band spectra of TCNQ° was obtained using XPS [5.125] and shows predominantly σ-levels bound by $\leqq 5$ eV (below E_F). UPS valence spectra of solid TCNQ appear in [5.126–128, 48], while [5.129] gives a vapor phase spectrum. All workers agree on the spectrum, shown for both solid and vapor phase TCNQ, for $h\nu = 21.2$ eV as given in Fig. 5.10b. (The solid TCNQ spectrum of [5.126] has the wrong energy scale, the spectrum being stretched in energy by ~ 20–30%, and should not be used. Other spectra in this reference are correct and agree with those in [5.127, 128, 48].) The position of the highest-lying orbital of the solid with respect to the Fermi level of the metallic substrate varies somewhat depending on the substrate [5.127] or method of preparation, but is always at ~ 3 to 4 eV below E_F.

The uppermost two emission features (at 3.5 and 5.0 eV of binding energy in Fig. 5.10b) are identified by X_α [5.130], ab initio [5.131, 132], and spectroscopy-parameterized CNDO calculations [5.133] as being composed of orbitals of b_{1u}, b_{3g}, and b_{2g} character (π-orbitals). CNDO/2 and MINDO/2 calculations [5.132] give a different ordering, with σ-orbitals included among the least bound orbitals. One important feature of such calculations is that they provide also affinity level wave functions (when the calculation is performed for the negative ion) which can be used as a tight-binding basis function for modeling the conduction electron wave functions in TTF-TCNQ.

Similarly, the photoemission measurements on TTF solid [5.126, 127], and TTF vapor [5.134, 135] provide information which, in conjunction with calculations of the TTF electronic structure [5.134–137] generally predict high-lying π-orbitals of b_{1a}, b_{2g}, b_{1u}, $2a_u$, and b_{3g} character [5.134, 135]. These calculations are in agreement even though they include three different approaches (extended Huckel [5.134, 135], X_α [5.136], and semiempirical [5.137]).

Figure 5.10 compares solid and vapor phase measurements of neutral TTF and TCNQ and demonstrates several important general features. One is that the good donor TTF, in the form of a thin solid film on Au, has the onset of emission from its highest-lying orbital within a few tenths of an eV of E_F; the strong acceptor TCNQ has the corresponding feature positioned ~ 3 eV below E_F in Fig. 5.10b (~ 2 eV in [5.126]). In [6.127] it is noted that the precise value of this gap is dependent on substrate, and obviously varies somewhat for different sample preparation conditions (compare [5.126] and [5.127] values). Nevertheless the conclusion, that a good electron donor shows little or no gap between E_F and the valence band edge, while a good electron acceptor has a several eV gap, is a general one.

Also, the strong correspondence between solid and vapor spectra is a very general feature [5.126, 133] and demonstrates the important point, already seen previously in this chapter, that in van der Waals solids, the interaction between molecules is very weak so that the electronic structure of the individual molecules (e.g., as seen in vapor phase spectra) totally dominates the solid-state spectrum.

We now refer to the K^+ $TCNQ^-$ spectrum (Fig. 5.10c), and the information on charge transfer which one can obtain from valence spectra. The importance of the K^+ $TCNQ^-$ spectrum (also reported in [5.127], but at a low-photon energy) is that it gives a good representation of the $TCNQ^-$ spectrum, since K^+ has no valence electrons. As the comparison between Figs. 5.10c and d shows, peaks A, B, and D in the TTF-TCNQ spectrum correspond exactly to peaks in the K^+ $TCNQ^-$ spectrum. On the other hand, superposition of the neutral TTF and TCNQ spectra does not give a reasonable representation of the TTF-TCNQ spectrum, even if one attempts to achieve a fit by a relative shift of the spectra of the neutral species. On this basis, in [5.126] it was concluded that the spectrum of TTF-TCNQ is best understood if peaks D and B (and 1/2 of peak A) are ascribed to $TCNQ^-$, and that peak C and 1/2 of peak A come from TTF^+.

Similar types of pictures are drawn in [5.48, 49, 127–129] (experimental) and [5.138] (theoretical), although in all of these papers the authors emphasize: I) that a significant amount of neutral molecule contribution to the spectra is consistent with either $h\nu \leqq 11$ eV or $h\nu = 21.2$ eV data; and II) that the surface layers which photoemission probes are possibly different than the bulk in their degree of charge transfer. Point I), on which there seems to be general agreement, was also made in [5.126]. The problem of the character of the surface layers probed by photoemission will be discussed further at this point.

The authors of [5.48, 49, 127, 128] have presented various types of data collected with $h\nu \leqq 11$ eV (e.g., surface cesiation [5.48, 128] layer by layer deposition of films [5.140], and measurement of the inelastic photoelectron attenuation length $\lambda_e(E)$ [5.140]); on the basis of these data they have argued that a short escape depth [$\lambda_e(E) \leqq 10$ Å [5.48] or $\lambda_e(E) \approx 4$–5 Å [5.49]] for $E < 10$ eV above E_F makes an analysis of bulk charge transfer difficult or impossible to base solely on UPS data. These arguments are well presented in the references mentioned, and we will here present arguments which lead to the opposite conclusion. We feel that the resolution of the disagreement in this area will occur when more and different types of data are available, but that there is value at this time in presenting the arguments in favor of a significant bulk-like contribution to the $h\nu = 21.2$ eV TTF-TCNQ spectrum of Fig. 5.10d.

The arguments against the interpretation that $\lambda_e(E)$ is so short that Fig. 5.10d is not typical of the bulk electronic structure are of two kinds. The first argument notes that Fig. 5.10d is obtained for $h\nu \approx 21.2$ eV so that $\lambda_e(E)$ might be greater than for spectra obtained for $h\nu \leqq 11$ eV. That $\lambda_e(E)$ may be longer at $E - E_F \sim 20$ eV than for $E - F_F \sim 5$ to 10 eV is suggested in [5.139] in which a plot of $\lambda_e(E)$ vs E already shows $\lambda_e(E)$ increasing at E–$E_F \approx 10$ eV. Also, it is well known that saturated overlayers of organic molecules on clean substrates (for example, see [5.53, 54]) yield UPS spectra for $h\nu \approx 21.2$ eV which show only an ~ 20–30% attenuation of the substrate emission by the molecular overlayer. Such data support the suggestion that $\lambda_e(E)$ for electrons excited ~ 20 eV above E_F in organic molecular films is *not* ~ 1 monolayer as suggested by [5.49], an experiment whose data are obtained for excited electrons only ~ 5 eV above E_F.

A second argument that the $h\nu = 21.2$ eV spectrum of TTF-TCNQ (Fig. 5.10d) is typical of the bulk, comes from a comparison of this spectrum with the XPS valence orbital spectrum ($h\nu \approx 1487$ eV) of TTF-TCNQ (from [5.127]) shown in Fig. 5.10e. The UPS and XPS spectra are remarkably similar, including the relative amplitude of the uppermost four orbitals. Since the relative amplitude of these orbitals is determined by the ratio of charged to neutral species in TTF-TCNQ, and since the XPS escape depth in organic solids is $\gtrsim 20$ Å (see [5.50–52]), the similarity of these spectra is a strong argument that bulk-like emission is dominating the 21.2 eV spectrum.

We now turn to photoemission data on tetraselenofulvalene (TSeF), the analogue of TTF which has Se in place of the four sulfurs of TTF. This molecule is especially interesting because one can compare the properties of the "organic metal" TSeF-TCNQ with those of TTF-TCNQ to better elucidate the

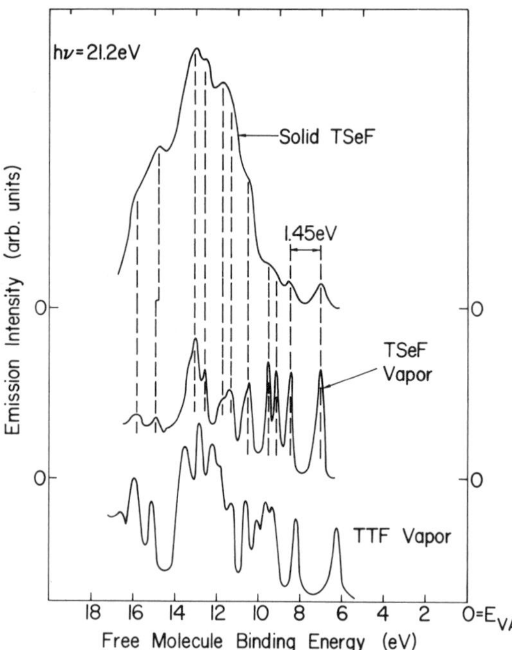

Fig. 5.11. The UPS spectra of solid phase from [5.142] and vapor phase TSeF are shown to be in close correspondence, as in the case of TTF° and TCNQ° in the previous figure, even though TSeF is a bent molecule in the solid state [5.142]. The solid phase spectrum is a UPS spectrum with an estimated background subtracted. For comparison, we also show a spectrum of TTF vapor, obtained using a constant energy analyzer pass energy. This latter spectrum should be compared to that in Fig. 5.10a, taken with the usual 127° deflection-type energy analyzer

nature of the electronic structure of these materials (see [5.145a, b]). For this reason, we present in Fig. 5.11 a comparison between the vapor phase spectrum of TTF and of TSeF [5.141], and of solid TSeF [5.142]. One first notes that TSeF is not as good a donor as TTF – the minimum ionization potential of TSeF being ~6.8 eV compared to ~5.9 eV for TTF. On this basis one might expect a smaller degree of charge transfer in TSeF-TCNQ than in TTF-TCNQ. Using electrochemical, spectroscopic, and mass spectroscopic measurements, the authors of [5.143] also obtained a greater ionization potential for TSeF than for TTF, although the absolute value of their result for the ionization potential of TSeF (7.21 ± 0.05 eV) is ~0.4 eV too high.

In Fig. 5.11 we also present an "optical density of states" (UPS spectrum with the secondary electron background subtracted) for solid TSeF. This spectrum has been arbitrarily shifted to show that, as in previous compounds, the solid and vapor phase spectrum are identical except for the polarization shift of the solid-state spectrum. This similarity of vapor and solid phase spectra is especially interesting since X-ray diffraction data [5.142] show that the unit cell of TSeF contains two "bent" molecules and one planar one. The similarity of the experimental solid and vapor spectra then may indicate that this molecule is rather "soft", so that bending does not change its electronic structure, or may simply imply that the escape depth even with $h\nu = 21.2$ eV excitation is smaller than our previous arguments would indicate. A more detailed discussion of UPS spectroscopy of TSeF is given in [5.142].

5.6.2 Valence Bands of $(SN)_x$

Due to the importance of $(SN)_x$ as the first true metallic [5.7] (and superconducting [5.8]) polymer, photoemission density of states overviews are of great interest. Such overviews enable one to check the accuracy of currently available theoretical band structures, e.g. [5.144–147], which can then be used to help us understand many electronic properties of $(SN)_x$ in more detail. The most complete theoretical work of *Salahub* and *Messmer* [5.145] (3-dimensional extended *Hückel* calculations) and *Rudge* and *Grant* [5.146] (non-self-consistent OPW method) agree qualitatively, predicting states of predominantly π-character at the top of the valence bands, mixed $\pi+\sigma$ character at around 3 to 5 eV binding energy and essentially σ-character at lower binding energies. These features have been labeled in the theoretical density of states curve in Fig. 5.12a. Also, both calculations predict strong dispersion of the energy bands for the direction ΓZ and along other lines parallel to the ΓZ chain axis of the polymer in the Brillouin zone, whereas almost completely flat bands are calculated for all directions perpendicular to the ΓZ direction.

The first photoemission valence band overviews from polycrystalline films were obtained using XPS [5.148, 149], while more recently UPS spectra have been obtained [5.150] from single crystals. Also, some UPS studies at low energies ($h\nu \lesssim 11.5$ eV) have been reported in [5.151], and *Stolz* et al. [5.152] recently succeeded in measuring angle-resolved EDC's from single crystals.

We combine, in Fig. 5.12, the spectra of [5.148–150] and [5.152], and the theoretical density of states from [5.146]. A detailed discussion of the spectra in comparison to the band-structure calculations appears in [5.150, 152], and we summarize the results here.

In Fig. 5.12a we have designated the various parts of the theoretical density of states spectrum by π, $\pi+\sigma$, and σ, depending on whether the state density receives its main contribution from the upper most π-, the mixed $\pi+\sigma$, or the σ-bands (for the correct ordering of the bands see, e.g., [5.145b]). One sees from Fig. 5.12a that the general features of the theoretical density of states correspond well to the general features of the $h\nu = 40.8$ eV UPS density of states overviews, and that the XPS spectra give a less detailed picture of the upper π and $\pi+\sigma$ valence bands, although the features due to the σ bands at 7 to 11 eV and 11 to 18 eV binding energy are seen more clearly in XPS. The 40.8 eV single-crystal spectrum in Fig. 5.12a is obtained in [5.150] by adding spectra taken with two orientations of an $(SN)_x$ crystal with respect to the analyzer. In this experiment the surface plane contains the b axis (fiber axis), which is placed either parallel to or perpendicular to the axis of the cylindrical mirror analyzer. In both cases the excitation light is incident normal to the crystal surface.

Differences in the XPS spectra and the 40.8 eV spectra have to be attributed to the energy-dependent cross section for s- and p-electrons, apart from possible contamination effects (note that the spectra in [5.148–151] have been obtained from samples prepared outside the spectrometer). Spectra from in situ prepared polycrystalline films or from in situ cleaved single crystals should be

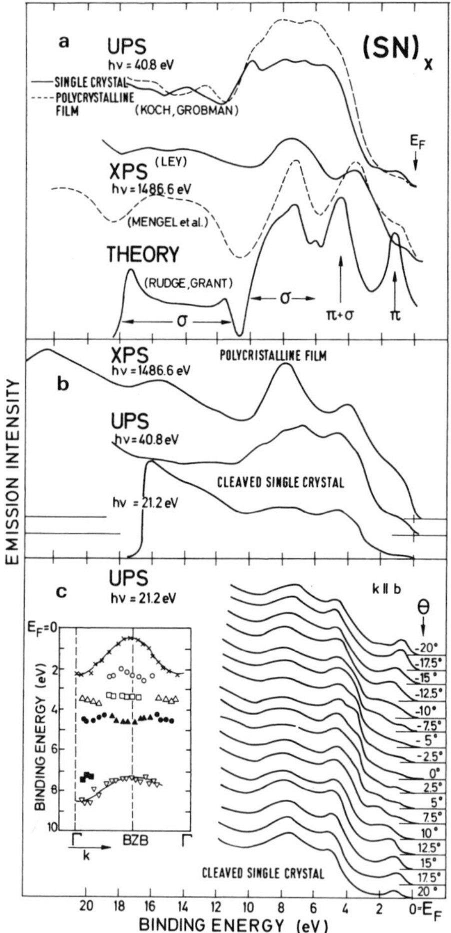

Fig. 5.12. **a** XPS, UPS, and theoretical valence band densities of states for $(SN)_x$ are compared, and show the general agreement between theory and the experimental position of prominent valence band state density features. The distortion of this state density by photon-energy-dependent optical cross sections is also apparent. [5.155]
b XPS spectra from in situ prepared polycrystalline films and UPS spectra at $h\nu = 40.8$ and 21.2 eV from in situ cleaved $(SN)_x$ single crystals. [5.152]
c Angle-resolved EDC's at 21.2 eV from in situ cleaved single crystals with the direction of the emitted photoelectrons parallel to the fiber axis b. θ denotes the polar angle. The insert shows a plot of peak positions as a function of $|k_{\parallel}|$ derived from the spectra [5.152]. $k_{\parallel}$ is the electron momentum component parallel to the crystal surface

free from possible distortions due to surface contamination. Such spectra as obtained by *Stolz* et al. [5.152] are displayed in Fig. 5.12b. Some differences between UPS and XPS spectra remain.

Further detailed information about the energy band dispersion of $(SN)_x$ was obtained from angle-resolved EDC's by the same authors [5.152]. Generally spectra with $k \parallel b$ (where k denotes the direction of the emitted photoelectron and b the fiber axis in the sample surface) showed strong dispersion, whereas for k perpendicular to the b axis no dispersion (within the limits of the experimental accuracy) was observed. Spectra for $k \parallel b$ are displayed in Fig. 5.12c. They have been obtained with the light source (angle of incidence $\theta_{\omega} = 45°$), the sample, and the entrance slit of the electron analyzer in a coplanar geometry. Since the wave vector parallel to the surface is conserved in photoemission, a variation of the polar angle θ changes the absolute value of $k_{\|}$ according to $k_{\|}=\sqrt{2E_{\mathrm{kin}}}\cdot\sin\theta$ where E_{kin} is the kinetic energy of the emitted electron. A collection of spectra with $k\|b$ for various angles θ appears in Fig. 5.12c, together with a plot of peak positions as a function of $|k_{\|}|$ derived from the spectra (insert). Assuming that final-state effects can be neglected, the latter plot can be related directly to the dispersion of the energy bands. In the range from 3 to 10 eV binding energy one observes a general downward trend of the upper bands and an upward dispersion of the lower bands, when going from Γ to the zone boundary. Above 2.5 eV the averaging of the properly weighted calculated conduction bands over the lines ΓZ, DB, CY and EA of the reciprocal lattice yields a dispersion which follows almost exactly the experimental one. Further, a conduction band width (averaged) of 2.4 ± 0.4 eV is deduced, which coincides well with more complete band-structure calculations [5.145, 146]. The important conclusion of this work is that the XPS and UPS spectra confirm the general features of the orthogonalized plane wave (OPW) band-structure calculation, and support the idea that this calculation can form a reliable basis for understanding further many features of the band structure of this important material.

5.6.3 The Absence of a Fermi Edge in Photoemission Spectra of Organic "Metals"

A remarkable feature common to photoemission studies of both TTF-TCNQ and $(SN)_x$ is the absence of a sharp emission edge (limited only by kT broadening and analyzer resolution) at E_F, as is observed in *all* inorganic metals that have been studied to date (see the discussion in [5.150]). This property may simply be due to an extremely low density of states at E_F as suggested in [5.152]. However, it seems to us that the reason might be more fundamental, namely that this qualitatively new feature of these organic "metals" is related to their lowered dimensionality, at least as far as the "fast" ($\tau \sim 10^{-16}$ s) photoemission experiment is concerned. The picture which emerges is, that on the short time scale of photoemission the hole left behind is rather localized, and the associated localized charge deforms the crystal (i.e., causes bond length changes) so that the electron excited from near E_F does not gain energy $h\nu$, since a small amount is left behind in bond vibrations (see the discussion of this picture in [5.126, 127]).

This unique feature of emission from states near E_F does not necessarily mean that "slower" phenomena (e.g., dc or microwave conductivity) obtain an anomalous, nonmetallic character in these materials. However, the fact that this striking qualitative difference between these "metals" and all other metals is seen in photoemission warns, that caution should be exercised in applying to these materials concepts which have been developed on the basis of experience with all previously studied, inorganic metals.

5.7 Core Orbital Spectroscopy of Organic Molecular Crystals

The application of XPS spectroscopy to the study of the core levels of organic molecules has been extensively reviewed in the past. Early studies were first comprehensively reported in [5.153] by *Siegbahn* and co-workers, and a more detailed review of the relation of core level binding energies to chemistry (i.e., to the charge distribution on molecules) was given in [5.14, 154, 155]. A recent general review of the field is found in [5.156] and [Ref. 1.1 Chap. 1], while the application of this technique specifically to structure and bonding in polymers is given in several work (see [5.157] and references therein). This brief list of references will serve to lead the interested reader to the wide variety of topics in this field which have been reviewed in the past, including the application of core-level spectroscopy to core-level binding energies (and charge distribution on molecules), linewidths and their relation to single- and many-electron broadening phenomena, and satellite structure and its relation to elementary excitations of molecules and solids.

In the present, necessarily short section, we wish to emphasize the new concepts developing in the application of core-level spectroscopy, concepts which can aid our understanding of charge transfer in organic crystals composed of molecules which are partially or fully ionized (molecular charge transfer salts). The studies listed above have emphasized that in neutral organic molecular crystals the relative binding energy of various core levels is little affected by solid-state effects. We show this in Fig. 5.13a ([Ref. 5.154, Fig. 5]) in which C1*s* core orbital chemical shifts in free molecules are plotted vs. the corresponding shifts for the same molecules condensed in the solid. Solid-state effects due to the charges in the solid acting on the core levels of a particular molecule cause at most a 10% correction to the shifts already present due to the distribution of valence orbital charge *within* a given molecule. This observation reiterates a dominant theme of the present review article which holds for valence orbital spectroscopy – namely that the electronic structure of molecules in a solid is totally dominated by that of the free molecules.

5.7.1 Solid-State Effects on Core Levels in Charge Transfer Salts

In charge transfer salts a new phenomenon arises. Due to the gross charges on individual molecules, a large spatially-varying crystal potential is produced which may cause additional XPS chemical shifts. These shifts can be used to understand the nature and amount of charge transfer from donor to acceptor molecules.

To put the problem in more detailed perspective, we turn to the "potential model" of XPS chemical shifts discussed in [5.14, 154, 155] [Ref. 1.1 Chap. 1]. In this model the atomic core-level binding energy is the reference energy, and

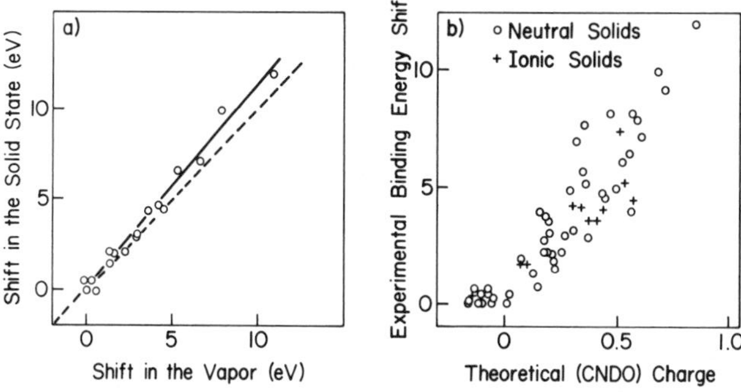

Fig. 5.13. **a** (Taken from [5.154]) XPS C 1*s* binding energy shifts (from the free atom value) for carbon atoms in organic molecules in the vapor state vs the shift for the corresponding quantity, measured when the sample is in the solid state. The least squares straight line fit to the data (solid) is very close to the unity slope line (dashed) passing through the origin. **b** (Taken from [5.154]) The experimental C1*s* core orbital shift shows the same dependence on theoretical atomic charge (in the molecule) for neutral (circles) and ionic (crosses) molecular solids

the shift from this energy in a molecule is due simply to the electrostatic potential due to three sources: the extra valence charge on the atom in question, the net charges (atomic charges) on the other atoms of the molecule of interest, and the atomic charges of the rest of the crystal. This can be written

$$\Delta E_i^A = E_i^A\,(\text{molecule}) - E_i^A\,(\text{atom}) = \frac{Q_i^A}{R_{ii}^{AA}} + \sum_{j \neq i} \frac{Q_j^A}{R_{ij}^{AA}} + \sum_{j,B \neq A} \frac{Q_j^B}{R_{ij}^{AB}}. \tag{5.1}$$

The first term on the right-hand side of (5.1) gives the core orbital shift due to the extra valence charge Q_i^A on atom i (beyond that of the neutral atom) of molecule A. The constant R_{ii}^{AA} in the denominator can be adjusted to give agreement between calculated values of Q_i^A and experimental values of ΔE_i^A as has been done, for example, in [5.154]. The physical interpretation of R_{ii}^{AA} is that it is a measure of the distance of the "shell" of valence charge from the nucleus of atom i. The second term gives the electrostatic potential at the center of atom i due to the charge on atom j, both being part of molecule A and separated by a distance R_{ij}^{AA}. Finally, the third term in (5.1) gives the electrostatic potential at the center of atom i of molecule A due to the individual atomic charges on all the other molecules of the crystal.

In a classical series of studies, the group at Uppsala under *Siegbahn* performed theoretical calculations of the Q_j^A, measured the values of ΔE_i^A, and thereby demonstrated the linear relationship between the C1*s* core-level binding energy shifts and the atomic charges predicted by (5.1). Figure 5.13b, which is [Ref. 5.154, Fig. 11], shows this linear correlation for a wide variety of organic molecular crystals (both neutral and ionic). In this study, the theoretical atomic charges Q_i^A were obtained using the semiempirical CNDO/2 molecular orbital calculation technique described by *Pople* and *Beveridge* [5.18]. Note that the linear relation between the atomic charges and binding energy shifts is a test not only of the "potential model" ((5.1) and [5.154]), but also of the theoretical molecular orbital model. Other types of theoretical models, including ab initio ones, also give good agreement with (5.1), see [5.155]. However, in all of these papers, the third, "solid-state" term, is replaced by a *constant* for all atoms of the crystal.

This fact is the central one of interest to us here, for model calculations of the third term in (5.1) (e.g., [5.158, 159]) predict, that in ionic crystals there should be large ($\sim$1 eV) shifts induced between core levels of atoms which are crystallographically inequivalent. On the other hand, based on Fig. 5.13a, *Gelius* et al. observed that "... Evidently the detailed effects of the crystal potential to some extent are averaged out for ionic (molecular) crystals as well as for nonionic compounds ..." [Ref. 5.154, p. 78].

A proposal for understanding in detail the way in which solid-state effects average out in ionic crystals has been offered in [5.160]. There it is noted that one must distinguish in calculating the Q_i^A used in (5.1) between those obtained from a molecular orbital calculation for the free molecular ions, and the new values of these atomic charges which result if the molecular orbital charges are permitted to shift in response to the solid-state potentials. Figure 5.14 illustrates the nature of the various effects which enter into the determination of the binding energy shifts in a particular model crystal – fully charge-transferred tetracyanoquinodymethane tetrathiafulvalene (TTF^+-$TCNQ^-$). If one considers the N-1*s* core orbitals on the $TCNQ^-$ molecule in this crystal (Fig. 5.14a) and schematically illustrates the atomic potentials vs. position x along a line joining the two nitrogen atoms shown in Fig. 5.14a, then one obtains a potential vs. position curve as shown in Fig. 5.14b for the free molecular ion. If one now "freezes" the atomic charges Q_i on the atoms of this molecular ion (as well as those of TTF^+) as one places these ions in the crystal, one obtains a potential vs. position curve as illustrated in Fig. 5.14c. Indeed, using such a procedure – calculating the atomic charges Q_i for *free* TTF^+ and $TCNQ^-$ using CNDO/2, and then doing a Madelung sum over these charges with the TTF^+ and $TCNQ^-$ placed in the crystal structure given by [5.161], a solid-state splitting of $\sim$1.4 eV is predicted for the N-1*s* core level in [5.158–160] since they are not crystallographically equivalent.

However, if the charges are now "unfrozen" and are permitted to flow on the molecule in response to this large crystal potential, the effect illustrated in Fig. 5.14d occurs. The nitrogen atom which resides at the most attractive crystal potential site gains valence charge. This *additional* valence charge causes an upward shift of the core level on this atom. Similarly, the nitrogen atom which lies at a high potential loses valence charge and its core level shifts down. Thus the effect of polarization of the molecular valence charge in the crystal field is in such a direction as to reduce the core level splitting caused by that field. Paper [5.154] shows that a loss of only 1/20 of an electron from a carbon atom causes about 1 eV of binding energy shift of the C-1*s* core level on that atom. This number emphasizes the importance of the effect illustrated in Fig.

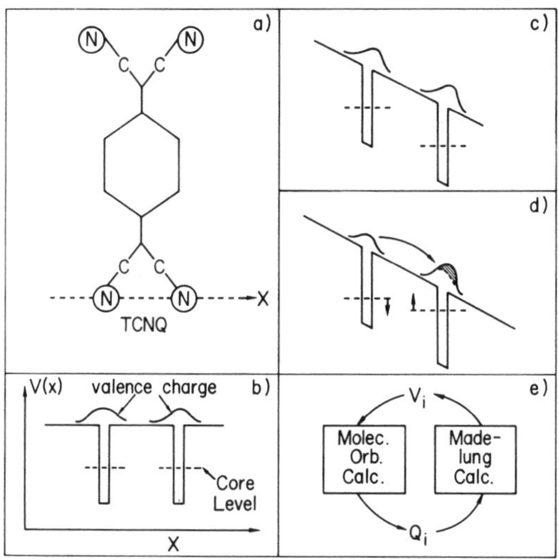

Fig. 14. a Shows the structure of the TCNQ molecule, and an axis (labeled "X") along which the schematic electrostatic potential will be shown in panels b–d.

b In neutral TCNQ, the electrostatic potential along X shows equal core orbital potential wells and binding energies for the N1*s* core orbital.

c In TTF^+–$TCNQ^-$, the potential along X varies as shown schematically, due to the crystalographic inequivalence of the two N atoms lying on axis X.

d In this panel, the valence charge on the solid-state molecular ion is "unfrozen" so that nitrogen atomic charge can flow in response to the solid-state fields.

e This panel schematically illustrates the way in which the processes in panels b–d can be calculated, using a self-consistent charge and potential calculation for the crystal, as described in [5.160]

5.14. Even if the polarization of molecular valence charge on a molecular ion in a crystal is small, it can have a large effect on crystal-induced core level splitting, and is always in such a direction as to reduce the splitting due to the crystal field itself.

In order to quantitatively investigate these effects, a self-consistent calculation of the charges in the model crystal TTF^+-$TCNQ^-$ was performed [5.160], as illustrated in Fig. 5.14e. First charges Q_i on the "free" molecular ions were calculated by using a molecular orbital calculation in which crystal potentials V_i at each atomic site were assumed to be equal to zero. The resulting Q_i were then used in a Madelung calculation to get a set of crystal potential values V_i at the atomic sites, due to these free ion charges. These atomic potentials were then input to the molecular orbital calculation to get a new set of Q_i, those which arose in response to this crystal potential. The new Q_i were then used to recalculate the V_i in the crystal, and the process repeated until the atomic potentials and charges reached self-consistency. The results of this detailed calculation, as well as those from a cruder one used to illustrate the physics more clearly, showed that charge flow on molecules in the crystal reduced solid-state splittings to $\sim 1/5$ of their original value, although the actual Madelung potential variations remained large. In other words, molecular polarization was a small effect and could not screen out the crystal fields. However, due to the great sensitivity of core-level binding energies even to small amounts of valence charge flow, molecular valence charge polarization in the crystal caused crystal-induced chemical shifts of the core levels which were opposite to and nearly equal to those due to the crystal field alone.

5.7.2 Core Level Spectroscopy and Charge Transfer in TTF-TCNQ

Armed with the above discussion, we now review briefly the recent attempts at relating core level spectra of TTF-TCNQ to the amount of charge transfer in this salt. The experimental spectra of the N-1*s* core level are simplest to interpret, for there is only one chemical specie of N in TTF-TCNQ and no intrinsic spin-orbit splitting as in the sulfur 3*p* core level.

High-resolution XPS spectra [5.126] of the N-1*s* level show two chemical states of N in the compound, and [5.126] interpreted this as evidence that one sees either $TCNQ^{\circ}$ or $TCNQ^{-}$ on the short XPS time scale, giving two core-level binding energies – the two emission peak areas then being in the ratio of the number of neutral to negatively charged TCNQ molecules in the crystal.

Following this work, [5.158, 159, 162] proposed that in fact the two core-orbital binding energies were due to a Madelung splitting of the binding energies of the two crystallographically inequivalent nitrogen atoms in TCNQ in the crystal.

In [5.160] a discussion summarized above was given, arguing that such Madelung splittings are largely cancelled by the XPS chemical shifts associated with molecular valence charge polarization in the crystal. Also it was argued that since the two N-1*s* peaks of different binding energies have different integrated areas and peak widths, one must be seeing the presence of different kinds of molecules rather than simply a Madelung splitting.

A separate question arises also: If two types of molecules are being seen, are they simply a bulk and surface species due to the photoemission sampling depth? Recently *Swingle* et al. [5.163] reported in an angle-resolved XPS experiment that a separate surface species is seen, added to the bulk emission spectrum. Further, *Ritsko* et al. [5.164] have even more recently probed optical transitions from N-1*s* to the empty valence states using inelastic electron scattering. This experiment observes predominantly the bulk of a film of TTF-TCNQ and sees only one strong and sharp ($\sim$0.7 eV wide) optical transition from N-1*s*. These two experiments are tentative evidence that the previous XPS results saw a bulk and a surface specie of TCNQ rather than two intrinsic charge states of TCNQ in the compound.

However, the evidence for this point of view is not yet conclusive for the following reasons. Of the two spectra shown in [5.163], the more "bulk-sensitive" one (in which electron emission was normal to the crystal surface) does not have the two primary N-1*s* peaks in the ratio reported in [5.126]. However, one of the authors of [5.126] could get the ratio seen by *Swingle* et al. [5.163] after heating the crystal [5.165]. (Such heating naturally occurs in the type of spectrometer used in [5.163].) Thus the data of [5.163] might show a surface of a different nature than the bulk due to fractional distillation of TTF from the surface, leaving a $TCNQ^{\circ}$-rich surface. In other words, in the region of the sample probed by XPS, the samples of [5.126, 163] may be different due to the preparation technique. (Indeed, in a table, the authors of [5.163] present two very different peak area ratios for the two N-2*p* peaks, obtained by them for two different samples. Thus, for these authors, there is obviously something nonreproducible about different samples.)

In the experiment reported in [5.164], only one optical transition is seen, although there are many empty valence final states. *Ritsko* et al. explain this observation by claiming that optical transition rules preclude most core-level transitions from N-1*s*. If this interpretation is correct, then it is conceivable that, due to the severe optical selection rules, low-energy transitions from N-1*s* to the empty orbitals of one charge specie may all be forbidden. In that case the one transition observed does not unambiguously prove that only one charge state of TCNQ exists in TTF-TCNQ. Indeed, we note that recent X-ray diffraction studies [5.166] strongly suggest that less than one electron is transferred from TTF to TCNQ per formula unit in this crystal.

5.7.3 Conclusions

The above discussion shows that the determination of solid-state properties, such as the amount of charge transfer in organic molecular compounds, is an extremely complicated matter. The quantum mechanical nature of the description of partial charge transfer and its relation to the "time scale" of the photoemission experiment is very poorly understood. The escape length (and thus surface sensitivity) in XPS of organic solids is not yet well understood. The question of Madelung splittings is not simple and can probably only be understood using self-consistent field calculations for the entire crystal. The models for this calculation described above were performed for fully charge-transferred crystals. However, in the case of partially charge-transferred crystals, one does not yet understand the electronic structure of the ground state well, so that self-consistent calculations are presently out of the question.

For these reasons, we feel that XPS core level spectra must be used with great care if we are to understand many solid-state properties. The information content of this spectroscopic technique is quite large, but this information may be related mostly to free molecule properties rather than to solid-state properties in many instances, and the solid-state information may perhaps only be extracted by using extremely sophisticated theoretical techniques.

Acknowledgments. We are grateful to a number of colleagues who have contributed to this article with useful discussions and comments. In particular we thank A.I. Belkind, J. Berkowitz, R. Caudano, J.E. Demuth, A.J. Epstein, D. Haarer, N. Karl, L. Ley, N. Lipari, P. Mengel, J.J. Pireaux, W. Pong, B. Schechtman, and A. Schweig for generously making available to us results of experiments and calculations prior to publication. Stimulating discussions with them, as well as with D.E. Eastman, are gratefully acknowledged. One of us (E.E.K.) would like to thank the IBM Corporation for the opportunity to work at the Thomas J. Watson Research Center as a visiting scientist.

References

5.1 F. Gutmann, L.E. Lyons: *Organic Semiconductors* (Wiley and Sons, New York 1967)

5.2 H. Meier: *Organic Semiconductors* (Verlag Chemie, Weinheim 1974)

5.3 N. Karl: "Organic Semiconductors". In *Festkörperprobleme XIV*, (Vieweg, Braunschweig 1974) p. 261

5.4 J. Ferraris, D.O. Cowan, V. Walatka, Jr., J.H. Perlstein: J. Am. Chem. Soc. **94**, 670 (1973)

5.5 L.B. Coleman, M.J. Cohen, D. Sandman, F.G. Yamagishi, A.F. Garito, A.J. Heeger: Solid State Commun. **12**, 1125 (1973)

5.6 J.P. Ferraris, D.O. Cowan, V.V. Walatka, J.H. Perlstein: J. Am. Chem. Soc. **95**, 948 (1973)

5.7 V.V. Walatka, J.H. Perlstein, M.M. Labes: Phys. Rev. Lett. **31**, 1139 (1973)

5.8 R.L. Greene, G.B. Street, L.J. Suter: Phys. Rev. Lett. **34**, 577 (1975) and references therein

5.9 J.F. Shchegolev: Phys. Status Solidi. (a) **12**, 9 (1972)

5.10 A.J. Heeger, A.F. Garito: In *Proc. of the NATO Summer Institute on Low Dimensional Conductors*, ed. by H.J. Keller (Plenum Press, New York 1975) p. 89

5.11 M.B. Robin: *Higher Excited States of Polyatomic Molecules*, Vols. 1 and 2 (Academic Press, New York 1974, 1975)

5.12 E.E. Koch, A. Otto: Int. J. Radiat. Phys. Chem. **8**, 113 (1976)

5.13 D.W. Turner, C. Baker, A.D. Baker, C.R. Brundle: *Molecular Photoelectron Spectroscopy* (Wiley and Sons, New York 1970)

5.14 K. Siegbahn et al.: *ESCA Applied to Free Molecules* (North-Holland, Amsterdam 1969)

5.15 J. Berkowitz: "High Temperatures UPS Studies", *Electron Spectroscopy* Vol. I, ed. by C.R. Brundle, A.D. Baker (Akademic Press, London 1977) p. 355

5.16 J.N. Murrell: *The Theory of the Electronic Spectra of Organic Molecules* (Methuen, London 1963)

5.17 L. Salem: *The Molecular Orbital Theory of Conjugated Systems* (Benjamin, New York 1966) and references therein

5.18 J.A. Pople, D.L. Beveridge: *Approximate Molecular Orbital Theory* (McGraw-Hill, New York 1970)

5.19 L.F. Lyons, G.C. Morris: J. Chem. Soc. (London) B, 5192 (1960)

5.20 A.P. Marchetti, D.R. Kearns: Mol. Cryst. Liq. Cryst. **6**, 299 (1970)

5.21 M. Kochi, Y. Harada, T. Hirooka, H. Inokuchi: Bull. Chem. Soc. Jpn. **43**, 2690 (1970)

5.22 F. Carswell, H. Iredale: Aust. J. Appl. Sci. **4**, 329 (1953)

5.23 F.I. Vilesov, A.A. Zagrubskii, D.F. Garbuzov: Fiz. Tverd. Tela **5**, 2000 (1963) [English transl.: Sov. Phys. Solid State **5**, 1460 (1964)]

5.24 J. Sworabowski: Phys. Status Solidi (a) **13**, 381 (1972); (a) **14**, K129 (1972); (a) **22**, K73 (1974)

5.25 J. Aihara: Phys. Status Solidi (a) **17**, K37 (1973)

5.26 A.I. Belkind, S.B. Aleksandrov, V.V. Aleksandrov, V.V. Grechov: "Micromechanism of Photoemission from Organic Semiconductors", In *Proceedings of the Karpacz Summer School* (1974), p. 8 and references therein

5.27 B.H. Schechtman, W.E. Spicer: Chem. Phys. Lett. **2**, 207 (1968)

5.28 A.I. Belkind, S.B. Aleksandrov: Phys. Status Solidi (a) **9**, 105 (1972)

5.29 S. Pfister, P. Nielsen: J. Appl. Phys. **43**, 3104 (1972)

5.30 H. Fujihira, T. Hirooka, H. Inokuchi: Chem. Phys. Lett. **19**, 584 (1973)

5.31 T. Hirooka, K. Tanaka, K. Kuchitsu, M. Fujihira, H. Inokuchi, Y. Harada: Chem. Phys. Lett. **18**, 390 (1973)

5.32 O.H. LeBlanc, Jr.: In *Physics and Chemistry of the Organic Solid State*, Vol. III, ed. by D. Fox, M.M. Labes, A. Weissberger (Interscience, New York 1967) p. 747; see also A. Bergman, J. Jortner: Phys. Rev. B**9**, 4560 (1974) and references therein

5.33 G. Johansson, J. Hedman, A. Berndtsson, M. Klasson, R. Nielsson: J. Electron Spectrosc. **2**, 295 (1973)

5.34 T. Dickinson, A.F. Povey, P.M.A. Sherwood: J. Electron Spectrosc. **2**, 441 (1973)

5.35 M.F. Ebel, H. Ebel: J. Electron Spectrosc. **3**, 169 (1974)

5.36 D.A. Huchital, R.T. McKeon: Appl. Phys. Lett. **20**, 158 (1972)

5.37 W.D. Grobman, E.E. Koch: To be published
5.38 N. Schwentner, F.J. Himpsel, V. Saile, M. Skibowski, W. Steinmann, E.E. Koch: Phys. Rev. Lett. **34**, 528 (1975)
5.39 G.M. Bancroft, I. Adams, H. Lampe, T.K. Sham: J. Electron Spectrosc. **9**, 191 (1976)
5.40 D.E. Eastman: Phys. Rev. B**8**, 6027 (1973)
5.41 I. Lindau, W.E. Spicer: J. Electron Spectrosc. **3**, 409 (1974)
5.42 C.J. Powell: Surf. Sci. **44**, 29 (1974)
5.43 W. Pong, J.A. Smith: J. Appl. Phys. **44**, 174 (1973)
5.44 S. Hino, N. Sato, H. Inokuchi: Chem. Phys. Lett. **38**, 494 (1976)
5.45 W.B. Berry: J. Electrochem. Soc.: Solid State Sci. **118**, 597 (1971)
5.46 Y.C. Chang, W.B. Berry: J. Chem. Phys. **61**, 2727 (1974)
5.47 J.T.J. Huang, J.L. Magee: J. Chem. Phys. **61**, 2736 (1974)
5.47a S. Evans, R.G. Pritchard, J.M. Thomas: J. Phys. C**10**, 2483 (1977)
5.48 S.F. Lin, W.E. Spicer, B.H. Schechtman: Phys. Rev. B**12**, 4184 (1975)
5.49 P. Nielsen, D. Sandman, A. Epstein: Solid State Commun. **17**, 1067 (1975)
5.50 B.L. Henke: J. Phys. (Paris) C**4**, 115 (1971)
5.51 B.L. Henke: *Advances in X-ray Analysis*, Vol. 13 (Plenum Press, New York 1970) p. 1
5.52 W.W. Riggs, R.C. Wendt: In *Proc. Int. Conf. on Elect. Spectrosc.* (Namur, Belgium 1974)
5.53 J.E. Demuth, D.E. Eastman: Phys. Rev. Lett. **32**, 1123 (1974)
5.54 G.W. Rubloff, H. Luth, W.D. Grobman: Chem. Phys. Lett. **39**, 493 (1976)
5.55 J.J. Pireaux, S. Svensson, E. Basilier, P.Å. Malmqvist, U. Gelius, R. Caudano, K. Siegbahn: Phys. Rev. A**14**, 2133 (1976)
5.56 J.J. Pireaux, R. Caudano: Preprint
5.57 J.M. André, J. Delhalle, S. Delhalle, R. Caudano, J.J. Pireaux, J.J. Verbist: Chem. Phys. Lett. **23**, 206 (1973)
5.58 J. Delhalle, J. M. André, S. Delhalle, J.J. Pireaux, R. Caudano, J.J. Verbist: J. Chem. Phys. **60**, 595 (1974) and references therein
5.59 U. Gelius, P.F. Hedén, J. Hedman, B.J. Lindberg, R. Manne, R. Nordberg, C. Nordling, K. Siegbahn: Phys. Scr. **2**, 70 (1970)
5.60 J.E. Demuth, D.E. Eastman: J. Vac. Sci. Technol. **13**, 283 (1976) and references therein
5.61 M.R. Philpott: Adv. Chem. Phys. **23**, 221 (1973)
5.62 W. von Niessen, L.S. Cederbaum, W.P. Krumer: J. Chem. Phys. **65**, 1378 (1976)
5.63 R.J. Buenker, S. Peyerimhoff: Chem. Phys. Lett. **3**, 37 (1969)
5.64 L. Åsbrink, C. Fridth, E. Lindholm: In *Chemical Spectroscopy and Photochemistry in the Vacuum-Ultraviolet*, ed. by C. Sandorfy, P.J. Ausloos, NATO Adv. Studies Inst. Ser. No. C-8 (1974)
5.65 A.J. Kitaigorodski: *Organic Chemical Crystallography* (Consultants Bureau, New York 1961)
5.66 K.Y. Yu, J.C. McMenamin, W.E. Spicer: Surf. Sci. **50**, 149 (1975)
5.67 U. Asaf, I.T. Steinberger: Chem. Phys. Lett. **33**, 563 (1975)
5.68 J.E. Demuth: Phys. Rev. Lett. **40**, 409 (1978)
5.69 E.E. Koch, W.D. Grobman: To be published
5.70 K. Sehi, H. Inokuchi, Y. Harada: Chem. Phys. Lett. **20**, 197 (1973)
5.71 K. Sehi, Y. Harada, K. Ohno, H. Inokuchi: Bull. Chem. Soc. Jpn. **47**, 1608 (1974)
5.72 N. Keno, Y. Hayasi, S. Kiyono: Chem. Phys. Lett. **35**, 31 (1975)
5.73 S. Seki, T. Hirooka, Y. Kamura, H. Inokuchi: Bull. Chem. Soc. Jpn. **49**, 904 (1976)
5.74 E.A. Silinsh, A.E. Belkind, D.R. Balode, A.J. Biseniece, V.V. Grechov, L.F. Taure, M.V. Kurik, J.I. Vertzymacha, I. Bok: Phys. Sol. a**25**, 339 (1974)
5.75 S. Hino, T. Hirooka, H. Inokuchi: Bull. Chem. Soc. Jpn. **48**, 1133 (1974)
5.76 R. Boschi, J.N. Murrell, W. Schmidt: Discuss. Faraday Soc. **54**, 116 (1972)
5.77 P.A. Clark, F. Brogli, E. Heilbronner: Helv. Chim. Acta **55**, 1415 (1972)
5.78 R. Boschi, E. Clar, W. Schmidt: J. Chem. Phys. **60**, 4406 (1974)
5.79 D.G. Streets, T.A. Williams: J. Electr. Spectrosc. **3**, 71 (1974)
5.80 N.O. Lipari, C.B. Duke: J. Chem. Phys. **63**, 748, 1748 (1975); **63**, 1768 (1975); N.O. Lipari: Private communication

5.81 J. Jager: Phys. Status Solidi. **35**, 731 (1969)
5.82 A.W. Potts, W.C. Price, D.G. Streets, T.A. Williams: Discuss. Faraday Soc. **54**, 168 (1972); and the following discussion by E. Lindholm (p. 200)
5.83 U. Gelius: J. Electron Spectrosc. **5**, 1039 (1974)
5.84 P.M. Williams: Preprint (1976); F.R. Shepherd, P.M. Williams: In *Vacuum Ultraviolet Radiation Physics*, ed. by E.E. Koch, R. Haensel, C. Kunz (Vieweg, Braunschweig 1974) p. 508
5.85a G.M. Bancroft, T.K. Sham, D.E. Eastman, W. Gudat: J. Am. Chem. Soc. **99**, 1752 (1973)
5.85b G.M. Bancroft, I. Adams, D.K. Creber, D.E. Eastman, W. Gudat: Chem. Phys. Lett. **38**, 83 (1976)
5.86a S. Evans, J.C. Green, P.J. Joachim, A.F. Orchard, D.W. Turner, J.P. Maier: J. Chem. Soc. Faraday Trans. II, 905 (1972)
5.86b W.C. Price, A.W. Potts, D.G. Streets: In *Electron Spectroscopy*, ed. by D.A. Shirley (North-Holland, Amsterdam 1972) p. 187
5.86c R.A. Pollak, S. Kowalczyk, L. Ley, D.A. Shirley: Phys. Rev. Lett. **29**, 274 (1972)
5.86d M. Cardona, J. Tejeda, N.J. Shevchik, D.W. Langer: Phys. Status Solidi (b) **58**, 483 (1973)
5.87 S. Hoste, H. Willemen, D. Van de Vondel, G.P. Van der Kelen: J. Electron Spectrosc. **5**, 227 (1974)
5.88 G.M. Bancroft, I. Adams, D.K. Cocker, D.E. Eastman, W.Gudat: Chem. Phys. Lett. **38**, 83 (1976)
5.89 R.F. Fenske: In *Progress in Inorganic Chemistry*, Vol. 21, ed. by S.J. Lippard (Wiley and Sons, New York 1975) p. 179
5.90 E.E. Koch, A. Otto: Chem. Phys. **3**, 370 (1974)
5.91 R.R. Chance, C.L. Braun: J. Chem. Phys. **64**, 3573 (1976); L.E. Lyons, K.A. Milne: J. Chem. Phys. **65**, 1474 (1976) and references therein
5.92 J.M. Caywood: Mol. Cryst. Liq. Cryst. **12**, 1 (1970)
5.93 H.J. Gaers, F. Willig: Chem. Phys. Lett. **32**, 300 (1975); F. Willig: Chem. Phys. Lett. **40**, 331 (1976) H. Killesreiter, H. Baessler: Phys. Status Solidi (b) **51**, 657 (1972) J. Singh, H. Baessler: Phys. Status Solidi (b) **62**, 147 (1974)
5.94 J. Cazaux: Opt. Commun. **3**, 221 (1971)
5.95 B.E. Cook, P.G. LeComber: J. Phys. Chem. Solids **32**, 1321 (1971)
5.96 M. Pope, H. Kallmann, J. Giachino: J. Chem. Phys. **42**, 2540 (1965)
5.97 M. Pope, J. Burgos: J. Mol. Cryst. **1**, 395 (1966)
5.98 G.T. Pott, D.F. Williams: J. Chem. Phys. **51**, 203 (1969)
5.99 D. Haarer, G. Castro: Chem. Phys. Lett. **12**, 277 (1971)
5.100 R. Wache, N. Karl: Unpublished R. Wache: Diplomarbeit, Universität Stuttgart (1973)
5.101 D.C. Northrop, O. Simpson: Proc. R. Soc. (London) A**244**, 377 (1958)
5.102 S. Choi, S.A. Rice: J. Chem. Phys. **38**, 366 (1963)
5.103 M. Silver, D. Olness, M. Swicord, R.C. Jarnagin: Phys. Rev. Lett. **10**, 12 (1963)
5.104 C.L. Braun: Phys. Rev. Lett. **21**, 215 (1968); see also [Ref. 5.91]
5.105 F.C. Strame, Jr.: Phys. Rev. Lett. **20**, 3 (1968)
5.106 E.G.E. Courtens, A. Bergman, J. Jortner: Phys. Rev. **156**, 948 (1967)
5.107 P. Holzman, R. Morris, R.C. Jarnagin, M. Silver: Phys. Rev. Lett. **19**, 506 (1967)
5.108 J. Fourny, G. Delacote, M. Schott: Phys. Rev. Lett. **21**, 1085 (1968)
5.109 M. Pope, J. Burgos, J. Giachino: J. Chem. Phys. **43**, 3367 (1965)
5.110 E.E. Koch, W.D. Grobman: Unpublished
5.111 N. Schwentner, E.E. Koch: Phys. Rev. B**14**, 5226 (1976)
5.112 G.W. Robinson: "Solid State Concepts in Radiation Chemistry and Biology", in *Computational Methods for Large Molecules and Localized States in Solids*, ed. by F. Herman, A.D. McLean, R.K. Nesbet (Plenum Press, New York 1973) p. 29
5.113 R.C. Hughes, Z.G. Soos: J. Chem. Phys. **63**, 1122 (1975)
5.114 A. Henriksson, M. Sandbom: Theor. Chim. Acta **27**, 213 (1972)

5.115 A.K. Ghosh, D.L. Morel, T. Feng, R.F. Shaw, C.A. Rowe: J. Appl. Phys. **45**, 230 (1974)
5.116 A. Broomberg, C.W. Tang, A.C. Albrecht: J. Chem. Phys. **60**, 4058 (1974)
5.117 For a detailed study of the near-infrared to vacuum ultraviolet absorption spectra and the optical constants, see B.H. Schechtman, W.E. Spicer: J. Mol. Spectrosc. **33**, 28 (1970)
5.118 B.F. Hoskins, S.A. Mason, J.C.B. White: J. Chem. Soc. London D 554 (1969)
5.119 M.V. Zeller, R.G. Hayes: J. Am. Chem. Soc. **95**, 3855 (1973)
5.120 A.J. Signorelli, R.G. Hayes: J. Chem. Phys. **64**, 4517 (1976)
5.121 D. Bettridge, J.C. Carver, D.M. Hereules: J. Electron Spectrosc. **2**, 327 (1973)
5.122 T. Kawai, M. Soma, Y. Matsumoto, T. Onishi, K. Tamaru: Chem. Phys. Lett. **37**, 378 (1976)
5.123 W. Pong, C.S. Inouye: J. Appl. Phys. **47**, 3444 (1976)
5.124 E. Subertova, J. Bok, P. Rihak, V. Prosser, E. Silinsh: Phys. Status Solidi (a) **18**, 741 (1973)
5.125 I. Ikemoto, J.M. Thomas, H. Kuroda: Faraday Discuss. Chem. Soc. **54**, 208 (1972)
5.126 W.D. Grobman, R.A. Pollak, D.E. Eastman, E.T. Maas, Jr., B.A. Scott: Phys. Rev. Lett. **32**, 534 (1974)
5.127 P. Nielsen, A.J. Epstein, D.J. Sandman: Solid State Commun. **15**, 53 (1974)
5.128 B.H. Schectman, S.F. Lin, W.E. Spicer: Phys. Rev. Lett. **34**, 667 (1975)
5.129 I. Ikemoto, K. Samizo, T. Fujikawa, K. Ishii, T. Ohta, H. Kuroda: Chem. Lett. **7**, 785 (1974)
5.130 F. Herman, I. Batra: Phys. Rev. Lett. **33**, 94 (1974)
5.131 T. Ladik, A. Karpfen, G. Stollhoff, P. Fulde: Chem. Phys. **7**, 267 (1975)
5.132 H.T. Jonkman, G.A. Van der Welde, W.C. Niewport: Chem. Phys. Lett. **25**, 62 (1974)
5.133 N.O. Lipari, P. Nielsen, J.J. Ritsko, A.J. Epstein, D.J. Sandman: Phys. Rev. B**15**, (1976)
5.134 R. Gleiter, E. Schmidt, D.O. Cowan, J.P. Ferraris: J. Electron Spectrosc. **2**, 207 (1973)
5.135 R. Gleiter: To be published
5.136 B.I. Bennett, F. Herman: Chem. Phys. Lett. **32**, 334 (1975)
5.137 A.J. Berlinsky, J.F. Carolan, L. Weiler: Solid State Commun. **15**, 795 (1974)
5.138 I.P. Batra, B.I. Bennett, F. Herman: Phys. Rev. B**11**, 4927 (1975)
5.139 B.H. Schechtman: "*Photoemission and Optical Studies of Organic Solids: Pthalocyanines and Porphrins*", Stanford Electron. Labs. Tech. Rpt. 5207-2 (1968)
5.140 Y. Tomkiewicz, E.M. Engler, T.D. Schultz: Phys. Rev. Lett. **35**, 456 (1975)
5.141 A. Schweig: Private communication
5.142 W.D. Grobman, S. Laplaca, E. Engler: To be published
5.143 E.M. Engler, F.B. Kaufman, D.C. Green, C.E. Klots, R.N. Compton: J. Am. Chem. Soc. **97**, 2921 (1975)
5.144a D.E. Parry, J.M. Thomas: J. Phys. C**8**, L45 (1975)
5.144b M. Schlüter, J.R. Chelikowsky, M.L. Cohen: Phys. Rev. Lett. **35**, 869 (1975); **36**, 452 (1976)
5.145a D.R. Salahub, R.P. Messmer: Phys. Rev. B**14**, 2592 (1976)
5.145b R.P. Messmer, D.R. Salahub: Chem. Phys. Lett. **41**, 73 (1976)
In Ref. [145a] a brief critical review of the available band-structure calculations for $(SN)_x$ may be found
5.146 W.E. Rudge, P.M. Grant: Phys. Rev. Lett. **35**, 1799 (1975)
5.147 H. Kamimura, A.M. Glazer, A.J. Grant, Y. Natsume, M. Schriber, A.D. Yoffe: J. Phys. C**9**, 291 (1976). However, the numerical values of overlap integrals in this paper are possibly incorrect. P. Grant: Private communication
5.148 L. Ley: Phys. Rev. Lett. **35**, 1796 (1975)
5.149 P. Mengel, P.M. Grant, W.E. Rudge, B.H. Schechtman, D.W. Rice: Phys. Rev. Lett. **35**, 1803 (1975)
5.150 E.E. Koch, W.D. Grobman: Solid State Commun. **23**, 49 (1977)
5.151 P. Mengel, W.D. Grobman, I.B. Ortenburger, P.M. Grant, B.H. Schechtman: Bull. Am. Phys. Soc. **21**, 254 (1976)
5.152 H.J. Stolz, L. Ley, T. Grandke, J. Azoulay: "*Angle resolved Photoelectron Spectroscopy on* $(SN)_x$" (to be published); see also
H.J. Stolz: Thesis, Stuttgart (1977)
5.153 K. Siegbahn: *ESCA* (Almqvist-Wiksells, Stockholm 1967)
5.154 U. Gelius, P.F. Heden, J. Hedman, B.J. Lindberg, R. Manne, R. Nordberg, C. Nordling, K. Siegbahn: Phys. Scr. **2**, 70 (1970)

5.155 M.E. Schwartz, J.D. Switalski, R.E. Strenski: "Core-level Binding Energy Shifts from Molecular-orbital Theory", in *Electron Spectroscopy*, ed. by D.A. Shirley (North-Holland, Amsterdam 1972) pp. 605–627
5.156 T.A. Carlson: *Photoelectron and Auger Spectroscopy* (Plenum Press, New York 1975)
5.157 D.T. Clark: "Structure and Bonding in Polymers as Revealed by ESCA", in *Proceedings of the NATO Advanced Study Institute*, Namur, Belgium (1974), ed. by J. Ladik, J.M. Andre (Plenum Press, New York 1975)
D.T. Clark: "The Application of ESCA to Studies of Structure and Bonding in Polymers", in *Spectroscopic Methods of Polymer Characterization*, ed. by K.J. Ivin (Wiley and Sons, New York 1976)
Refs. [5.57] and [5.58] above
5.158 A.J. Epstein, N.O. Lipari, P. Nielsen, D.J. Sandman: Phys. Rev. Lett. **34**, 914 (1975)
5.159 A.J. Epstein, N.O. Lipari, D.J. Sandman, P. Nielsen: Phys. Rev. B**13**, 1569 (1976)
5.160 W.D. Grobman, B.D. Silverman: Solid State Commun. **19**, 319 (1976)
5.161 T.J. Kistenmacher, T.E. Phillips, D.O. Cowan: Acta Crystallogr. Sect. B**30**, 763 (1974)
5.162 R.M. Metzger, A.N. Bloch: Bull. Am. Phys. Soc. **20**, 415 (1975); J. Chem. Phys. (in press)
A.J. Epstein, N.O. Lipari, D.J. Sandman, P. Nielsen: Bull. Am. Phys. Soc. **20**, 465 (1975)
5.163 R.S. Swingle, R.P. Groff, B.M. Monroe: Phys. Rev. Lett. **35**, 452 (1975)
5.164 J.J. Ritsko, N.O. Lipari, P.C. Gibbons, S. Schnatterly: Phys. Rev. Lett. **37**, 1068 (1976)
5.165 R.A. Pollack: Private communication
5.166 F. Denoyer, R. Comes, A.F. Garito, A.J. Heeger: Phys. Rev. Lett. **35**, 445 (1975)

6. Synchrotron Radiation: Overview

C. Kunz

With 33 Figures

The application of synchrotron radiation from electron synchrotrons and storage rings to photoemission spectroscopy during the last few years is one of the most important technical developments in this field. While above ~10 eV photon energy, only a few resonance lines from gases and solids provide sufficient intensity at discrete energies; synchrotron radiation (SR) covers the spectral range from the visible through the vacuum ultraviolet into the X-ray region with its intense, highly polarized, continuous spectrum. SR is generated at several laboratories as a by-product of high-energy physics; in addition, there is a small, now growing, number of storage rings, which are specifically built as radiation sources. SR finds various applications in the investigation of electronic states of atoms, molecules, solids and biological materials with different methods (see Table 6.1), its application to photoemission being among the most successful ones. In addition, synchrotron radiation is used for structural analysis by means of X-ray diffraction and soft X-ray microscopy, as well as for several purposes in the fields of applied research. A series of reviews covers the fundamentals of synchrotron radiation [6.1–18] and the results obtained from its application [6.19–38]. For information concerning photoemission see especially [6.25, 31, 37, 38].

Table 6.1. Different techniques used with synchrotron radiation. The number of crosses is a subjective measure of the relative importance in the vaccum ultraviolet (VUV) and X-ray regions of the spectrum

Technique	VUV ($h\nu < 1000$ eV)	X-ray ($h\nu > 1000$ eV)
1. Absorption	+ + +	+ + +
2. Reflection	+ + +	—
3. Interferometry	+	+
4. Fluorescence	+ +	+
5. Photoemission	+ + +	+
6. Mass-spectrometry	+	—
7. Time resolution	+ +	+ +
8. X-ray diffraction	—	+ + +
9. Raman and Compton	—	+
10. Nuclear fluorescence	—	+
11. Topography	—	+ +
12. Microscopy	+ + +	+
13. Dosimetry	—	+
14. Radiometry	+ +	—

6.1 Overview

For photoemission the availability of a tunable light source instead of only a few lines at fixed photon energies (with a large gap between ~40 and ~1000 eV) allows for several new techniques in addition to the commonly measured energy distribution curves (EDC). These are: 1) yield spectroscopy (PEYS), where the photoelectric yield is measured as a function of photon energy; 2) constant final-state spectroscopy (CFS), where the electron energy analyzer is set to a fixed energy while the photon energy is scanned, and 3) constant initial-state spectroscopy (CIS), where electron and photon energies are scanned synchronously. This chapter is especially concerned with experimental results obtained with these new techniques, since they provide considerable insight (that was not possible before) into the electronic structure of matter. Quite generally, the possibility of varying the photon energy allows for a measurement of core electron excitation near threshold, variation of partial excitation cross sections, tuning-in at resonance, and variation of the electron escape depth through a variation of their kinetic energy. The latter technique also provides a method to change the relative strength of bulk to surface features in the spectra. The linear polarization of synchrotron radiation is especially useful in combination with angular dependent photoemission. This aspect, however, is treated separately in [Ref. 1.1, Chap. 6].

The scope of this chapter is to give first a short introduction to the properties of SR, followed by a comparison of SR with other sources, and by some aspects of the development of so-called dedicated storage rings which are optimized as light sources from the outset (Sect. 6.2.3). Some technical aspects of the set-up of laboratories and the instrumentation used are explained in Sect. 6.3. Section 6.4 gives a description of the various spectroscopic techniques applied in connection with SR. In the last four sections a selection of experiments, which have been performed using these techniques, is reviewed. The selection of the experimental material, which constitutes the main body of this chapter, was guided by the following ideas: Since investigation of semiconductors, rare earths and transition metals are treated in separate chapters in this book, primarily results on insulators like rare-gas solids and alkali halides were chosen. These materials also serve extremely well to exemplify the power of the techniques applied. Measurements on some other materials like Se, metal alloys and simple metals, however, are also included, especially for demonstrating the power of yield spectroscopy (Sect. 6.5). Section 6.6 deals with the investigation of valence and conduction band states, while Sect. 6.7 is concerned with relaxation processes and excitons. Although this monograph is not devoted to surface photoemission (see, e.g., Sect. 2.9, [6.39, 40] and [Ref. 6.37, p. 319]), two examples of the investigation of surface states on clean samples and on adsorbate covered samples should serve to stress the importance of SR for this branch of photoemission spectroscopy in Sect. 6.8. Quite naturally the author draws heavily from material obtained at the DESY SR

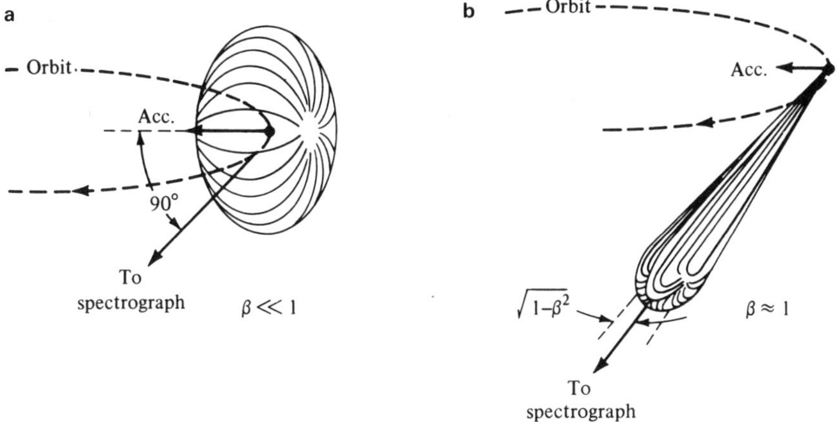

Fig. 6.1a and b. Angular distribution of emitted intensity **a** from a slow electron on a circular orbit, **b** from a relativistic electron moving with the velocity βc. The dipole pattern **a** is distorted into a narrow cone **b** in the instantaneous direction of motion of the electron [6.7]

laboratories. He hopes, however, to have done justice to the investigators at other places by giving ample reference to their work. While the important methods applying synchrotron radiation should be thoroughly documented in this chapter, it is far from being an exhaustive review.

6.2 Properties of Synchrotron Radiation

6.2.1 Basic Equations

The properties of SR can be calculated by applying the methods of classical electrodynamics to relativistic electrons and positrons on circular orbits. While already *Schott* [6.41] treated this problem in connection with classical models of the atom in the beginning of the century, *Ivanenko* and *Pomeranchuk* [6.1], followed by *Schwinger* [6.2], were the first who predicted its importance for circular particle accelerators. Now the main equations are derived in standard textbooks, like those of *Sommerfeld* [6.8] and *Jackson* [6.9].

The important properties of synchrotron radiation are qualitatively understandable from the picture of an oscillating dipole (the accelerated electron in orbit) and the relativistic transformation of the dipole radiation pattern into the laboratory frame (Fig. 6.1), this transforms the dipole pattern of Fig. 6.1a into the narrow beam of Fig. 6.1b. This beam hits an observer only during a very short time each revolution. This explains the production of very high harmonics of the revolution frequency. The properties of SR are listed as follows:

1) Continuous spectrum from the infrared into the region of hard X-rays,

2) strong collimation in the instantaneous direction of flight (typically 1 mrad),

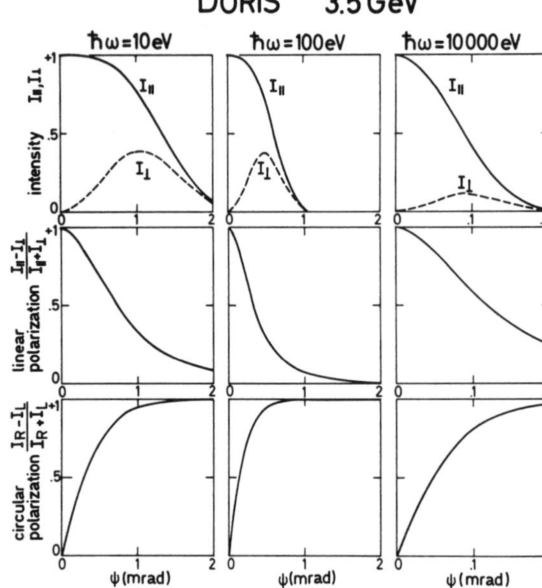

Fig. 6.2. Angular distribution of the radiation intensity of the two components $I_{\parallel}$, parallel and $I_{\perp}$, perpendicular to the plane of the orbit. In addition the linear and circular degrees of polarization are given. I_R and I_L are the intensities of the right-hand and left-hand components, respectively [6.33]

3) linear polarization in the plane of the orbit,

4) circular polarization in the "wings" above and below the plane of the orbit,

5) pronounced time structure, which is a copy of the pulse structure of the electron beam (pulses as short as 100 ps),

6) absolute calculability of all the properties of the source, once parameters of the particle beam are given,

7) cleanliness and stability (in particular with storage rings) of the source, which, in contrast to gas-discharge sources, exists in an extremely good vacuum.

The power of radiation $P(\psi, \lambda)$, in erg s^{-1} rad^{-1}, of a single electron with energy E on a circular orbit with radius R at an azimuth angle ψ relative to the plane of the orbit and per unit wavelength interval in λ (the wavelength) is given (see, e.g., [6.7, 13, 28]) by

$$P(\psi, \lambda) = \frac{8}{3} \frac{\pi R e^2 c}{\lambda^{-4}} \gamma^{-4} (1 + x^2)^2 \left[K_{2/3}^2(\xi) + \frac{x^2}{1 + x^2} K_{1/3}^2(\xi) \right] \tag{6.1}$$

$$x = \gamma \psi, \quad \xi = [2\pi R/(3\lambda)] \gamma^{-3} (1 + x^2)^{3/2}, \quad \gamma = E/(mc^2).$$

In (6.1) $K_{2/3}$ and $K_{1/3}$ are modified Bessel functions of the second kind. It is sometimes more convenient to deal with the power emitted from a unity particle current since this eliminates the geometrical deviations from a circular orbit in which case the right hand side of (6.1) has to be multiplied by $2\pi R/(ec)$. The additivity of radiation emitted by an ensemble of electrons is granted, provided the distribution of electrons is random on the scale of the wavelengths emitted.

The first and the second term in square brackets in (6.1) describe the two components of polarization with the electric vector in the plane of the orbit and perpendicular to it, respectively. This is illustrated for several parameters in Fig. 6.2. The two terms also serve to calculate the degrees of linear and circular polarization since the amplitudes of the two linearly polarized components are out of phase by $\pm\pi/2$.

If $P(\psi,\lambda)$ according to (6.1) is integrated over all vertical angles ψ, this results in a spectral distribution of the total power given by

$$P(\lambda)=\frac{3^{5/2}}{16\pi^2}\frac{e^2c}{R^3}\gamma^7\left(\frac{\lambda_c}{\lambda}\right)^3\int_{\lambda_c/\lambda}^{\infty}K_{5/3}(\eta)d\eta\,. \tag{6.2}$$

This distribution has its spectral maximum at $\lambda=0.42\,\lambda_c$, where

$$\lambda_c=\frac{4\pi R}{3}\gamma^{-3} \tag{6.3}$$

is called the critical wavelength and characterizes the short wavelength end of the spectral distribution of a specific SR source. For practical purposes the following equations are helpful:

$$\begin{aligned}\lambda_c\,(\text{Å})&=5.59\,R[\text{m}]\cdot(E[\text{GeV}])^{-3}\\ \varepsilon_c\,(\text{eV})&=2.22\;10^3(E[\text{GeV}])^3\cdot(R[\text{m}])^{-1}\,,\end{aligned} \tag{6.4}$$

where ε_c is the corresponding critical photon energy. Accelerators and storage rings provide useful intensities up to photon energies $\varepsilon\approx4\times\varepsilon_c$. Instead of $P(\psi,\lambda)$ and $P(\lambda)$ it is sometimes more useful to plot the number of photons available per unit interval of photon energy. This is done, for example, in Fig. 6.3 for realistic acceptance angles, which are quite different for small and large machines. A useful approximation to the number of photons available at photon energies $\varepsilon\ll\varepsilon_c$ in a horizontal section 1 mrad wide and of sufficient vertical height to accept all the radiation is given by

$$N\left[\frac{\text{photons}}{\text{s}\cdot\text{eV}\cdot\text{mA}\cdot\text{mrad}}\right]=4.5\;10^{12}j[\text{mA}]\cdot(R[\text{m}])^{1/3}\cdot(\varepsilon[\text{eV}])^{-2/3}\,, \tag{6.5}$$

where j is the electron current. It should be noted that for a fixed photon energy $\varepsilon\ll\varepsilon_c$ N is mainly determined by the current of particles j. Its dependence on R is much weaker, the intensity per mrad *increases* like $R^{1/3}$.

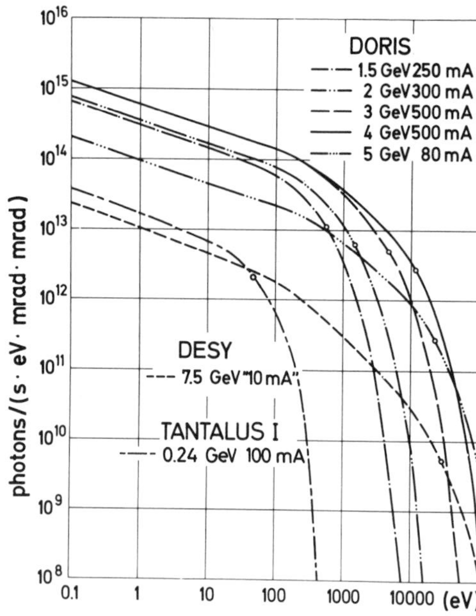

Fig. 6.3. Distribution of intensity into an aperture 1 mrad wide and 1 mrad high around the tangential direction. The open circles indicate ε_c [6.4]. While such a small aperture is typically accepted by an experiment at a large machine, at a small storage ring like TANTALUS I a 10 mrad wide bundle could easily be accepted

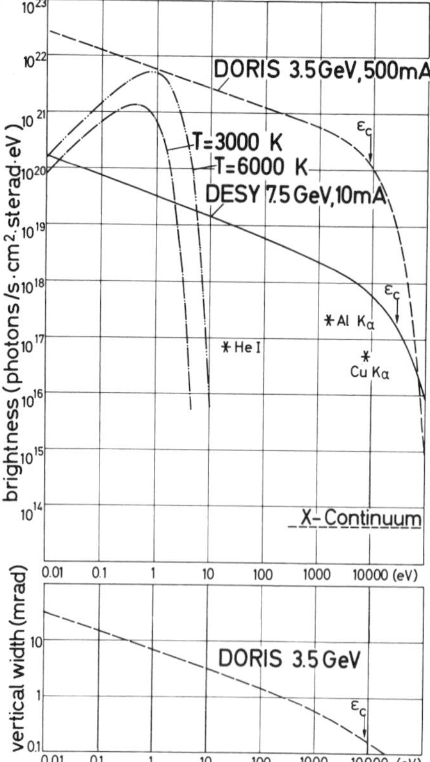

Fig. 6.4. Brightness of SR sources in the plane of the orbit compared to the brightnesses of classical sources: black-body radiators at different temperatures, gas-discharge lines and X-ray sources. The vertical angular width of DORIS is also plotted as a function of photon energy [6.15]

The angular spread $\langle\psi\rangle$ can be approximated for $\varepsilon \ll \varepsilon_c$ by

$$\langle\psi\rangle \approx \left(\frac{\varepsilon_c}{\varepsilon}\right)^{1/3} \cdot \gamma^{-1} = (\tfrac{3}{2}\, hc)^{1/3} R^{-1/3} \varepsilon^{-1/3} \,. \tag{6.6}$$

Note that for a fixed photon energy the angular spread is reduced with increasing R like $R^{-1/3}$.

Let us now inspect the brightness of a SR source. The spectral brightness $\eta(\psi, \varepsilon)$ is the averaged number of photons per unit area of the source per unit solid angle. A synchrotron or a storage ring is a source with a strong variation of the brightness with angle ψ in contrast to most other sources, e.g., black-body radiators or X-ray tubes. In the plane of the orbit ($\psi=0$) the brightness has a maximum. The dependence of this maximum brightness on the machine parameters for $\varepsilon \ll \varepsilon_c$ can be obtained from (6.5, 6):

$$\eta(0, \varepsilon) \propto j \cdot R^{2/3} \cdot A^{-1} \cdot \varepsilon^{-1/3} \,, \tag{6.7}$$

where A is the area of the electron beam.

In Fig. 6.4 the brightness $\eta(0, \varepsilon)$ is plotted for DESY and DORIS for the full range of ε. In addition $\langle\psi\rangle$ is given as a function of ε.

6.2.2 Comparison with Other Sources

Figure 6.4 can also serve as a basis for comparing a SR source with a classical source. A SR source has a large brightness η_s, but a small emittance e' (source area times usable solid angle). A classical source with its nearly isotropic brightness η_c usually fills the acceptance a' (e.g., entrance slit area times solid angle accepted by a grating) of a monochromator. In this case the useful intensity I_c at the experiment is given by $I_c \propto \eta_c \cdot a'$. If, however, as with SR sources, $a' > e'$ then $I_s \propto \eta_s \cdot e'$.

In the vacuum ultraviolet the continuous emission from rare-gas discharges, summarized [6.17] in Fig. 6.5, is available [6.42]. No single one of these sources covers a large spectral range; in addition, the peak intensities are lower by approximately one order of magnitude compared to a SR source. The comparison is based on measured intensities behind the exit slits of the monochromators and is to be understood only as giving the order of magnitude in view of the different types of such lamps available. The He *I* resonance line at 21.2 eV (584 Å), with its natural width <20 meV, plays a special role. For many purposes in photoemission a He source producing this line can be used without a monochromator. In this case up to $5 \cdot 10^{11}$ photons have been obtained [6.43] at the useful area of the sample. This intensity surpasses that of a synchrotron like DESY in combination with a monochromator by two orders of magnitude. The He *I* intensity is still somewhat higher than that of monochromatized SR at the storage rings which are presently in use, but should be surpassed at dedicated storage rings with appropriate monochromators.

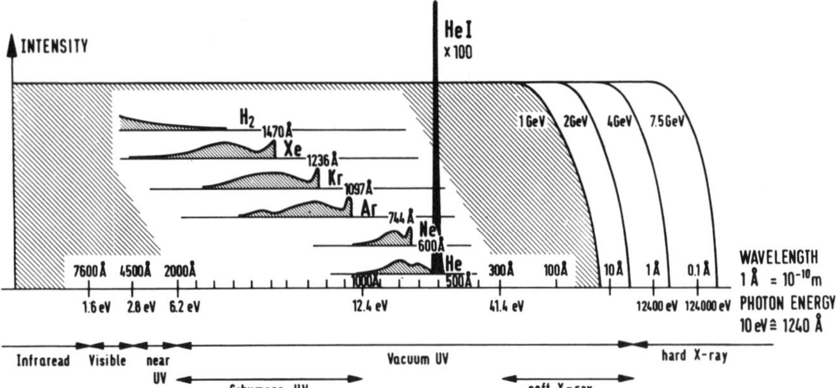

Fig. 6.5. Schematic comparison of the spectral emission of radiation from a typical large synchrotron (e.g., DESY) with that of classical discharge sources. The intensities are roughly on the same relative scale [6.17]

In the regime of X-rays (above 1000 eV) SR from presently available sources cannot compete with characteristic X-ray lines as a source for ordinary photoemission experiments. Although, according to Fig. 6.4, the brightness of these lines is lower by one and three orders of magnitude compared to a typical synchrotron or storage ring respectively, bent crystal monochromators [6.44] can make up for this difference in brightness by very large acceptances which cannot be filled from a SR source. When, however, tunability is required, SR has to be compared with the bremsstrahlung continuum and is superior in brightness by about three orders of magnitude. Nonetheless, the intensity is fairly low [6.45] for routine applications. At large dedicated SR sources, especially with the availability of wigglers (see below), the situation will change and make synchrotron radiation also an attractive source for photoemission spectroscopy in the X-ray region.

6.2.3 Evolution of Synchrotron Sources

Blewett [6.3], in about 1945, was the first to look for SR from a 100 MeV betatron, but he found only its effects on the electron orbit. In 1947 a technician working at the General Electric Research Laboratory in Schenectady was the first man to directly observe synchrotron light from the 70 MeV synchrotron with the aid of a mirror [6.4, 46]. In the years since 1953 *Tomboulian* and co-workers [6.5–7] made extensive investigations of SR from the 300 MeV synchrotron at Cornell University in Ithaca and showed that it was applicable to spectroscopy. In the early sixties *Codling* and *Madden* [6.10, 23, 28] began to use the 180 MeV synchrotron at the NBS in Washington, DC mainly for atomic spectroscopy. Now, more than twenty synchrotrons and storage rings are used as radiation sources all over the world. Table 6.2 summarizes the most important properties of these machines.

If we consider different generations of these machines we can rank them according to their usefulness as light sources in the following way:

1) parasitically used synchrotrons,
2) parasitically used storage rings,
3) storage rings dedicated to synchrotron radiation work,
4) storage rings equipped with periodic wigglers.

Conflict of interest occurs in the joint operation of the machines in cathegories 1) and 2) with the interests of high-energy physics at several points:

1) *Beam energy.* This according to (6.4) determines ε_c.

2) *Beam current.* This should usually be maximized for spectroscopists, while high-energy physicists wish to maximize "luminosity" (collision probability) with storage rings.

3) *Design parameters.* For a colliding beam storage ring the beam conditions are optimized at the interaction point, while synchrotron radiation requires an optimization in the deflecting magnets. Moreover, synchrotron radiation can live with one single stored beam, rather than two colliding ones. This can result in higher current and more stability. Also, down times are reduced if the complexity is reduced by use of a single beam.

4) *Planning ahead.* Scheduling of machine parameters in high-energy machines is usually on short terms and subject to the outcome of the most recent high-energy experiments; these two factors cause difficulties for the planning of the spectroscopists.

All these problems, together with the strong recent interest in SR, have led to a series of projects to build dedicated storage rings (third generation light sources – see Table 6.3). Detailed considerations lead to the conclusion that two different storage rings should be built, one for the spectral region below $\sim$1000 eV (vacuum ultraviolet) and one for the region above 1000 eV (X-rays). The small storage ring would have an energy of $\sim$700 MeV and radius of curvature in the magnets of roughly 1 m with the consequence of short distances to the experiment. Further, such a small ring would allow for easy access to the instrumentation during operation since no hard radiation is produced. The operation would have low running costs.

A large X-ray machine must be operated at energies of at least 2 to 3 GeV. Building and operation of such a source involves a large group of accelerator physicists and operators. Two projects of this kind are under way: NINA II at Daresbury is in the stage of construction, and the Brookhaven storage ring has been financially granted. Table 6.3 gives a survey on the available and proposed storage rings and their properties as light sources.

The generation of radiation sources (see Item 4 above) would involve multiple wigglers [6.47, 48]. These consist of straight sections with alternating

Table 6.2. Synchrotrons (SY) or storage rings (ST) used as light sources. E, particle energy; R, magnet radius; I, max current (during acceleration period for SY); ε_c, characteristic photon energy; "Exp. Stations" gives only an approximate number of experimental stations (if no number is given we lack information on this point); SR, synchrotron radiation

Name and Location	Type	E[GeV]	R[m]	I[mA]	ε_c[eV]	Exp. stations	Remarks
Group I, $\varepsilon_c = 1$–60 eV							
PTB (Braunschweig)	SY	0.14	0.46		13	1	
Tantalus I (Stoughton)	ST	0.24	0.64	100	48	10	Dedicated
Surf II (Washington)	ST	0.24	0.83	~ 30	37	5	Dedicated
INS-SOR II (Tokyo)	ST	0.3	1.1	~200	54	(~ 7)	Dedicated
Group II, $\varepsilon_c = 60$–$2{,}000$ eV							
Bonn I (Bonn)	SY	0.5	1.7	30	163	3	SR Lab
ACO (Orsay)	ST	0.55	1.11	35	333	5	Dedicated
C-60 (Moscow)	SY	0.68	2	10	349		
VEPP-2M (Novosibirsk)	ST	0.67	~ 2	~100	350	~ 4	SR Lab
Frascati (Frascati)	SY	1.1	3.6	15	821	2	
LUSY (Lund)	SY	1.2	3.65	10	1,050	1	
SIRIUS (Tomsk)	SY	1.3	4.2	20	1,160	~ 1	SR Lab
PACHRA (Moscow)	ST	1.3	4.0	100	1,220	4	Dedicated SR Lab
INS-SOR I (Tokyo)	SY	1.3	4.0	50	1,220	2	
Adone (Frascati)	ST	1.5	5.0	60	1,550	(1)	SR Lab
Group III, $\varepsilon_c = 2$–30 keV							
DCI (Orsay)	ST	1.8	3.82	400	3.390	4	SR Lab
VEPP3 (Novosibirsk)	ST	2.2	6.15	80–500	3.800		SR Lab
Bonn II (Bonn)	SY	2.5	7.65	30	4,530	4	SR Lab
DORIS (Hamburg)	ST	3.5	12.12	500	7,850	8	2 SR Labs
SPEAR (Stanford)	ST	4	12.7	60	11,200	~ 8	SR Lab
NINA I (Daresbury)	SY	5.0	20.8	50	13,300	8	SR Lab
ARUS (Erevan)	SY	6.0	24.65	20	19,500	3	SR Lab
DESY (Hamburg)	SY	7.5	31.7	30	29,500	10	2 SR Labs
Group IV, $\varepsilon_c \geqq 30$ keV							
Cornell III (Ithaca)	SY	12	120	2	32,000	1	SR Lab

Table 6.3. Storage rings (most of them dedicated as light sources). *E*, particle energy; *R*, magnetic radius; *I*, maxium current; SR, synchrotron radiation. The numbers on the proposed storage rings are usually subject to changes

Name	Location	E[GeV]	R[m]	I[mA]	ε_c[eV]	State
Group I $\varepsilon_c \leqq 60$ eV						
Tantalus I	Stoughton	0.24	0.64	100	48	Dedicated
Surf II	Washington	0.24	0.83	30	37	Dedicated
INS-SOR II	Tokyo	0.30	1.1	200	54	Dedicated
Group II $\varepsilon_c = 60$–2,000 eV						
ACO	Orsay	0.55	1.11	35	333	Dedicated
VEPP-2M	Novosibirsk	0.67	~ 2	~ 100	~ 350	SR Lab
Brookhaven I	Upton	0.70	~ 2	1,000	~ 400	Dedicated, proposed add. wigglers
ALADDIN	Stoughton	0.75	2.08	1,000	450	Dedicated, proposed add. wigglers
Adoné	Frascati	1.5	5.0	60	1,500	SR Lab
PAMPUS	Amsterdam	1.5	4.17	500	1,800	Dedicated, proposed add. wigglers
Group III $\varepsilon_c = 2$–30 keV						
Tantalus 2.5	Stoughton	1.76	4.5	100	2,690	Dedicated, proposed add. wigglers
SR (Nina II)	Daresbury	2.0	5.55	1,000	3,200	Dedicated, add. wigglers under construction
Brookhaven II	Upton	2.0	~ 8.0	1,000	2,200	Dedicated, proposed add. wigglers
DCI	Orsay	1.8	3.82	400	3,390	SR Lab
Moscow	Moscow	2	5	1,000	3,500	Dedicated, proposed
VEPP 3	Novosibirsk	2.2	6.15	80–500	3,800	SR Lab, wiggler under construction
Photon Factory	Japan	2.5	8.0	500	~ 4,300	Dedicated, proposed
DORIS	Hamburg	3.5	12.12	500	7,850	2 SR Labs
SPEAR	Stanford	4	12.7	60	11,200	SR Labs
VEPP 4	Novosibirsk	6	33	100	14,500	Under constr., SR Labs
Group IV $\varepsilon_c \geqq 30$ keV						
Cornell IV	Ithaca	8	32	100	35,000	Proposed, SR lab
PEP	Stanford	15	170	100	44,000	Under construction
PETRA	Hamburg	19	200	90	75,000	Under construction

(or rotating) transversal magnetic fields. The number of periods η would be $n=10$ to $n=100$ and a periodicity length λ_ω of a few cm might be achieved. Under optimum conditions the amplitude of radiation from the different wigglers adds coherently, which would produce a spike in the spectrum at [6.47]

$$\lambda_{\mathrm{w}} \approx \frac{\lambda_\omega}{\gamma^2} \tag{6.8}$$

with an n^2-fold increase of intensity. As an example one could take $\lambda_{\mathrm{w}} \approx 10$ Å for $\lambda_\omega = 9$ cm, $\gamma = 10^4$ ($E \approx 5$ GeV). A gain in intensity of 100–10^4 would open new possibilities for synchrotron radiation work, e.g., photoemission in the X-ray region. It should, however, be mentioned that such wigglers have not yet been realized in circular machines and, as a first step, a development program needs to be carried through.

6.3 Arrangement of Experiments

6.3.1 Layout of Laboratories

A typical arrangement of experiments at small storage rings is the arrangement at TANTALUS I (see Fig. 6.6). A 240 MeV machine which, though not originally built as such, is now being operated as a successful dedicated storage ring light source [6.49]. At such a small machine beam lines can be short; the experimental area needs only to be evacuated by the personnel during injection, and the adjustment of instruments can be made by using the visible part of the spectrum just like with ordinary laboratory sources. Typically, one can accept 10 mrad of the horizontal width of SR or more. The installation of a large number of beam ports is not very expensive, and thus usually not more than two experiments have to share the same beam port. This adds to the flexibility of scheduling and operation which is of particular importance.

The typical installation at big machines is quite different. Figure 6.7 shows an artist's view of the laboratory at the accelerator DESY which in principle is representative also for the arrangement at storage ring laboratories at large machines like DORIS [6.50] or SPEAR [6.51]. A single long beam line takes radiation into the laboratory building, which, due to radiation safety considerations, has to be located at a distance of between 20 and 40 m. Different portions of the beam are deflected by grazing incidence mirrors to form secondary beams. Several experiments share the same beam and have to be arranged in such a way that those which are nearer to the source can let the beam pass on to the next instrument by removing a grating or another optical component. Such instruments are used alternatively and have to share the available beam time.

At a storage ring the first mirrors have to take the full load of up to 100 W of SR which could make water cooling a necessity [6.52]. In addition, at

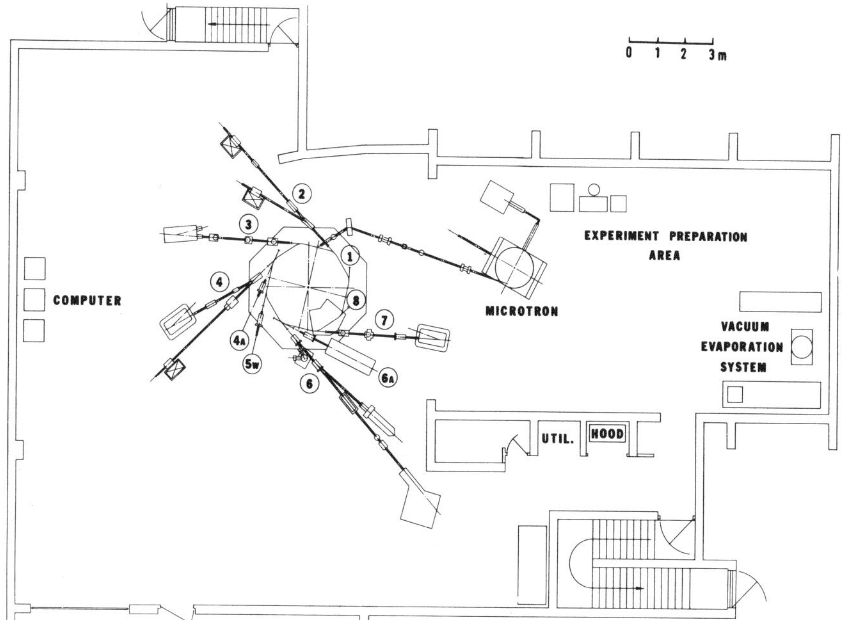

Fig. 6.6. Arrangement of beam ports and experiments at the 0.24 GeV storage ring TANTALUS I at Stoughton, Wisconsin. This can be taken as the model for an operation at a small storage ring. (With permission by E.M. Rowe)

storage rings the vacuum system must have an ultrahigh vacuum layout. This is not only a requirement to protect the storage ring from contaminants, but also to protect the mirror surfaces from cracked hydrocarbons. Only sample chambers which are buffered by differential pumping may be operated at pressures above 10^{-9} Torr. Further, a fast closing valve near the machine, which is activated by vacuum sensors at the experimental area, protects the ring in case of a vacuum breakdown.

Radiation safety considerations require that instrumentation near the direct beam be operated by remote control. In the case of the laboratory shown in Fig. 6.7 this holds for the lower level of the building. Access to the experiments is only allowed when the beam shutters (lead blocks) are closed. More details on the different laboratories can be found in users' handbooks which are available from several laboratories.

6.3.2 Monochromators

Synchrotron radiation in the energy region 5 to 1000 eV is monochromatized by grating monochromators. Either commercially available monochromators [6.42] are adapted to the special conditions of this source or new types of

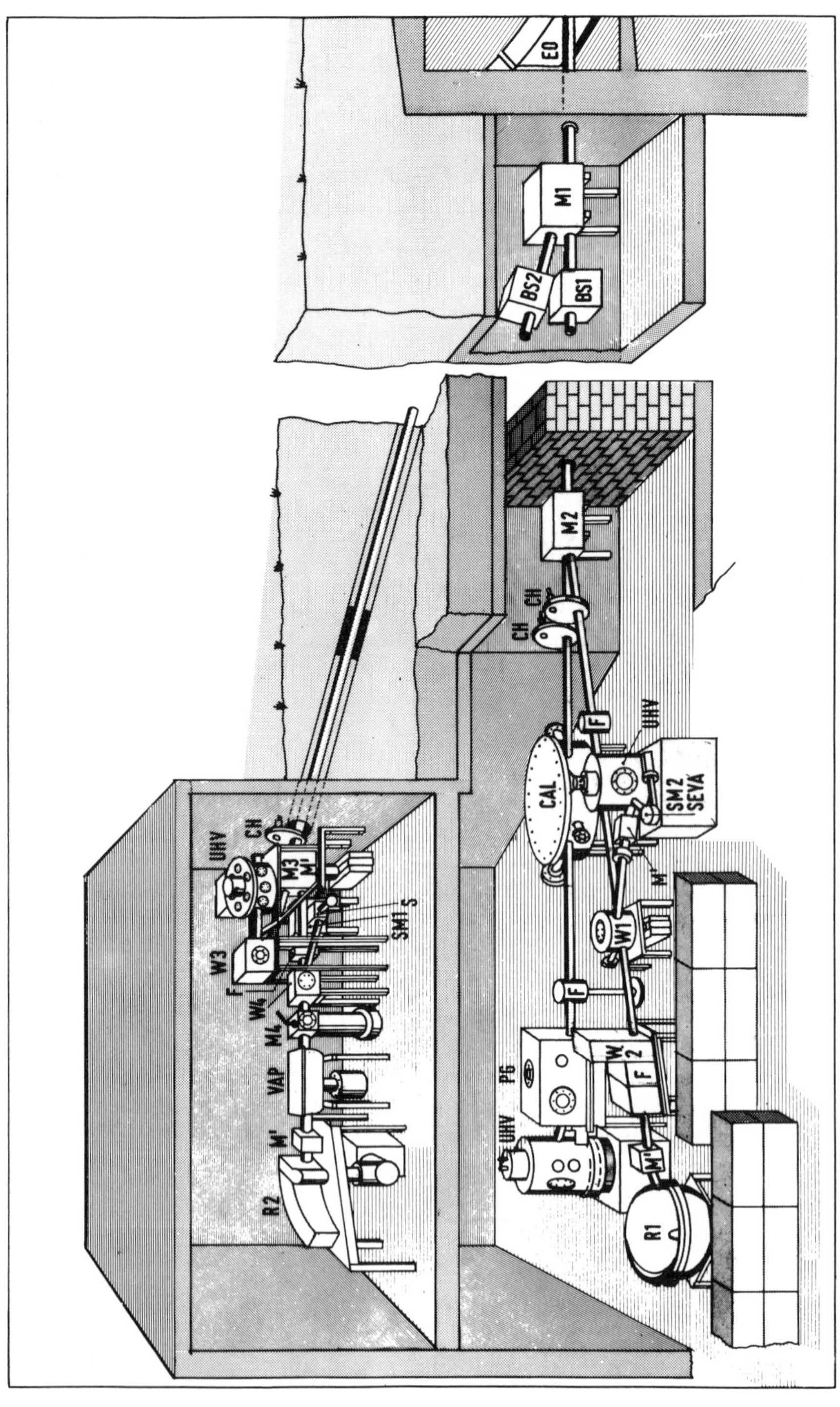
EO
M1
BS2
BS1
M2
CH
CH
F
UHV
CAL
SM2
SEVA
M'
W1
F
PG
W
2
F
UHV
M'
R1
UHV
CH
M3
M'
S
SM1
W3
F
W4
M4
VAP
M'
R2

monochromators are developed for this purpose [6.53]. In adapting a monochromator with an entrance slit, a focusing mirror has to image the source point into the entrance slit, e.g., grazing incidence toroidal mirrors have been used [6.53–61]. Advantages are gained when the monochromator is arranged in such a way that the dispersing plane is vertical. In this case the natural polarization of synchrotron radiation results in the higher reflectivities and efficiencies of *s*-polarization (rather than *p*). Moreover, the shape of the electron beam source is usually better matched to the shape of the entrance slit.

Two types of monochromators are used: normal incidence instruments, which are useful from 5 to 40 eV, and grazing incidence instruments which are operated typically in the range 30 to 600 eV. It is not the intention here to discuss different monochromators in detail, but rather to refer to a series of publications [e.g., 6.42, 53–61] which discuss the monochromators more fully.

6.4 Spectroscopic Techniques

6.4.1 Spectroscopy of Directly Excited Electrons

When measuring photoelectrons which are excited with a continuously tunable light source, the experimentalist has the option of varying either the pass energy E_{kin} of the electron analyzer and keeping the photon energy $\hbar\omega$ fixed, or of varying $\hbar\omega$ and setting E_{kin} to a fixed value. There are further possibilities for combined variations of $\hbar\omega$ and E_{kin}. There is a need for guidelines for the interpretation of the spectra obtained in these uncommon modes. For this purpose we go back to the simplified one-electron density of states picture (see [Ref. 1.1, eq. (1.89)].

The number of photoelectrons excited to a final-state energy characterized by E_{kin} is given by (6.62)

$$N(E_{\mathrm{kin}}, \hbar\omega) \propto \int d^3\boldsymbol{k} M_{\mathrm{if}}(\boldsymbol{k})\,\delta[E_{\mathrm{f}}(\boldsymbol{k}) - E_{\mathrm{i}}(\boldsymbol{k}) - \hbar\omega]\,\delta[E_{\mathrm{f}}(\boldsymbol{k}) - E_{\mathrm{kin}}]\,. \tag{6.9}$$

The escape probability through the surface has been assumed to be independent of all variables. The indices i, f represent initial and final states with energies E_{i} and E_{f}, respectively. $M_{\mathrm{if}}(\boldsymbol{k})$ is the matrix element for dipole transitions between these states at the point $\boldsymbol{k}$ in the first Brillouin zone. Under the simplifying assumption that $M_{\mathrm{if}}(\boldsymbol{k})$ is also constant throughout the

Fig. 6.7. Arrangement of primary and secondary beam lines at the 7.5 GeV synchrotron DESY. The secondary beams are split off by grazing incidence mirrors M1, M2. The beam can be blocked by beam shutters BS1, BS2. Normal incidence (W1–W4) and grazing incidence (R1, R2, PG) monochromators serve for wavelength selection. Sample chambers are F, VAP, UHV, S. Two secondary monochromators SM1 and SM2 serve for the decomposition of fluorescence radiation. Rotating chopper wheels (CH) cut out appropriate time intervals [6.34]

Brillouin zone and further that we deal with flat initial-state bands or have a breakdown of the $\boldsymbol{k}$-selection rule we can rewrite (6.9) and obtain

$$N(E_{\rm kin}, \hbar\omega) \propto M \cdot \varrho_{\rm i}(E_{\rm kin} - \hbar\omega) \cdot \varrho_{\rm f}(E_{\rm kin}), \tag{6.10}$$

where $\varrho_{\rm i}$ and $\varrho_{\rm f}$ are initial- and final-state densities of states, respectively. With realistic band structures, however, the $\boldsymbol{k}$-selection rule, which allows only for direct transitions, will reduce the number of states contributing to $\varrho_{\rm i}$ and $\varrho_{\rm f}$ in (6.10).

The three types of spectroscopies which have been used can be interpreted depending on how the parameters $E_{\rm kin}$ and $\hbar\omega$ are varied in (6.10).

1) EDC (Energy Distribution Curves) variation of $E_{\rm kin}$ for a constant $\hbar\omega$ gives joint information on $\varrho_{\rm i}$ and $\varrho_{\rm f}$.
2) CFS (Constant Final-state Spectroscopy), variation of $\hbar\omega$ gives $\varrho_{\rm i}$.
3) CIS (Constant Initial-state Spectroscopy), synchronous variation of $E_{\rm kin}$ and $\hbar\omega$ such that $E_{\rm kin} - \hbar\omega = \text{const}$ gives $\varrho_{\rm f}$.

These spectroscopies will be discussed in more detail in the following sections.

6.4.2 Energy Distribution Curves (EDC)

EDC is the common way and the only way to measure photoelectron spectra when using light sources of fixed (discrete) photon energies. With synchrotron radiation a series of EDC spectra at arbitrary photon energies can be measured. Since matrix elements are varying slowly compared to variations in the density of occupied states, but are usually determined in a characteristic way by the atomic parentage of the electron state in a solid (see Sect. 2.4.4), such measurements can serve to disentangle different contributions to a valence band. Indeed, it was recently demonstrated that such typical atomic features like the *Cooper* minimum [6.63, 64] in the 4*d* cross section of Ag [6.65] is clearly borne out in the EDC spectra of solids (Fig. 6.8). Figure 6.8 also shows the delayed onset of the 4*f* electron excitations in Au [6.66] and the 5*d* cross section of Au [6.67]. Another aspect of a variation of photon energy is concerned with rapid changes of structure in valence band spectra, which can be followed easily when spectra at photon energy intervals of about 1 eV are taken. The origin of these structures is explainable from (6.9) and is a consequence of direct transitions between valence and conduction bands. Only transitions from certain parts of the valence band contribute to the EDC at a certain photon energy, and intensity shifts back and forth as a function of $\hbar\omega$ in the energy range limited by the lower and upper bound of the valence band. In other words, the energy distribution of the joint density of states (EDJDOS) rather than the simple density of states (DOS) plays the dominant role. Such measurements serve for a detailed check and improvement of band-structure calculations. They have

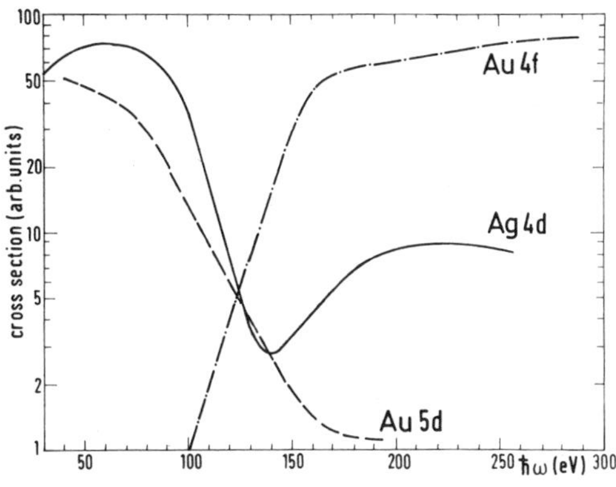

Fig. 6.8. Measured variation of subshell cross sections of the Au 5*d* [6.67], Au 4*f* [6.66] and the Ag 4*d* [6.65] excitations. The so-called Cooper minimum at 140 eV in the Ag cross section is of a purely atomic origin. The cross section of different shells relative to each other are not on scale

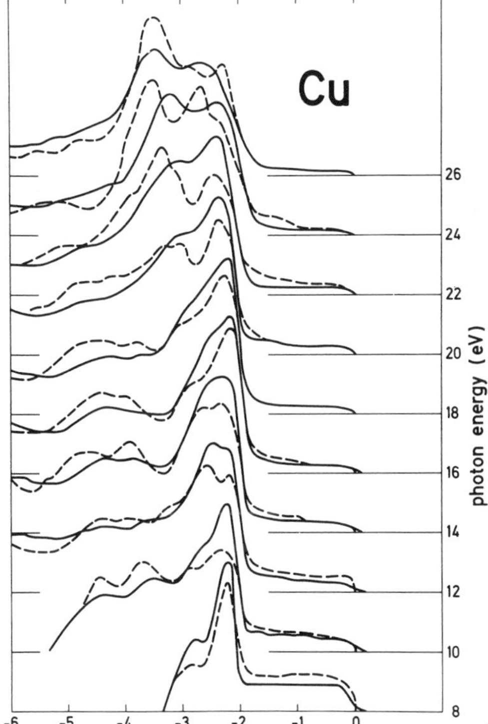

Fig. 6.9. Energy distribution curves for the copper valence band, experimental (solid), theoretical (dashed) [6.69]

been very successful with semiconductors, eg., Ge [6.68, 62] and metals, eg., Cu [6.69] (see Fig. 6.9).

The energy region where the EDJDOS dominates the spectra lies below 30 to 50 eV. Above, there is a gradual change to a DOS picture. This change is explained by *Feibelman* and *Eastman* (6.70) by the decrease of the mean free path of electrons with a minimum at kinetic energies around 100 eV, where the mean free path is typically of the order of a few lattice constants. Consequently, the component the momentum of the final state *perpendicular to the surface* $k_\perp$ is uncertain by an amount of the order of the reciprocal mean free path. Because of the high density of final states in this energy range, there occurs a merging of final states, so that there is always one available into which an initial-state electron with momentum $\boldsymbol{k}$ can be excited by arbitrary photon energies $\hbar\omega$. These final states can be regarded as inverted LEED states ([see Ref. 1.1, Chap. 2]). *Lindau* et al. [6.66] showed, however, that even at energies above 100 eV some features undergo considerable changes in relative intensity. The ratio of the lower to the higher valence band maximum in Au, which are only ~3 eV apart in binding energy, varies by more than 50% in the energy range 100 to 200 eV. This might throw some doubt on the "theorem" that above 100 eV such spectra are essentially unchanged unless atomic matrix elements are involved. The reason for the observed changes in the spectra of Au is not clear at the moment. An explanation of this effect might be of an atomic origin rather than of a solid-state origin [6.71].

The variation of the mean free path with excitation energy can further be applied to the discrimination between volume and surface effects. Since, however, the excitation cross section is varied at the same time, this method must be used with great caution. Also the polarization with respect to the surface normal, and with respect to crystallographic directions in single crystals, is variable with SR, unlike with other sources. Polarization-dependent matrix elements have to be used to identify volume states of specific symmetries and also to clarify the geometry of surface states [6.39].

6.4.3 Constant Final-State Spectroscopy (CFS)

The CFS technique (6.72) is visualized in Fig. 6.10 by showing how the measured spectrum originates from the excitation of valence or core electrons into a fixed energy interval at $E_{\mathrm{kin}} = E^0_{\mathrm{kin}}$. Positions of levels are obtained in just the same way as with EDC's. There are, however, differences between EDC's and CFS's which can be explained from (6.10). Since E_{kin} is fixed, ϱ_{i} is measured directly (in the $\boldsymbol{k}$-nonconservation DOS approximation). Any variation of the shape of ϱ_{i} with different $\hbar\omega$ can immediately be interpreted as a breakdown of the simplifying assumptions underlying (6.10).

For all values of E_{i} the same range below the surface is sampled in CFS (provided the concept that the mean free path is only a function of E_{kin} is valid). For a comparison of relative intensities, say of core excitations with valence excitations, the relative partial excitation cross sections for the same E_{kin} but different $\hbar\omega$ come into play. Further, the intensity variation of the photon flux on the sensitive part of the sample needs to be determined as a function of $\hbar\omega$. The latter is the counterpart to the transmission function of the electron analyzer for a quantitative evaluation of EDC curves. By choosing different values for the parameter E^0_{kin} the probed depth can be varied, and more or less surface sensitivity is obtained.

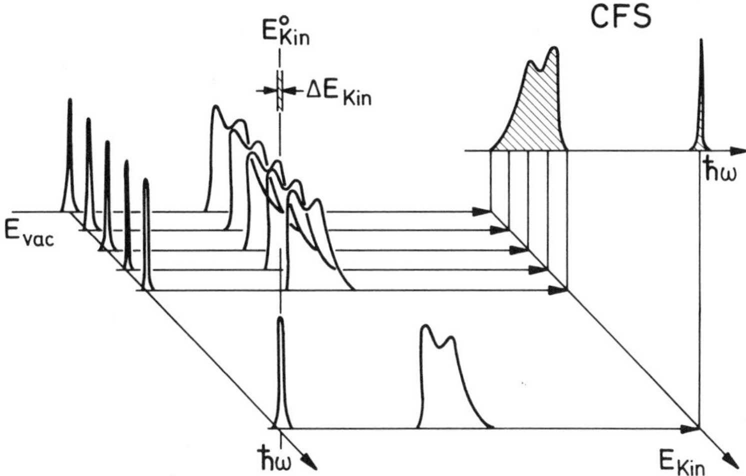

Fig. 6.10. Schematic explanation of constant final-state (CFS) spectra as generated from a series of EDC spectra at different $\hbar\omega$. E^0_{kin} is kept constant while $\hbar\omega$ is varied. In the upper right corner the shaded curve is the thus generated CFS spectrum [6.39]

6.4.4 Constant Initial-State Spectroscopy (CIS)

The CIS technique was first applied by *Lapeyre* and co-workers [6.72–74]. The principle is visualized in Fig. 6.11. By a synchronous scanning of E_{kin} and $\hbar\omega$ electrons from the same initial-state energy are excited to different final-state energies [see (6.10)]. Thus a mapping of the final-state density of states is obtained. On the basis of (6.10) a series of spectra, in which different initial states are excited, should show the same ϱ_{f}. If this is not the case, this is an indication of deviations from the simplified picture underlying (6.10).

In Fig. 6.12 a set of CIS spectra, as obtained by *Lapeyre* et al. [6.73] on KCl, shows features A, B, and C, which persist when E_{i} is varied. According to the discussion above, these features are attributed to density of states structure in ϱ_{f}. This interpretation is confirmed by the upper curve, in which scattered electrons which originate from primary excitations at higher energies are shown. These scattered electrons acquire a distribution which is essentially proportional to the available density of states ϱ_{f}. This method was also applied to KI (6.75). Scattered electrons will be discussed in more detail in Sect. 6.4.6. The small peak D in the spectra of Fig. 6.12 shifts with photon energy as is demonstrated in more detail in [6.73]. Peak D is always excited at a fixed photon energy and thus could be attributed to a resonance transition, e.g., an exciton related to a higher conduction band of KCl. Such structures, however, have to be carefully checked against structure in the primary spectrum of the monochromator and also against structure in the reflectivity of the material under investigation in order to avoid false interpretations.

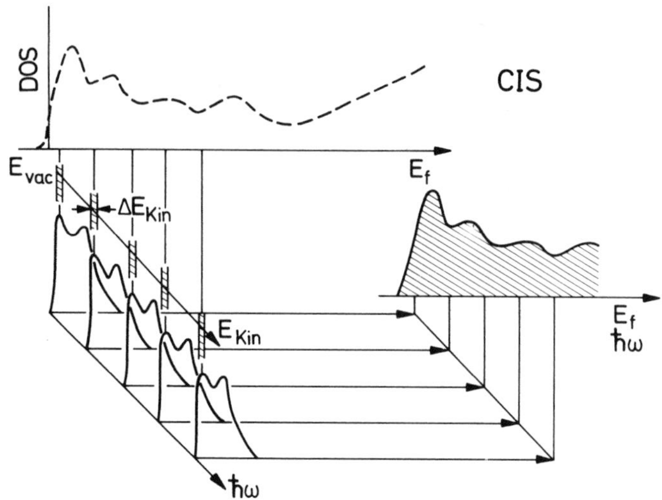

Fig. 6.11. Schematic explanation of constant initial-state (CIS) spectra. The energy interval, which accepts electrons from a fixed initial state is shifted synchronously with $\hbar\omega$. Thus the final-state, DOS represented by the dashed curve in the upper part of the figure is reproduced as visualized in the right part of the figure [6.39]

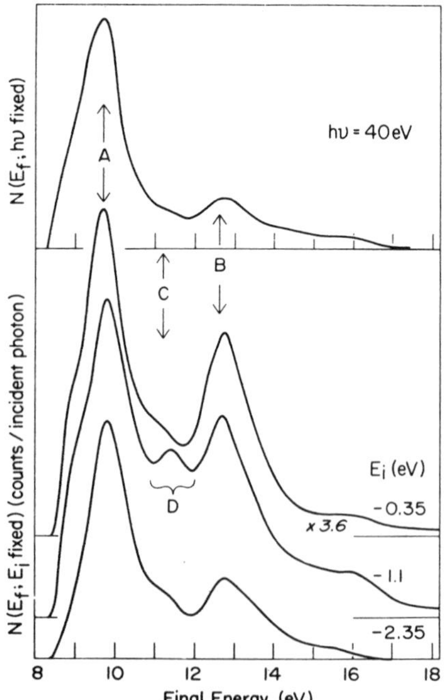

Fig. 6.12. *Bottom:* A set of CIS spectra for KCl taken at several initial-state energies E_i plotted as a function of final-state energy. *Top:* Portion of an EDC containing inelastically scattered electrons. Peaks *A* and *B* are common to both types of spectra and are due to peaks in the conduction-band density of states of KCl [6.73]

6.4.5 Angular Resolved Photoemission (ARP, ARPES)

A combination of these new techniques of spectroscopy with angular resolved photoemission opens up very attractive new possibilities for the study of single crystals. It has been amply demonstrated, using classical radiation sources (see, e.g., [6.76, 77]), that photoemission normal to a crystal surface scans the Brillouin zone along a straight line going through its center. This can be chosen to coincide with a crystallographic axis along which band calculations usually are available. Variability of $\hbar\omega$ allows for an extension of such investigations.

At a nonzero fixed polar angle, the conservation of $k_{\|}$ the parallel component of $\boldsymbol{k}$ to the surface will result in scanning another portion of the Brillouin zone in an EDC spectrum. In layered compounds bands can be mapped (see, e.g., [6.77]). This is caused by varying the magnitude of $k_{\|}$.

In CFS spectroscopy $k_{\|}$ is fixed for a fixed polar angle. This allows now for scanning along a straight line (or several lines) in $\boldsymbol{k}$-space which does not necessarily intersect the center of the Brillouin zone. *Lapeyre* demonstrated for *Tungsten* [6.78] that by a proper choice of angle and final-state energy, spectra are obtained which represent the band structure along a line on the surface of the Brillouin zone. Portions of the Brillouin zone which are not accessible with classical sources thus can be inspected in a well-defined manner. In addition, advantage is taken of the selection rules for transitions excited with different polarizations of the incoming SR light beam.

In angle-integrated spectroscopy usually no band gaps are seen other than the fundamental band gap in semiconductors and insulators. Along well-defined lines in $\boldsymbol{k}$-space, however, other band gaps do exist. With CFS spectroscopy a final state in such a band gap can be specifically selected and only excitations near the surface will be seen since electrons cannot propagate inside the crystal in such states. At very large values of E_{kin} the uncertainty in $\boldsymbol{k}$ due to the decreasing mean free path comes into play (see Sect. 6.4.2), and probably destroys this concept, which could be called "band gap spectroscopy". No actual results obtained by this method are published as yet. For further applications of ARP in combination with synchrotron radiation the reader is refered to [Ref. 1.1, Chap. 6].

6.4.6 Secondary Processes

Figure 6.13 gives a schematic decomposition of the primary and secondary processes which contribute to a typical EDC. A density of states $\boldsymbol{k}$-non-conserving model of an insulator is used.

The most important process is electron–electron scattering. In a metal this process is of importance for all possible kinetic energies, while in an insulator the electron kinetic energy must be high enough to excite a valence band electron into a bound (or free) state. The electron–electron scattering cross

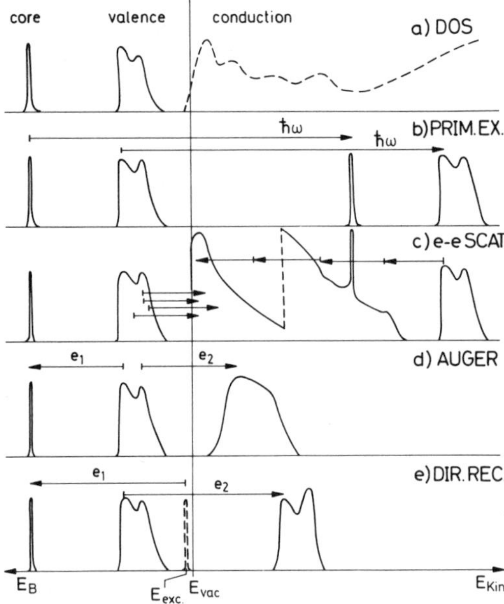

Fig. 6.13a–e. Contribution of primary and secondary processes to an EDC spectrum (schematic): **a** as an example a DOS of an insulator is assumed with one core and one valence band; **b** primary excitations with photons of energy $\hbar\omega$; **c** electron-electron scattering of energetic electrons; **d** Auger decay of a hole in the core state; **e** direct recombination of a bound core exciton state (autoionization) [6.39]

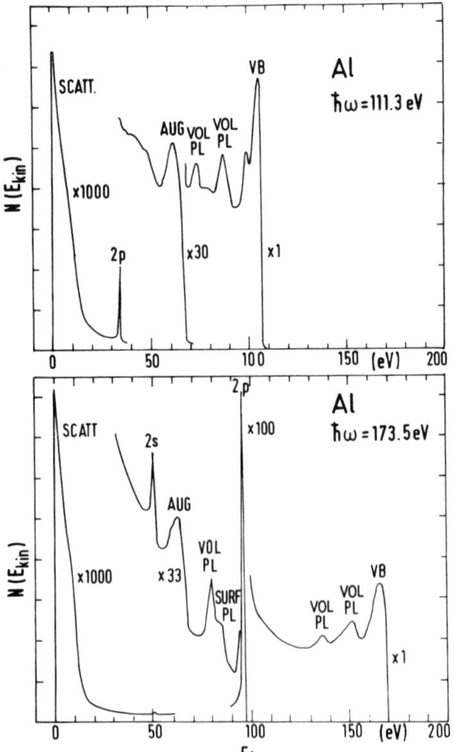

Fig. 6.14. EDC curves of evaporated Al plotted against the kinetic energy of the electrons E_{kin} for two values of $\hbar\omega$. Volume and surface plasma (*PL*) excitations and direct excitations from the valence band *VB* and the Al 2*p* and 2*s* levels are observed. The Auger edge (*AUG*) is always located at the same E_{kin}. The intensity ratios are given as measured with a double cylindrical mirror analyzer [6.79]

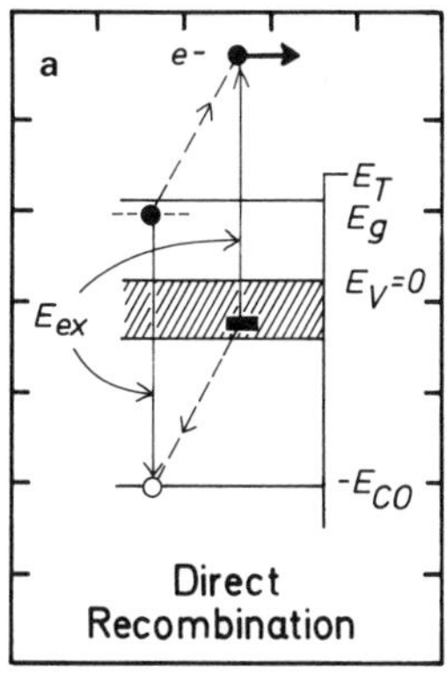

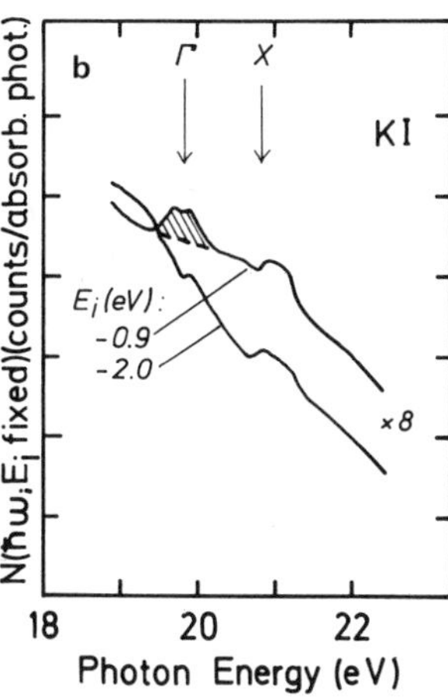

Fig. 6.15. **a** Energy diagram for radiationless direct recombination decay of a core exciton. **b** CIS spectra for KI for two initial-state energies E_i. One of the enhanced peaks due to direct recombination decay of a core exciton is shaded [6.72]

section, proportional to the inverse of the mean free path, leads to a strong multiplication of electrons and thus to the actual accumulation of electrons at low kinetic energies where the scattering cross section is small. This is demonstrated in Fig. 6.14 by the spectrum of Al [6.79]. One of the main energy loss processes, the excitation of surface and volume plasma oscillations at 10 and 15 eV respectively, is clearly identified.

Another important process producing secondary electrons is Auger decay (Fig. 6.13). This leads, e.g., to the strong broad maximum in the spectra of Fig. 6.14. This peak occurs at a fixed E_{kin} since Auger decay is independent of the excitation energy of a core level. Thus a variation of $\hbar\omega$ is an easy way of finding the Auger contribution to a spectrum.

Special processes occur when resonance excitations (e.g., core excitons) are involved. A core exciton can decay and give all its energy to a valence electron. In a EDC such a decay mode cannot be distinguished from a direct valence band excitation by a photon of the same energy as that of the exciton. CIS spectra, however, clearly show this decay process by setting the CIS energy to an appropriate valence band state and scanning $\hbar\omega$. Then exciton peaks show up only if direct recombination takes place. This process was discussed by *Haensel* et al. [6.80] for the NaCl, and proven for potassium halides by *Blechschmidt* et al. [6.81]. Figure 6.15 shows an experimental proof with CIS by *Lapeyre* et al. [6.72] which indicates that this process occurs with the $3p$ exciton in KI. Enhancement above the background of directly excited valence band excitations is observed for different values of E_i. Some differences in the spectra for different E_i which are observed at the Γ exciton are an indication of different selection rules for this decay into different hole states. *Lapeyre* et al. [6.67] give in a rough estimate the ratio of probabilities for direct recombination to that of Auger decay as 1 to 2. Clearly, these contributions are small enough to escape detection in a family of EDC's.

6.4.7 Photoelectron Yield Spectroscopy (PEYS)

Yield spectroscopy can be performed for any photon energies above the photoemission threshold. Many investigations were made in the region of the fundamental absorption on alkali halides [6.81–83], pure and doped rare-gas solids [6.84–90] and pure and oxidized Al [6.91–94]. For aluminum the polarization dependence of the plasma excitation was investigated.

Yield spectroscopy in the region of core electron excitations is quite different from that in the fundamental absorption region. Historically the first series of such measurements for several materials in the soft X-ray region was carried out by *Lukirskii* and co-workers using the bremsstrahlen continuum of a soft-X-ray tube [6.95–97]. These results showed already a very close resemblance of PEYS and absorption spectra. It was demonstrated [6.80, 82] that this similarity exists as long as core excitations are involved. *Gudat* and *Kunz* [6.98] showed that SR is the ideal tool for such investigations. A simple theoretical interpretation leads to a satisfactory understanding of PEYS. The method is exploited to investigate the absorption coefficient of liquid metals [6.99, 100], single crystals [6.98, 101], alloys [6.101], and surface states [6.102–105].

According to the preceding section each photoexcitation is followed by a series of scattering and secondary processes, which lead to an accumulation of electrons at low kinetic energies (see Fig. 6.14). These slow electrons are the dominant contribution to the yield. Their mean free path determines an average depth L which contributes to the yield and is typically in the order of 30 to 50 Å for metals and about one order of magnitude larger for insulators. In fact, for insulators, L can be determined by the film thickness. Usually L is much smaller than the penetration depth of light, which is the reciprocal of the absorption coefficient. At high enough photon energies, where reflectivity can be neglected, the yield $Y(\omega)$ is proportional to the number of photons absorbed in this layer, namely to $\mu(\omega)\cdot L$, $\mu(\omega)$ is the absorption coefficient. A smoothly varying multiplication factor $F(\omega)$ has to be introduced for which we assume: $F(\omega)\propto\hbar\omega$, since the number of slow electrons created should be proportional to the energy initially deposited in the material. We thus obtain

$$Y(\omega)\propto F(\omega)\cdot\mu(\omega)\cdot L\,. \tag{6.11}$$

In this model all structure, which occurs in $Y(\omega)$, originates from $\mu(\omega)$. Over limited spectral regions the $Y(\omega)$ and the $\mu(\omega)$ spectra are practically proportional to each other.

Figure 6.16 shows a comparison between a yield spectrum and the absorption coefficient of Al. The directly measured yield spectrum, uncorrected for the spectral dependence of the incident intensity, is shown. It has to be corrected by a smooth slowly varying function of ω, representing the spectral distribution of the incident intensity, in order to obtain the absolute yield.

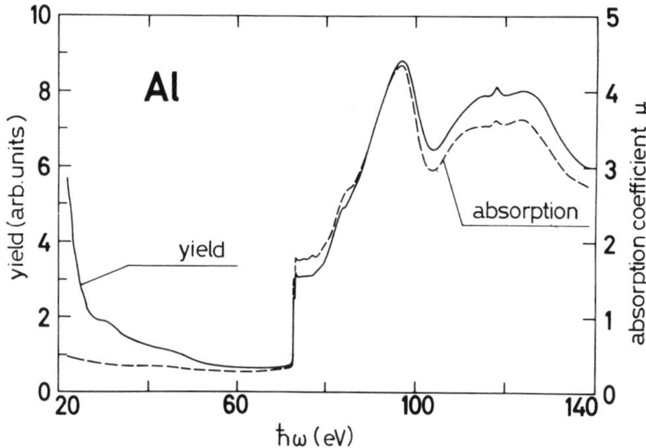

Fig. 6.16. Comparison of photoelectric yield spectroscopy (PEYS) and absorption spectroscopy from a clean evaporated Al sample. The yield curve is shown as measured, e.g., it has to be corrected by a smooth slowly varying instrument response function [6.101]

Therefore the variation of $F(\omega)$ in (6.11) cannot be deduced from Fig. 6.16. The exact correlation of structures in both spectra, however, is clearly borne out.

A technique related to PEYS is the so-called partial yield spectroscopy. Two versions have been applied, either an energy dispersive electron analyzer selects scattered electrons, which fall into a fixed energy window, or a retarding field analyzer selects all electrons, which have a kinetic energy above a certain threshold. This method has a reduced signal level compared to PEYS, but allows for varying the depth probed through a variation of E_{kin}. It is nowadays a standard technique for the investigation of empty surface states and surface excitons [6.102–105].

6.4.8 Yield Spectroscopy at Oblique Incidence

It is quite useful to consider PEYS at nonnormal incidence of the light beam since many experiments are performed with such a geometry and, further, since in principle, information could be derived from such experiments which goes beyond the normal incidence measurements.

In going to more and more grazing angles the penetration depth of light is reduced. This can be accounted for in the frame of classical optics by calculating an effective absorption coefficient $\mu(\omega,\theta)$ with θ the angle of incidence

$$\mu(\omega,\theta)=\frac{2\omega}{c}\,\mathrm{Im}\,\{\sqrt{\varepsilon_1(\omega)-\sin^2\theta+\mathrm{i}\varepsilon_2(\omega)}\}\,. \tag{6.12}$$

$\varepsilon_1(\omega)$ and $\varepsilon_2(\omega)$ are the real and imaginary parts of the frequency-dependent dielectric constant. This description may become invalidated through nonlocal

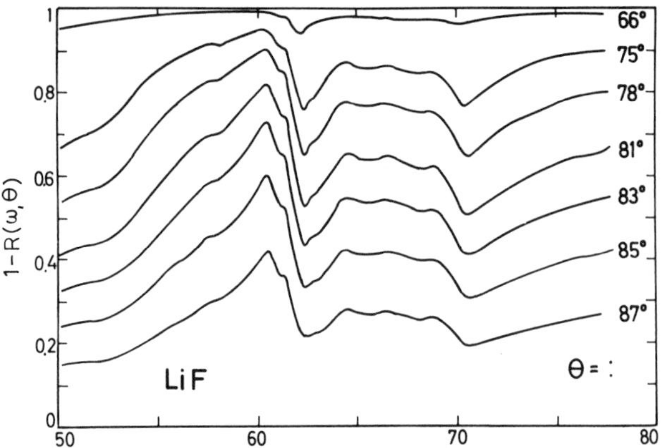

Fig. 6.17. Calculated spectra of $1-R(\omega,\theta)$, where $R(\omega,\theta)$ is the angle-dependent reflectivity of LiF in the region near the onset of Li^+ 1*s* transitions. The optical constants for these calculations were generated by a Kramers-Kronig analysis of absorption measurements [6.101]

effects which could be of importance with *p*-polarized light [6.106]. We will not consider such effects here, because according to existing evidence they usually are weak in the energy region of core electron excitations.

Since at oblique angles the penetration depth of light, namely $1/\mu(\omega,\theta)$, can be of the same magnitude as L, we have to generalize (6.11) also for this case. The result is obtained under the assumption that the escape probability of an electron excited at a depth x below the surface can be described by an exponential $\exp(-x/L)$. We then obtain [see Ref. 1.1, Eq. (6.15)]

$$Y(\omega,\theta)\propto F(\omega)\cdot\frac{\mu(\omega,\theta)}{\mu(\omega,\theta)+1/L}\cdot[1-R(\omega,\theta)]\,, \tag{6.13}$$

where $R(\omega,\theta)$ is the angular dependent reflectivity which cannot be neglected any more with very oblique angles of incidence θ.

Gudat [6.10] made a careful study of $Y(\omega,\theta)$ from a 3000 Å thick layer of LiF evaporated in situ under ultrahigh vacuum conditions. The angle of incidence θ was varied in such a way that the light was *s*-polarized. The yield was investigated for $\hbar\omega$ between about 55 and 75 eV which is the region of strong core exciton structures [6.107, 108] at the onset of the $Li^+ 2p$ electron transitions. Optical constants were generated from a Kramers–Kronig analysis of a combination of absorption coefficients from several sources in the spectral region 10 to 400 eV. Then these optical constants were used to calculate the different factors determining $Y(\omega,\theta)$ according to (6.13). $F(\omega)$ was set constant because of the small spectral interval involved. The result of the calculation for $[1-R(\omega,\theta)]$ and for $\mu(\omega,\theta)$ is shown in Figs. 6.17, 18. The individual contributions to the distortion of the original $\mu(\omega,0)$ spectrum of these factors becomes evident.

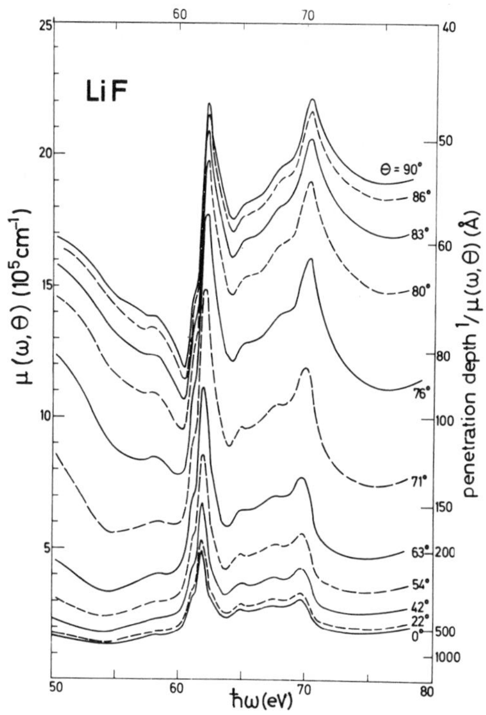

Fig. 6.18. Spectral distribution of the effective absorption coefficient $\mu(\omega, \theta)$ and the corresponding penetration depth for different angles of incidence θ for LiF according to (6.12) [6.101]

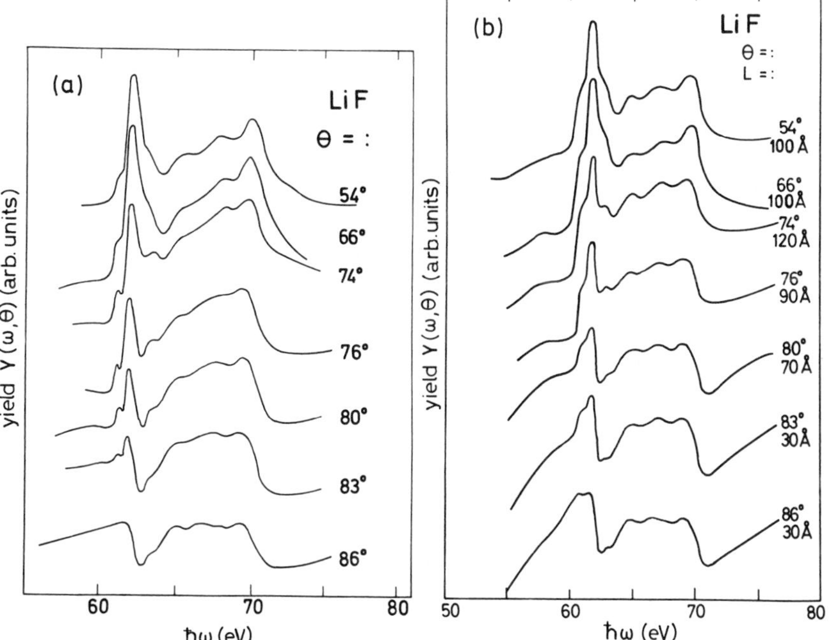

Fig. 6.19. a Experimental photoelectric yield spectra $Y(\omega, \theta)$ as a function of the angle of incidence onto an evaporated film which was not prepared in situ; **b** calculated yield spectra $Y(\omega, \theta)$ according to (6.13). The electron escape depth L was fitted by calculating a series of spectra and selecting those, which gave the best agreement with the experiment [6.101]

Figure 6.19 shows the result of the calculation of $Y(\omega, \theta)$ according to (6.13), in which the parameter L was varied to obtain optimum agreement with experiment. It is interesting to note that the details of the shapes of the experimental structures are reproduced. Up to an angle of 80° the values of $L(L \approx 100\,\text{Å})$ are consistent within their estimated errors of about $\pm 30\,\text{Å}$. For $\theta > 80°$, however, the shape of the curves is very sensitive to a variation of L as can be understood from (6.13) since there $\mu(\omega, \theta)$ and $1/L$ are of the same order of magnitude. For L at 83° and 86° a fit of the experimental data requires $L = 30 \pm 10\,\text{Å}$. There are several possible reasons for this discrepancy and more investigations are needed to clarify this point: roughness of the film, wrong optical constants, change of optical constants in the surface layer, surface scattering, etc. In summarizing we can state that experimental results for angles of incidence smaller than 70° appear to be quite well described by (6.13) and the intensity enhancement due to the increase in the effective absorption coefficient can profitably be used.

6.5 Applications of Yield Spectroscopy

6.5.1 Anisotropy in the Absorption Coefficient of Se

Anisotropy effects on the excitation spectra of core electrons have not yet been extensively studied since they encounter considerable experimental difficulties. An investigation on CdS was carried out on small thin flakes in absorption near the onset of the sulphury $L_{2,3}$ transitions in the range 160 to 170 eV [6.109]. Small anisotropy effects were identified. Core electron transitions involve the well-defined atomic symmetries of the hole and the empty band states. In the frame of a one-electron picture, selection rules for transitions into the different states of the conduction band can be obtained from group theory. Thus, such anisotropy measurements should provide information on the band states and might be a means for improving the understanding of core electron transitions.

Yield spectroscopy is well suited for such investigations since cleaved single crystals can be used. Trigonal Se is a good candidate: the c-axis is contained in a natural cleavage surface. *Gudat* [6.101] measured the yield spectra of the $3d$ electron transitions in the region from 54 eV on with the electric vector $\boldsymbol{E}$ parallel and perpendicular to the c-axis (Fig. 6.20). The spectra show richer and sharper structure than previously obtained from amorphous and crystalline evaporated films [6.110, 111]. There is a clearly detectable difference between the spectra taken with the two different directions of polarization for the transitions into the low-lying parts of the conduction band, namely the first large peak between 54 and 58 eV. The next higher band complex with peaks E to G is much weaker and shows less pronounced anisotropy effects. Altogether the anisotropy effects are far from being dramatic, but its details should be interpretable from a band-structure picture.

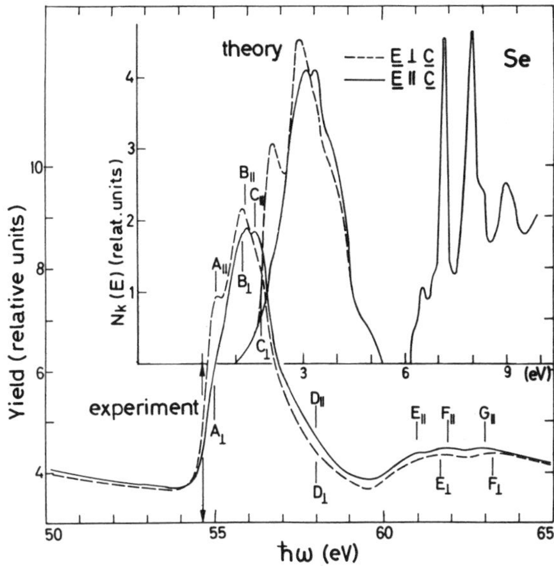

Fig. 6.20. The experimental photoelectric yield from a Se cleavage surface measured with two directions of polarization when exciting electrons from the $3d_{5/2}$ and $3d_{3/2}$ states. The upper part contains theoretical spectra which are obtained from a conduction band DOS, taking into account selection rules and the spin-orbit splitting of the d-level [6.101]

Gudat [6.101] took the density of states calculation by *Kramer* et al. [6.112] and modified it by applying the selection rules of the two different geometries $\boldsymbol{E} \| c$ and $\boldsymbol{E} \perp c$. A qualitative density of states for the two cases, modified to include the superpositions of the two hole levels $d_{5/2}$ and $d_{3/2}$ with a spin orbit splitting of 0.83 eV, was obtained. The absolute position of the calculated absorption structure was adjusted to the experimental curves. This leads to the densities of states curves as given in the upper part of Fig. 6.20. There is a satisfactory agreement with experiment in the region below 58 eV. A modification of the density of states above 60 eV was not included since the symmetry of the corresponding bands was not known. These bands, however, are mainly formed from $5s$ and $5d$ orbitals into which transitions from $3d$ states are forbidden from atomic selection rules. This explains why they show up so weakly in the spectra.

Although the band picture appears to be quite adequate for a description of the results shown above, the influence of the electron-hole interaction on these spectra might not be neglected. *Shevchik* et al. [6.113] determine from XPS measurements the energy separation between the $3d_{5/2}$ level and the top of the valence band as 54.64 eV. This implies a separation of $3d_{5/2}$ to the first peak in the density of states as 56.8 eV while the experimental value is 55.0 eV, therefore 1.8 eV lower. As in equivalent investigations of this kind [6.107, 114, 115] on insulators (see Sect. 6.1.2) a fairly large exciton binding energy of 1.8 eV must be the reason for this shift. The relatively satisfactory interpretation of the observations based on the one-electron picture are therefore not sufficient for a complete understanding of the observed anisotropy and the shape of the band states.

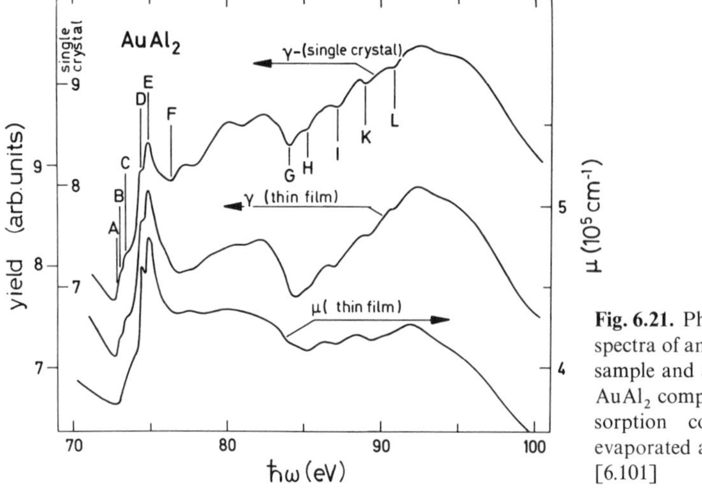

Fig. 6.21. Photoelectron yield spectra of an evaporated alloy sample and a single crystal of $AuAl_2$ compared with the absorption coefficient of an evaporated alloy film [6.118], [6.101]

6.5.2 Investigation of Alloys

Alloys of some transition metals of the series Ti to Ni and $CuAl_2$, $PrAl_2$ [6.116–118] and $AuAl_2$ [6.118, 119] and CuNi [6.120] have been investigated in absorption with synchrotron radiation. Direct absorption measurements necessitate the preparation of these alloys as very thin films by simultaneous evaporation from two sources. The subsequent annealing process is restricted to temperatures at which these films are not destroyed. Furthermore, there is always the danger of oxidation; even a slight oxidation can be of importance with films in the thickness range 200 to 500 Å. There is, therefore, a desire to investigate bulk crystalline samples of alloys, whose composition and phase structure is known from investigations with complementary methods.

The absorption of $AuAl_2$ was first investigated by *Gudat* et al. [6.119] in thin films; quite unexpected results occur with structures in the spectra near the Al $L_{2,3}$ threshold (see Fig. 6.21). The onset of the $L_{2,3}$ transitions shows up weakly at an energy 0.3 eV higher than with pure Al but without the characteristic spin-orbit splitting of the p-states, while at a 1.3 eV higher energy a very steep and prominent edge occurs with a clearly detectable spin-orbit split counterpart. This result was later essentially confirmed by *Hagemann* [6.118]. Since at that time there was some ambiguity about phase homogeneity of the thin film samples, and since such a steep rise in the density of states at an energy 1.3 eV above the Fermi energy was without precedent, this system was reinvestigated by *Gudat* [6.101] with PEYS on two well-defined crystalline samples from different sources.

The bulk $AuAl_2$ alloy samples were cleaned by scraping in ultrahigh vacuum of $2 \cdot 10^{-10}$ Torr. Figure 6.21 shows the result of PEYS together with the absorption measurement [6.118] and a PEYS result on a evaporated sample which was prepared on a Ta strip by allowing a Au/Al sandwich to alloy at roughly 280 °C. The PEYS result essentially confirms the absorption measurements near the onset while there are some differences in the region above 80 eV. The yield spectra also show the spin-orbit splitting of the threshold which might be observable because of a better crystalline ordering of the bulk samples, or because of a reduction of impurities due to the evaporation under UHV conditions in the case of the alloying sandwich.

The conclusion drawn from these experiments was the following: The strong rise at ~1.3 eV above threshold must be a density of states effect. In fact, such a rise in the density of states, although less pronounced, is borne out in theoretical calculations [6.121]. In later studies a rise of the absorption coefficient of comparable prominence was observed in the absorption spectra of the intermetallic compounds NiAl and FeAl [6.116] which is in much better agreement with band calculations [6.122] than the results on $AuAl_2$.

6.5.3 Investigation of Liquid Metals

It is extremely difficult to obtain liquids in the form of a thin film for transmission experiments and can be successful only under very favorable conditions for a few substances. With yield spectroscopy substances can be investigated whose vapor pressure is sufficiently low above the melting point. In this case partial PEYS must be applied in order to separate the thermally emitted electrons.

Li has been investigated at different temperatures up to the liquid state near the K threshold [6.123] as well as to higher excited states [6.100]. Further studies were made of Na and Al for a wide range in the region of the $L_{2,3}$ transitions [6.99]. In these investigations (Fig. 6.22) it was demonstrated by *Petersen* and *Kunz* [6.99] that the spike at the Na $L_{2,3}$ absorption edge, which is interpreted as due to a many-body phenomenon prompted by the sudden creation of a localized hole in a metal ([6.33, 124]; see also [Ref. 1.1, Chap. 5]), persists in the spectrum of the liquid. This proves that it could not be possibly due to any effect which depends on the crystalline order, like, e.g., band-structure density of states. If we consider now the structure at higher energies above the edge in Fig. 6.22 we see that peaks C and D are completely washed out while the long-range modulation of the absorption coefficient, which manifests itself in peaks E and F, is considerably weakened. The spectrum becomes very similar to the result of atomic calculations [6.125], except for the residual E-F peaks. The long-range modulation which is quite pronounced also in the $L_{2,3}$ spectrum of solid Al (while absent with liquid Al [6.99]) is explained as extended absorption fine structure (EXAFS) [6.33, 126, 127]. EXAFS is usually investigated above the K edges in the X-ray region and originates from the interference of the outgoing electron wave with the backscattered waves from the neighboring atoms. This interference builds up the amplitude of the final-state wave function at the core of the excited atom. Its amplitude

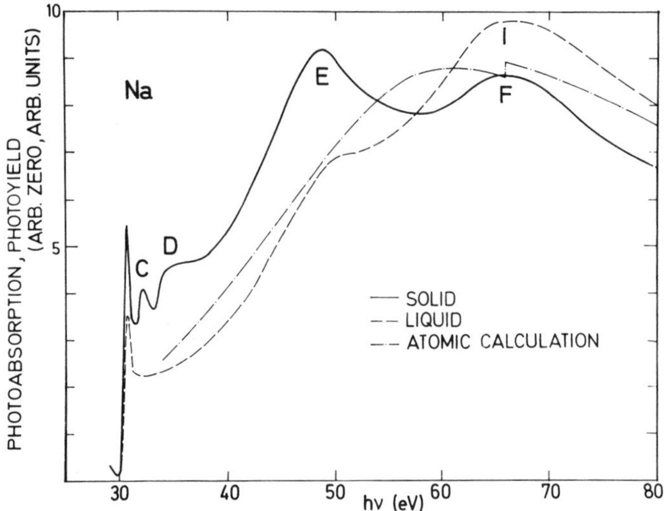

Fig. 6.22. Photoelectric yield from liquid Na compared to the absorption coefficient of solid Na in the region of the Na $2p$ transitions. The calculated atomic cross section [6.125] is also shown [6.99]

modulation reflects itself through the matrix element in the absorption coefficient. When the crystalline order is destroyed on melting such interference structures are weakened although they need not disappear completely: short-range order is still present in the liquid phase with nearest-neighbor distances similar to those found in the solid. Hence residual EXAFS effects may still be observed in the molten phase.

6.6 Experiments Investigating Occupied and Empty States

6.6.1 Valence Bands in Rare-Gas Solids

The number of SR investigations of valence bands, mainly of metals and semiconductors, on polycrystalline samples and on single-crystal samples by methods which average over fairly large emission angles of photoelectrons, is continuously growing (e.g., [6.128–131]; see also the other chapters of this volume). These investigations, as mentioned above, take advantage of the possibilities to vary matrix elements for the different types of contributing atomic orbitals at will, e.g., [6.132–134], comparing both valence and conduction band structures with theoretical bands by applying the concept of direct transitions, e.g., [6.62], and finally investigating the transition to the so called X-ray limit where final-state structure does not interfere (e.g., [6.66]). Such investigations are presently being repeated by means of angular resolved photoemission from single crystals (see, e.g., [6.67, 78, 135, 136] and also [Ref.

1.1, Chap. 6]). Such measurements yield by far more specific information than angularly averaged photoemission.

We shall concentrate here on the investigation of a class of substances, the rare-gas solids, which because of their van der Waals binding belong to the simplest solids. There are many band calculations published in the literature, which in spite of the simplicity of these substances differ considerably (see e.g., *Rössler* [6.137]). Their investigation ideally takes advantage of the cleanliness of the source, and of the variability of the spectrum. The experiments using reflection and absorption techniques have been extensively reviewed by *Sonntag* [6.32]. While these spectra yield considerable insight into the formation and behavior of strongly bound excitons, almost no information could be obtained on the actual band structure [6.137]. EDC measurements on thin polycrystalline films which provide this essential information were taken by *Schwentner* et al. [6.138–140].

Figure 6.23 shows a series of EDC curves of the four rare-gas solids Ne, Ar, Kr, and Xe measured on thin films of typical thicknesses 20 to 100 Å in order to avoid charging effects [6.138]. There is a large difference in the width of the valence band between the different rare-gas solids, which mainly reflects the different spin-orbit splitting which in the gas state varies between 0.14 eV for Ne and 1.31 eV for Xe. The single peak to the left (higher binding energy) which appears clearly for Xe and Kr corresponds to the bands originating from the $P_{1/2}$ state of the hole while the broader, structured peak to the right originates from $P_{3/2}$ derived bands. A quite fortunate situation compared to semiconductors and metals occurs: all these spectra are obtained by exciting into final states which lie below the electron–electron scattering threshold.

The authors compared their results for Xe with the available relativistic band-structure calculations [6.137, 141, 142]. As shown in Fig. 6.24, these bands are separated at the Γ point practically by the atomic spin-orbit splitting. EDC's calculated by *Rössler* [6.143] under the assumption of a constant matrix element according to (6.9) from his band calculation [6.142] show pronounced deviations from the measurements (Fig. 6.24). The theoretical band is 40% too narrow. The comparison has also been made at the other photon energies for Xe [6.139]. Although the general topological appearance of a narrow high-binding energy band and a split low-binding energy band are borne out by both the theoretical and experimental curves, the quantitative agreement is poor and asks for an improved theoretical treatment. Another comparison between theory and experiment was attempted by *Kunz* et al. [6.144].

Schwentner et al. [6.138, 139] have also performed matrix isolation experiments in which they investigated the EDC of samples with 1% Xe in Ar and Ne matrices (Fig. 6.24). In this case only two lines are observed which are separated by the spin-orbit splitting of the Xe gas. This demonstrates that the width of the bands in the EDC's of Fig. 6.23 originates from genuine band effects and is not determined by any artifacts, e.g., electron-phonon scattering. In a very recent investigation by *Nürnberger* et al. [6.145] the continuous evolution of the Xe and Ar valence bands was observed by investigating

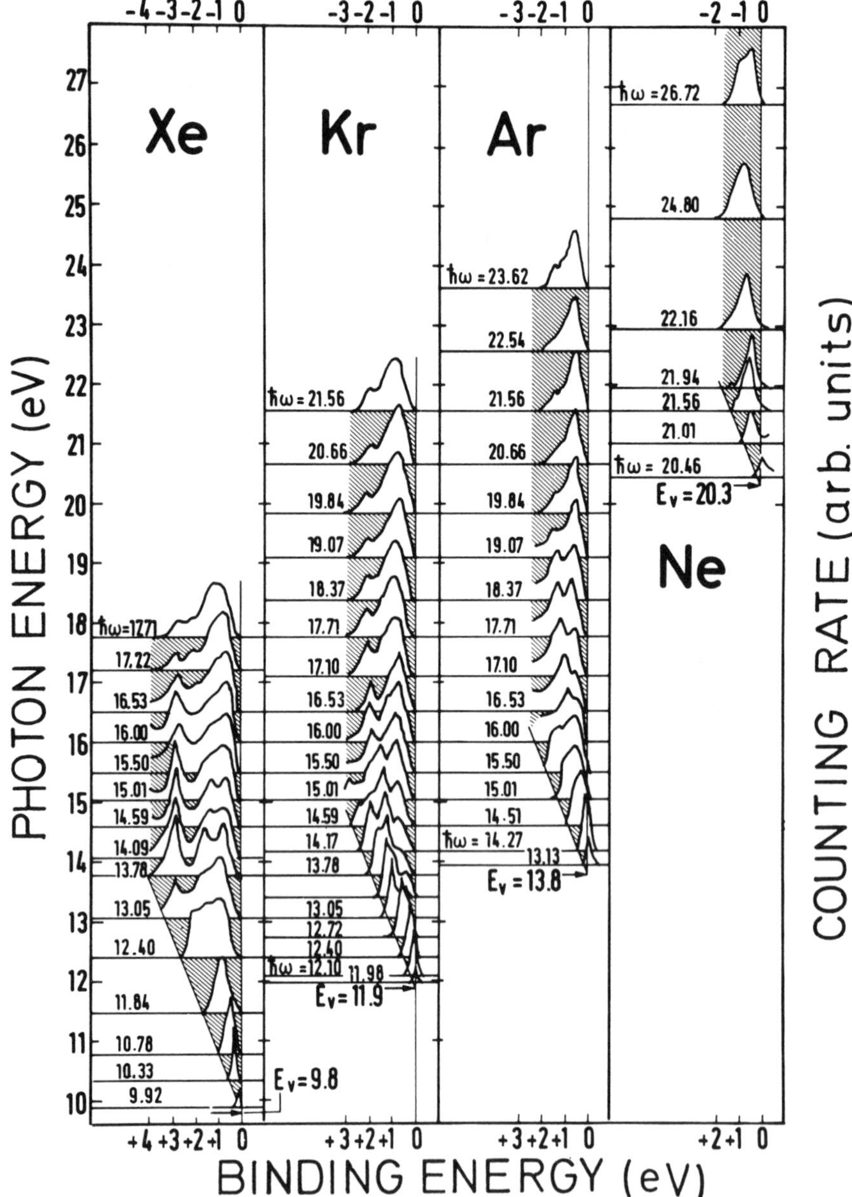

Fig. 6.23. EDC's of solid Ne, Ar, Kr, and Xe for various photon energies below the onset of electron-electron scattering. Except near threshold the spectra are normalized to the same height. The zero-count line for each spectrum is shifted upwards proportionally to $\hbar\omega$ [6.138]

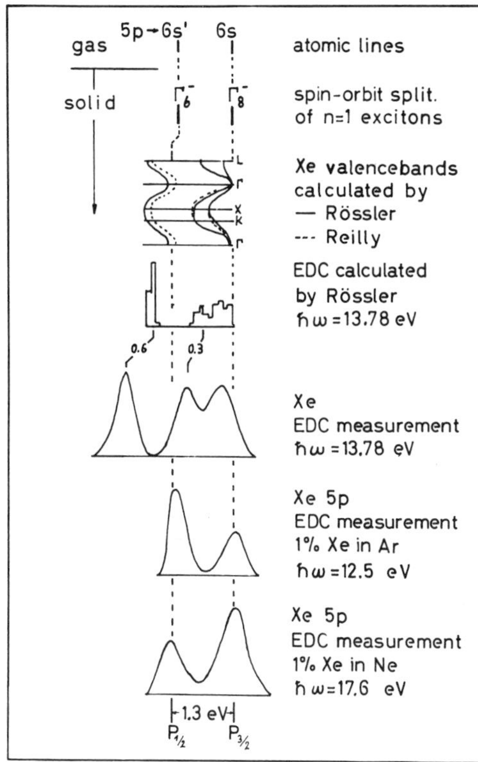

Fig. 6.24. A representative EDC of pure Xe is compared with EDC's of the Xe $5p$ levels of Xe in Ar and Xe in Ne hosts. Also the spin-orbit splitting of the atomic lines and of the $n=1$ excitons of solid Xe is shown together with valence band calculations [6.141, 142] and the calculated EDC [6.143] [6.139]

rare-gase mixtures of Xe and Ar containing between 5 and 100 Xe. While the $P_{1/2}$ peak of Xe is almost unchanged the $P_{3/2}$ peak begins to broaden at a concentration of about 50% Xe; at 70% it approaches a square shape, and finally it splits at 100% Xe. It should be noted that this clear splitting into two distinct peaks is occurring most pronouncedly at $\hbar\omega = 13.8$ eV. At other photon energies (Fig. 6.23) different structures are observed due to the selection rules which allow only direct interband transitions. Energy transfer in Xe/Ar mixtures will be discussed below (see Fig. 6.31).

In addition, *Schwentner* [6.146] has published a careful analysis of the electron-electron scattering process which sets in at 17.8 eV for Xe, 21.9 eV for Kr, and 26.3 eV for Ar. He found that the mean free path decreases monotonically from about 1000 Å at threshold to 1 to 5 Å at 10 eV above threshold.

6.6.2 Conduction Band State from Angular Dependent Photoemission

While usually angular resolved photoemission is applied to probe the valence band density of states, see [Ref. 1.1, Chap. 6], or at most the joint density of states, there is one special type of experiments in which SR allows for an

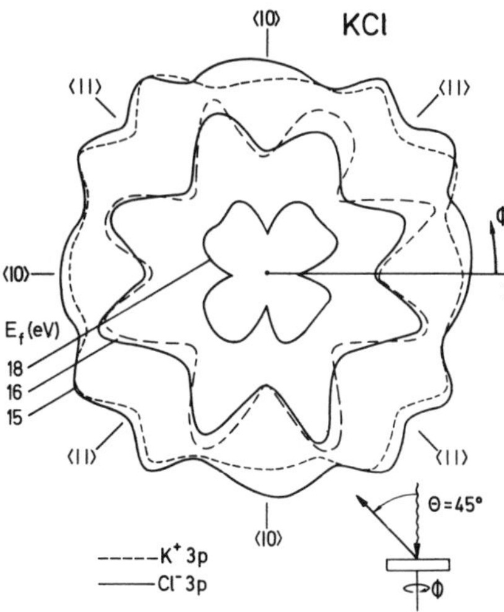

Fig. 6.25. Polar diagram of the photoemission intensity as a function of azimuth ϕ for a KCl (100) cleavage face at a fixed polar angle of 45° for three final-state energies (15, 16, and 18 eV above the top of the valence band) and two different initial-state levels (Cl^{-} $3p$ and K^{+} $3p$) [6.147]

investigation of the final states only. This technique is well suited for insulators, like NaCl and KCl, for which the energy region below the electron-electron scattering threshold is especially large. The final-state energy is fixed and $\hbar\omega$ is fixed, while the angle under investigation is varied. In its simplest form, which was first applied by *Himpsel* and *Steinmann* [6.147], the azimuth angle is changed by rotating the single crystals around an axis normal to the surface (see Fig. 6.25).

Figure 6.25 shows the angular pattern obtained for a single-crystal cleavage surface of KCl. The charging of the crystal is compensated for by means of a flood gun. The final-state energy is measured relative to the top of the valence band. The electron-electron scattering threshold sets in at about $E_f = 17.4$ eV. Therefore a large mean free path belongs to practically all the final states involved in Fig. 6.25. As it is clearly demonstrated, the angular pattern in these spectra is independent of the initial state involved. For the pattern at $E_f = 15$ eV (and also for that at $E_f = 16$ eV) the same angular symmetry is observed irrespective of whether K^{+} $3p$ electrons or Cl^{-} $3p$ electrons are excited into these final-state channels. Small changes of E_f, in some cases of only 0.5 eV, completely change the angular pattern. The authors could show for NaCl that from $E_f = 20$ eV on for NaCl (the electron-electron scattering threshold there lies at about 17.5 eV) the angular patterns are sensitive to both initial- and final-state energies [6.147–149].

The proposed interpretation is as follows. At low energies E_f the escape depth is very large as can be inferred from the high photoelectric yield of NaCl exceeding 0.5 electrons per photon [6.150]. Along their path the electrons undergo several electron-phonon scattering events by which all correlation with the initial $\boldsymbol{k}$ vector of excitation is lost, while at energies E_f above the electron-electron scattering limit the much smaller number of escaping electrons are still in the states belonging to the original **k** determined by the initial excitation process. After several phonon scattering processes all states with a certain energy E_f will be populated proportional to their density of states. (This assumption might be questioned, especially for the region of band-minima.) For a given combination of energy E_f, polar and azimuth angles, only states in the Brillouin zone which lie on several lines can contribute. This information is yet insufficient to plot band structures from the data obtained. *Himpsel* [6.149] therefore has tried to make a comparison between theory and experiment on the level of the observed angular distributions. He calculated photoemission from the pseudopotential bands matching Bloch waves inside the crystal to free electron waves outside. The overall symmetry of the theoretical distributions is in agreement with experiment, while many details still show considerable differences [6.148, 149].

It should be mentioned that the primary assumption underlying this interpretation of the angular patterns of the results discussed above is confirmed by a measurement in which the states at the energy E_f are indirectly populated [6.149]. Population is achieved by scattering of electrons which are primarily excited to states above the electron-electron scattering limit. The angular distributions obtained are the same. This investigation is thus analogous to the CIS spectroscopy mentioned in Sect. 6.4.4, of which it is an extension yielding more detailed information on the final states.

6.7 Experiments on Relaxation Processes and Excitons

6.7.1 Phonon Broadening of Core Lines

It was recognized quite early in photoemission experiments from core levels that the widths of core lines from most compounds of a metal are considerably larger than those of the pure metal (see, e.g., [6.151]). *Citrin* et al. [6.152] and *Matthew* and *Devey* [6.153], gave an explanation in terms of excitation of optical phonons. The model underlying their calculation is visualized in Fig. 6.26. We take the $Na^+ 2p$ excitation in NaCl as an example. When Na^+ is ionized during the photoemission process and changes to Na^{++} the surrounding Cl^- octahedron finds itself at an average distance q_0 with respect to Na^{++}, which is not any more the equilibrium distance. Thus a lattice relaxation takes place after the photoemission process. The corresponding relaxation energy is (see [Ref. 1.1, Eq. (0.75)])

$$E_R = e^2 \left(\frac{6}{\pi V}\right)^{1/3} \left(\frac{1}{\varepsilon_\infty} - \frac{1}{\varepsilon_0}\right), \tag{6.14}$$

where ε_∞ is the dynamic, ε_0 is the static dielectric constant, and V the volume of the primitive cell. In the case of NaCl we obtain $E_R = 1.73$ eV. If we now apply the Franck-Condon principle to the transition shown in Fig. 6.26, we learn that

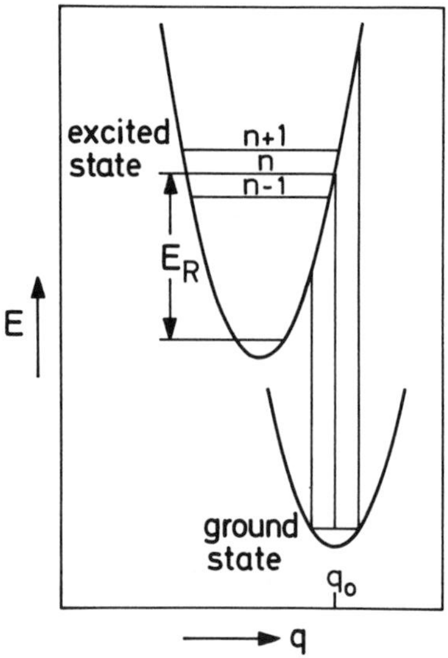

Fig. 6.26. Configuration diagram showing the ground state at the equilibrium position q_0 and the ionized state of Na^{2+} of NaCl. A band of phonons centered at the nth phonon state is excited after the photoelectron has left the lattice site [6.155]

the zero point (or finite temperature) oscillations of the Na^+ ion in its octahedral Cl^- cage is reflected into the steep part of the final-state potential curve. As a consequence, not a well-defined number of optical phonons but a whole band (about 50 on the average) is excited. In the experiment the individual phonon lines, which are separated by about 33 meV, are not resolved, and a finite envelope width (FWHM) of the photoemission line results as given by [Ref. 1.1, Eq. (1.76)]

$$\Delta E = 2.35\left(\hbar\omega_{\rm LO}E_{\rm R}\coth\frac{\hbar\omega_{\rm LO}}{2kT}\right)^{1/2}, \tag{6.15}$$

where $\omega_{\rm LO}$ is the frequency of the longitudinal optical phonons.

Figure 6.27 shows (6.15) compared to an experimental linewidth obtained from a 70 Å thick NaCl film as a function of temperature by *Iwan* and *Kunz* [6.154, 155]. The small thickness was chosen to avoid charging, but the experimental results were reproduced with films in a whole range of different thicknesses. The photon energy in this experiment was 60 eV, and the combined resolution of the monochromator and the electron energy analyzer was 0.3 eV. This together with the spin-orbit splitting of 0.17 eV was estimated to contribute only 16% to the experimental linewidth as plotted. Charging was shown to be negligible. The experimental width is about 0.5 eV larger than the width according to (6.15), while the slope with temperature is in satisfactory agreement. Not much importance should, however, be attributed to this discrepancy. The model underlying (6.14, 15) is crude; it does not account for different frequencies $\hbar\omega_{\rm LO}$ for the excited and for the ground state and further neglects dispersion of the LO phonon frequencies.

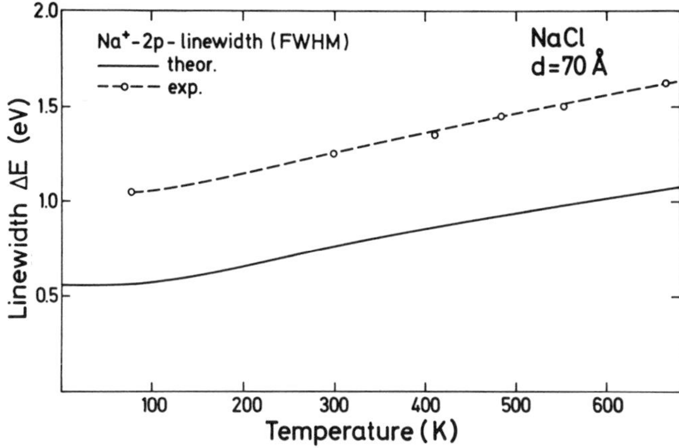

Fig. 6.27. Temperature dependence of the Na^+ $2p$ photoemission linewidth ΔE [6.155]

6.7.2 Exciton Effects with Core Excitations

One of the crucial questions for the interpretation of the absorption spectra of core electrons is concerned with separating bound electron-hole states from interband transitions if this is possible at all. From a combination of photoemission with absorption data it could be shown that the onset of the $Li^+ 1s$ excitations in LiF occurs at an 8 eV lower energy than the onset of the $Li^+ 1s$ interband transitions [6.107]. The identification of the lowest structure belonging to a $Li^+ 1s$ excitation has been recently questioned [6.157]: in this work a peak at 61 eV is identified as the onset of $Li^+ 1s$ transitions from results obtained with electron energy loss. This results in a quantitative change of the interpretation but has no influence on the essence of the argument. The method is the following. From an EDC the energy separation between the top of the valence band and the core level is determined. The separation from the top of the valence band to the bottom of the conduction band is obtained from an analysis of the fundamental absorption or reflection spectrum usually involving the determination of the limit of convergence of an exciton series. By adding the two energies, the binding energy of the core level relative to the bottom of the conduction band is obtained. Any structure which occurs at a lower photon energy is attributed to a bound electron-hole state.

Similar studies have also been performed on NaCl. Figure 6.28 shows as an example the absorption spectrum of NaCl [6.158] for Na-$2p$ excitations. The arrow shows the photon energy at which transitions into the conduction band should set in [6.115, 156, 159]. Thus peaks A–C are clearly identified as bound electron-hole states. Are these the only bound electron-hole states?

It is well known that the influence of the electron-hole interaction is not limited to the formation of bound exciton states below threshold which can be

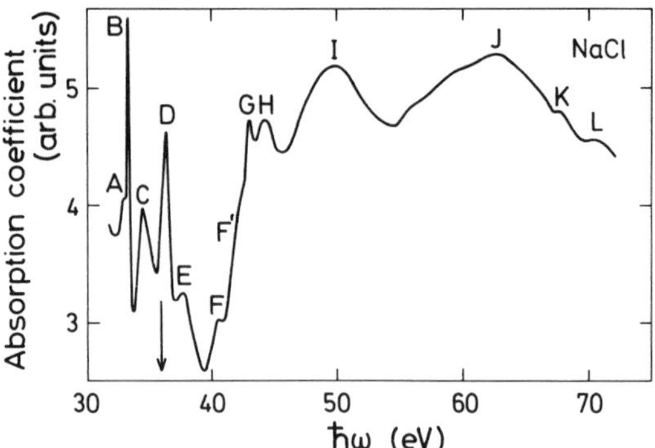

Fig. 6.28. Absorption spectrum of NaCl in the region of the Na^+ $2p$ excitations. The arrow marks the onset of interband transitions [6.159] [6.158]

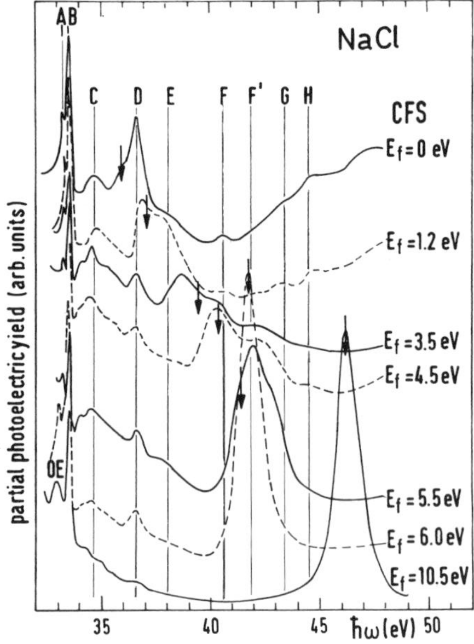

Fig. 6.29. CFS spectra of a 70 Å thick evaporated NaCl film in the region of the Na^+ $2p$ transitions. The vertical arrows in the spectra mark the position of the final-state channel. While the directly excited Na^+ $2p$ peak is well observed for $E_f = 6$ and 10.5 eV, it is considerably distorted in the excitonic region below $E_f = 6$ eV. The structures A to H are identical to structures in the absorption spectrum (Fig. 6.28); $0E$ is probably a surface excitation (see also Fig. 6.32) [6.159]

described within the effective mass approximation. *Altarelli* and *Dexter* [6.160] showed that for the $2p$ absorption spectrum of Si a dramatic buildup of oscillator strength for transitions to the low part of the conduction band is due to the electron-hole interaction. In alkali halides the electron-hole interaction is definitely stronger than in Si and probably will destroy any correlation of the absorption behavior with the density of states in the low parts of the conduction band. Therefore it appears to be justified to try a localized molecular approach rather than DOS's from band calculations in order to explain such narrow structures as D–G in the spectrum of NaCl of Fig. 6.28. *Dehmer* and *Åberg* [6.161] ascribed these structures qualitatively to resonance states within a local cage around the Na^{++} ion which is generated by the repulsive pseudopotential built up from the Cl^- ions. They also classify the states according to the octahedral symmetry. For a quantitative description, however, a knowledge of the potential is needed. Could photoemission experiments provide any experimental clue to these questions?

Iwan and *Kunz* [6.154, 156, 159] have performed CFS measurements on NaCl in which E_f, as measured from the bottom of the conduction band, was varied between 0 and 10.5 eV. Thus E_f covers the range of the sharp structures D–G, Fig. 6.29 shows the results. At energies $E_f \geqq 6$ eV a prominent peak of direct excitations from the Na^+2p level into the conduction band shows up. The width of this peak was already explained in Sect. 6.7.1. At most values of E_f also the structures A–H which are well known from the absorption spectrum (Fig. 6.28) are recognized. They contribute to these E_f channels because of the decay due to an Auger process or direct recombination followed by electron-electron scattering processes (see Sect. 6.4.6). The peak marked with the vertical arrow for $E_f = 10.5$ eV shifts with decreasing E_f towards lower photon energies and thus seems to correspond to a direct excitation. If the simple density of states approximation would hold, the peak of directly excited electrons must shift with the position of E_f as indicated by the arrows in the spectra of Fig. 6.29. The intensity is expected to be modulated by the absorption coefficient, but the width should be practically unchanged. Bound electron-hole states, due to their local nature, could give rise to lattice relaxation before they decay into single-particle states.

The behavior for $E_f < 6$ eV appears to differ completely from that for $E_f \geqq 6$ eV in Fig. 6.29. It can be described by a strong distortion of the direct peak mainly towards higher energy than that indicated by the arrow. In the spectrum taken at $E_f = 4.5$ eV a peak F′ centered 2 eV above the corresponding arrow still has a considerable amplitude. How do these electrons, which are primarily excited to $E_f' = 6.5$ eV, get into the final-state channel at $E_f = 4.5$ eV? An energy loss of 2 eV must occur. The suggested interpretation based on the model of *Åberg* and *Dehmer* is as follows. The primary excitation is into a quasi-bound state around the Na^{++} ion with a lifetime which allows for a relaxation process involving the emission of many phonons. Since the energy of these states lies above the bottom of the conduction band, eventually these states can autoionize, thus emitting the electron into a conduction band state. With such an interpretation the absorption structures D–G of Fig. 6.28 are explained as being due to quasi-bound inner well states. They cannot be attributed to any features in the conduction band density of states (e.g., [6.162]).

6.7.3 Energy Transfer Processes

Photoemission from insulators is not possible when $\hbar\omega$ is less than the separation between the top of the valence band and the vacuum level or less than the band gap energy E_g. For example in solid Ar the electron affinity is

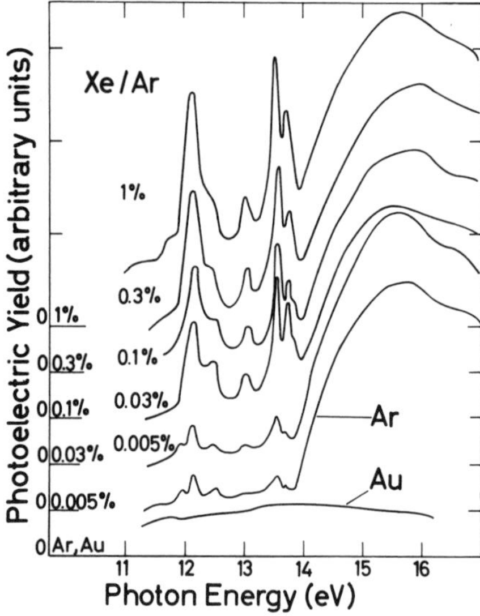

Fig. 6.30. Photoelectric yield of pure and Xe doped Ar films of 60 Å thickness on a Au substrate. The spectra are not corrected for the contributions of the hot electrons from the substrate and for reflectivity losses [6.89]

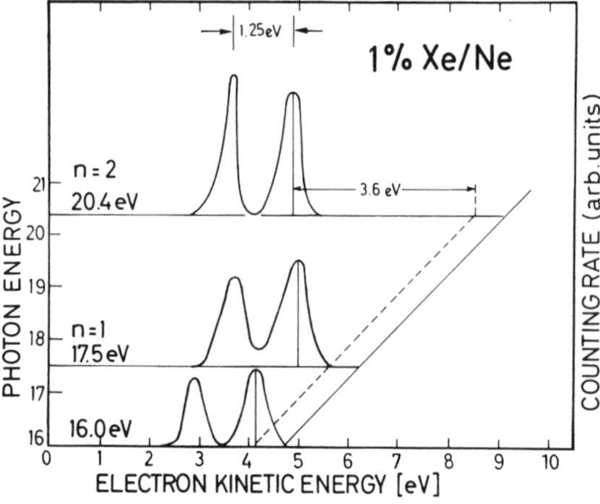

Fig. 6.31. Photoelectron distribution curves from 1% Xe in a Ne matrix for different excitation energies: 16 eV (below the first exciton line of the host), 17.5 eV (into the $n=1$ host exciton bands), and 20.4 eV (into the $n=2$ host exciton bands) [6.164]

negative (the vacuum level is located 0.4 eV below the bottom of the conduction band and thus photoemission sets in for $\hbar\omega = E_g$ (see Fig. 6.30), $E_g = 14.2$ eV for pure Ar. Optical excitations, however, set in with the $n=1$ exciton at 12.1 eV. Any impurities incorporated into solid Ar which have an ionization threshold below 12.1 eV can be ionized by colliding with an Ar exciton.

Figure 6.30 taken from an investigation by *Ophir* et al. [6.89] shows the photoelectric yield of a 60 Å thick Ar film as a function of the Xe concentration. The yield at the Ar excitons for pure Ar is either due to a process in which Ar excitons diffuse to the Au substrate and excite electrons from there [6.87, 163], or due to the presence of impurities other than Xe. When increasing the Xe concentration to 1% the yield in the exciton region rises dramatically. This is explained by the energy transfer from the Ar host to the impurities [6.89]. Such data have been analyzed in terms of an exciton diffusion model and the relevant rate constants for energy transfer and exciton lifetimes can be extracted. The spectra of Fig. 6.30 also show the abruptly rising yield at the onset of interband transitions in the Ar host which is independent of the impurity concentration. Such investigations are impossible without a tunable source.

The mechanism of energy transfer was further elucidated by *Schwentner* and *Koch* [6.164], who measured EDC's of the emitted photoelectrons. Several processes are possible: 1) excitation of an exciton with $n>1$ of the host and direct transfer of energy to the impurity atom; 2) first a relaxation to the $n=1$ exciton and thereafter transfer of energy to the impurity; 3) relaxation of the $n=1$ exciton to its "self-trapped" state (from which luminescence usually occurs) and then energy transfer to the impurity, if energetically possible. In Ar the $n=1$ exciton has a binding energy of 12.1 eV while the self-trapped exciton has a transferable energy of only 9.8 eV, which is too low to ionize a Xe impurity atom.

The results of [6.164] show a pronounced difference between Xe in an Ar or in a Ne matrix. In the first case, when exciting to the $n=2$ exciton, electron energies which correspond to the direct ionization from this exciton are observed and in addition some contribution of energy transfer from partly relaxed $n=2$ excitons occurs. For the case of direct excitation of the $n=2$ exciton of Ne with 1 at% Xe, Fig. 6.31 shows clear evidence that the first step is a fast transition to the $n=1$ exciton. Energy transfer occurs only on the level of the $n=1$ exciton even if excitons with $n>1$ are primarily excited. From such experiments time hierarchies for the different relaxation and transfer time constants can be deduced [6.164].

6.8 Surface States and Adsorbates

Although surface physics is outside the scope of this volume, a survey on the potential applications of synchrotron radiation to photoemission spectroscopy would be quite incomplete without at least a glimpse at this field. There are now surveys which cover the recent developments [6.165, 39, 40]. Since application of synchrotron radiation to surface physics is only a few years old there is no doubt that we are still at the beginning.

6.8.1 Surface Core Excitons on NaCl

One of the most genuine applications of SR that cannot be replaced by any other source is the investigation of empty surface states and of surface excitons by partial yield spectroscopy. Such investigations were carried out on Ge (111)

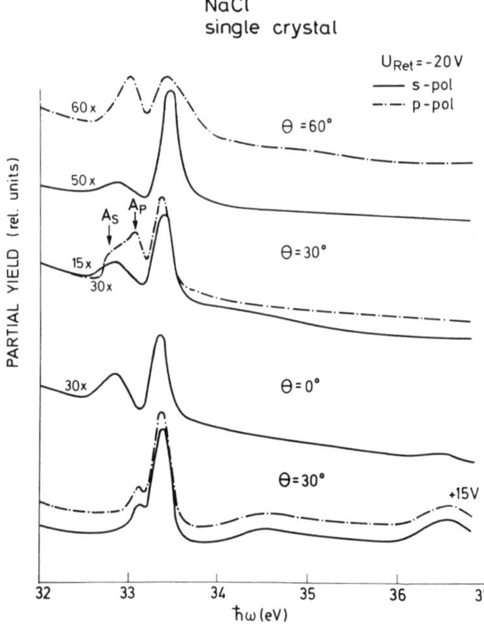

Fig. 6.32. Partial yield spectra from a cleaved NaCl (100) crystal surface. The spectra are obtained for polar angles 0° and 30° and for *s*- and *p*-polarized light. A retarding potential between +15 and −20 V was applied to allow for a variation of the probe depth. This varies from volume sensitivity at +15 V to maximum surface sensitivity at −20 V. Two surface-sensitive peaks emerge below the volume 2*p* core exciton, which is equivalent to peaks *AB* in Fig. 6.28 [6.167]

surfaces [6.102] on GaAs (110) surfaces [6.166] and then on a whole series of other semiconductors. For these semiconductors electron-electron scattering is large, because of the small gap, even at the photoemission threshold. Hence the mean free path is always small with uv excitation and the photoemission spectra are strongly influenced by surface properties (see Chap. 2) [6.102]. The situation is different, however, with insulators for which low-energy electrons can travel large distances in the volume, without suffering electron–electron scattering.

In a recent experiment by *Rehder* et al. [6.167] the $Na^+ 2p$ surface excitations could be investigated on freshly-cleaved surfaces of NaCl. A set of spectra is shown in Fig. 6.32. Alternate pulses of light from the *Desy* synchrotron and electrons from a charge compensation electron gun were impinging onto the highly insulating sample. A retarding field in front of the electron detector allowed for a rejection of stray electrons and for selecting only those photoelectrons which have energies above a certain energy threshold U_{Ref}. The spectrum at $U_{Ref} = +15\,eV$ is the total yield. The total yield is dominated by contributions from the bulk and constitutes a measurement of the absorption coefficient by means of PEYS. The first spin-orbit split $Na^+ 2p$ exciton, which was already discussed in Sect. 6.7.2, shows up. When decreasing the retarding potential U_{Ref} the escape depth of the contributing electrons is reduced more and more, and two peaks at the low-energy side of the bulk structures begin to appear at $U_{Ref} = -10\,V$. These peaks become even more pronounced when U_{Ref} is finally reduced to $-20\,V$.

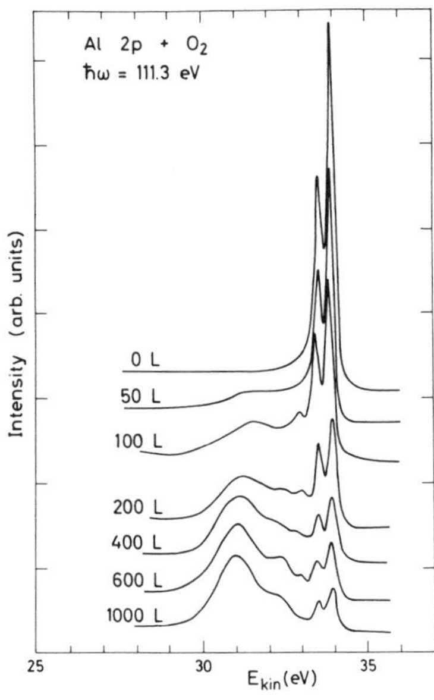

Fig. 6.33. Variation of the UPS spectrum in the region of the spin-orbit split K 2*p* lines of Al with oxidation by 50 *L* to 1000 *L* oxygen. The results were obtained on a polycrystalline evaporated Al film [6.173]

The angle of incidence chosen for these spectra was 30° and spectra were obtained for both types of polarization *s* and *p*. For *s*-polarized light the electric field vector parallel to the surface gives rise to the peak labelled A_s in Fig. 6.32. For *p*-polarized light there are two components of the electric field – one parallel, and one perpendicular to the surface. In this case a second peak, A_p, which must originate from excitation by the perpendicular component the field shows up at lower binding energy. These surface excitons are explained as localized excitations on a Na^+ ion split in the noncubic field on the surface of the crystal [6.168]. Two of the special properties of SR, its continuous spectrum and its linear polarization, are of importance for such investigations.

6.8.2 Adsorbates and Oxidation

Several studies have been performed on adsorbates like CO on Ni and Pd [6.169], and CO on Ir [6.170]. Especially the investigations making use of angular resolved photoemission of CO on Ni should be mentioned [6.171]. Further H on W was investigated [6.172]. Then a series of oxidation studies was carried out with Si, Ge, GaAs, GaP, InSb (see, e.g., [6.104, 165]). As an example we show in Fig. 6.33 the result of a measurement of the chemical shift of the Al 2*p* level on oxidation of a polycrystalline evaporated film of Al by *Eberhardt* et al. [6.173]. The high resolution of this measurement allows us to resolve the $P_{3/2}$ and $P_{1/2}$ levels of pure Al. With high exposure above 200 *L* a broad maximum shifted by 2.7 eV to higher binding energies occurs. The faint structures at lower binding energies are probably due to surface sites of different quality as they are expected to occur on the surface of evaporated films. The results are quite similar to those of *Flodström* et al. [6.174] at high exposures when the better resolution in the results of Fig. 6.33 is taken into consideration. There is, however, a remarkable difference of the results for exposure below 100 L. This demonstrates that evaporated Al films can differ considerably in quality. The measurements of *Flodström* et al., which seemed to indicate a surface phase transition, ought to be repeated on single-crystal surfaces, which are a better defined system.

Acknowledgment. The author wants to thank W. Gudat for a critical reading of the manuscript and many useful comments. He also is indebted to E.E. Koch and M. Skibowski for suggesting several improvements. He further tanks E. Thumann for her careful writing of the manuscript and V. Fischer, W. Knaut, K. Köhler, M. Sörensen, and J. Schmidt for preparing the figures.

References

6.1 D. Ivanenko, J. Pomeranchuk: Phys. Rev. **65**, 343 (1944)
6.2 J. Schwinger: Phys. Rev. **70**, 798 (1946); Phys. Rev. **75**, 1912 (1949)
6.3 J.P. Blewett: Phys. Rev. **69**, 87 (1946)
6.4 F.R. Elder, A.M. Gurewitsch, R.V. Langmuir, H.C. Pollock: Phys. Rev. **71**, 829 (1947); J. Appl. Phys. **18**, 810 (1947)
6.5 P.L. Hartman, D.H. Tomboulian: Phys. Rev. **87**, 233 (1952)
6.6 P.L. Hartman, D.H. Tomboulian: Phys. Rev. **91**, 1577 (1953)
6.7 D.H. Tomboulian, P.L. Hartman: Phys. Rev. **102**, 1423 (1956)
6.8 A. Sommerfeld: *Elektrodynamik* (Akademische Verlagsgesellschaft, Leipzig 1949)
6.9 J.D. Jackson: *Classical Electrodynamics* (Wiley and Sons, New York 1962) p. 481ff
6.10 K. Codling, R.P. Madden: J. Appl. Phys. **36**, 380 (1965)
6.11 R. Haensel, C. Kunz: Z. Angew. Phys. **23**, 276 (1967)
6.12 A.A. Sokolov, J.M. Ternov: *Synchrotron Radiation* (translated from the Russian) (Pergamon Press, Oxford 1968)
6.13 R.P. Godwin: *Synchroton Radiation as a Light Source*, Vol. 51, ed. by G. Höhler. Springer Tracts in Modern Physics (Springer, Berlin, Heidelberg, New York 1969) p. 1
6.14 R.E. Watson, M.L. Perlman: Brookhaven National Laboratory Rpt. BNL 50 381 (1973)
6.15 C. Kunz: In *Vacuum Ultraviolet Radiation Physics*, ed. by E.E. Koch, R. Haensel, C. Kunz (Vieweg-Pergamon, Braunschweig 1974) p. 753
6.16 R.W. Morse: In "An Assessment of the National Need for Facilities Dedicated to the Production of Synchrotron Radiation", Rpt. to the National Academy of Sciences (Washington D.C. 1976)
6.17 E.E. Koch: Proc. 8th All Union Conf. High Energy Part. Phys., Erevan, (1975); Vol. 2, Erevan (1976) p. 502
6.18 J.W. McGowan, E.M. Rowe (eds.): Proc. of Synchrotron Radiat. Facilities, Quebec Summer Workshop, Rpt. U. of West. Ontario (London, Ontario 1976)
6.19 G. Rosenbaum, K.C. Holmes: Nature **230**, 434 (1971)
6.20 W. Hayes: Contemp. Phys. **13**, 441 (1972)
6.21 G.V. Marr, I.H. Munro, J.C.C. Sharp: *Synchrotron Radiation: A Bibliography*, Daresbury, Nucl. Phys. Lab., Rpt. DNPL/R **24** (1972); and DL/TM **127** (1974)
6.22 G.V. Marr, I.H. Munro (eds.): Proc. Int. Symp. Synchrotron Radiat. Users, Daresbury, Nucl. Phys. Lab., Rpt. DNPL/R **26** (1973)
6.23 K. Codling: Rep. Prog. Phys. **36**, 541 (1973)
6.24 F.C. Brown: Solid State Phys. **29**, 1 (1974)
6.25 E.E. Koch, R. Haensel, C. Kunz (eds.): *Vacuum Ultraviolet Radiation Physics* (Vieweg-Pergamon, Braunschweig 1974)

6.26 E.E. Koch: In *Chemical Spectroscopy and Photochemistry in the* VUV, ed. by C. Sandorfy, F.J. Ausloos, M.B. Robin (Reidel, Dodrecht 1974) p. 559
6.27 M.L. Perlman, E.M. Rowe, R.E. Watson: Phys. Today **27**, 30 (1974)
6.28 R.P. Madden: In X-*Ray Spectroscopy*, ed. by L.V. Azároff (McGraw-Hill, New York 1974) p. 338
6.29 J. Taylor: In *Chemical Spectroscopy and Photochemistry in the Vacuum Ultraviolet*, ed. by C. Sandorfy, P.J. Ausloos, M.B. Robin (Reidel, Dodrecht 1974) p. 543
6.30 R. Haensel: In *Festkörperprobleme, Advances in Solid State Physics*, Vol. 15, ed. by H.-J. Queisser (Pergamon-Vieweg, Braunschweig 1975) p. 203
6.31 E.E. Koch, C. Kunz, B. Sonntag: Phys. Rpt. **29**C, 153 (1977)
6.32 B. Sonntag: In *Rare Gas Solids II*, ed. by M.L. Klein, J.A. Venables (Academic Press, New York 1977) p. 1021
6.33 C. Kunz: In *Optical Properties of Solids – New Developments*, ed. by B.O. Seraphin (North-Holland, Amsterdam 1976) p. 473
6.34 C. Kunz: Phys. Bl. **32**, 9, 55 (1976)
6.35 J. Barrington Leigh, G. Rosenbaum: Annu. Rev. Biophys. Bioeng. **5**, 239 (1976)
6.36 E.E. Koch: In *Interaction of Radiation with Condensed Matter*, Vol. II, ed. by L.A. Self (Trieste Center for Theor. Physics, IAEA, Wien 1976) p. 225
6.37 A.N. Mancini, I.F. Quercia (eds.): Proc. Intern. School on Synchrotron Radiation Research, Vol. I, Alghero 1976 (Intern. College on Appl. Phys., Instituto Nazionali di Fisica Nucleare, Catania 1976)
6.38 K.O. Hodgson, H. Winick, G. Chu (eds.): *Synchrotron Radiation Research*, SSSRP Rpt. No. 76/100 (1976)
6.39 B. Feuerbacher, B. Fitton, R.F. Willis (eds.): *Photoemission from Surfaces* (Wiley, London 1978)
6.40 B. Feuerbacher, B. Fitton, R.F. Willis (eds.): *Proceedings of an International Symposium on Photoemission, Noordwijk* ESA Rpt. SP 118
6.41 G.A. Schott: *Electromagnetic Radiation* (Cambridge University Press, Cambridge 1912)
6.42 J.A.R. Samson: *Techniques of Vacuum Ultraviolet Spectroscopy* (Wiley and Sons, New York 1967)
6.43 J.A. Kinsinger, W.L. Stebbings, R.A. Valenzi, J.W. Taylor: In *Electron Spectroscopy*, ed. by D.A. Shirley (North-Holland, Amsterdam 1972) p. 155
6.44 K. Siegbahn: J. Electron Spectrosc. **5**, 3 (1974)
6.45 I. Lindau, P. Pianetta, S. Doniach, W.E. Spicer: Nature **250**, 214 (1974)
I. Lindau, P. Pianetta, K. Yu, W.E. Spicer: Phys. Lett. **54**A, 47 (1975)
6.46 G.C. Baldwin: Phys. Today **28**, 9 (1975)
6.47 B. Kincaid: J. Appl. Phys. **48**, 2684 (1977)
6.48 J.P. Blewett, R. Chasman: J. Appl. Phys. **48**, 2692 (1977)
6.49 E.M. Rowe, F.E. Mills: Part. Accel. **4**, 221 (1973)
6.50 E.E. Koch, C. Kunz, E.W. Weiner: Optik **45**, 395 (1976)
6.51 H. Winick: In [Ref. 6.25, p. 776]
6.52 J.L. Stanford, V. Rehn, D.S. Kyser, V.O. Jones, A. Klugman: In [Ref. 6.25, p. 783]
6.53 M. Skibowski, W. Steinmann: J. Opt. Soc. Am. **57**, 112 (1967)
6.54 C. Kunz: In [Ref. 6.22, p. 68]
6.55 D.E. Eastman, W.D. Grobman, J.L. Freeouf, M. Erbudak: Phys. Rev. B**9**, 3473 (1974)
6.56 K. Codling, P. Mitchell: J. Phys. E: **3**, 685 (1970)
6.57 H. Dietrich, C. Kunz: Rev. Sci. Instrum. **43**, 434 (1972)
6.58 F.C. Brown, R.Z. Bachrach, S.B.M. Hagström, N. Lien, C.H. Pruett: In [Ref. 6.25, p. 785]
6.59 V. Saile, P. Gürtler, E.E. Koch, A. Kozevnikov, M. Skibowski, W. Steinmann: Appl. Opt. **15**, 2559 (1976)
6.60 P. Jaeglé, P. Dhez, F. Wuilleumier: In [Ref. 6.25, p. 788]
6.61 K. Thimm: J. Electron Spectrosc. **5**, 755 (1974)
6.62 See e.g. D.E. Eastman: In [Ref. 6.25, p. 417]
6.63 J.W. Cooper: Phys. Rev. **128**, 681 (1962)
6.64 U. Fano, J.W. Cooper: Rev. Mod. Phys. **40**, 441 (1968); **41**, 724 (1969)

6.65 P.S. Wehner, J. Stöhr, G. Apai, F.R. McFeely, R.S. Williams, D.A. Shirley: Phys. Rev. B**14**, 2411 (1976)
6.66 I. Lindau, P. Pianetta, K.Y. Yu, W.E. Spicer: Phys. Rev. B**13**, 492 (1976)
I. Lindau, P. Pianetta, S. Doniach, W.E. Spicer: Nature **250**, 214 (1974)
6.67 J. Stöhr, G. Apai, P.S. Wehner, F.R. McFeely, R.S. Williams, D.A. Shirley: Phys. Rev. B**14**, 5144 (1976)
6.68 W.D. Grobman, D.E. Eastman, J.L. Freeouf, J. Shaw: Proc. 12th Int. Conf. Phys. Semiconductors, Stuttgart (1974)
6.69 J.F. Janak, A.R. Williams, V.L. Moruzzi: Phys. Rev. B**11**, 1522 (1975)
6.70 P.J. Feibelman, D.E. Eastman: Phys. Rev. B**10**, 4932 (1974)
6.71 J.L. Dehmer, J. Berkowitz: Phys. Rev. A**10**, 484 (1974)
T.E.H. Walker, J. Berkowitz, J.L. Dehmer, J.T. Waber: Phys. Rev. Lett. **31**, 678 (1973)
6.72 G.J. Lapeyre, A.D. Baer, J. Hermanson, J. Anderson, J.A. Knapp, P.L. Gobby: Solid State Commun. **15**, 1601 (1974)
6.73 G.J. Lapeyre, J. Anderson, P.L. Gobby, J.A. Knapp: Phys. Rev. Lett. **33**, 1290 (1974)
6.74 G.J. Lapeyre, J. Anderson, J.A. Knapp, P.L. Gobby: In [Ref. 6.25, p. 380]
6.75 A.D. Baer, G.J. Lapeyre: Phys. Rev. Lett. **31**, 304 (1973)
6.76 N.E. Christensen, B. Feuerbacher: Phys. Rev. B**10**, 2349, 2373 (1974)
6.77 R.Z. Bachrach, M. Skibowski, F.C. Brown: Phys. Rev. Lett. **37**, 40 (1976)
6.78 G.J. Lapeyre: Nuovo Cimento **39**B, 693 (1977)
6.79 W. Eberhardt, G. Kalkoffen, C. Kunz: Unpublished
6.80 R. Haensel, G. Keitel, G. Peters, P. Schreiber, B. Sonntag, C. Kunz: Phys. Rev. Lett. **23**, 530 (1969)
6.81 D. Blechschmidt, M. Skibowski, W. Steinmann: Phys. Status Solidi **42**, 61 (1970)
6.82 D. Blechschmidt, M. Skibowski, W. Steinmann: Opt. Commun. **1**, 275 (1970)
6.83 H. Sugawara, T. Sasaki, Y. Iguchi, S. Sato, T. Nasu, A. Ejiri, S. Onari, K. Kojima, T. Oya: Opt. Commun. **2**, 333 (1970)
Y. Iguchi, T. Sasaki, H. Sugawara, S. Sato, T. Nasu, A. Ejiri, S. Onari, K. Kojima, T. Oya: Phys. Rev. Lett. **26**, 82 (1971)
T. Sasaki, Y. Iguchi, H. Sugawara, S. Sato, T. Nasu, A. Ejiri, S. Onari, K. Kojima, T. Oya: J. Phys. Soc. Jpn. **30**, 580, 581 (1971)
6.84 N. Schwentner, M. Skibowski, W. Steinmann: Phys. Rev. B**8**, 2965 (1973)
6.85 E.E. Koch, V. Saile, N. Schwentner, M. Skibowski: Chem. Phys. Lett. **28**, 562 (1974)
6.86 D. Pudewill, F.-J. Himpsel, V. Saile, N. Schwentner, M. Skibowski, E.E. Koch, J. Jortner: J. Chem. Phys. **65**, 5226 (1976)
6.87 E.E. Koch, B. Raz, V. Saile, N. Schwentner, M. Skibowski, W. Steinmann: Jpn. J. Appl. Phys. Suppl. **2**, 775 (1974)
6.88 V. Saile, N. Schwentner, E.E. Koch, M. Skibowski, W. Steinmann, Z. Ophir, B. Raz, J. Jortner: In [Ref. 6.25, p. 352]
6.89 Z. Ophir, B. Raz, J. Jortner, V. Saile, N. Schwentner, E.E. Koch, M. Skibowski, W. Steinmann: J. Chem. Phys. **62**, 650 (1975)
6.90 S.S. Hasnain, I.H. Munro, T.D.S. Hamilton: Submitted to: J. Phys. C
6.91 W. Steinmann, M. Skibowski: Phys. Rev. Lett. **16**, 989 (1966)
6.92 M. Skibowski, B. Feuerbacher, W. Steinmann, R.P. Godwin: Z. Phys. **211**, 342 (1968)
6.93 B. Feuerbacher, M. Skibowski, W. Steinmann, R.P. Godwin: J. Opt. Soc. Am. **58**, 137 (1968)
6.94 B.P. Feuerbacher, R.P. Godwin, M. Skibowski: Z. Phys. **224**, 172 (1969)
6.95 A.P. Lukirskii, I.A. Brytov: Fiz. Tverd. Tela **6**, 43 (1964) [English Transl.: Sov. Phys. – Solid State **6**, 32 (1964)]
6.96 A.P. Lukirskii, O.A. Ershov, T.M. Zimkina, E.P. Savinov: Fiz. Tverd. Tela **8**, 1787 (1966) [English Transl.: Sov. Phys. – Solid State **8**, 1422 (1966)]
6.97 A.P. Lukirskii, T.M. Zimkina: Izv. Akad. Nauk SSSR, Ser. Fiz. **28**, 765 (1964)
6.98 W. Gudat, C. Kunz: Phys. Rev. Lett. **29**, 169 (1972)
6.99 H. Petersen, C. Kunz: Phys. Rev. Lett. **35**, 863 (1975)
6.100 H. Petersen, C. Kunz: In [Ref. 6.25, p. 587] and to be published
6.101 W. Gudat: Thesis, University Hamburg (1974), Internal Rpt. DESY F41–74/10

6.102 D.E. Eastman, J.L. Freeouf: Phys. Rev. Lett. **33**, 1601 (1974)
6.103 D.E. Eastman, J.L. Freeouf: Phys. Rev. Lett. **34**, 1624 (1975)
6.104 W. Gudat, D.E. Eastman: [Ref. 6.39, Chap. 11]
6.105 W. Gudat, D.E. Eastman, J.L. Freeouf: J. Vac. Sci. Technol. **13**, 250 (1976)
6.106 See e.g. K.L. Kliewer: Phys. Rev. Lett. **33**, 900 (1974)
6.107 W. Gudat, C. Kunz, H. Petersen: Phys. Rev. Lett. **32**, 1370 (1974)
6.108 W. Gudat, C. Kunz: In [Ref. 6.25, p. 392]
6.109 M. Watanabe, H. Yamashita, Y. Nakai, S. Sato, S. Onari: Phys. Status Solidi (b) **43**, 631 (1971)
6.110 M. Cardona, W. Gudat, B. Sonntag, P.Y. Yu: Proc. 10th Int. Conf. Phys. of Semiconductors, ed. by S.P. Keller, J.C. Hensel, F. Stern, CONF-700801 (Cambridge, Mass. 1970) p. 209
6.111 B. Sonntag, T. Tuomi, G. Zimmerer: Phys. Status Solidi (b) **58**, 101 (1973)
6.112 B. Kramer, K. Maschke, L.D. Lande: Phys. Rev. B**8**, 5781 (1973)
6.113 N.J. Shevchik, M. Cardona, J. Tejeda: Phys. Rev. B**8**, 2833 (1973)
6.114 S.T. Pantelides, F.C. Brown: Phys. Rev. Lett. **33**, 2981 (1974)
6.115 S.T. Pantelides: Phys. Rev. B**11**, 2391 (1975)
6.116 H.J. Hagemann, W. Gudat, C. Kunz: Solid State Commun. **15**, 655 (1974)
6.117 H.J. Hagemann, W. Gudat, C. Kunz: Phys. Status Solidi (b) **74**, 507 (1976)
6.118 H.J. Hagemann: Diplomarbeit, University Hamburg (1974), Internal Rpt. DESY F41–74/4
6.119 W. Gudat, J. Karlau, C. Kunz: Proc. Int. Symp. X-Ray *Spectra and Electronic Structure of Matter*, Munich, 1972, Vol. 1, ed. by A. Faessler, G. Wiech (München 1973) p. 205
6.120 W. Gudat, C. Kunz: Phys. Status Solidi (b) **52**, 433 (1972)
6.121 A.C. Switendick: Nat. Bur. Stand. (U.S.) Spec. Publ. 323, p. 297 (1971)
A.C. Switendick, A. Narath: Phys. Rev. Lett. **22**, 1423 (1969)
6.122 J.W.D. Connolly, K.H. Johnson: Nat. Bur. Stand. (U.S.) Spec. Publ. 323, p. 19 (1971)
6.123 C. Kunz, H. Petersen, D.W. Lynch: Phys. Rev. Lett. **33**, 1556 (1974)
6.124 G.D. Mahan: Solid Sate Phys. **29**, 75 (1974)
6.125 e.g. E.J. McGuire: Sandia Lab. Res. Rpt. No. SC-RR-721 (unpublished)
6.126 D.E. Sayers, F.W. Little, E.A. Stern: In *Advances in X-Ray Analysis*, Vol. 13, ed. by B.L. Henke, J.B. Newkirk, G.R. Mallett (Plenum Press, New York 1970) p. 248
6.127 J.J. Ritsko, S.E. Schnatterly, P.C. Gibbons: Phys. Rev. Lett. **32**, 671 (1974)
6.128 Z. Hurych, J.C. Shaffer, D.L. Davies, T.A. Knecht, G.J. Lapeyre, P.L. Gobby, J.A. Knapp, C.G. Olson: Phys. Rev. Lett. **33**, 830 (1974)
6.129 Z. Hurych, D. Davies, D. Buczek, C. Wood, G.J. Lapeyre, A.D. Baer: Phys. Rev. **9**, 4392 (1974)
6.130 I.T. McGovern, R.H. Williams: J. Phys. C **9**, L337 (1976)
6.131 R.H. Williams, I.T. McGovern, R.B. Murray, M. Howells: Phys. Status Solidi (b) **73**, 307 (1976)
6.132 D.E. Eastman, J.L. Freeouf: Phys. Rev. Lett. **34**, 395 (1975)
6.133 W. Braun, M. Iwan, C. Kunz, H. Petersen: Proc. 8th Int. Conf. Phys. of Semiconductors, Rome, 1976 (North-Holland, Amsterdam, in press)
6.134 G.M. Bancroft, W. Gudat, D.E. Eastman: Phys. Rev. B **17**, 4499 (1978)
6.135 R.Y. Koyama, L.R. Hughey: Phys. Rev. Lett. **29**, 1518 (1972)
6.136 I.T. McGovern, A. Parke, R.H. Williams: J. Phys. C **9**, L511 (1976)
6.137 U. Rössler: In *Rare Gas Solids*, Vol. I, ed. by M.L. Klein, J.A. Venables (Academic Press, London 1975) p. 505
6.138 N. Schwentner, F.-J. Himpsel, V. Saile, M. Skibowski, W. Steinmann, E.E. Koch: Phys. Rev. Lett. **34**, 528 (1975)
6.139 N. Schwentner, F.-J. Himpsel, E.E. Koch, V. Saile, M. Skibowski: In [Ref. 6.25, p. 355]
6.140 N. Schwentner: Thesis, University Munich (1974)
6.141 M.J. Reilly: J. Phys. Chem. Solids **28**, 2067 (1967)
6.142 U. Rössler: Phys. Status Solidi **42**, 345 (1970)
6.143 U. Rössler: Unpublished
6.144 A.B. Kunz, D.J. Mickish, S.K.V. Mirmira, T. Shima, F.-J. Himpsel, V. Saile, N. Schwentner, E.E. Koch: Solid State Commun. **17**, 761 (1975)

6.145 R. Nürnberger, F.-J. Himpsel, E.E. Koch, N. Schwentner: Rpt. DESY SR-77/01, and submitted to Phys. Status Solidi
6.146 N. Schwentner: Phys. Rev. B **14**, 5490 (1976)
6.147 F.-J. Himpsel, W. Steinmann: Phys. Rev. Lett. **35**, 1025 (1975)
6.148 F.-J. Himpsel, W. Steinmann: In [Ref. 6.40, p. 137]
6.149 F.-J. Himpsel: Thesis, University Munich (1976)
Internal Rpt. DESY F41-77/01 (1977)
6.150 P.H. Metzger: J. Phys. Chem. Solids **26**, 1879 (1965)
6.151 P.H. Citrin, P.M. Eisenberger, W.C. Mara, T. Åberg, J. Utriainen, E. Källne: Phys. Rev. B **10**, 1762 (1974)
6.152 P.H. Citrin, P. Eisenberger, D.R. Hamann: Phys. Rev. Lett. **33**, 965 (1974)
6.153 J.A.D. Matthew, M.G. Devey: J. Phys. C **7**, L335 (1974)
6.154 M. Iwan, C. Kunz: In [Ref. 6.40, p. 127]
6.155 M. Iwan, C. Kunz: Phys. Lett. **60** A, 345 (1977)
6.156 M. Iwan: Diplomarbeit, University Hamburg (1976)
Internal Rpt. DESY F41-76/09 (1976)
6.157 J.R. Fields, P.C. Gibbons, S.E. Schnatterly: Phys. Rev. Letters **38**, 430 (1977)
6.158 R. Haensel, C. Kunz, T. Sasaki, B. Sonntag: Phys. Rev. Lett. **20**, 1436 (1968)
6.159 M. Iwan, C. Kunz: J. Phys. C **11**, 905 (1978)
6.160 M. Altarelli, D.L. Dexter: Phys. Rev. Lett. **29**, 1100 (1972)
6.161 T. Åberg, J.L. Dehmer: J. Phys. C **6**, 1450 (1973)
6.162 N.O. Lipari, A.B. Kunz: Phys. Rev. B **3**, 491 (1971)
6.163 Z. Ophir, N. Schwentner, B. Raz, M. Skibowski, J. Jortner: J. Chem. Phys. **63**, 1072 (1975)
6.164 N. Schwentner, E.E. Koch: Phys. Rev. B **14**, 4687 (1976)
6.165 I. Lindau: In [Ref. 6.37, p. 319]
6.166 G.J. Lapeyre, J. Anderson: Phys. Rev. Lett. **35**, 117 (1975)
6.167 U. Rehder, W. Gudat, R.G. Hayes, C. Kunz: Proc. 7th Intern. Vacuum Congress and 3rd Conf. on Solid Surfaces, ed. by R. Dobrozemsky, F. Rüdenauer, F.P. Viehböck, A. Breth (Wien 1977) p. 453
6.168 V.E. Henrich, G. Dresselhaus, H.J. Zeiger: Phys. Rev. Lett. **36**, 158 (1976)
6.169 T. Gustafsson, E.W. Plummer, D.E. Eastman, J.L. Freeouf: Solid State Commun. **17**, 391 (1975)
6.170 T.N. Rhodin, C. Brucker, G. Brodén, Z. Hurych, R. Benbow: Solid State Commun. **18**, 105 (1976)
6.171 R.J. Smith, J. Anderson, G.J. Lapeyre: Phys. Rev. Lett. **37**, 1081 (1976)
6.172 J. Anderson, G.J. Lapeyre: Phys. Rev. Lett. **36**, 377 (1976)
6.173 W. Eberhardt, G. Kalkoffen, C. Kunz: Surf. Sci. **75**, 709 (1978)
6.174 S.A. Flodström, R.Z. Bachrach, R.S. Bauer, S.B.M. Hagström: Phys. Rev. Lett. **37**, 1282 (1976)

7. Simple Metals

P. Steiner, H. Höchst, and S. Hüfner

With 10 Figures

The theory of the photoemission process is inspected taking into account those many-body effects, which manifest themselves as asymmetric line shapes and plasmons. Simple equations for the analysis of experimental spectra are deduced which allow the determination of the relative importance of the many-body plasmons (intrinsic plasmons) from the observed spectra. A commonly used formula for the background correction in XPS spectra is derived from an inspection of the photoemission process. An analysis of the core level spectra of Be, Na, Mg, and Al metal yields the magnitude of the extrinsic and intrinsic contribution to the plasmon creation rate. The valence band spectra of Be, Na, Mg, and Al have been analysed in terms of existing band structures. It is found empirically that good agreement with existing band structures can be obtained if many-body effects of the same magnitude as observed in the core level spectra and different photoionization cross sections for the partial bands of electrons with different angular momenta are included in the analysis of the valence band spectra.

7.1 Historical Background

For the understanding of the properties of metals the investigation of the simple $s-p$ band metals like Li, Na, Mg ... has always played a prominent role because of the seemingly simple model (free-electron gas) that can be used for their description. This holds also for photoemission experiments. In particular in the XPS regime the spectra of these metals show some features which do not appear to be so well defined in other cases: the zero-loss "line" (this may be a core line or a valence band) is accompanied by a series of equal distant satellite lines with decreasing intensity, due to electrons which have lost a discrete part of their energy by creation of plasmons [7.1–9]. This spectral part contains, in principle, a lot of information. On the other hand, the presence of these plasmon satellites makes the extraction of the zero-loss features a problem especially in the case of the valence band.

In the simplest approximation, the valence band of a simple metal is parabolic, and therefore its density of states (DOS) shows an $E^{1/2}$ dependence. The first X-ray emission experiments in solids, some decades ago, tried to recover this parabolic shape [7.10]. Since then many investigations of these valence bands by various methods (such as soft X-ray emission spectroscopy

Table 7.1. Width of the valence band of Be, Na, Mg, and Al estimated from XPS, AES, and SXE experiments compared with the theoretical results

	XPS	AES	SXE	Theory
Be	11.9(3) [7.20]		$K_\beta \sim 6.0$ [7.76] $K_\beta \sim 6.0$ [7.77]	12.1 [7.70] 11.9 [7.69] 16.75 [7.75]
Na	3.1(1) [7.20] 3.0(2) [7.59] 2.5(1) [7.4]	$KL_{2,3}V$ 4.0(2) [7.63] KL_1V 4.0(2) [7.63]	$L_{2,3}$2.7 [7.60] $L_{2,3}$3.05 [7.61]	3.19 [7.58] 3.23 [7.95] 3.2 [7.71]
Mg	6.9(2) [7.20] 6.5(1) [7.62]	$KL_{2,3}V$ 7.5 [7.62] KL_1V 8.5 [7.62] $KL_{2,3}V$ 7.5 [7.88] KL_1V 8.2 [7.88]	$L_{2,3}$7.0 [7.90] $L_{2,3}$6.86 [7.91] $L_{2,3}$6.4 [7.92]	6.9 [7.79] 9.4 [7.75] 7.1 [7.96]
Al	11.5(3) [7.20] 12.0 [7.1]	$L_{2,3}VV \sim 11$ [7.89]	$L_{2,3}$11.67(14) [7.93] $L_{2,3}$10.6 [7.92] $L_{2,3}$11.2 [7.94]	11 [7.97] 11.5 [7.94]

(SXE) [7.11], Auger electron spectroscopy (AES) [7.12, 13], photoelectron spectroscopy [7.14–19]) have appeared, but as can be seen from Table 7.1, the width of the valence bands in these simple metals cannot yet be considered to have been accurately determined. This can have its reason in the fact that different techniques measure different bandwidths; it can, however, also be connected with the problem of recovering the band-structure information from the measured spectra.

Since the electronic structure of the simple metals can be described by a simple model, these substances lend themselves to the attempt of describing the photoelectron spectra in a closed form and thus to come to a better understanding of them. Such attempts have been successfully performed within the last year for XPS spectra, and the results of this work will be presented here [7.9, 20–23]. The theoretical treatment employed by these authors reproduces the core level and the valence band spectra with a surprising degree of accuracy and allows the determination of the "background" of the spectra, the rate for intrinsic and extrinsic plasmon creation, the valence bandwidth, and the relative strength of the components of different angular momentum of which the valence band DOS is composed [7.20].

The photoelectron spectrum of a simple metal contains besides a given zero-loss core line or valence band contributions due to two other effects. The first is the coupling of the conduction electrons to the positive hole, produced by the photoemission process, which manifests itself in an asymmetric line shape in the core level spectra (and a similar tailing out of the valence band spectra) [7.24–27]. This effect is extensively dealt with by G.K. Wertheim in [Ref. 1.1, Chap. 5] and we shall not elaborate on this matter. The sudden creation of the positive hole also gives rise to a series of excitations called intrinsic plasmons because they are inherently connected with the photoemission process

[7.28–34]. The sum of the electrons in these spectral features will be called the primary spectrum. Secondly, on the way to the surface, and during the escape through the surface (three-step model, see [Ref. 1.1, Chaps. 1, 2]) the photoexcited electron does undergo inelastic scattering processes thereby exciting plasmons (extrinsic plasmons) and a smooth background. The quantitative description of all these effects seems now possible. *Höchst* et al. [7.9] showed that intrinsic plasmons contribute significantly to the total plasmon spectrum of the Be 1s core level, a point which had been of long-standing controversy. These findings were then substantiated by an analysis of the spectra of Na, Mg, and Al [7.21, 23]. Secondly, the successful description of the total spectrum makes it also possible to give a method to account for the smooth background in the photoemission spectrum which can be justified on theoretical grounds [7.35] and which is easy to perform for an experimentalist. The analysis shows within certain approximations that over a limited energy range the background at energy ω is always proportional to the total number of electrons emitted from the sample with higher kinetic energy $\omega' > \omega$, that is, in the experimental spectra the smooth background at ω is proportional to the total spectral area above ω[1]. If the procedures applied for the analysis of core lines in simple metals are also used to analyze the valence bands, one comes to a surprising agreement between synthetic (theoretical) and experimental photoemission spectra. This means that for this class of materials the many-body effects for core states and valence band states seem to be quite similar – a somewhat surprising and not yet fully understood result [7.36, 37].

7.2 Theory of the Photoelectron Spectrum

The theory of the photoelectron spectra has been worked out by a number of authors and we shall give here only the results which are necessary for a numerical evaluation of the spectra [7.38–44], a general discussion is found in [Ref. 1.1, Chaps. 2, 3, 5]. The basis for our interpretation of a photoemission spectrum is the three-step model developed by *Berglund* and *Spicer* [7.38], which brakes the photoemission process up into the absorption of the impinging photon with photoexcitation of an electron, the transport of this electron to the surface, and the escape of this electron through the surface [Ref. 1.1, Chaps. 1, 2]. The first step gives the primary spectrum (zero-loss line plus intrinsic plasmons) whereas the second two steps produce the inelastic tail (plasmons plus unstructured smooth background).

The use of the three-step model has been challenged by *Šunjić* and *Šokčević* [7.45]. They argue that photoemission is a many-body process and thus the breakdown into intrinsic and extrinsic effects is artificial. They show that a negative interference term exists, which can be viewed as reducing the intrinsic plasmon contribution. This interference effect, however, seems to be small,

[1] In the following energies E and frequencies ω are given in units of eV.

judging from their results making the breakdown into intrinsic and extrinsic effects not unreasonable, and the use of the three-step model justifiable.

The primary spectrum can be described within the theory given by *Mahan* [7.24], and *Nozières* and *DeDominics* [7.26]. The application to the core line has been given by *Doniach* and *Šunjić* [7.27], and *Mahan* [7.25], and the inclusion of the intrinsic plasmons has been derived by *Langreth* [7.46], *Hedin* et al. [7.47], and *Minnhagen* [7.48]. These theories show that the primary spectrum consists of an asymmetric core line at binding energy ω_0 and a series of plasmon-aided lines at $\omega_0-\omega_p$, $\omega_0-2\omega_p, \ldots$.

The primary photoelectron spectrum $F(\omega)$ can be described by [7.46, 25]

$$F(\omega)=\frac{1}{2\pi}\int_{-\infty}^{+\infty} dt\,\mathrm{e}^{-\mathrm{i}\omega t}\cdot f(t) \tag{7.1}$$

with the Fourier transform of $F(\omega)$, $f(t)$ given by

$$f(t)=\exp[-m(t)]$$

and

$$m(t)=\int_0^\infty d\omega\,\frac{\varrho(\omega)}{\omega}(1-\mathrm{e}^{-\mathrm{i}\omega t}),$$

where in the random-phase approximation (RPA) the dynamic form factor of the electron, $\varrho(\omega)$, is

$$\varrho(\omega)=\frac{\theta(\omega)}{\pi\omega}\sum_{\boldsymbol{q}}\frac{|V_q|^2}{v_q}\,\mathrm{Im}\left\{\frac{-1}{\varepsilon(q,\omega)}\right\} \tag{7.2}$$

with

$$v_q=\frac{4\pi e^2}{q^2}\quad\text{and}\quad V_q=\int d^3r\,\mathrm{e}^{\mathrm{i}\boldsymbol{q}\boldsymbol{r}}\,V(r). \tag{7.3}$$

$V(r)$ is the sudden change in Coulomb potentials through the photoexcitation process, as seen by the conduction electrons, and $\Theta(\omega)$ the step function.

$\varrho(\omega)$ peaks sharply at the plasma frequency ω_p and therefore $m(t)$ contains an oscillatory component of frequency ω_p. This oscillatory component when inserted into the exponential giving $f(t)$ yields components of frequencies ω_p, $2\omega_p$, $3\omega_p \ldots$ and thus the possibility of intrinsic multiple plasmon production. Under the assumption of a Lorentzian of width $2\gamma_p$ for $\varrho(\omega)$ we find [7.46]

$$f(t)=\sum_{n=0}^{\infty}\exp(-b)\,b^n\exp(-\mathrm{i}n\omega_p t-\mathrm{i}n\gamma_p t)/n! \tag{7.4}$$

The asymmetric core line is often described by a line shape given by *Doniach* and *Šunjić* [7.27]. This line shape, although it fits the lines of simple metals very well, has the disadvantage that its area diverges. This problem does not occur in a modified line shape given by *Mahan* [7.25] yielding

$$\varrho_0(\omega)=\alpha \mathrm{e}^{-\omega/\zeta}, \tag{7.5}$$

ζ = cut off parameter of oder $E_F/\hbar$, with E_F equal to the Fermi energy

$$f_0(t)=(1+\mathrm{i}t\zeta)^{-\alpha}, \tag{7.6}$$

$$F_0(\omega)=\frac{\zeta^{-1}}{\Gamma(\alpha)}\frac{\mathrm{e}^{+\omega/\zeta}}{|\omega/\zeta|^{1-\alpha}}\cdot\Theta(-\omega), \tag{7.7}$$

where the index zero indicates that we are dealing with the zero-loss line at ω_0. $\Gamma(\alpha)$ is the gamma function.

In the region of the core line ($\omega \approx 0$) one finds using a screened Coulomb potential

$$\varrho(\omega)=\frac{2e^2}{\pi\omega}\int dq\frac{q^4}{(q^2+k_F^2T)^2}\varepsilon_2(q,0)\,c^2, \tag{7.8}$$

where $k_FT=\left(\frac{6\pi n e^2}{E_F}\right)^{1/2}$ is the Thomas–Fermi screening wave vector and $\frac{V_q}{\varepsilon(q,0)}=c\frac{4\pi e^2}{q^2+k_F^2T}$, where c is the effective charge of the photohole seen by the conduction electrons.

For $\omega=0$ this yields

$$\varrho(0)=0.083\cdot r_s\cdot c^2=\alpha,$$

where α is the exponent in the *Mahan* line shape (7.7). Inserting the experimental values for α one finds $c\approx 0.75$ for Na, Mg, and Al – a not unreasonable result.

Equations (7.5, 7) actually differ from those proposed by *Mahan* for the X-ray anomaly at the absorption edge for core transitions in that α_l is replaced by $-\alpha$, the orthogonality contribution to α_l (for a discussion see [Ref. 1.1, Chap. 5). In order to obtain the core line shape, (7.6) must be convoluted with a Lorentzian to include core hole lifetime effects. The effects of (7.1, 4, 6) can be obtained by simple multiplication of the Fourier transforms in time-space, (7.4, 6). One obtains [7.30]

$$f(t)=\sum_{n=0}^{\infty}\mathrm{e}^{-b}\frac{b^n\exp[-\mathrm{i}n\omega_p t-(n\gamma_p+\gamma_0)t]}{n!(1+\mathrm{i}t\zeta)^{\alpha}}, \tag{7.9}$$

where $b \approx 2\alpha$, ω_p is the plasmon energy, and $2\gamma_p$ its linewidth. This equation shows that besides the zero-loss core line we find an infinite number of satellite lines at positions $\omega_0 - n\omega_p$, with widths $2(\gamma_0 + n\gamma_p)$ and relative intensities $b^n/n!$

Since the total escape depth in the XPS regime is of the order of 20 Å, the intrinsic contribution of surface plasmons to the primary spectrum should be small. This contribution will therefore be neglected in the following. The transport to the surface can be described by the transport equation given by *Wolff* [7.35]. Assuming predominant forward scattering this yields [7.22, 23]

$$P(\omega) = F(\omega)\cdot\lambda_{tot}(\omega) + \int_{\omega'>\omega} d\omega' \frac{\lambda_{tot}(\omega)\cdot P(\omega',\omega)}{v(\omega')} P(\omega'). \tag{7.10}$$

$F(\omega)$ is the primary spectrum, $P(\omega)$ the photoelectron spectrum inside the solid, $\lambda_{tot}(\omega)$ the total escape depth, $v(\omega')$ the velocity of an electron with energy ω', and $P(\omega',\omega)$ the probability per unit time that an electron of energy ω' is scattered to ω with [7.22]

$$1/\lambda_{tot}(\omega) = \int_{\omega_F}^{\omega} d\omega' \cdot P(\omega,\omega')/v(\omega'). \tag{7.11}$$

For the contribution from extrinsic plasmon creation one thus obtains:

$$P(\omega) = [\lambda_{tot}(\omega)/\lambda_{ep}(\omega)] \cdot \int d\omega' f(\omega' - w_p, \gamma_p)\, P(\omega + \omega'), \tag{7.12}$$

where $\lambda_{ep}(\omega)$ is the attenuation length due to plasmon creation and $f(\omega' - \omega_p, \gamma_p)$ the normalized lineshape of the plasmon resonance described by a single Lorentzian at energy ω_p with linewidth $2\gamma_p$. For the smooth background one estimates after some algebraic manipulations

$$P_b(\omega) = \frac{3}{2\bar{\omega}} \int_{\omega'>\omega} d\omega' P(\omega'), \tag{7.13}$$

where $\bar{\omega}$ is the mean energy of the photoelectrons for the spectrum under consideration.

From (7.12) one finds that the intensity of the *n*th extrinsic plasmon is given by the intensity of the (*n*-1st) total (extrinsic plus intrinsic) plasmon [7.30], times the constant $\lambda_{tot}(\omega)/\lambda_{ep}(\omega)$ which should be only weakly dependent on the energy.

The smooth background has a simple feature, namely it is at the frequency ω proportional to the total number of electrons with energy $\omega' > \omega$. This equation produces the background usually observed in photoemission experiments rising slowly behind a zero-loss line. Empirically this sort of background was found in earlier XPS work and approximated by a straight line with a fixed slope, a good approximation if compared with real data. The escape through the surface leads to the creation of surface plasmons and an additional small contribution to the unstructured background.

The dispersion of the plasmons is taken into account by using in (7.12) a function f proportional to

$$\mathrm{Im}\{1/\varepsilon(q,\omega)\} = \omega_\mathrm{p}^2\omega\gamma_q/[(\omega^2-\omega_q^2)^2+(\omega\gamma_q)^2] \tag{7.14}$$

with

$$\omega_q = \omega_\mathrm{p}[1+A(q/q_\mathrm{c})^2]$$

and

$$\gamma_q = \gamma_\mathrm{p}[1+B(q/q_\mathrm{c})^2].$$

Here A and B are constants, determined in energy-loss experiments, and q_c is a cut-off wave vector.

To incorporate this into the treatment, one has to multiply this equation with the wave vector dependent plasmon creation rate [7.49] and then to integrate this expression up to the cut-off wave vector. This produces an asymmetric intensity distribution with a tail to higher binding energies. If, from a practical point of view, one wants to incorporate this into a fitting routine, one finds it very time consuming. It was therefore tried to approximate the bulk plasmon intensity distribution by a simple function. *Steiner* et al. [7.50] found that a superposition of two Lorentzians of different intensity and width with an adjustable energy separation ΔE when substituted into two equations of the type of (7.9) approximate the photoelectron spectra with plasmon losses very well. This is a very useful approximation because now all the necessary convolutions can be performed in closed form, which simplifies the actual handling of the problem considerably. It can indeed be seen in the next section, that a solution using the approximate form for the plasmon intensity gives results which are as good as those obtained by using the exact form [7.23]. The surface plasmons contribute only weakly to the total spectrum and were therefore approximated by a single Lorentzian.

From a practical point of view one can decompose the photoelectron spectrum therefore into four contributions

$$P(\omega) = P_0 + \sum_{n=1}^{\infty} P_n + P_\mathrm{b} + P_\mathrm{s}, \tag{7.15}$$

where P_0 is the zero-loss line and has a *Doniach–Šunjić* or modified *Mahan* shape [7.27, 25]. The plasmon contribution P_n at $\omega_0 - n\omega_\mathrm{p}$ has an integrated strength given by

$$P_n = P_0\left(b_n + a\frac{P_{n-1}}{P_0}\right); \tag{7.16}$$

$$b_n = \mathrm{e}^{-b} b^n / n!$$

$$a = \lambda_{\mathrm{tot}} / \lambda_{\mathrm{ep}}$$

and the same line shape as the zero-loss line with an enlarged linewidth $2(\gamma_0 + n\gamma_{\mathrm{p}})$.

The smooth background is given by (7.13).

Finally, we include the surface plasmons through

$$P_{\mathrm{s}} = \sum_{n=0}^{\infty} a_{\mathrm{s}} \cdot p_n \tag{7.17}$$

with a relative intensity a_{s} at the energy $\omega_0 - n\omega_{\mathrm{p}} - \omega_{\mathrm{s}}$ and linewidth $2(\gamma_0 + n\gamma_{\mathrm{p}} + \gamma_{\mathrm{s}})$, where ω_{s} and $2\gamma_{\mathrm{s}}$ are the energy and linewidth of the surface resonance and approximately $\omega_{\mathrm{s}} = \omega_{\mathrm{p}}/\sqrt{2}$. Equation (7.17) assumes that only surface plasmon can accompany each bulk plasmon satellite.

To obtain the experimentally measured photoelectron spectra, the total spectrum obtained in this way has to be convoluted with the resolution function of the photoelectron spectrometer, which for the results represented later is a Gaussian of 0.60 eV full width at half maximum (FWHM).

This form of decomposing the spectrum can easily be used to generate synthetic photoelectron spectra. *Penn* [7.23] has included in his treatment the exact dispersion of the bulk plasmons and of the surface plasmons. His results, as will be shown in the next section, are as gratifying as those of *Steiner* et al. [7.21] justifying the tailoring of the theory to the needs of the experimentalist as performed by the latter authors.

The treatment so far applies "rigorously" only to core levels. For an analysis of valence band spectra the validity of the Mahan–Nozières–DeDominicis theory has to be reconsidered. In principle, one can argue that the recoil transmitted to the photohole upon photoionization will make the photohole "move" in the valence band and thus reduce the many-body effects. Thus in the treatment of *Doniach* [7.36], who takes the hole mobility into account, only in cases with large effective masses or narrow bands – which does not apply to the simple metals – can one expect that the core-hole approach is also valid for valence bands. On the other hand *Poole* [7.37] calculated the relaxation energies for the valence bands of some alkali metals with a semiempirical method. The result of this work is, that the model of the localized photohole gives a better agreement with the experimental data than the model using a complete nonlocalized photohole in the valence band. Since there is no

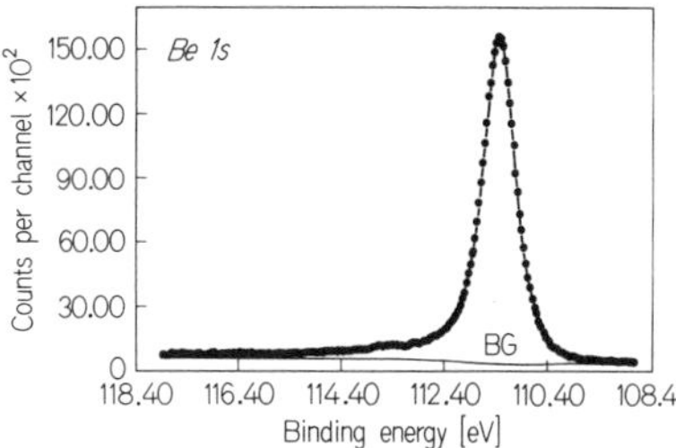

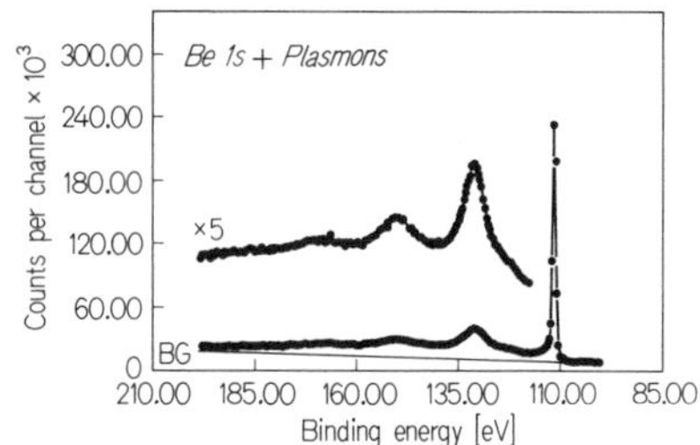

Fig. 7.1. Core level spectra of Be 1*s*. The points are raw data. The full lines are the result of a fit using the theory outlined in the text. In this and subsequent figures BG labels the smooth background

established method to deal with the many-body effects in simple metal valence bands, one can take the opposite view and apply the core level approach to find by analyzing any appearing discrepancies the effects of the recoil on the photohole.

7.3 Core Level Spectra

The question whether the plasmon satellites seen in the core level spectra are intrinsic or extrinsic in nature has aroused much controversy. The most elegant experiment to settle this question has been designed by *Fuggle* et al. [7.8]. They compared the core level spectra of Mn and Al in overlayer samples of Al on Mn. Whereas the Al spectra always showed their plasmon structure, even for thick (40 Å) layers of Al on Mn, no Al plasmons could be seen on the Mn core levels, which should have been the case if the plasmons were predominantly extrinsic in nature. On the basis of this experiment the authors judged that, at least in Al, the plasmon satellites are mostly *intrinsic.* in origin. We believe that this otherwise rather nicely designed experiment produced however the wrong results. One can speculate that the Al overlayers had large islands, making the missing plasmons in the spectra of the overlayer samples understandable. *Pardee* et al. [7.7] analyzed the core level spectra of Mg, Al, and Na in terms of (7.16): the intensities of the various intrinsic plasmon satellites should follow a Poisson distribution as a function of n, while the n-th extrinsic contribution is obtained by multiplying the total $n-1$ strength by the factor a of (7.16). Thus *Pardee* et al. came semiquantitatively to the conclusion of small intrinsic plasmon contributions in Al and Mg and a measurable one in Na. These findings were later substantiated by more quantitative analyses [7.21, 23]. The 1*s* spectrum of Be was the first for which *Höchst* et al. [7.9] could give a quantitative analysis of the total spectrum, also proving definitely the existence of a large intrinsic contribution to the plasmon spectrum. Figure 7.1 shows core level spectra of the Be 1*s* line [the points are the raw data and the full lines are the fits to the data with (7.15)]. Figure 7.2 shows the 2*s* and 2*p* spectra of Na,

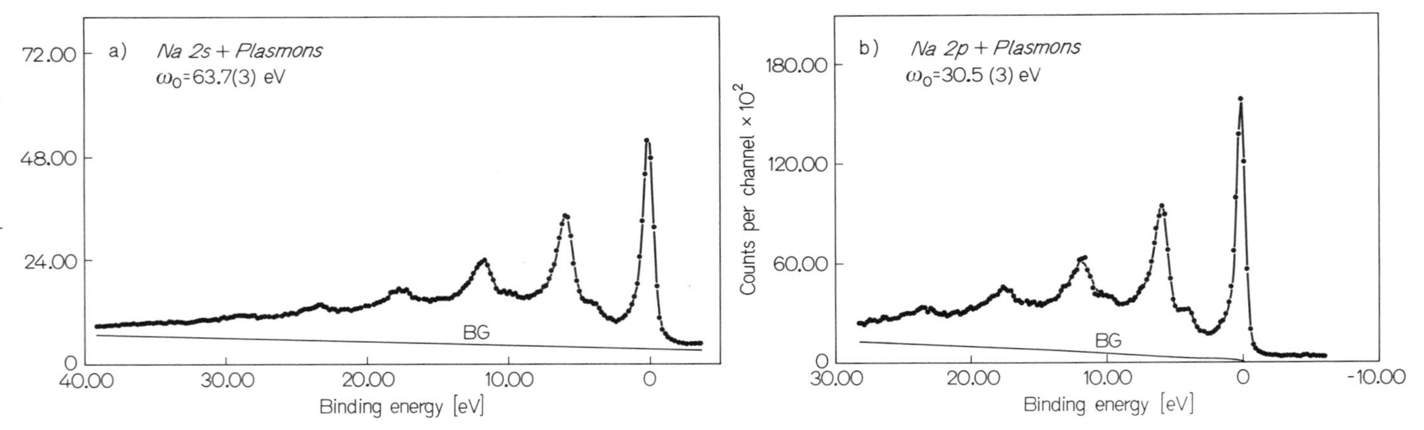

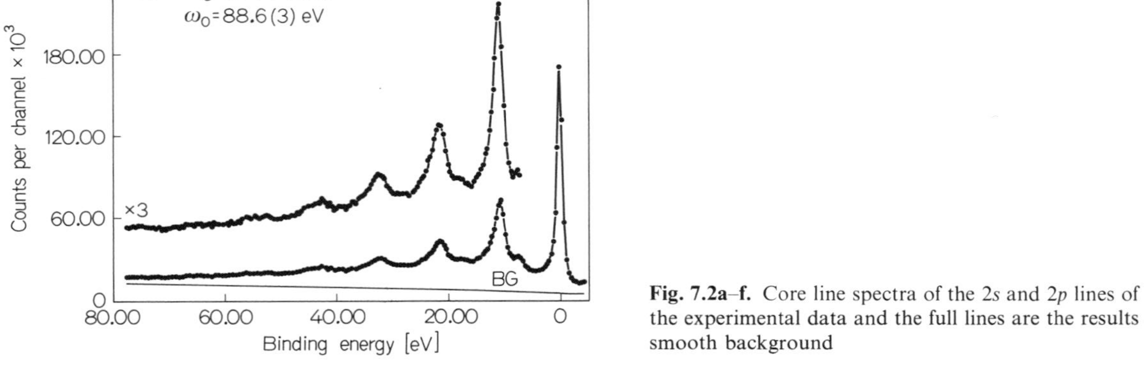

Fig. 7.2a–f. Core line spectra of the 2*s* and 2*p* lines of Na, Mg, and Al. The points are the experimental data and the full lines are the results obtained from the fit. BG is the smooth background

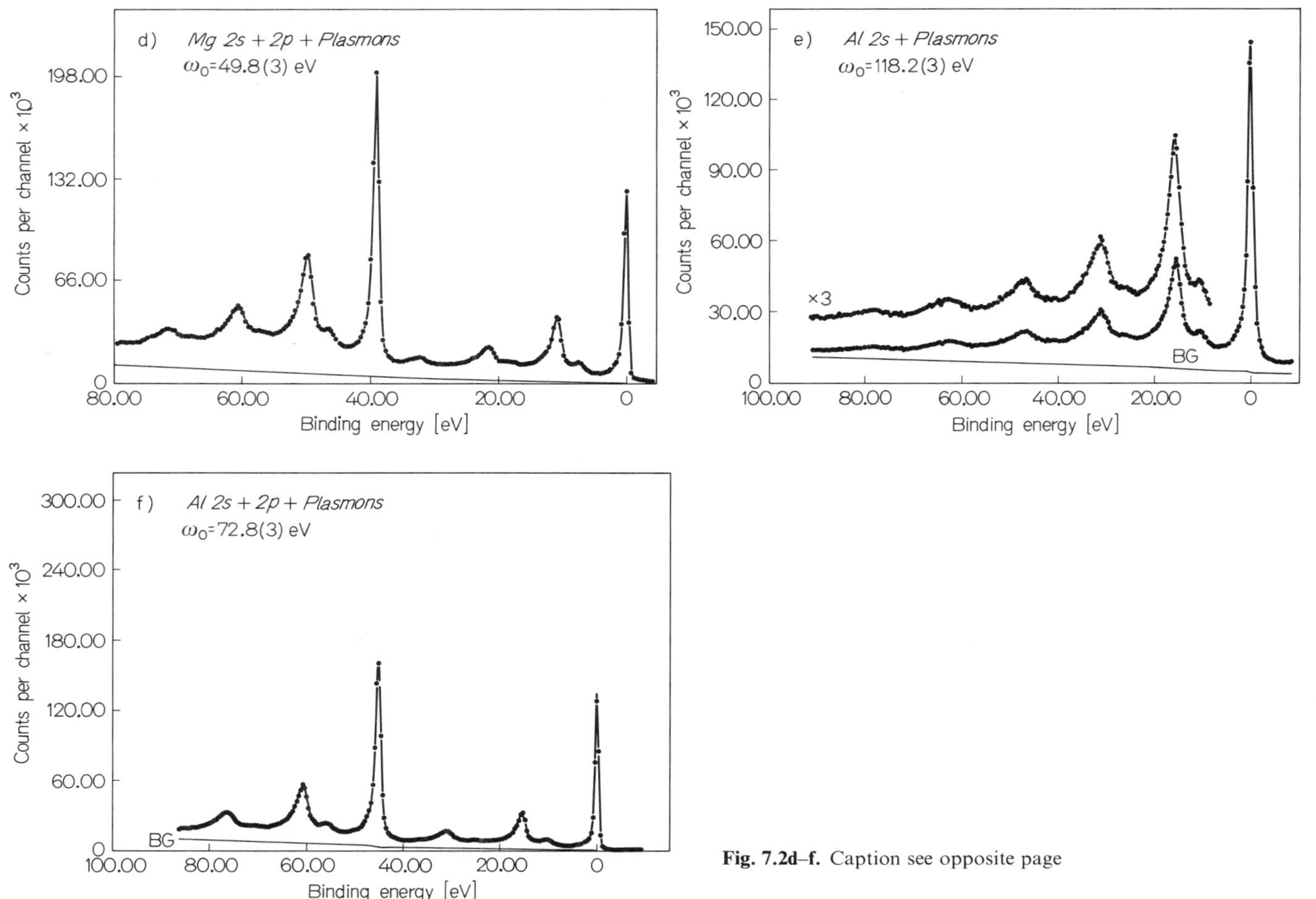

Fig. 7.2d–f. Caption see opposite page

Table 7.2. Relevant experimental parameters determined from a fit of XPS data for core levels and plasmon satellites.
ω_p = bulk plasmon energy from energy loss measurements [7.49]; ω_s = surface plasmon energy (this work); $2\gamma_p$ = bulk plasmon width from energy loss measurements; $2\gamma_s$ = surface plasmon width (this work); E_1 = Energy of main Lorenzian used to simulate the plasmon dispersion, the second Lorenzian has an energy which is 20% larger and its intensity is about 20% of that at E_1. The width of both these Lorenzians are about twice $2\gamma_p$

	E_0 [eV]	$2\gamma_0$ [eV]	α	ζ [eV]	Bulk plasmons [eV]	Surface plasmons [eV]	a in (7.16)	b in (7.16)
Be 1*s*	111.3(3)	0.35(10)	0.05(1)	15(5)	$\omega_p = 19.2$	$\omega_s = 12.3$	0.55(10)	0.38(10)
					$2\gamma_p = 5.0$			
$E_F = 11.9$ eV					$E_1 = 19.4$	$2\gamma_s = 11$		
Na 2*p*	30.5(3)	0.05(10)	0.19(1)	3.2(3)	$\omega_p = 5.7$		0.67(5)	0.53(5)
2*s*	63.7(3)	0.14(10)						
1*s*	1070.8(3)	0.50(10)			$2\gamma_p = 0.4$	$\omega_s = 3.9$		
$E_F = 3.2$ eV					$E_1 = 5.9$	$2\gamma_s = 0.4$		
Mg 2*p*	49.8(3)	0.05(10)	0.12(1)	15(5)	$\omega_p = 10.4$		0.65(5)	0.27(5)
2*s*	88.6(3)	0.48(10)			$2\gamma_p = 0.7$	$\omega_s = 7.3$		
$E_F = 6.9$ eV					$E_1 = 10.8$	$2\gamma_s = 0.8$		
Al 2*p*	72.8(3)	0.05(10)	0.10(1)	20(5)	$\omega_p = 15.0$		0.66(5)	0.11(5)
2*s*	118.2(3)	0.76(10)			$2\gamma_p = 0.5$	$\omega_s = 10.5$		
$E_F = 11.4$ eV					$E_1 = 15.6$	$2\gamma_s = 1.4$		

Included spin-orbit splitting [7.98] $\Delta = E(2p_{1/2}) - E(2p_{3/2})$.
Na: $\Delta = 0.18$ eV; Mg: $\Delta = 0.29$ eV; Al: $\Delta = 0.44$ eV.

Mg, and Al. Fits of similar quality have been obtained for Na, Mg, and Al by *Penn* [7.23] and are shown in Fig. 7.3. In comparing the last two figures, one finds that the description of the photoelectron spectra by the theoretical curves is of similar quality in both cases. This is a good justification for the simplifications made by *Höchst* et al., and *Steiner* et al., namely approximating the plasmon intensity distribution by the sum of two Lorentzians (instead of integrating over the imaginary part of the inverse dielectric function) and approximating the background by (7.13) instead of solving exactly for the dielectric constant.

The relevant parameters obtained from the fits are collected in Table 7.2. For analyzing the core lines *Höchst* et al. [7.9], and *Steiner* et al. [7.21], employed the line shape (7.7) given by *Mahan* [compare (7.9)]. The singularity exponent α obtained in this way does not differ from the one obtained in the core-line analysis by *Citrin* et al. [7.51] using the simpler *Doniach-Šunjić* line shape [7.27], see [Ref. 1.1, Chap. 5]. The cut-off parameter ζ is also obtained from the fits, but of course with quite large uncertainties; it is, however, gratifying to realize that these numbers are of the order of the Fermi energy (and the bulk plasmon energy), as they should be. The b values give a measure of the strength of the intrinsic plasmon creation rate [7.52]. From the a and b values, one can

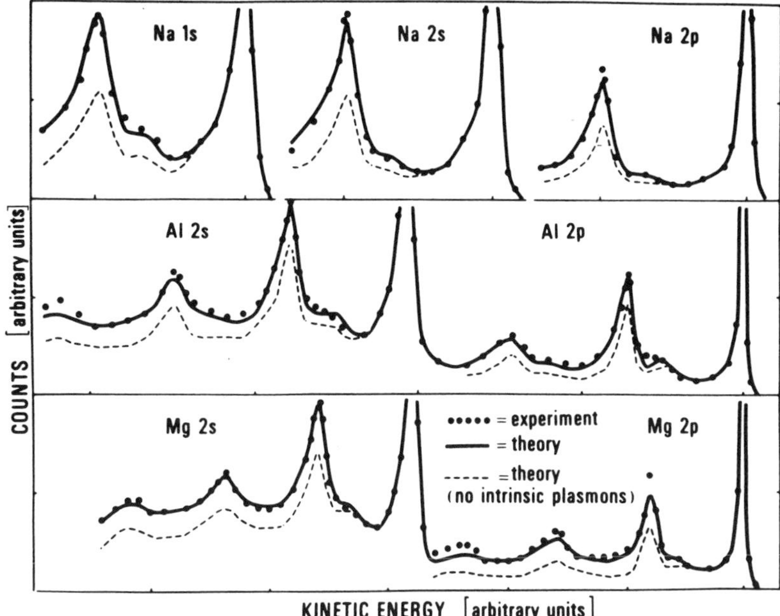

Fig. 7.3. Core level spectra of Na, Mg, and Al with fits to theory, as obtained by *Penn* [7.23]

calculate an intrinsic contribution to the first plasmon of 44, 29, and 14% in Na, Mg, and Al, respectively numbers that agree reasonably well with the estimates of 41, 36, and 26% given by *Penn* [7.23]. One may argue that the higher theoretical values found in Mg and Al are caused by the neglect of interference effects between the intrinsic and extrinsic plasmon production. These interference effects should amount to a reduction in the plasmon satellites by 50, 35, and 20% for the 2*s* and 2*p* levels of Al, Mg, and Na, respectively [7.53], thus giving a correction in the right direction. However, they should also give a much larger correction, namely 50% for the 1*s* level in Na, and within the quoted limits of error one finds the same *a* and *b* values for the 1*s*, 2*s*, and 2*p* level in Na, indicating very small interference effects.

The plasmon linewidths and energies also follow the anticipated behavior, as shown in Fig. 7.4 for the 2*s* plasmons in Mg metal, namely they are a linear function of the excitation number *n*. This shows that there is no plasmon–plasmon coupling present. The linewidth plotted in Fig. 7.4 is of course not the true plasmon linewidth, but an augmented one due to the plasmon dispersion. If this is subtracted from the measured width, the linewidth data are in good agreement with those from direct energy-loss measurements. The same statement holds for the measured plasmon energies. They come out slightly larger than those from the direct energy-loss measurements, which is again a result of averaging over the dispersion in the XPS experiment.

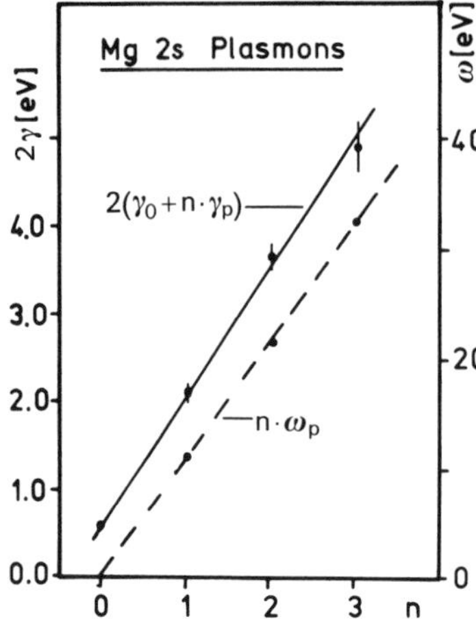

Fig. 7.4. Measured energy shifts ω and linewidth 2γ of the plasmons accompanying the 2s line in Mg. The measured linewidths are not the true linewidths, because the dispersion introduces an additional broadening (see text for details)

Table 7.3. Electron attenuation length λ_{tot}^{exp}[Å] obtained from the bulk plasmon creation rate a and calculated values λ_{ep}[Å] of the plasmon scattering lenth. λ_{ei}[Å] is the contribution of the electron–electron and electron–ion interaction to λ_{tot}. λ_{tot}^{theory} are values obtained from theoretical calculations of [7.54]

	E_{kin} [eV]	a	λ_{ep} [Å]	λ_{tot}^{exp} [Å]	λ_{tot}^{theory} [Å]	λ_{ei} [Å]
Be 1s	1376	0.55	32	18	24	41
Na 1s	416	0.67	31	21	18	65
Na 2s	1423	0.67	86	58	49	178
2p	1456		86	58	50	158
Mg 2s	1398	0.65	52	34	32	98
2p	1437		53	35	33	103
Al 2s	1369	0.66	39	25	26	70
2p	1414		40	26	27	74

The data in Table 7.2 can be used to determine the electron mean free paths. The constant a is given by $a=\lambda_{tot}/\lambda_{ep}$ where λ_{tot} is the total mean free path of an electron of kinetic energy ω and λ_{ep} is the mean free path due to extrinsic plasmon creation. Now $\lambda_{ep}=4r_0(\omega/\omega_p)/\ln(\omega/E_F)$, where r_0 is the first Bohr radius, ω_p is the bulk plasmon energy and E_F is the Fermi energy [7.49]. From these relations the total mean free paths given in Table 7.3 are calculated, and they show good agreement with theoretical numbers from *Penn* [7.54].

The intensity of the surface and bulk plasmon peaks accompanying a zero-loss line are a function of the total mean free path of the electrons in the solid. It is intuitively apparent that the bulk plasmon intensity will increase with increasing mean free path, starting out with zero intensity for zero mean free

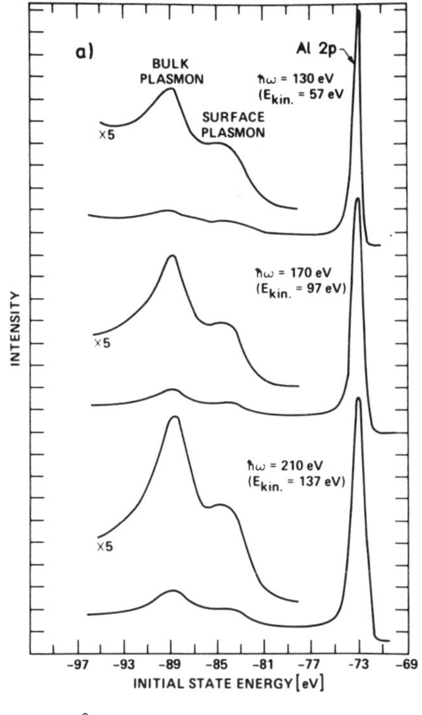

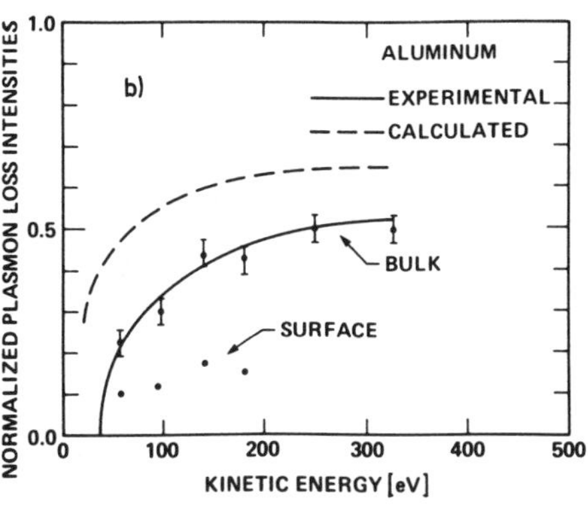

Fig. 7.5. (a) Al $2p$ spectrum from *Flodström* et al. [7.55] taken with three different exciting photon energies. **(b)** Intensity of the first bulk plasmon loss accompanying the $2p$ core level spectrum of Al [7.55] as a function of the electron energy

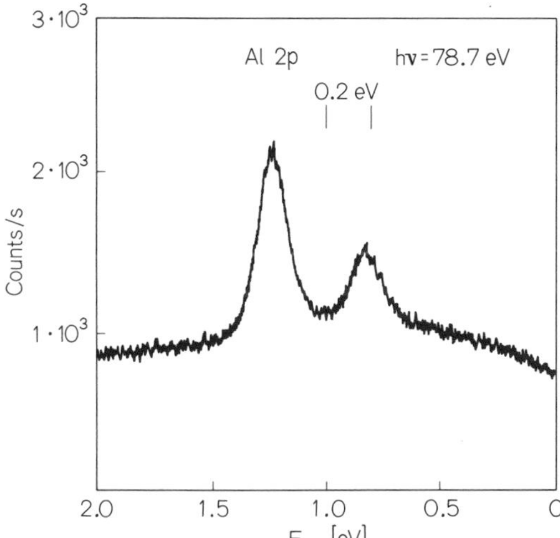

Fig. 7.6. Al $2p$ spectrum measured by *Eberhardt* et al. [7.57] with synchrotron radiation near the threshold for photoexcitation to obtain maximum resolution for the electron spectrum

path. On the other hand, the surface plasmon intensity will be approximately constant for almost all values of the mean free path – and only show a relative increase for very small escape depths. These predictions have been confirmed in experiments performed by *Flodström* et al. [7.55], who studied the $2p$ plasmon intensity in Al with synchrotron radiation. A few spectra from this work are shown in Fig. 7.5a and the relative plasmon intensities are shown in Fig. 7.5b.

The full curve in Fig. 7.5b derives from a calculation of *Kleinman* [7.56] who found that the bulk plasmon intensity should go as $1/(1+\lambda_{ep}/\lambda_{tot})^2$. Although this gives the correct trend, one finds that the calculations overestimate the actual plasmon intensity.

Finally we show in Fig. 7.6 the Al $2p$ spectrum measured at threshold with synchrotron radiation by *Eberhardt* et al. [7.57] displaying clearly the spin-orbit splitting of the $2p_{1/2}$ and $2p_{3/2}$ line. This shows the best resolution which can be obtained these days in photoelectron spectra in solids. One notes from Figs. 7.2, 3 that under conditions commonly used in XPS the Al $2p_{1/2}$ and $2p_{3/2}$ lines are not resolved.

7.4 Valence Band Spectra

Theoretical and experimental work on DOS curves of simple metals is available now over a period of almost 40 years. However, if one compares for the specific element the various experimental results among each other and also with theoretical predictions (see Table 7.1), one finds surprising disagreements. The experimental bandwidth of Na metal is 4.0 to 2.5 eV depending on the experimental technique, whereas the corresponding theoretical value is 3.2 eV [7.58]. Even the results obtained with the same technique differ among each other: XPS data give a width of 3.0 and 2.5 eV [7.59, 4], and X-ray emission measurements give widths of 2.7 and 3.05 eV [7.60, 61]. Similar disagreements can be found for other metals [7.62]. In view of the fact that X-ray emission and photoelectric method are used to determine DOS of a great number of often very complicated compounds, it seems advisable to test with the simple metals in how far these methods actually can produce reliable data or which type of correction has to be applied. In order to review the situation in a typical case, namely Na, we present in Fig. 7.7 a number of DOS curves obtained by an methods, SXE [7.60], AES [7.63], XPS, and UPS [7.64]. SXE with different initial state (one-hole state), different from XPS or UPS, produces, the same final state as UPS or XPS (one-hole state), but the $L_{2,3}$ spectrum shown measures only the s contribution (which in Na should be the largest!) to the total DOS. In addition, at the Fermi energy, the Manan-Nozière-DeDominicis effect causes a spike. AES always produces a two-hole state – and therefore the interpretation of these spectra in terms of a DOS is not yet established.

In principle, photoelectron measurements do yield the total DOS. However, a number of corrections have to be applied to the photoelectron spectrum in order to compare it directly with a theoretical DOS curve. These corrections are the different transition probability of electrons of different angular momentum [7.65, 66, Ref. 1.1, Chap. 3], the correction introduced by the photohole-conduction electron coupling and the correction produced by the inelastic scattering of the photoexcited electron. In performing an actual experiment and analyzing the data, it is very hard to recover the DOS function from the

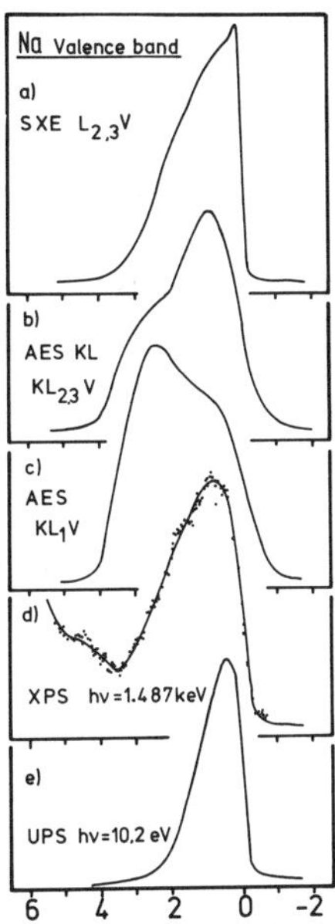

Fig. 7.7a–e. Valence band DOS measurements of Na obtained by different experimental methods (see text for details)

measured spectrum. Rather one applies the mentioned corrections to a theoretical DOS spectrum in order to come to a meaningful comparison with the experimental data.

It has been shown by various authors that matrix elements for states with different orbital momentum influence the spectra severely [7.43, 67, 68]. So far this problem has, however, not been studied in metals. The photohole-conduction electron coupling gives rise to the just discussed effects (an asymmetry of the line and a multiplasmon tail) in core level XPS spectra in metals. The magnitude of this effect in valence bands is still not established. Therefore *Höchst* et al. [7.20] have used the experimental core level spectra, which implicitly contain this effect (and the plasmon structure due to direct energy loss), and used them to fold the theoretical DOS curves, in order to simulate the photohole-conduction electron coupling correction (and the energyloss plasmon). The smooth electron scattering background was corrected for by use of (7.13). *Höchst* et al. [7.20] applied all these corrections to calculated DOS curves and then compared these modified DOS curves with the measured XPS spectra.

As an example of the procedure, the results of Be metal will be discussed. Figure 7.8b shows an XPS spectrum of Be metal and as an inset the total DOS curve from [7.69]. Several authors [7.70–75] have calculated these DOS curves, but the deviations between the various calculations are not too large. The agreement between the XPS spectrum and the calculated total EDC curve is not good. One notes especially that the XPS curve shows a peak at about 8 eV below the Fermi energy, whereas the calculated total DOS curve shows a peak

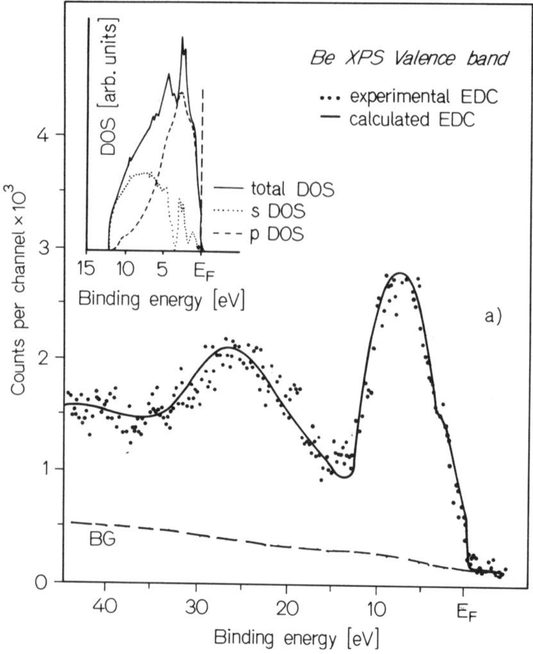

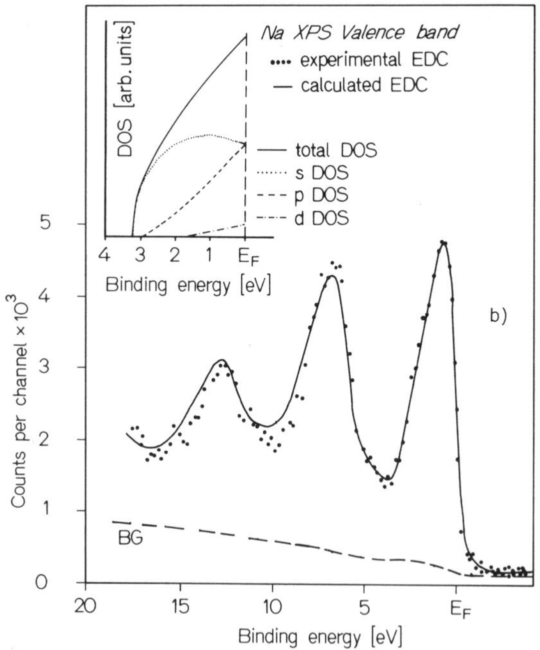

Fig. 7.8. (a) Density of states (DOS) information on Be. The points represent raw XPS data, and the insert is a total DOS, decomposed into the *s* and *p* partial DOS [7.76, 69]. The full curve is a synthetic spectrum obtained after folding the partial DOS with the core level plasmon spectrum and the resolution function of the spectrometer. BG is the smooth background. **(b)** Total and partial densities of states for Na [7.58] and valence band spectrum (points) of Na including the plasmon structure. The full line is the synthetic spectrum obtained by the procedure described in the text. BG is the smooth background

around 2 eV below the Fermi energy. A clue to this puzzling difference is obtained from the SXE spectrum of Be metal [7.76, 77]. Since this spectrum originates from an *s* level, it measures only the *p* partial DOS curve of this material. If one decomposes the total calculated DOS curves of Be metal by using this *p* contribution [7.76, 74], one obtains the *s* partial DOS as shown in the inset of Fig. 7.8a. We see that the peak in the *s* partial DOS curve roughly coincides with the peak in the experimental DOS curves. This is an indication for the fact that the XPS spectrum is dominated by the *s* DOS curve. The separation of the maxima in the *s* and *p* partial DOS curves is 6 eV in agreement with conclusions reached from Auger electron spectra [7.78]. The theoretical *s* and *p* partial DOS was folded with the function that has an asymmetry known from the core hole spectrum and also its surface and bulk plasmon tail. Then the *s* and *p* partial DOS modified in this way were added in order to produce the "theoretical" XPS spectrum. The only free parameter in this procedure was the relative intensity of the *p* and *s* partial DOS. Applying these corrections to the calculated DOS, the full curve in Fig. 7.8b was obtained with a ratio of the *s* to *p* photo excitation cross sections of $\sigma_s:\sigma_p=5:1$. The agreement between the measured and the generated curve is quite good, indicating the validity of the procedure and the accuracy of the calculated DOS curve. The theoretical bandwidth which is in agreement with the experimental spectrum is 11.9 eV. The theoretical DOS curve of *Herring* and *Hill* [7.70], which was obtained in 1940, gives results not very much different from those shown in Fig. 7.8a.

Figure 7.8b shows the result of an analysis for Na similar to that described just for Be. The calculated EDC was constructed by using the partial DOS obtained in an APW calculation [7.58] and by folding them with the experimental spectrum from core levels; cross section ratios of $\sigma_s:\sigma_p:\sigma_d=1:5.1:5$ were obtained. Note that the intensity of the first plasmon is reproduced correctly. This plasmon has about 50% of its intensity from intrinsic processes in the core level spectra. Since it has the same relative height in the valence band spectra, we have to assume large intrinsic inelastic processes also in the valence bands. Figures 7.9a–c show experimental and theoretical EDC's for Na, Mg, and Al on a wider scale in order to show in greater detail the valence band region. In each case the theoretical total and partial DOS are also shown, which are from *Gupta* and *Freeman* [7.79], and *Leonard* [7.80], for Mg and Al respectively. The cross section ratios are $\sigma_s:\sigma_p:\sigma_d=1:1.4:2.2$ and $\sigma_s:\sigma_p:\sigma_d=1:1:1.6$ for Mg and Al, respectively.

There are also a number of UPS studies on other simple metals available [7.64, 81–87, 55]. The most interesting fact that emerges from these studies is a very small electron escape depth (~ 1 Å at 10 eV) in the uv regime as compared to other metals [7.86]. This indicates that UPS spectra can contain a mixture of bulk and surface DOS.

A quantitative interpretation of UPS spectra has so far not been given, even for Na or K. Figure 7.10 shows results of *Koyama* and *Smith* [7.82] for K (which are very similar to those obtained on Na), and one can see that the measured EDCs (full line) do not agree with the theoretical ones (dashed line). It is unclear at this point whether this is a consequence of the neglect of many-body effects in the calculations or of the application of the direct model for the calculation of the EDC's. If indeed the escape length is as small as indicated above, one might expect a breakdown of the direct transition model, because then the escape length is of the order of variation of the crystal potential at the surface.

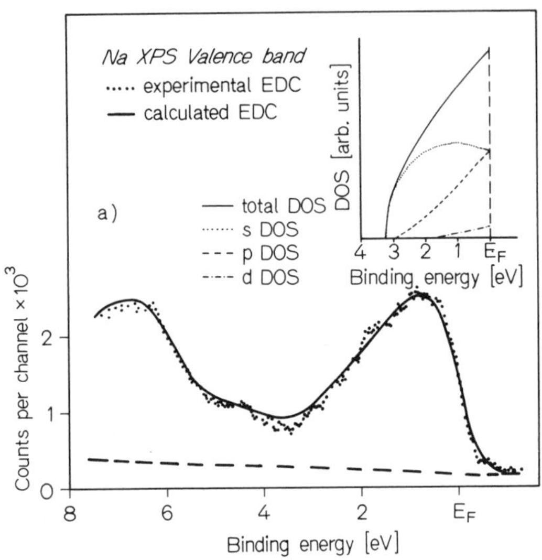

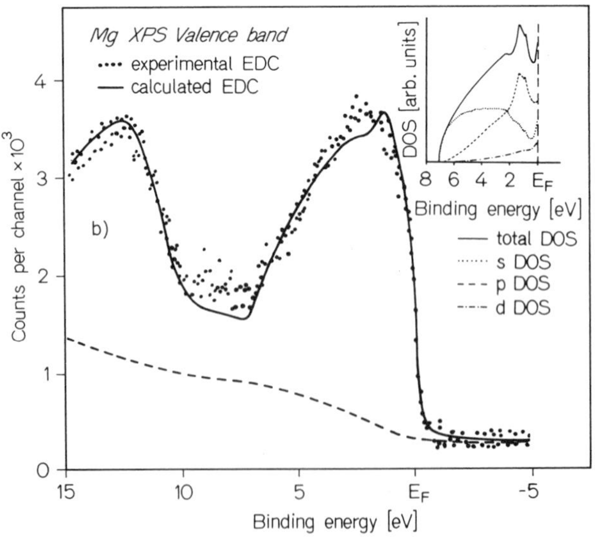

Fig. 7.9. (a) Theoretical and experimental EDC's of the valence band region of Na. The theoretical total and partial DOS is from [7.58]. **(b)** Same as Fig. 7.9a for Mg; theoretical total and partial DOS is from [7.79]

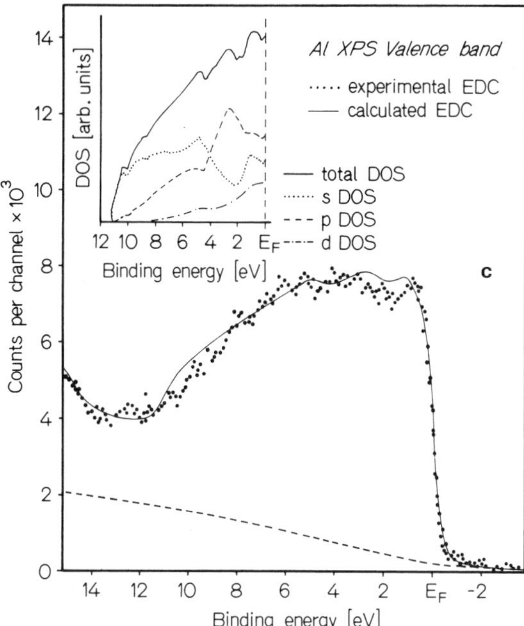

Fig. 7.9. (c) Same as Fig. 7.9a for Al; theoretical total and partial DOS is from [7.80]

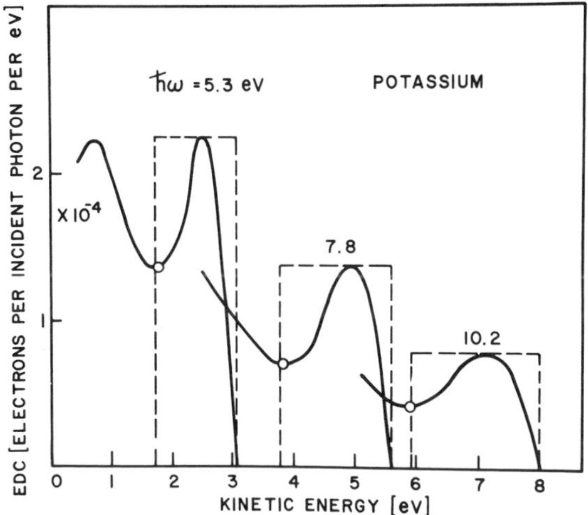

Fig. 7.10. EDC of K measured by *Koyama* and *Smith* [7.82] (full curve) compared with the free electron EDJDOS (dashed rectangles)

7.5 Summary

For the simple metals a long-standing problem in photoemission experiments has been solved at least for the XPS regime: namely the full and quantitative description of an electron energy distribution curve (EDC) by a theory. This theory relies on the three-step model and takes into account the plasmon dispersion as well as the creation of plasmons by intrinsic and extrinsic processes.

The analysis of the core lines gives quantitatively the relative magnitude of the intrinsic and extrinsic plasmon creation rate. For Be and Na they are of similar magnitude, for Mg and Al the intrinsic contributions are small. A form for the background correction in the XPS spectra has been derived: it is proportional at energy ω to the number of electrons between ω and ω_0 (where ω_0 is the zero-loss energy) which should be applicable also to the analysis of spectra of other metallic samples.

With respect to the valence bands the results are twofold. First of all, as demonstrated most clearly with the data in Be metal, photoelectric data cannot be directly compared with DOS curves but have to be analyzed in terms of the partial DOS for the electrons with different angular momentum. In order to use theoretical DOS curves partial DOS are mandatory for the comparison with experimental results.

Secondly, the interaction of the hole and the photoexcited electron with the electrons of the solid can be described with the same parameters for the valence band states as for the core levels. This is demonstrated by the fact that the parameters obtained from the analysis of the core levels also describe very accurately the valence band spectra. This seems to hold for the asymmetry of the lines as well as for the strength of the surface and bulk plasmons. This means especially that the recoil of the hole in the valence band has no dramatic effect on the shape of the XPS valence band spectrum. A strong localization of the photohole in the valence band has been found recently in a calculation of the relaxation energy for the simple metals by *Poole* [7.37]. It is well realized that this is a quite unexpected result, because due to the large bandwidth in the simple metals, one might have anticipated a weakening of the many-body effect due to the recoil of the hole [7.27, 36]. However, the finding rests on the analysis of valence band spectra for four metals. In addition, the many-body effects manifest themselves by two effects: by the occurrence of intrinsic contributions to the plasmons and by asymmetric line shapes. One might argue that the asymmetric line-shape function has spuriously been corrected for with the background. However, the fact that the plasmon intensities especially in the case of the first plasmon of Na, where the intrinsic contribution amounts to 45% of the total intensity, are correctly reproduced is a strong argument in favor of the findings of *Höchst* et al. [7.20].

Acknowledgment. This work was supported by the "Deutsche Forschungsgemeinschaft".

References

7.1 Y. Baer, G. Busch: Phys. Rev. Lett. **30**, 280 (1973)
7.2 A. Barrie: Chem. Phys. Lett. **19**, 109 (1973)
7.3 J. Tejeda, M. Cardona, N.J. Shevchik, D.W. Langer, E. Schönherr: Phys. Status Solidi B **58**, 189 (1973)
7.4 S.P. Kowalczyk, L. Ley, F.R. McFeely, R.A. Pollak, D.A. Shirley: Phys. Rev. B **8**, 3583 (1973)
7.5 R.A. Pollak, L. Ley, F.R. McFeely, S.P. Kowalczyk, D.A. Shirley: J. Electron Spectrosc. **3**, 381 (1974)

7.6 L. Ley, F.R. McFeely, S.P. Kowalczyk, J.G. Jenkin, D.A. Shirley: Phys. Rev. B **11**, 600 (1975)
7.7 W.J. Pardee, G.D. Mahan, D.E. Eastman, R.A. Pollak, L. Ley, F.R. McFeely, S.P. Kowalczyk, D.A. Shirley: Phys. Rev. B **11**, 3614 (1975)
7.8 J.C. Fuggle, D.J. Fabian, L.M. Watson: J. Electron Spectrosc. **9**, 99 (1976)
7.9 H. Höchst, P. Steiner, S. Hüfner: Phys. Lett. **60** A, 69 (1977)
7.10 H.W.B. Skinner: Phil. Trans. R. Soc. London A **239**, 95 (1940)
7.11 D.J. Fabian, L.M. Watson, C.A.W. Marshall: Rep. Prog. Phys. **34**, 601 (1971)
7.12 C.C. Chang: In *Characterisation of Solid Surfaces*, ed. by P.F. Kane, G.B. Larabee (Plenum Press, New York, London 1974)
7.13 T.A. Carlson: *Photoelectron and Auger Spectroscopy* (Plenum Press, New York, London 1975)
7.14 K. Siegbahn: *ESCA Applied to Free Molecules* (North-Holland, Amsterdam 1969)
7.15 N.V. Smith: CRC Critical Rev. in Sol. State Sci. **2**, 45 (1971)
7.16 S.B. Hagström: In *Electron Spectroscopy*, ed. by D.A. Shirley (North-Holland, New York 1972) p. 515
7.17 W.E. Spicer: J. Phys. (Paris) **34**, Suppl. C **6**, 6 (1973)
7.18 D.E. Eastman: In *Vacuum Ultraviolet Radiation Physics*, ed. by E.E. Koch, R. Haensel, C. Kunz (Pergamon-Vieweg, Braunschweig 1973) p. 417
7.19 I. Lindau: In *International College on Applied Physics*, Istituto Nazionale Di Fisica Nucleare, Course on Synchrotron Radiation Physics, Vol. 1, ed. by A.N. Mancini (I.F. Quercia 1976) p. 321
7.20 H. Höchst, P. Steiner, S. Hüfner: J. Phys. F **7**, L 309 (1977)
7.21 P. Steiner, H. Höchst, S. Hüfner: Phys. Lett. **61** A, 410 (1977)
7.22 D.R. Penn: J. Vac. Sci. Technol. **14**, 300 (1977)
7.23 D.R. Penn: Phys. Rev. Lett. **38**, 1429 (1977)
7.24 G.D. Mahan: Phys. Rev. **163**, 612 (1967)
7.25 G.D. Mahan: Phys. Rev. B **11**, 4814 (1975)
7.26 P. Nozières, C.T. DeDominicis: Phys. Rev. **178**, 1097 (1969)
7.27 S. Doniach, M. Šunjić: J. Phys. C **3**, 285 (1970)
7.28 J.J. Chang, D.C. Langreth: Phys. Rev. B **5**, 3512 (1972)
7.29 R.G. Cavell, S.P. Kowalczyk, L. Ley, R.A. Pollak, B. Mills, D.A. Shirley, W. Perry: Phys. Rev. B **7**, 5313 (1973)
7.30 D.C. Langreth: "Theory of Plasmon Effects in High Energy Spectroscopy" (presented before the NOBEL SYMPOSIUM, Sweden 1973) in *Nobel Foundation Series "Nobel Symposia, Medicine and Natural Sciences"*, Vol. 24 (Academic Press, New York, London 1974)
7.31 S.Q. Wang, G.D. Mahan: Phys. Rev. B **6**, 4517 (1972)
7.32 G.D. Mahan: Phys. Status Solidi b **55**, 703 (1973)
7.33 M. Šunjić, D. Šokčević: Solid State Commun. **15**, 165 (1974)
7.34 A.M. Bradshaw, W. Wyrobisch: J. Electron Spectrosc. **7**, 45 (1975)
7.35 P.A. Wolff: Phys. Rev. **95**, 56 (1954)
7.36 S. Doniach: Phys. Rev. B **2**, 3898 (1970)
7.37 R.T. Poole: Chem. Phys. Lett. **42**, 151 (1976)
7.38 C.N. Berglund, W.E. Spicer: Phys. Rev. **136**, A 1030 (1964)
7.39 I. Adawi: Phys. Rev. **134**, A 788 (1964)
7.40 W.I. Schaich, N.W. Ashcroft: Solid State Commun. **8**, 1959 (1970)
7.41 W.I. Schaich, N.W. Ashcroft: Phys. Rev. B **3**, 2452 (1971)
7.42 G.D. Mahan: Phys. Rev. B **2**, 4334 (1970)
7.43 C. Caroli, D. Lederer-Rozenblatt, B. Roulet, D. Saint-James: Phys. Rev. B **8**, 4552 (1973)
7.44 P.J. Feibelman, D.E. Eastman: Phys. Rev. B **10**, 4932 (1974)
7.45 M. Šunjić, D. Šokčević: Solid State Commun. **18**, 373 (1976)
7.46 D.C. Langreth: Phys. Rev. B **1**, 471 (1970)
7.47 L. Hedin, B.I. Lundquist, S. Lundquist: J. Res. Nat. Bur. Stand. **74** A, 417 (1970)
7.48 P. Minnhagen: Phys. Lett. **56** A, 327 (1976)
7.49 H. Raether: In *Solid State Excitations by Electrons*, Springer Tracts in Modern Physics, Vol. 38, ed. by G. Höhler (Springer, Berlin, Heidelberg, New York 1965) p. 84
7.50 P. Steiner, H. Höchst, S. Hüfner: Z. Physik B **30**, 129 (1978)

7.51 P.H. Citrin, G.K. Wertheim, Y. Baer: Phys. Rev. Lett. **35**, 885 (1975)
7.52 D.C. Langreth: Phys. Rev. Lett. **26**, 1229 (1971)
7.53 W.J. Gadzuk: J. Electron Spectrosc. **11**, 355 (1977)
7.54 D.R. Penn: J. Electron Spectrosc. **9**, 29 (1976)
7.55 S.A. Flodström, R.Z. Bachrach, R.S. Bauer, J.C. McMenamin, S.B.M. Hagström: J. Vac. Sci. Technol. **14**, 303 (1977)
7.56 L. Kleinman: Phys. Rev. B **3**, 2982 (1971)
7.57 W. Eberhardt, G. Kalkoffen, C. Kunz: DESY Rt. SR78/15 (1978)
7.58 R.P. Gupta, A.J. Freeman: Phys. Lett. **59** A, 223 (1976)
7.59 P.H. Citrin: Phys. Rev. B **8**, 5545 (1973)
7.60 R.S. Crisp, S.E. Williams: Philos. Mag. **6**, 625 (1961)
7.61 D.H. Tomboulian: In *Handbuch der Physik*, Vol. XXX, ed. by S. Flügge (Springer, Berlin, Göttingen, Heidelberg 1957) p. 246
7.62 S.P. Kowalczyk: Thesis, University of California at Berkeley (January 1976) (LBL-4319) p. 552
7.63 A. Barrie, F.J. Street: J. Electron Spectrosc. **7**, 1 (1975)
7.64 N.V. Smith, W.E. Spicer: Phys. Rev. **188**, 593 (1959)
7.65 V.V. Nemoshkalenko, V.G. Aleshin, Yu.N. Kucherenko: Solid State Commun. **20**, 1155 (1976)
7.66 V.G. Aleshin, Yu.N. Kucherenko, V.V. Nemoshkalenko: Solid State Commun. **20**, 913 (1976)
7.67 G.K. Wertheim, L.F. Mattheis, M. Campagna, T.P. Pearsall: Phys. Rev. Lett. **32**, 997 (1974)
7.68 A. Goldmann, J. Tejeda, N.J. Shevchik, M. Cardona: Phys. Rev. B **10**, 4388 (1974)
7.69 S.T. Inoue, J. Yamashita: J. Phys. Soc. Jpn. **35**, 677 (1973)
7.70 C. Herring, A.G. Hill: Phys. Rev. **58**, 132 (1940)
7.71 J.F. Cornwell: Proc. R. Soc. London A **261**, 551 (1961)
7.72 T.L. Loucks, P.H. Cutler: Phys. Rev. **133** A, 819 (1964)
7.73 J.H. Terrell: Phys. Rev. **149**, 526 (1966)
7.74 P.O. Nilsson, G. Arbman, T. Gustafsson: J. Phys. F **4**, 1937 (1974)
7.75 S. Chatterjee, P. Sinha: J. Phys. F **5**, 2089 (1975)
7.76 G. Wiech: In *Soft X-Ray Band Spectra and the Electronic Structure of Metals and Materials*, ed. by D.J. Fabian (Academic Press, New York 1968) pp. 59–70
7.77 G. Dräger, O. Brümmer: Phys. Status Solidi b **78**, 729 (1976)
7.78 H.G. Maguire, P.D. Augustus: Philos. Mag. **30**, 15 (1974)
7.79 R.P. Gupta, A.J. Freeman: Phys. Rev. Lett. **36**, 1194 (1976)
7.80 P. Leonard: Private communication
7.81 N.V. Smith, G.B. Fisher: Phys. Rev. B **3**, 3662 (1971)
7.82 R.Y. Koyama, N.V. Smith: Phys. Rev. B **2**, 3049 (1970)
7.83 K.A. Kress, G.J. Lapeyre: Solid State Commun. **9**, 827 (1971)
7.84 K.A. Kress, G.J. Lapeyre: Phys. Rev. Lett. **28**, 1639 (1972)
7.85 C.R. Helms, W.E. Spicer: Phys. Rev. Lett. **28**, 565 (1972)
7.86 C.R. Helms, W.E. Spicer: Phys. Lett. **57** A, 369 (1976)
7.87 P.O. Nilsson, G. Arbman, D.E. Eastman: Solid State Commun. **12**, 627 (1973)
7.88 J.C. Fuggle, L.M. Watson, P.R. Norris, D.J. Fabian: J. Phys. F **5**, 590 (1975)
7.89 C.J. Powell: Phys. Rev. Lett. **30**, 1179 (1973)
7.90 H. Neddermeyer: Verh. Dtsch. Phys. Ges. VI **10**, 527 (1975)
7.91 L.M. Watson, R.K. Dimond, D.J. Fabian: In *Soft X-Ray Band Spectra and Electronic Structure of Metals and Materials*, ed. by D.J. Fabian (Academic Press, New York 1968) p. 45
7.92 W.M. Cady, D.H. Tomboulian: Phys. Rev. **59**, 381 (1941)
7.93 H. Neddermeyer: Z. Phys. **271**, 329 (1974)
7.94 G.A. Rooke: J. Phys. C **2**, 767 (1968)
7.95 S. Ham: Phys. Rev. **128**, 826 (1962)
7.96 J.C. Kimball, R.W. Stark, F.M. Mueller: Phys. Rev. **162**, 600 (1967)
7.97 L. Smrčka: Czech. J. Phys. B **21**, 683 (1971)
7.98 K. Schwarz: Private communication

Appendix: Table of Core-Level Binding Energies

This table lists binding energies (up to ~1500 eV) of core levels obtained from elements in their natural form using photoemission spectroscopy. The binding energies are given in electron volts [eV] relative to the vacuum level for the rare gases and H_2, N_2, O_2, F_2, Cl_2; relative to the Fermi level for the metals; and relative to the top of the valence bands for semiconductors. Errors in the last digit(s) are given parenthetically as they have been quoted by the authors. Since these errors are in almost all cases (except for [40]) a mere measure of the *precision* of the measurements, we have tried to list whenever possible more than one binding energy to convey a feeling for the *accuracy* of the binding energies. In a number of elements only a few binding energies were obtained under UHV conditions from clean surfaces; we have then used the energy differences of *Bearden* and *Burr* [19] to derive the missing energies. For the elements P, Cl, Zr, Nb, Ru, I, Hf, Os, and the radioactive elements Po through Pa we had to rely entirely on the compilation by *Siegbahn* et al. [22] because no new trustworthy data seemed to be available. These values are set in parentheses. Electrons contributing to the valence bands or molecular orbits of a solid or molecule are marked “VE” (valence electrons). The spin-orbit splitting of levels, which can be measured more accurately than the absolute binding energies of the doublet components, are sometimes given behind the initials s.o.

	$1s$	$2s$	$2p_{1/2}$	$2p_{3/2}$	$3s$	$3p_{1/2}$	$3p_{3/2}$	$3d$	$4s$
^{1}H	16.0 [1]								
^{2}He	24.59 [23]								
^{3}Li	54.9 (1) [10] 54.8 (1) [2] 54.3 (3) [8]	VE							
^{4}Be	111.7 [2] 111.4 (3) [8]	VE							
^{5}B	(188) [22]								
^{6}C	284.3 (3) [3] 284.7 [2] (graphite) 283.5 (3) [29] (diamond)	VE	VE						
^{7}N	409.9 (1) [4]	37.3 [4] (σ_g 2s)	VE						
^{8}O	543.1 (2) [4, 5]	41.6 [4] (σ_g 2s)	VE						
^{9}F	696.71 (5) [6]	VE	VE						
^{10}Ne	870.2 (1) [4]	48.42 (5) [4] 48.47 [7]	21.661 [7]	21.564 [7]					

	$1s$	$2s$	$2p_{1/2}$	$2p_{3/2}$	$3s$	$3p_{1/2}$	$3p_{3/2}$	$4s$
^{11}Na	1071.7 (1) [2] 1070.8 (2) [10] 1070.8 (3) [8]	63.4 (1) [2] 63.6 (2) [10] 63.7 (3) [8]	30.6 (1) [2]	30.3 (1) [2] 30.5 (2) [10]	VE			
			30.5 (3) [8] s.o. = 0.17 [10]					
^{12}Mg	1303.0 (1) [2]	88.55 (10) [2] 88.7 (2) [10] 88.6 (3) [8]	49.5 (1) [2]	49.2 (1) [2]	VE			
			49.7 (2) [10]; 49.8 (3) [8]; s.o. = 0.28 [10]					
^{13}Al	1562.3 (5) [2]	117.5 (2) [2] 117.9 [9] 118.2 (2) [8]	72.6 (2) [2] 72.9 (2) [10]	72.2 (2) [2] 72.5 (2) [10]	VE	VE		
			72.8 (3) [8, 9]; 73.0 (1) [13]	s.o. = 0.35 [10] s.o. = 0.42 (5)[13]				
^{14}Si		149.5 (7) [19] 149.8 (5) [20]	99.2 (1) [20a]	99.8 (1) [20a]	VE	VE		
		$(2s-2p)$ = 51.1 (1) [48]	99.4 (2) [11]; s.o. = 0.6 (2) [20] 98.4 (5) [20]; s.o. = 0.62 (3) [20a]					
^{15}P		(189 [22])	(136 [22])	(135 [22])	VE	VE		
^{16}S		230.9 (7) [19]	163.6 (3) [42]; 162.5 (3) [42] s.o. = 1.15 (5) [42]		VE	VE		
^{17}Cl		(270 [22])	(202 [22])	(200 [22])	VE	VE		
^{18}Ar	3205.9 (5) [4]	326.3 (1) [4]	250.56 (7) [4]	248.45 (7) [4]	29.3 (1) [4]	15.94 (1) [7]	15.76 (1) [7]	
^{19}K	(3608.4 (2) [2])	(378.6 (3) [2])	(297.3 (3) [2])	(294.6 (3) [2]	(34.8 (3) [2])			
						(18.3 (1) [2])		
^{20}Ca	(4038.5 (4) [2])	(438 (1) [2]) 439.0 (3) [8]	(350.4 (5) [2]) 350.3 (3) [8]	(346.8 (5) [2]) 346.5 (3) [8]	(44.1 (6) [2]) 44.0 (3) [8]	26.5 (3) [8]	25.1 (3) [8]	
						(24.8 (2) [2])		

	$2s$	$2p_{1/2}$	$2p_{3/2}$	$3s$	$3p_{1/2}$	$3p_{3/2}$	$3d_{3/2}$	$3d_{5/2}$	$4s$
^{21}Sc	498.0 (3) [2]	403.58 (10) [2]	398.65 (10) [2]	51.1 (1) [2]	28.3 (1) [2]		VE		VE
^{22}Ti	561.4 (3) [2]	461.0 (2) [2]	454.9 (2) [2]	58.4 (2) [2]	32.6 (2) [2]		VE		VE
^{23}V	627.2 (4) [2]	521.07 (8) [2]	513.41 (8) [2]	66.4 (2) [2]	37.2 (2) [2]		VE		VE
^{24}Cr	697.8 (3) [2, 5]	585.35 (23) [2]	576.04 (17) [2]	75.2 (3) [2, 5]	43.1 (3) [2]		VE		VE
^{25}Mn	769.4 (2) [2, 5]	650.6 (3) [2] 649.7 (3) [8]	639.4 (2) [2] 638.2 (3) [8]	82.4 (3) [2, 5]	47.3 (2) [2] 47.0 (3) [8]		VE		VE
^{26}Fe	848.7 (2) [2, 5]	720.65 (20) [2] 720.3 (3) [8]	707.55 (20) [2] 706.82 (9) [40] 707.2 (3) [8]	91.6 (2) [2, 5]	53.0 (2) [2]		VE		VE
^{27}Co	926.6 (2) [2, 5, 14]	796.2 (4) [2, 14] 793.4 (3) [8]	781.0 (4) [2, 14] 778.4 (3) [8]	105.4 (1) [2, 5, 14] 101.0 (3) [5, 8]	59.4 (1) [2, 14]		VE		VE
^{28}Ni	1010.4 (5) [2, 5]	871.5 (2) [2] 870.8 (3) [8]	854.2 (1) [2] 853.3 (3) [8]	110.9 (2) [2, 5] 110.2 (3) [5, 41] 111.5 (3) [5, 8]	66.5 (2) [2] 67.4 (3) [41] 67.0 (3) [8]	65.7 (3) [41] 65.7 (3) [8]	VE		VE
^{29}Cu	1098.6 (7) [2] 1096.4 (3) [39]	952.6 (2) [2] 952.7 (3) [8] 952.5 (2) [17] 952.1 (1) [39] 952.35 (9) [40]	932.8 (2) [2] 932.9 (3) [8] 932.7 (2) [17] 932.2 (1) [39] 932.53 (9) [40]	122.5 (1) [2] 122.4 (2) [17] 122.4 (1) [39]	77.23 (10) [2] 77.1 (1) [39] 77.3 (2) [17]	15.07 (10) [2] 15.2 (1) [39] 14.9 (2) [17]	VE		VE
^{30}Zn	1200.7 (3) [2]	1045.1 (2) [2] 1044.7 (2) [17]	1022.0 (2) [2] 1021.6 (2) [17]	139.9 (2) [2] 139.6 (2) [17]	91.31 (15) [2] 91.4 (2) [17]	88.70 (15) [2] 88.5 (2) [17]	9.77 (10) [16]	9.23 (10) [16]	VE
					s.o. = 2.86 (2) [17]		10.2 (2) [15]; 10.08 [17] ΔE [18] = 0.55 [17]		

	$2s$	$2p_{1/2}$	$2p_{3/2}$	$3s$	$3p_{1/2}$	$3p_{3/2}$	$3d_{3/2}$	$3d_{5/2}$	$4s$	$4p$
^{31}Ga	1299.0 (7) [19]	1143.6 (7) [19]	1116.7 (7) [19]	159.4 (2) [20]	107.3 (2) [20]	104.2 (2) [20]			VE	VE
							18.7 (2) [20]; 18.34 (10) [16]			
^{32}Ge	1414.6 (7) [19]	1248.1 (7) [19]	1217.0 (7) [19]	180.1 (2) [20]	124.9 (2) [20]	120.8 (2) [20]	29.9 (1) [46]	29.3 (1) [46] 29.1 (1) [47]	VE	VE
							29.0 (2) [20]; s.o. = 0.53 (6) [46]			
^{33}As	1527.0 (7) [19]	1359.1 (7) [19]	1323.6 (7) [19]	204.7 (2) [20]	146.2 (2) [20]	141.2 (2) [20]			VE	VE
							41.7 (2) [20]			
^{34}Se	1652.0 (7) [19]	1474.3 (7) [19]	1433.9 (7) [19]	229.6 [21]	166.5 [21]	160.7 [21]	55.47 [21]	54.64 [21]	VE	VE
^{35}Br	(1782 [22])	(1596 [22])	(1550 [22])	(257 [22])	(189 [22])	(182 [22])	(70 [22])	(69 [22])	VE	VE
^{36}Kr		1730.9 (5) [4]	1678.4 (5) [4]	292.8 (2) [4]	222.2 (2) [4]	214.4 (2) [4]	94.9 (2) [4] 95.04 [23]	93.7 (2) [4] 93.83 [23]	27.4 (2) [4] 27.51 [23]	14.08 [4] s.o.=0.65 [23]
		s.o. = 52.5 (3) [4]			s.o. = 7.8 (1) [4]					
	$3s$	$3p_{1/2}$	$3p_{3/2}$	$3d_{3/2}$	$3d_{5/2}$	$4s$	$4p_{1/2}$	$4p_{3/2}$	$5s$	$4d$
^{37}Rb	326.7 (10) [24]	248.7 (5) [24]	239.1 (5) [24]	113.0 (5) [24]	112.0 (5) [24]	30.5 (5) [19]	16.1 (1) [24] 16.4 (3) [25]	15.2 (1) [24] 15.3 (3) [25]	VE	
^{38}Sr	358.0 (3) [8]	279.1 (3) [8]	269.6 (3) [8]	135.8 (3) [8]	134.1 (3) [8]	38.8 (3) [8]	21.2 (3) [8]	20.1 (3) [8]	VE	
^{39}Y	392.0 (8) [19]	310.6 (3) [45]	298.8 (3) [45]	157.7 (3) [45]	155.8 (3) [45]	43.8 (8) [19]	24.4 (3) [45]	23.1 (3) [45]	VE	VE
							s.o. = 1.21 (5) [45]			
^{40}Zr	(431 [22])	(345 [22])	(331 [22])	(183 [22])	(180 [22])	(52 [22])	24.6 (5) [44]		VE	VE
							(29 [22])			

	$3s$	$3p_{1/2}$	$3p_{3/2}$	$3d_{3/2}$	$3d_{5/2}$	$4s$	$4p_{1/2}$	$4p_{3/2}$	$4d_{3/2}$	$4d_{5/2}$
^{41}Nb	(469 [22])	(379 [22])	(363 [22])	(208 [22])	(205 [22])	(58 [22])	28.7 (5) [44] (34 [22])		VE	VE
^{42}Mo	504.9 (8) [19]	410.0 (8) [19]	392.5 (6) [19]	230.6 (2) [16]	227.4 (2) [16]	62.1 (6) [19]	35.1 (6) [19]		VE	VE
				s.o. = 3.17 (18) [16]						
^{43}Tc	(544 [22])	(445 [22])	(425 [22])	(257 [22])	(253 [22])	(68 [22])	(39 [22])		VE	VE
^{44}Ru	(585 [22])	(483 [22])	(461 [22])	(284 [22])	(279 [22])	(75 [22])	(43 [22])		VE	VE
^{45}Rh	627.4 (6) [19]	521.3 (6) [19]	496.2 (3) [41]	312.3 (5) [20]	307.2 (3) [41] 307.5 (5) [20]	81.4 (6) [19]	48.3 (6) [19]		VE	VE
				s.o. = 5.27 (12) [16]						
^{46}Pd	671.7 (6) [19]	560.0 (3) [8]	532.2 (2) [39] 532.5 (3) [8]	340.7 (5) [20] 340.9 (2) [16] 340.4 (3) [8] 340.49 (5) [40]	335.0 (5) [20] 335.6 (2) [16] 335.1 (3) [8] 335.20 (5) [40] 335.2 (1) [39]	87.1 (6) [19]	51.8 (6) [19]		VE	VE
^{47}Ag	719.1 (6) [19]	604.0 (6) [19]	573.0 (1) [39] 573.0 (3) [41]	374.5 (5) [20] 375.8 (3) [8] 374.23 (5) [40]	368.5 (5) [20] 368.2 (2) [39] 369.8 (3) [8] 368.23 (5) [40]	96.8 (6) [19]	64.2 (6) [19]	57.5 (6) [19]	VE	VE
^{48}Cd	771.5 (6) [19]	652.0 (6) [19]	618.0 (3) [41]	411.8 (6) [19]	404.9 (3) [41]	108.9 (6) [19]	[27]	[27]	11.5 (3) [41] 11.46 (9) [26] 11.15 (10) [16]	10.6 (3) [41] 10.47 (9) [26] 10.20 (10) [16]
									ΔE [18] = 0.99 (5) [26]; 0.95 (3) [16]	
^{49}In	826.4 (8) [19]	703.0 (6) [19]	665.1 (3) [41]	451.6 (6) [19]	443.9 (6) [19]	122.7 (6) [19]	[27]	[27]	17.64 (9) [26] 17.26 (10) [16]	16.74 (9) [26] 16.40 (10) [16]
									ΔE [18] = 0.90 (1) [26]; 0.86 (3) [16]	
^{50}Sn	883.9 (4) [19]	756.5 (3) [8]	714.5 (3) [8] 714.8 (3) [41]	493.1 (3) [8]	484.9 (3) [8]	136.5 (5) [19]	[27]	[27]	24.76 (9) [26] 24.8 (3) [8]	23.68 (9) [26] 23.8 (3) [8]
									s.o. = 1.08 (3) [26]	

	$3s$	$3p_{1/2}$	$3p_{3/2}$	$3d_{3/2}$	$3d_{5/2}$	$4s$	$4p_{1/2}$	$4p_{3/2}$	$4d_{3/2}$	$4d_{5/2}$	$4p, 4s$		
^{51}Sb	944 (1) [19]	812.6 (7) [19]	766.3 (7) [19]	537.7 (5) [20]	528.2 (5) [20]	152.7 (7) [19]	[27]	[27]	33.44 (9) [26]	32.14 (9) [26]	VE		
									s.o. = 1.25 (4) [26]				
^{52}Te	1006 (1) [19]	869.5 (7) [19]	818.5 (7) [19]	582.2 (5) [20]	572.5 (5) [20]	169.8 [43]	[27]	[27]	41.80 (9) [26]	40.31 (9) [26]	VE		
									s.o. = 1.51 (1) [26]				
^{53}I	(1072 [22])	(931 [22])	(875 [22])	(631 [22])	(620 [22])	(186 [22])					VE		
							(123 [22])		(50 [22])				
	$3s$	$3p_{1/2}$	$3p_{3/2}$	$3d_{3/2}$	$3d_{5/2}$	$4s$	$4p_{1/2}$	$4p_{3/2}$	$4d_{3/2}$	$4d_{5/2}$	$5s$	$5p_{1/2}$	$5p_{3/2}$
^{54}Xe	1148.7 (5) [4]	1002.1 (3) [4]	940.6 (2) [4]	689.0 (2) [4]	676.4 (1) [4]	213.2 (2) [4]	[27]	145.5 (2) [4]	69.5 (1) [4]	67.5 (1) [4]	23.3 (1) [4] 23.39 [23]	13.4 (1) [4] 13.43 [23]	12.13 [4] 12.13 [23]

	$3s$	$3p_{1/2}$	$3p_{3/2}$	$3d_{3/2}$	$3d_{5/2}$	$4s$	$4p_{1/2}$	$4p_{3/2}$	$4d_{3/2}$	$4d_{5/2}$	$5s$	$5p_{1/2}$	$5p_{3/2}$	
^{55}Cs	1211 (1) [19]	1071 (1) [24]	1003 (1) [24]	740.5 (2) [24]	726.6 (2) [24]	232.3 (10) [24]	172.4 (5) [24]	161.3 (5) [24]	79.8 (2) [24]	77.5 (2) [24]		14.1 (2) [24] 14.2 (2) [25]	12.1 (2) [24] 12.3 (2) [25] 11.8 (4) [28]	
^{56}Ba	1293 (1) [19]	1137 (1) [19]	1063 (1) [19]	795.7 (3) [8]	780.5 (3) [8]	252.6 (3) [8]	[27]	178.7 (3) [8]	92.6 (3) [8]	90.0 (3) [8]	30.1 (3) [8]	16.8 (3) [8]	14.6 (3) [8]	
^{57}La	1362 (1) [19]	1209 (1) [19]	1128 (1) [19]	852.9 (3) [8] 853.2 (3) [20]	836.1 (3) [8] 836.0 (3) [20]	274.7 (3) [8]	[27]	196.0 (3) [8]	105.3 (3) [8]	102.5 (3) [8]	34.3 (3) [8]	19.3 (3) [8]	16.8 (3) [8]	
^{58}Ce	1436 (1) [19]	1274 (1) [19]	1187 (1) [19]	902.7 (3) [5, 8] 902.1 (5) [5, 20]	884.2 (3) [5, 8] 883.5 (5) [5, 20]	291.0 (3) [5, 8] $\Delta E = 1.4$ (3) [36]	[27]	206.5 (3) [8]	109.0 (3) [8, 34] 109.0 (3) [34, 36]		[5] $\Delta E = 1.0$ (3) [36]	19.8 (3) [8]	17.0 (3) [8]	
	$3p_{1/2}$	$3p_{3/2}$	$3d_{3/2}$	$3d_{5/2}$	$4s$	$4p_{1/2}$	$4p_{3/2}$	$4d_{3/2}$	$4d_{5/2}$	$5s$	$5p_{1/2}$	$5p_{3/2}$	$4f_{5/2}$	$4f_{7/2}$
^{59}Pr			[34] 948.3 (5) [20]	[34] 928.8 (5) [20]	[5] $\Delta E = 2.0$ (3) [36]			[34] 115.1 (5) [20]		[5] $\Delta E = 1.4$ (2) [36]			[34]	[34]
^{60}Nd			[34] 1003.3 (8) [20]	[34] 980.4 (5) [20]	319.2 (8) [5, 20] $\Delta E = 2.7$ (3) [36]			[34] 120.5 (5) [20]		[5] $\Delta E = 1.6$ (2) [36]			[34]	[34]
^{61}Pm													[34]	[34]
^{62}Sm			[34] 1110.9 (8) [20]	[34] 1083.4 (8) [20]	347.2 (8) [5, 20] $\Delta E = 5.4$ (3) [36]			[34] 129.0 (5) [35]		[5] $\Delta E = 2.9$ (2) [36]			[34] 5.2 (2) [37]	

	$3d_{3/2}$	$3d_{5/2}$	$4s$	$4p_{1/2}$	$4p_{3/2}$	$4d_{3/2}$	$4d_{5/2}$	$5s$	$5p_{1/2}$	$5p_{3/2}$	$4f_{5/2}$	$4f_{7/2}$
^{63}Eu	[34] 1158.6 (8) [20]	[34] 1127.5 (8) [20]	[5] $\Delta E = 7.4$ (3) [36]			[34]	[34] $^9D_6 = 127.7$ (2) [35]	[5] $\Delta E = 3.8$ (3) [20]			[34]	[34]
^{64}Gd	[34] 1221.9 (8) [20]	[34] 1189.6 (8) [20]	378.6 (8) [5, 20] $\Delta E = 7.8$ (2) [36]			[34]	[34] $^9D = 142.6$ (4) [35]	43.5 (6) [5, 20] $\Delta E = 3.6$ (2) [20]			[34]	[34]
											8.6 (1) [38]	
^{65}Tb	[34] 1275.0 (2) [33] 1278.8 (8) [20]	[34] 1239.1 (2) [33] 1243.2 (8) [20]	396.0 (2) [5, 33] $\Delta E = 6.7$ (2) [33]	322.4 (2) [5, 33]	284.1 (2) [5, 33]	[34]	[34]	45.6 (2) [5, 33] $\Delta E = 3.2$ (3) [36]; 3.0 (2) [33]	28.7 (2) [5, 33]	22.6 (2) [5.33]	[34] 7.8 (1) [38] 7.5 (2) [37]	[34] 2.6 (1) [38] 2.2 (2) [37]
						150.5 (5) [35]						
^{66}Dy	[34]	[34] 1292.6 (8) [20]	414.2 (2) [5, 33] $\Delta E = 5.8$ (2) [33]	333.5 (2) [5, 33]	293.2 (4) [5, 33]	[34]	[34]	49.9 (2) [5, 33] $\Delta E = 2.8$ (2) [36]; 2.4 (2) [33]	29.5 (2) [5, 33]	23.1 (2) [5, 33]	[34]	[34]
						153.6 (8) [20]					^{5}L = 7.7 (2) [37]; 8.2 (1) [38] ^{7}F = 4.0 (2) [37]; 4.5 (1) [38]	
^{67}Ho			432.4 (2) [5, 33] $\Delta E = 4.4$ (2) [33]		308.2 (4) [5, 33]	[34]	[34]	49.3 (2) [5, 33] $\Delta E = 2.4$ (2) [36]; 2.0 (2) [33]	30.8 (2) [5, 33]	24.1 (2) [5, 33]	[34] ^{4}K = 8.6 (2) [37]	[34] ^{6}H$_{11/2}$ = 5.2 (2) [37]
						160 (2) [20]						
^{68}Er			449.8 (2) [5, 33] $\Delta E = 3.2$ (2) [33]		320.2 (2) [5, 33]	[34]	[34]	50.6 (2) [5, 33] $\Delta E = 1.4$ (2) [33]	31.4 (2) [5, 33]	24.7 (2) [5, 33]	[34]	[34]
						167.6 (5) [20]						^{5}I$_8$ = 4.7 (2) [37]
^{69}Tm			470.9 (4) [33]		332.6 (2) [5, 33]	[34]	[34]	54.7 (4) [5, 33]	31.8 (2) [5, 33]	25.0 (2) [5, 33]	[34]	[34] ^{4}I$_{15/2}$ = 4.6 (2) [37]
						175.5 (5) [20]						
^{70}Yb			480.9 (2) [33] 480.0 (3) [8]	388.9 (4) [33] 388.4 (3) [8]	339.9 (2) [33] 339.5 (3) [8]	191.7 (2) [33] 190.8 (3) [8]	182.7 (2) [33] 182.0 (3) [8]	52.7 (4) [33] 51.3 (3) [8]	30.5 (2) [33] 30.0 (3) [8]	24.5 (2) [33] 23.7 (3) [8]	2.5 (2) [33] 2.5 (3) [8] 2.5 (2) [37]	1.3 (2) [33] 1.3 (3) [8] 1.2 (2) [37]

	$4s$	$4p_{1/2}$	$4p_{3/2}$	$4d_{3/2}$	$4d_{5/2}$	$5s$	$5p_{1/2}$	$5p_{3/2}$	$4f_{5/2}$	$4f_{7/2}$	$5d_{3/2}$	$5d_{5/2}$
^{71}Lu	506.8 (4) [33]	412.4 (4) [33]	359.2 (4) [33]	206.2 (2) [33] 206.0 (3) [8]	196.3 (2) [33] 196.2 (3) [8]	57.3 (2) [33]	33.5 (2) [33] 33.7 (3) [8]	26.6 (2) [33] 26.8 (3) [8]	8.9 (2) [33] 8.8 (3) [8]	7.6 (2) [33] 7.3 (3) [8]	VE	VE
^{72}Hf	(538 [22])	(437 [22])	(380 [22])	(224 [22])	(214 [22])	(65 [22])	(38 [22])	(31 [22])	(19 [22])	(18 [22])	VE	VE
^{73}Ta	563 (1) [19]	462.3 (8) [19]	402.0 (8) [19]	238.6 (5) [20]	227.1 (5) [20]	68.6 (8) [19]	42.4 (8) [19] 42.1 (5) [44]	33.9 (8) [19] 31.4 (5) [44]	24.8 (2) [20]	23.0 (1) [20]	VE	VE
^{74}W	592 (1) [19]	489 (1) [19]	422 (1) [19]	255.2 (5) [20]	242.9 (5) [20]	74 (1) [19]	44 (1) [19] 46.5 (5) [44]	33 (1) [19] 34.7 (5) [44]	33.6 (2) [20]	31.5 (2) [20]	VE	VE
^{75}Re	625.0 (8) [19]	517.9 (8) [19]	444.4 (8) [19]	273.7 (5) [20]	260.2 (5) [20]	82.8 (8) [19]	45.6 (7) [19] 50.4 (5) [44]	34.6 (6) [19]	42.9 (2) [20]	40.5 (2) [20]	VE	VE
^{76}Os	(655 [22])	(547 [22])	(469 [22])	(290 [22])	(273 [22])	(84 [22])	(58 [22])	(46 [22])	(52 [22])	(50 [22])	VE	VE
^{77}Ir	690.6 (8) [19]	577.6 (8) [19]	494.8 (6) [19]	311.5 (5) [20]	295.8 (5) [20]	95.2 (8) [19]	63.0 (8) [19]	50.5 (8) [19]	63.6 (2) [20]	60.7 (2) [20]	VE	VE
^{78}Pt	725 (1) [19]	608.4 (8) [19]	519.9 (8) [19]	331.6 (5) [20]	314.8 (5) [20] 314.5 (1) [39]	101.7 (8) [19]	65.3 (8) [19]	51.0 (8) [19]	74.5 (1) [20] 74.4 (3) [8]	71.2 (1) [20] 71.1 (3) [8] 71.1 (1) [39]	VE	VE
^{79}Au	759 (1) [19]	643.5 (3) [8]	546.5 (3) [8]	353.0 (8) [19]	335.1 (1) [39]	107.2 (8) [19]	71.2 (8) [19]	58 (1) [19]	87.6 (1) [20] 87.7 (3) [8] 87.74 (2) [40]	84.0 (1) [20] 84.0 (1) [39] 84.0 (3) [8] 84.07 (2) [40]	VE	VE
^{80}Hg	800 (2) [19]	677 (2) [19]	571 (2) [19]	378 (2) [19]	360 (2) [19]	120 (2) [19]	81 (2) [19]	65 (1) [19]	103.9 (3) [20]	99.8 (3) [20]	9.5 (1) [30]	7.7 (1) [30]

	$4s$	$4p_{1/2}$	$4p_{3/2}$	$4d_{3/2}$	$4d_{5/2}$	$4f_{5/2}$	$4f_{7/2}$	$5s$	$5p_{1/2}$	$5p_{3/2}$	$5d_{3/2}$	$5d_{5/2}$	$6s$		
^{81}Tl	846 (2) [19]	721 (2) [19]	609 (2) [19]	407 (2) [19]	386 (2) [19]	123 (1) [19]	119 (1) [19]	136 (1) [19]	100 (1) [19]	73 (1) [19]	14.53 (7) [31] 15.64 (10) [16]	12.30 (7) [31] 13.43 (10) [16]	VE		
											s.o. = 2.21 (7) [16]				
^{82}Pb	893 (1) [19]	763 (1) [19]	644 (1) [19]	434.2 (5) [20]	412.0 (5) [20]	141.2 (3) [20]	136.4 (3) [20]	147 (1) [19]	104 (1) [19]	82 (1) [19]	20.32 (7) [31] 20.18 (10) [16]	17.70 (7) [31] 17.52 (10) [16]	VE		
											s.o. = 2.66 (9) [16]				
^{83}Bi	939 (1) [19]	805.8 (8) [19]	679.4 (8) [19]	464.2 (5) [20]	440.5 (5) [20]	162.2 (3) [20]	159.3 (3) [20]	159.3 (7) [19]	116.8 (7) [19]	93.1 (7) [19]	26.94 (7) [31] 26.9 [32] 27.4 (1) [16]	23.90 (7) [31] 23.8 [32] 24.3 (1) [16]	VE		
											s.o. = 3.10 (12) [16]				
^{84}Po	(995 [22])	(851 [22])	(705 [22])	(500 [22])	(473 [22])	(184 [22])		(177 [22])	(132 [22])	(104 [22])	(31 [22])		VE		
^{85}At	(1042 [22])	(886 [22])	(740 [22])	(533 [22])	(507 [22])	(210 [22])		(195 [22])	(148 [22])	(115 [22])	(40 [22])		VE		
^{86}Rn	(1097 [22])	(929 [22])	(768 [22])	(567 [22])	(541 [22])	(238 [22])		(214 [22])	(164 [22])	(127 [22])	(48 [22])		(26 [22])	(11 [22])	
^{87}Fr	(1153 [22])	(980 [22])	(810 [22])	(603)22])	(577 [22])	(268 [22])		(234 [22])	(182 [22])	(140 [22])	(58 [22])		(34 [22])	(15 [22])	
^{88}Ra	(1208 [22])	(1958 [22])	(879 [22])	(636 [22])	(603 [22])	(299 [22])		(254 [22])	(200 [22])	(153 [22])	(68 [22])		(44 [22])	(19 [22])	
^{89}Ac	(1269 [22])	(1080 [22])	(890 [22])	(675 [22])	(639 [22])	(319 [22])		(272 [22])	(215 [22])	(167 [22])	(80 [22])				
^{90}Th	(1330 [22])	(1168 [22])	(968 [22])	(714 [22])	(677 [22])	(344 [22])	(335 [22])	(290 [22])	(229 [22]	(182 [22])	(95 [22])	(88 [22])	(60 [22])	(45 [22])	
	$4s$	$4p_{1/2}$	$4p_{3/2}$	$4d_{3/2}$	$4d_{5/2}$	$4f_{5/2}$	$4f_{7/2}$	$5s$	$5p_{1/2}$	$5p_{3/2}$	$5d_{3/2}$	$5d_{5/2}$	$6s$	$6p_{1/2}$	$6p_{3/2}$
^{91}Pa	(1387 [22])	(1224 [22])	(1007 [22])	(743 [22])	(708 [22])	(371 [22])	(360 [22])	(310 [22])	(232 [22])		(94 [22])				
^{92}U	1439 (1) [19]	1271 (1) [19]	1042 (1) [19]	778.5 (3) [8]	736.5 (3) [8]	388.2 (3) [8]	377.3 (3) [8]	321 (1) [19]	257 (1) [19]	192 (1) [19]	102.7 (3) [8]	93.7 (3) [8]	45.0 (3) [8]	26.0 (3) [8]	17.0 (3) [8]

References

1 This is the vertical ionization potential. The adiabatic value is 15.45 eV. See D. H. Turner: *Molecular Photoelectron Spectroscopy* (Wiley-Interscience, New York 1970)

2 D.A.Shirley, R.L.Martin, S.P.Kowalczyk, F.R.McFeely, L.Ley: Phys. Rev. B **15**, 544 (1977)

3 G.Johansson, J.Hedman, A.Berndtsson, M.Klasson, R.Nilsson: J. Electr. Spectr. **2**, 295 (1973)

4 K.Siegbahn, C.Nordling, G.Johansson, J.Hedman, P.F.Hedén, K.Hamrin, U.Gelius, T. Bergmark, L.O.Werme, R.Manne, Y.Baer: *ESCA Applied to Free Molecules* (North-Holland, Amsterdam 1971)

5 This line shows multiplet splitting ΔE; the energy given is that of the most intense component

6 T.X.Carroll, R.W.Shaw, Jr., T.D.Thomas, C.Kindle, N.Bartlett: J. Amer. Chem. Soc. **96**, 1989 (1974)

7 W.Lotz: J. Opt. Soc. Am. **57**, 873 (1967); **58**, 236 (1968); **58**, 915 (1968); from optical data

8 S.Hüfner: Private communication

9 J.C.Fuggle, E.Källne, L.M.Watson, D.J.Fabian: Phys. Rev. B **16** ,750 (1977)

10 P.H.Citrin, G.K.Wertheim, Y.Baer: Phys. Rev. B **16**, 4256 (1977)

11 R.S.Bauer, R.Z.Bachrach, J.C.McMenamin, D.E.Aspnes: Nuovo Cimento **39** B, 409 (1977)

12 F.C.Brown, Om P.Rustgi: Phys. Rev. Lett. **28**, 497 (1972)

13 S.A.Flodstrom, R.Z.Bachrach, R.S.Bauer, S.B.M.Hagström: Phys. Rev. Lett. **37**, 1282 (1976)

14 The Co binding energies quoted by Shirley et al. [2] appear to be consistently too high by ~ 2 eV. The Co2$p_{3/2}$ binding energy deviates by $\sim +1.5$ eV from the trend observed for the series Ti through Ni. [Compare Y.Fukuda, W.T.Elam, R.L.Park: Phys. Rev. B **16**, 3322 (1977)]

15 L.Ley, S.P.Kowalczyk, F.R.McFeely, R.A.Pollak, D.A.Shirley: Phys. Rev. B **8**, 2392 (1973)

16 R.T.Poole, P.C.Kemeny, J.Liesegang, J.G.Jenkin, R.C.G.Leckey: J. Phys. F: Metal Phys. **3**, L 46 (1973)

17 G.K.Wertheim, M.Campagna, S.Hüfner: Phys. Cond. Matter **18**, 133 (1974)

18 The splitting is larger than the free atom spin-orbit splitting due to crystal field effects; see L.Ley, R.A.Pollak, F.R.McFeely, S.P.Kowalczyk, D.A.Shirley: Phys. Rev. B **9**, 600 (1974)

19 Obtained by combining the photoemission binding energies with energy differences from J.A.Bearden, A.F.Burr: Rev. Mod. Phys. **39**, 125 (1967)

20 S.P.Kowalczyk, Ph.D.Thesis, University of California, Berkeley (1976) unpublished

20a W. Eberhardt, G. Kalkofen, C. Kunz, D. Aspnes, M. Cardona: Phys. Stat. Sol. (b) **88**, 135 (1978)

21 N.J.Shevchik, M.Cardona, J.Tejeda: Phys. Rev. B **8**, 2833 (1973)

22 K.Siegbahn, C.Nordling, A.Fahlman, R.Nordberg, K.Hamrin, J.Hedman, G.Johansson, T.Bergmark, S.E.Karlsson, I.Lindgren, B.Lindberg: Nova Acta Regiae Soc. Sci. Ups. Ser. **IV**, Vol. 20 (Uppsala 1967)

23 C.E.Moore: *Atomic Energy Levels*, Washington, Nat. Bureau of Standards, Circ. 467 (1949, 1952, 1958)

24 G.Ebbinghaus: Ph. D. Thesis, Stuttgart (1977) unpublished

25 R.G.Oswald, T.A.Callcott: Phys. Rev. B **4**, 4122 (1971).

26 R.A.Pollak, S.P.Kowalczyk, L.Ley, D.A.Shirley: Phys. Rev. Lett. **29**, 274 (1972)

27 Broadened beyond recognition due to multielectron effects. See for example, U.Gelius: J. Electr. Spectr. **5**, 985 (1967)
S.P.Kowalczyk, L.Ley, R.L.Martin, F.R.McFeely, D.A.Shirley: Farad. Disc. Chem. Soc. **60**, 7 (1975)

28 H.Petersen: Phys. Stat. Sol. (b) **72**, 591 (1975)

29 F.R.McFeely, S.P.Kowalczyk, L.Ley, R.G.Cavell, R.A.Pollak, D.A.Shirley: Phys. Rev. B **9**, 5268 (1974)

30 S.P.Kowalczyk, L.Ley, R.A.Pollak, D.A.Shirley: Phys. Lett. **41** A, 455 (1972)

31 L.Ley, R.A.Pollak, S.P.Kowalczyk, D.A.Shirley: Phys. Lett. **41** A, 429 (1972)

32 Z.Hurych, R.L.Benbow: Phys. Rev. Lett. **38**, 1094 (1977)

33 B.D.Padalia, W.C.Lang, P.R.Norris, L.W.Watson, D.J.Fabian: Proc. Roy. Soc. London A **354**, 269 (1977)

34 Complex multiplet structure; for details see [33] and also
Y. Baer, G. Busch: Phys. Rev. Lett. **31**, 35 (1973)
Y. Baer, G. Busch: J. Electr. Spectr. **5**, 611 (1974)
S. P. Kowalczyk, N. Edelstein, F. R. McFeely, L. Ley, D. A. Shirley: Chem. Phys. Lett. **29**, 491 (1974)
F. R. McFeely, S. P. Kowalczyk, L. Ley, D. A. Shirley: Phys. Lett. **45** A, 227 (1973)
M. Campagna, G.K. Wertheim, Y. Baer: "Unfilled Inner Shells: Rare Earths and Their Compounds", Chap. 4 of this volume
If a binding energy is given, it is that of the most intense peak or a member of the multiplet that is identified
35 S. P. Kowalczyk, N. Edelstein, F. R. McFeely, L. Ley, D. A. Shirley: Chem. Phys. Lett. **29**, 491 (1974)
36 F. R. McFeely, S. P. Kowalczyk, L. Ley, D. A. Shirley: Phys. Lett. **49** A, 301 (1974)
37 Y. Baer, G. Busch: J. Electr. Spectr. **5**, 611 (1974)
38 F. R. McFeely, S. P. Kowalczyk, L. Ley, D. A. Shirley: Phys. Lett. **45** A, 227 (1973)
39 G. Schön: J. Electr. Spectr. **1**, 377 (1972/73)
40 K. Asami: J. Electr. Spectr. **9**, 469 (1976)
41 S. Hüfner, G. K. Wertheim, J. H. Wernick: Sol. State Commun. **17**, 417 (1975)
42 Obtained for a solid film of S_8; from W. R. Salaneck, N. O. Lipari, A. Paton, R. Zallen, K.S. Liang: Phys. Rev. **12** B, 1493 (1975); the binding energies have been corrected for a Au$4f_{7/2}$ energy of 84.0 below E_F
43 S. Svensson, N. Martensson, E. Basilier, P. A. Malmquist, U. Gelius, K. Siegbahn: Physica Scripta **14**, 141 (1976)
44 From characteristic electron energy loss measurements; B. M. Hartley: Phys. Stat. Sol. **31**, 259 (1969)
45 J. Azoulay: Private communication
46 M. Cardona, J. Tejeda, N. J. Shevchik, D. W. Langer: Phys. Stat. Sol. (b) **58**, 483 (1973)
47 W. D. Grobman, D. E. Eastman, J. L. Freeouf: Phys. Rev. B **12**, 4405 (1975)
48 B. von Roedern: Private communication

Additional References with Titles

Chapter 2

A.A. Akhayan, A.N. Brozdnichenko, E.V. Bursion: Influence of the spontaneous polarization on the photoelectric emission from $LiNbO_3$. Sov. Phys.–Solid State **20**, 912 (1978)

P.M.Th.M. van Atteleum, J.M. Trooster: Bulk and surface plasmon loss intensities in photoelectron Auger and electron energy loss spectra of Al metal. Phys. Rev. B**18**, 3872 (1978)

J. Baars, D. Basselt, M. Schultz: Metal-semiconductor barrier studies of PbTe. Phys. Status Solidi a**46**, 489 (1978)

R.Z. Bachrach, A. Bianconi: Interface States at the Ga-GaAs interface. J. Vac. Sci. Technol. **15**, 525 (1978)

R.L. Benbow, Z. Hurych: Angle resolved photoemission from layered Bi_2Te_3: theory and experiment. Solid State Commun. **28**, 641 (1978)

K. Berndt, U. Marx, O. Brümmer: Electronic structure of FeNi alloys by means of photoelectron spectroscopy. Phys. Status Solidi b**90**, 487 (1978)

D.W. Bullett: Electronic band structure and bonding in transition metal layered dichalcogenides. J. Phys. C**22**, 4501 (1978)

D.J. Chadi: (110) Surface states of GaAs: sensitivity of electronic structure to surface structure. Phys. Rev. B**18**, 1800 (1978)

P.W. Chye, I. Lindau, P. Pianetta, W.E. Spicer, C.M. Garner: Evidence for a new type of metal-semiconductor interaction on GaSb. Phys. Rev. B**17**, 2682 (1978)

M. Cini: Theory of the Auger effect in solids: plasmon effects in electron spectroscopies of valence states. Phys. Rev. B**17**, 2486 (1978)

O.B. Dabbousi, P.S. Wehner, D.A. Shirley: Temperature independence of the angle-resolved X-ray photoemission spectra of Au, Pt, valence bands. Solid State Commun. **28**, 227 (1978)

E. Dietz, D.E. Eastman: Symmetry method for the absolute determination of energy band dispersions $E(\boldsymbol{k})$. Phys. Rev. Lett. **41**, 1674 (1978)

S. Evans, R.G. Pritchard, J.M. Thomas: Relative differential subshell photoionisation cross-sections ($\mathrm{Mg}\,K_\alpha$) from lithium to uranium. J. Electr. Spectrosc. **14**, 341 (1978)

C.M. Garner, W.E. Spicer: New phenomena in adsorption of O_2 on Si. Phys. Rev. Lett. **40**, 403, (1978)

S.M. Goldberg, C.S. Fadley, S. Kono: Photoelectric cross sections for fixed orientation atomic orbitals: relationship to the plane-wave final state approximation and angle resolved photoemission. Solid State Commun. **28**, 459 (1978)

T. Grandke, L. Ley, M. Cardona: Angle-resolved uv photoemission and electronic band structure of the lead chalcogenides. Phys. Rev. B**18**, 3847 (1978)

H.W. Haak, G.A. Sawatzky, T.D. Thomas: Auger photoelectron coincidence measurements in Cu. Phys. Rev. Lett. **41**, 1825 (1978)

G.V. Hansson, S.A. Flodström: Photoemission from surface states and surface resonances on the [100], [110], and [111] crystal faces of aluminium. Phys. Rev. B**18**, 1562 (1978)

G.V. Hansson, S.A. Flodström: Photoemission study of the bulk and surface electronic structure of single crystals of gold. Phys. Rev. B**18**, 1572 (1978)

F.J. Himpsel, W. Steinmann: Angle resolved photoemission from NaCl (100) face. Phys. Rev. B**17**, 2537 (1978)

H. Ihara, H. Abe, S. Endo, I. Irie: Valence band densities of states of $CdIn_2S_4$ and In_2S_3 from X-ray photoelectron spectroscopy. Solid State Commun. **28**, 563 (1978)

5th Annual Conference on the Physics of Compound Semiconductor Interfaces. J. Vacuum Sci. Technol. **15** (1978)

N.D. Land, A.R. Williams: Core holes in chemisorbed atoms. Phys. Rev. **16**, 2408 (1977)

A.I. Larkin, V.I. Molnikov: Energy distribution of X-ray photoelectrons. Sov. Phys.–Solid State **20**, 74 (1978)

M.G. Mason, L.J. Gerenser, S.T. Lee: Electronic structure of catalytic mutal clusters studied by X-ray photoemission spectroscopy. Phys. Rev. Lett. **39**, 288 (1977)

L. Mihich: Self-consistent calculation of energy band structure of $(SN)_x$ by the intersecting spheres model. Solid State Commun. **28**, 521 (1978)

Y. Mizokawa, H. Iwasaki, R. Nishitani, S. Nakamura: ESCA studies of Ga, As, GaAs, Ga_2O_3, As_2O_3, and As_2O_5 (core and Auger spectra only). J. Electr. Spectrosc. **14**, 129 (1978)

I.D. Moore, J.B. Pendry: Theory of spin polarized photoemission from Ni. J. Phys. C**22**, 4615 (1978)

V.V. Nemoshkalenko, V.G. Aleshin, Yu. N. Kucherenko: Theoretical investigation of the structure of X-ray photoelectron spectra of crystals (diamond, Si, GaAs). J. Electr. Spectrosc. **13**, 361 (1978)

J.A. Nicholson, J.D. Riley, R.C.G. Leckey, J.G. Jenkin, J. Liesegang, J. Azoulay: Ultraviolet photoelectron spectroscopy of the valence bands of some Au alloys. Phys. Rev. B**18**, 2561 (1978)

A.W. Parke, A. McKinley, R.H. Williams: Surface states and the surface structure of cleaved silicon. J. Phys. C**24**, L993 (1978)

P. Pianelta, I. Lindøn, C.M. Garner, W.E. Spicer: Chemisorption and oxidation studies of the (110) surface of GaAs, GaSb, InP. Phys. Rev. B**18**, 2792 (1978)

A. Platau, S.E. Karlsson: Valence band of γ-Ce studied by UPS and XPS. Phys. Rev. **18**, 3820 (1978)

E.W. Plumer, W.R. Salaneck, J.S. Miller: Photoelectron spectra of transition metal carbonyl complexes. Comparison with the spectra of adsorbed CO. Phys. Rev. B**18**, 1673 (1978)

W. Pong, C.S. Inouye, S.K. Okada: Ultraviolet photoemission studies of BaF_2 and $BaCl_2$. Phys. Rev. B**18**, 4422 (1978)

M. Sagurton, D. Liebowitz, N.J. Shevchik: Oxidation-induced breakdown of the conservation of perpendicular momentum in angle resolved photoemission spectra of Cu(111). Phys. Rev. Lett. **42**, 274 (1979)

W.R. Salaneck, H.R. Thomas: Energy-Gain Satellite in the C(1 s) X-ray photoemission spectra of organic macromolecules. Solid State Commun. **27**, 685 (1978)

B. Schröder, W. Grobman, W.L. Johnson, C.C. Tsuei, P. Chaudari: A comparative study of amorphous and crystalline superconducting molybdenum films by ultraviolet photoelectron spectroscopy. Solid State Commun. **28**, 631 (1978)

N.J. Shevchik, D. Liebowitz: Theory of angle resolved photoemission from the bulk bands of solids I. Formalism. Phys. Rev. **16**, 1618 (1978)

N.J. Shevchik, D. Liebowitz: Application to Ag (111). Phys. Rev. **18**, 1630 (1978)

J. Szajman, J. Liesegang, R.C.G. Leckey, J.G. Jenkin: Photoelectron determination of mean free paths of 200–1500 electrons in KI. Phys. Rev. B**18**, 4010 (1978)

P. Thiry, Y. Petroff, R. Pinchaud, C. Guillot, Y. Ballu, J. Lecante, J. Paigne, F. Levy: Experimental band structure of GaSe obtained by angular resolved photoemission. Solid State Commun. **22**, 685 (1977)

S. Tougard: Surface relaxation of zincblende (110). Phys. Rev. B**18**, 3799 (1978)

S.L. Weng, E.W. Plummer, T. Gustafsson: Experimental and theoretical studies of the surface resonances on the (100) faces of W and Mo. Phys. Rev. B**18**, 1718 (1978)

F. Werfel, G. Bräger, B. Brümmer, H. Jurisch: XS (X-ray emission) and XPS investigation of V_3Si electron structure. Phys. Status Solidi b**80**, KSJ (1977)

A.R. Williams, N.D. Lang: Core level binding energy shifts-metals. Phys. Rev. Lett. **40**, 954 (1978)

K.C. Woo, F.C. Brown: Infrared study of superlattice formation in $Ti_{1+x}Se_2$. Solid State Commun. **28**, 341 (1978)

Chapter 5

D.T. Clark: "ESCA Applied to Organic and Polymeric Systems", in: *Handbook of X-ray and Ultraviolet Photoelectron Spectroscopy*, ed. by D. Briggs (Heyden, London 1977) Chap. 6, p. 211

J.A. Connor: "XPS Studies of Inorganic and Organometallic Compounds", in: *Handbook of X-ray and Ultraviolet Photoelectron Spectroscopy*, ed. by D. Briggs (Heyden, London 1977) Chap. 5, p. 183

D.T. Clark: Some experimental and theoretical aspects of structure, bonding and reactivity of organic and polymeric systems as revealed by ESCA. Phys. Scr. **16**, 307 (1977)

S. Hashimoto, S. Hino, K. Seki, H. Inokuchi: Anisotropic photoemission from oriented polyethylene. Chem. Phys. Lett. **40**, 279 (1976)

K. Seki, S. Hashimoto, N. Sato, Y. Harada, K. Ishii, H. Inokuchi, J. Kanbe: Vacuum-ultraviolet photoelectron spectroscopy of hexatriacontane (n-$C_{36}H_{74}$) polycrystal: a model compound of polyethylene. J. Chem. Phys. **66**, 3644 (1977)

J.J. Pireaux, J. Riga, R. Caudano, J.J. Verbist, J. Delhalle, S. Delhalle, J.M. André, Y. Gobillon: Polymer primary structures studied by ESCA and EHCO methods. Phys. Scr. **16**, 329 (1977)

M. Fujihira, H. Inokuchi: Photoemission from polyethylene. Chem. Phys. Lett. **17**, 554 (1972)

D. Betteridge, D.J. Joyner, F. Gruming, N.R. Shoko, M.E.A. Cudby, H.A. Willis, T.E. Attwood, L. Henriksen: The analysis of polymer degradation products by UV-photoelectron spectroscopy. Phys. Scr. **16**, 339 (1977)

J. Riga, J.J. Pireaux, J.J. Verbist: An ESCA study of the electronic structure of solid benzene. Valence levels, core level, and shake-up satellites. Mol. Phys. **34**, 131 (1977)

J. Riga, J.J. Pireaux, R. Caudano, J.J. Verbist: Y comparative ESCA study of the electronic structure of solid acenes: benzene, naphthalene, anthracene, and tetracene. Phys. Scr. **16**, 346 (1977)

F.E. Fischer, S.R. Kelemen, H.P. Bonzel: Adsorption of acetylene and benzene on the Pt(100) surface. Surf. Sci. **64**, 157 (1977)

J.E. Demuth: Initial-state shift of the carbon 2s levels of chemisorbed hydrocarbons on nickel. Phys. Rev. Lett. **40**, 409 (1978)

W.R. Salaneck: Intermolecular relaxation energies in anthracene. Phys. Rev. Lett. **40**, 60 (1978)

W.R. Salaneck, H.R. Thomas: Energy-gain satellite structure in the C (1s) X-ray photoemission spectra of organic macromolecules. Solid State Commun. **27**, 685 (1978)

S. Hino, K. Seki, H. Inokuchi: Photoelectron spectra of p-Terphenyl in gaseous and solid state. Chem. Phys. Lett. **36**, 335 (1975)

K. Seki, H. Inokuchi, N. Sato, K. Ishii: "VUV-Photoelectron Spectroscopy of Hexatriacontane (n-$C_{36}H_{74}$) in Solid and Gaseous Phases", in: 8th Mol. Cryst. Symp. Santa Barbara Calif., May 29–June 2, 1977

L. Nemec, H.J. Gaers, L. Chia, P. Delahay: Photoelectron spectroscopy of liquids up to 21.2 eV. J. Chem. Phys. **66**, 4450 (1977)

J. Knecht, H. Bässler: An ESCA-study of solid 2,4-hexadiyne-1,6-diol-bis-(toluenesulfonate) and its constituents before and after polymerization. Chem. Phys. **33**, 179 (1978)

J.J. Ritsko, P. Nielsen, J. S. Miller: Photoemission from ferrocene, decamethylferrocene, and decamethylferrocene bis-(7,7,8,8-tetracyano-p-quinodimethane). J. Chem. Phys. **67**, 687 (1977)

N.S. Hush, A.S. Cheung: Study of valence level splitting in a porphin type π-cation dimer by He I photoelectron spectroscopy. Chem. Phys. Lett. **47**, 1 (1977)

S. Muralidharan, R.G. Hayes: XPS studies of the valence electron levels of metallophorphyrins. Chem. Phys. Lett. **57**, 630 (1978)

H. Höchst, A. Goldmann, S. Hüfner, H. Malter: X-ray photoelectron valence band studies on phthalocyanine compounds. Phys. Status Solidi (b) **76**, 559 (1976)

H. Malter: On the electronic molecular structure of the organic semiconductor copper phthalocyanine. Phys. Status Solidi (b) **74**, 627 (1976)

F.L. Battye, A. Goldmann, L. Kasper: Ultraviolet photoelectron valence band studies on phthalocyanine compounds. Phys. Status Solidi (b) **80**, 425 (1977)

M. Iwan, W. Eberhardt, G. Kalkoffen, E.E. Koch, C. Kunz: Photoemission studies on phthalocyanine compounds: cross section dependence of outer core levels. Chem. Phys. Lett., in press (DESY preprint SR-78/04)

J. Berkowitz: Photoelectron spectroscopy of phthalocyanine vapors. J. Chem. Phys., in press (preprint Nov. 1978)

M. Iwan, E.E. Koch: 3*p*-core threshold effects in photoemission from quasiatomic Ni in nickel-phthalocyanine. (preprint Jan. 1979)

R.J. Dam, C.A. Burke, O.H. Griffith: Photoelectron quantum yields of the amino acids. Biophys. J. **14**, 467 (1974)

R.J. Dam, K.F. Kongslie, O.H. Griffith: Photoelectron quantum yields of hemin, hemoglobin, and apohemoglobin. Biophys. J. **14**, 933 (1974)

C.A. Burke, G.B. Birrelli, G.H. Lesch, O.H. Griffith: Depth resolution in photoelectron microscopy of organic surfaces. The photoelectric effect of phthalocyanine thin films. Photochem. Photobiol. **19**, 29 (1974)

R.J. Dam, K.F. Kongslie, O.H. Griffith: Photoelectron quantum yields and photoelectron microscopy of chlorophyll and chlorophyllin. Photochem. Photobiol. **22**, 265 (1975)

R.J. Dam, K.K. Nadakavukaren, O.H. Griffith: Photoelectron microscopy of cell surfaces. J. Microsc. **3**, 211 (1977)

D. Bloor, G.C. Stevens, P.J. Page, P.M. Williams: Photoelectron spectra of single crystal diacetylene polymers. Chem. Phys. Lett. **33**, 61 (1975)

G.C. Stevens, D. Bloor, P.M. Williams: Photoelectron valence band spectra of diacetylene polymers. Chem. Phys. **28**, 399 (1978)

C.B. Duke, A. Paton, W.R. Salaneck, H.R. Thomas, E.W. Plummer, A.J. Heeger, A.G. Macdiramid: Electronic structure of polyenes and polyacetylene. Chem. Phys. Lett. **59**, 146 (1978)

W.R. Salaneck, J.W. Lin, A.J. Epstein: X-ray photoemission spectroscopy of the core levels of polymeric sulfurnitride $(SN)_x$. Phys. Rev. B**13**, 5574 (1976)

H.J. Stolz: „Gitterdynamische und elektronische Eigenschaften von Polyschwefelnitrid $(SN)_x$". Dissertation, Universität Stuttgart 1977

J. Sharma, Z. Iqbal: X-ray photoelectron spectroscopy of brominated $(SN)_x$ and S_4N_4. Chem. Phys. Lett. **56**, 373 (1978)

K. Seki, Y. Kamura, J. Shirotani, H. Inokuchi: Absorption spectra and photoemission of amonium-TCNQ salt evaporated films. Chem. Phys. Lett. **35**, 513 (1975)

J.J. Ritsko, A.J. Epstein, W.R. Salaneck, D.J. Sandman: Surface electronic structure of tetrathiafulvalene-tetracyanoquinodimethane. Phys. Rev. B**17**, 1506 (1978)

P. Nielsen: Substrate dependent ionization and polarization energies of molecules: dibenz-tetrathiafulvalene. Solid State Commun. **26**, 835 (1978)

J.A. Riga, J.J. Verbist, F. Wudl, A. Kruger: The electronic structure and conductivity of tetrathiotetracene, tetrathionaphtalene, and tetraselenotetracene studied by ESCA. J. Chem. Phys. **69**, 3221 (1978)

Subject Index

Page numbers in *italics* refer to **Photoemission in Solids I: General Principles,** Topics in Applied Physics, Vol. 26, ed. by M. Cardona, L. Ley (Springer, Berlin, Heidelberg, New York 1978)

Absorption coefficient 41
– edge 14, 41
– index 41
Acenes 268
–, molecular orbitals 270
Adenine 280
Adsorbates, alkali metals *43*
–, synchrotron radiation experiments 341, 343, 344
Ag 194–201
–, core line asymmetry *225, 228*
–, 4*d* subshell-photoionization cross sections 315
–, photoionization cross section 68
–, UPS spectra 199, 209
–, valence band spectra (XPS) 196
–, work function 19
AgBr 21, 67–72
AgCl 21, 67–72
–, band structure 22
–, partial density of states 71
AgI 23, 67–72
–, band structure 22
–, partial density of states 69
–, valence band spectra (XPS and UPS) 69, 70
Ag-O-Cs *6*
AgPd alloys 210
– –, valence band spectra (XPS) 207
– –, virtual bound state parameters 208
AgPt alloys, virtual bound state parameters 208
Ag_2S 13
Al 9, 350, *149*
–, core level spectrum 359, 363
–, oxidized UPS spectra 343
–, photoabsorption coefficient *149*
–, plasmons 359–361, 363
–, SXPS spectra (synchrotron radiation) 320
–, valence band spectrum 369
–, work function *39*
–, yield spectrum and absorption spectra 323
Alkali halides 124, *74, 76, 178*
– metals 365, 366, *5*
Alloys, concentrated 210
–, dilute 206
–, minimum polarity model 206
– of transition metals 206
–, virtual bound state model (Friedel-Anderson) 206
AlN 23, 120
AlSb 21
–, amorphous 101
–, critical points 59
–, work function *49*
–, XPS spectrum 57
Amorphous III–V compounds 100–104
– group V semiconductors 104, 108
– group VI semiconductors 111–114
– semiconductors 41
Analysis, elemental concentration through core level intensities *80*
Angular asymmetry parameter (cross sections) *81*
Angular resolution *242*
Angular resolved photoemission (ARP, ARPES) 319, *237*
– – –, conduction band states 333–335
– – – in metals *258–262*
– – – in semiconductors *249, 254–259*
– – – of surface states 139
– – – orbital information *249*
– – –, valence bands of semiconductors 80–85
Anodes *52*
Anthracene 272–277
–, absorption spectrum 276
–, Frenkel exciton 272, 273
–, MO calculations 273
–, photoemission spectrum 269, 273, 274, 276
–, – –, angle resolved 274
Antifluorite structure 24
Ar, photoionization cross section 68
–, solid UPS spectra (synchrotron radiation) 332
Aromatic hydrocarbons 267
As, 104
–, amorphous 105–107

As, orthorhombic 107
–, photoabsorption cross section *154, 155*
–, Raman spectrum 105
–, valence band spectrum 106
As_2S_3 11, 31, 32, 86
As_4S_4, valence band spectrum (XPS) 110
As_2Se_3 86, 111
As_2Te_3 32, 86, 111
A7 structure 96, 104, 107
Asymmetry, core lines 352, 353, *15*
Au 194–200, 202
–, $5d$ and $4f$ subshell photoionization cross sections 315
–, angular resolved PES *251*
–, core lines *207*
–, photoionization cross section (photoabsorption) *45, 146, 147, 153, 154*
– standard *13*
AuAg alloy 210
AuAl 212
Au_2Al 212
$AuAl_2$ 212, *75*
–, yield spectrum 328
AuCu alloys 210
Auger decay *78–80*
– processes, interatomic 245–249, *80*
– spectroscopy *9, 15, 60*
– spectrum, Na 365
AuPd alloys, virtual bound state parameter 208
$Au_{0.1}Pt_{0.9}$ *75*
AuSn *75*
$AuSn_4$ *75*

Back-bound, Si 140, 142
–, Si(111):H 153
Background in photoemission spectra (inelastic tail) 193, 354
Band bending 128, 133, 156, *24*
– gap spectroscopy 319
– structure calculations 15–36
– – –, augmented plane waves (APW) 35, *44*
– – –, bond, orbital model (BOM) 18, 22
– – –, empirical pseudopotential method (EPM) 16, 19, 22, 25, 26, 29, 30
– – –, empirical tight-binding model (ETBM) 17
– – –, orthogonalized plane waves (OPW) 32
– –, complex 90, 98
– – of semiconductors 15
– – regime in photoemission, Ge 51
– –, two-dimensional 32–39, *255, 256*
– tailing 115
– –, photoemission spectrum a-Si 116
– width, $3d$ electrons 191

Ba, photoabsorption cross section *157–159*
–, photoionization *187–189*
Be 350, 358
–, core level spectrum 357
–, density of states 366
–, plasmon 357, 360
–, valence band spectrum 366
Benzene, UPS spectrum 269, 271
Bethe lattice 94, 95
Beyond the one-electron picture *165*
Bi 104, 105
–, amorphous 105
–, photoabsorption cross section *147, 148, 153, 154*
–, Raman spectrum 105
–, spin-orbit splitting 105
–, valence band spectrum (XPS) 106
BiI_3 77
Binary alloys, stability *51*
Binding energies, $4f$ and $4d$ electrons in rare earths 253, 254
– –, core levels 373, *60–70, 265*
– – in ionic solids *73*
– – in semiconductors 126–129
– –, $5s$ and $5p$ electrons in rare earths 236
Bi_2Se_3 32
Bi_2Te_3 32
Black phosphorus structure 29, 107
Bond orbital model (BOM) 17, 18, 93
Bonding charge 118, 130
Born-Oppenheimer approximation *177*
Brillouin zone, fcc lattice 83
Bulk incoming wave state *112*
– outgoing wave components *111, 121, 123*
Butane, UPS spectrum 267

CaB_6 245
Calibration, energy *13*
Catalysis, heterogeneous 153
Cd, core line asymmetry *228*
Cd_3As_2 24
CdI_2 33
$CdIn_2S_4$ 26
CdS 23
–, band structure 23
CdSe 23
$CdSnAs_2$, valence band spectrum (XPS) 60
Ce 235, 240, 252
–, $4f$ orbitals 235
–, α-phase 237
–, γ-phase 237
–, $\gamma \rightarrow \alpha$ transition 235
–, halides 238
–, photoionization cross sections 68, *157*
–, XPS spectrum 230, 237

CeAs, XPS spectra 241, 251, 252
CEL 43
see Electron energy losses
CeF_3 252
Central field approximation *136, 140*
CeSb, XPS spectra 241, 251–253
CF_4 *179*
CH_4 56, *179*
–, valence band spectrum (XPS) 267
Chalcopyrite compounds 24, 61, 70
Channeltron, channel plate *56*
Charge density waves (CDW) 36–38
– – –, amplitude 38
– – –, commensurate 37
– – –, effect on core levels 38
– – –, first-order phase transitions 38
– – –, incommensurate 37, 39
– – –, phase 38
– – –, Raman effect 38
– transfer 126
Charging, in organic compounds 262
–, in photoemission 262, *13, 17*
Chemical potential *33*
– shift *14, 60–75*
– – of core levels of rare gases, implanted in noble metals *70–73*
– shift in alloys *74, 75*
Chemisorption 151, 154, *57*
Cleaning by milling, filing, brushing *59*
Cleavage face, polar, nonpolar 148
Cleaving *58*
Clusters, finite *98*
Co 200, *179*
–, valence band spectrum (XPS) 201
Coherent potential approximation 210, 211
Cohesive energy *35, 36*
Compatibility relations, zincblende-diamond 20
Compounds, I–III–VI_2 24
–, II–IV–V_2 24
–, II_3–V_2 24
–, II–VI 19, 23
–, II–VI: valence band spectra (XPS) 57
–, III–V 19
–, III–V: valence band spectra (XPS) 57
–, III–VII 28
–, III_2–VI_3 24
–, IV–VI 62
–, IV–VI: valence band spectra 63
–, V_2VI_3 30, 31
Configuration interaction *14, 170, 182–186*
– – final state (FSCI) *182–186*
– – in the continuum (CSCI) *156, 182, 184, 187*
– –, initial state (ISCI) *182, 184, 189*
Conservation of $k_{||}$ 53, *121, 239, 254–257*
Constant final state spectroscopy (CFS) 300, 314, 316, 317, *240, 260, 262*
– initial state spectroscopy (CIS) 2, 79, 300, 314, 317, 318
Contact potential *4, 13, 22*
– – difference 150
Contamination 265, *57, 58*
Continuous random network 87, 99
CoO 183
–, UPS spectrum and partial *d*-, *p*-components 182
–, valence band spectrum (XPS) 188
Cooper minimum 314, 315, *145, 156*
Core excitons 337–339, *9*
– levels *60*
– –, cross sections *80*
– – lifetime *79, 80*
– –, line asymmetry 353, *201, 202, 205*
– – line shape 353, *197–229*
– – relaxation *141, 152*
– – shifts 121, 126, 127, 129, *60–75*
– – –, effect of molecular polarization 290, 291
– – – in charge transfer salts 288, 289
– – – in organic molecules 288–293
– – –, potential model 288, 289, *61, 64–70*
– –, singularity index 353, 354, *202, 204, 226*
– –, spectra of simple metals 357–364, *210–224*
– – width *76–80, 208–217*
– – –, vibrational contribution 335, *76*
– polarization *167*
Correlation 16, *156, 181–186*
– energy 176, 191, 224, *35, 36*
– –, Ce 235
– –, intraatomic 257
–, intershell 250
–, intrashell 250
Covalent gap 121
Critical points 52, 59, 65, 84, *8*
– –, interband 41
Cr_2O_3 180
Cr_2O_3, UPS spectra and partial *d*-, *p*-components 182, 189
Cr_2O_3, valence band spectrum (XPS) 181
Cross section, partial 68, 219, 271, 367
– –, photoabsorption (photoionization) *82, 83, 136–160*
– –, –, accurate calculations *149–159*
– –, –, C, Si, Ge 55, 56
Crystal field splitting 179
Cs coverage 87, *5, 17, 42, 43*
$CsPbBr_3$ 28
$CsPbCl_3$ 28

Cs_3Sb photocathode 12, *6–8*
Cu 194–201, 315, *8, 87–89*
–, angular resolved photoemission 199, *258*
–, core line *223*
–, density of states 175, 195
–, photoionization cross section 68
–, UPS spectrum 175, 195, 315, *87, 89*
–, valence band spectrum (XPS) 175, 196, 197
–, work function *38*
CuBr 21, 67–72
–, temperature effect on EDC 72
–, valence band spectrum (XPS) 70
CuCl 21, 67–72
–, valence band spectrum (XPS) 70
CuI 21, 67–72
–, valence band spectrum (XPS) 70
CuNi alloys 206, 210
$Cu_{0.62}Ni_{0.38}$, UPS spectrum 211
$Cu_{0.62}Ni_{0.38}$, theoretical density of states 211
CuO 177, 192
–, XPS spectrum 178
Cu_2O 177, 179, 192
–, XPS spectra 178, 185
CuPd alloys, virtual bound state parameters 208
Cyclotron resonance 14
Cytosine 280

Dangling bond 48, 114, 131, 140, 142, 145
– – on GaAs 148–151
– – on Si charge density 144
– – on Si density of states 146
Debye-Waller factor 81
Dedicated storage rings 309
Defect tetrahedral structures 24
Delayed absorption maximum *144, 146, 147*
– onset of transitions 314
Density of conduction states 42, 78
– – states 18, *88, 140*
– – –, cross section weighted 193, 221, 367, 368
– – –, joint 369, *86*
– – –, one-dimensional 83, 198
– – –, optical 41, 42
– – –, partial 46, 47, 50, 68–71, 73, 366–368
– – –, surface 137, 140–143, 145–147, 150, 194
DESY, experimental layout 312
– synchrotron, intensity compared with other sources 306
– –, intensity distribution and brightness 304
Detailed balance theorems *123, 125*
Diamond 11, *15*
–, valence band spectrum (XPS) 56, *15*
Dielectric constant 41
– –, longitudinal 44
Dipole acceleration *130, 139*
– approximation expression *137, 138*
Dipole layer (surface) *32, 33, 38*
– length expression *139, 141*
– matrix element *138, 142*
– velocity expression *139*
Direct transitions 53, 85, 87
Dispersion compensation 227, *12*
Doniach-Šunjić shape 232, 240, 246, 355, *200, 206, 232*
Doping 133
DORIS storage ring: intensity distribution and brightness 304
DOS *see* Density of states
Double quantum photoemission 276, 277
Dy, density of valence states 233
–, valence band spectrum (XPS) 228, 231, 233
DySb, valence band spectrum (XPS) 241

Eclipsed configuration 23, 26
EDJDOS *see* Energy distribution of joint density of states
Effective electromagnetic field *119, 127*
– independent particle sytem *110*
Effusion method (work function) *31*
Einstein's law *3, 135*
Electrochemical potential *16*
Electron affinity *17*
– –, for Si 133
Electron escape depth (mean free path) 354, 362, 367, *2, 3, 8, 55, 57, 81, 92, 122, 125, 192, 193, 247*
– – – in organic materials 264, 283
– – – for Si 49
– energy analyzers *9, 11, 55, 65, 241–244*
– –, losses (CEL) 12, 40, 43
– –, loss spectroscopy 132, 150
– mean free path, *see* Electron escape depth
– momentum parallel to surface 81, *239, 247*
– spectrometer, calibration *57*
– –, resolution 193, 227, 228, *56*
– storage rings 299
– synchrotrons 299
– transport term in photoemission 174, *85, 91*
Electronegativity 119, *48, 51*
see also inside cover
Electron-electron scattering *109*
Electron-hole excitations 193, 350, *201, 202, 204*
– interaction effects on core absorption 327
– pair production 53
Elemental analysis, composition determination by XPS *59, 60*

Energy band structure *see* Band structure
– distribution curves (EDC) 314, *2, 84–89*
– – of joint density of states (EDJDOS) 174, *88, 238*
– gap 11
– sum rule *175*
– transfer processes (excitons) 275, 339
Epitaxial films 63
Equivalent cores approximation *70, 177*
Er, UPS and XPS spectra 233, 252
ErB_n, valence band spectra (XPS) 248
ErSb, valence band spectra (XPS) 241
ESCA *10, 12*
Escape depth *see* Electron escape depth
– function *85*
Ethane, valence band spectrum (XPS) 267
Eu 225, 252
– chalcogenides 217, 238
– valence band spectrum (XPS) 232
EuO 218, 238, 254
–, valence band spectrum (XPS) 219, 242, *76, 172*
$EuPt_2$ 252
$EuRh_2$ 252
EuS 238
–, UPS spectra 73, 218
EuTe, XPS spectrum *172*
EXAFS *see* Extended X-ray absorption fine structure
Exchange energy *35, 36, 143*
–, Kohn-Sham-Gaspar *37*
–, Slater *37, 143*
– splitting *see* Multiplet splitting
Exciton annihilation (organics) 275–277
Excitonic shift in core hole absorption spectra 150, 337
Extended X-ray absorption fine structure (EXAFS) 86, 136, 329

f-levels in rare earths 217, 221
– – – –, Johansson scheme 237
– – – –, promotion energy 225, 236
FC-2 98
Fe 200–202, *169*
–, density of valence states 201
–, soft X-ray emission spectrum 201
–, valence band spectra 201
FeAl, absorption spectrum 329
FeAu alloys 210
FeCu alloys 210
FeF_2 181, *170*
–, valence band spectrum (XPS) 182
Fe_xO, UPS spectra and partial *d*-, *p*-components 182
Fermi edge, in organic metals 287
– level *14, 16, 46*
– –, pinning 134, 137, 154
– surface, two-dimensional 37
Ferromagnetic metals: Fe, Co, Ni 200
Field emission 132
– – microscope *30*
– –, photoassisted *4, 29, 30, 129*
Final state effects in photoemission 78, 177, 188, 317, 333, *165*
Flash evaporation *59*
Floodgun, electron *13*
Fluorescence yield *78*
Form factor 89
Fractional parentage coefficients 181, 221–223, 240, *167*
Frank-Condon diagram for NaCl 336
– principle 335, 336, *76, 77*
Frenkel exciton, in anthracence 271, 273, 277

GaAs 40, *48*
–, amorphous 100, 101
–, – densities of states of model structures 101
–, angular resolved PES *248, 261*
–, band structure 19, 20, *49*
–, critical points 59
–, density of states 150
–, electroreflectance 43
–, LEED 148
–, oxidation 343
–, photoabsorption cross section *155*
–, photoemission spectrum 150
–, reflectivity 42
–, surface 148–151
–, – relaxation 148, 149
–, valence band spectrum (XPS) 57, 58, 122
–, yield spectrum 150
Gallium, photoabsorption cross section *154, 155*
GaN 23, 120
GaP 21
–, amorphous 100, 101, 104
–, – EDC's 103
–, critical points 59
–, oxidation 343
–, work function *49*
Gap, indirect or direct 40
Gap states, photoemission spectrum of amorphous Ge 117
– –, amorphous semiconductors 114–118
– –, – Si, photoemission spectrum 116, 117
– –, metal induced 155
GaS 26, 75
GaSb, amorphous 100, 101
–, critical points 59
–, valence band spectrum (XPS) 57
–, work function *49*

GaSe 12, 26, 75
–, angular resolved PES *251, 255, 256*
–, band structure 26
–, charge distribution 26
Ga_2Te_3 24
Gd, XPS spectrum 232, 233
GdB_N 246
GdB_6 246
–, valence band spectrum (XPS) 247, 249
GdS 217, 238
–, UPS spectrum 73
GdSb, XPS spectrum 241, 247
Ge 14, 40, 315, *47*
GeIII 97, 98
GeIV 97, 98
–, amorphous 87–100
–, – density 89
–, – dielectric constant 90, 93
–, – UPS derivative spectrum 117
–, – valence band spectrum (XPS) 88
–, band structure 16, 18
–, chalcogenides 107
–, – amorphous 107, 108
–, density of states 88
–, oxidation 343
–, polytypes 97, 98
–, UPS cesiated 54
–, valence band spectrum (XPS) 56
–, work function *49*
GeH_4 56
GeO_2 86
GeS 67
GeS, nearest neighbor distance 125
GeSe 67
–, amorphous 109
–, nearest neighbor distance 125
GeTe 29
–, amorphous 108, 109
–, critical points 65, 66
–, nearest neighbor distance 125
– UPS spectrum 108
–, valence band spectrum (XPS) 109
Ge_xTe_{1-x} 86
Ge_xTe_{1-x}, valence band spectra (XPS) 109
Glasses 85, 86
Golden rule *109, 125, 140*
Graphite *13, 179*
–, angular resolved PES *255*
–, valence band spectrum (UPS) 270
Gray tin 17
Green's functions theory of photoemission *109, 115*
Group V elements (As, Sb, Bi) 28, 67
– – –, valence band spectra (XPS) 106

Hall effect 13
Hartree-Fock 187, *64, 65, 143, 150, 166, 174*
– -Slater central field wave functions *143*
Heat of formation *51*
He-source *9*
Heterojunctions *47*
Heteropolar gap 102, 121
HfC, calculated density of valence states 190
–, valence band spectra (XPS) 190
HfS_2 34, 35
–, valence band spectrum (XPS) 74
Hg, photoabsorption cross section *154, 155*
HgS, HgSe, HgTe, work function *49*
Ho, valence band spectrum (XPS) 234
HoB_6, valence band spectrum (XPS) 248
HoS, valence band spectrum (XPS) 244
HoSb, valence band spectrum (XPS) 241, 244
Hubbard gap, of NiO, CoO, MnO 188
– –, of VO_2 189
– model 176, 192
Hume-Rothery rule 104
Hund's rule 180, 181, 225, *173*
Husumi cactus lattice 94, 95
Hybridization, temperature dependence 72, *76*
Hydrogen chemisorbed on Si(111) 151–154
– – – – density of states 153
– – – – structure 152
– – – – UPS spectra 153
Hydrogenic atom *143*

Impurity scattering, phase shifts *227*
In, core level line shape *228*
– on Si(111) surface photoemission spectra 156
– – – band bending 156
–, photoabsorption cross section *155*
InAs, amorphous 101, 103, 104
–, critical points 59
–, valence band spectra 58
–, work function *49*
Independent particle reduction of photoemission theory *109, 117, 119*
Inelastic processes *see* Plasmons
Infrared catastrophe 179, *202*
InP, amorphous 100, 101
–, critical points 59
–, valence band spectra 58
–, work function *49*
InSb, amorphous 100, 101, 103, 104
–, critical points 59
–, oxidation 343
–, valence band spectra 58
–, work function *49*

InSe, angular resolved photoemission spectrum *255, 256*
Insulator 11
In_2Te_3 24
Intercalation 32
Interconfiguration fluctuations 235
Interface states 134, 154
– –, extrinsic 155
– –, metal-semiconductor 154
Interference terms in photoionization 50
Intermediate valence (IV) 250, 254
Intermetallic compounds of transition metals 212
Internal conversion *10*
International conferences on amorphous semiconductors 13
– – – semiconductors 13
Ioffe-Regel rule 93
Ion bombardment *59*
Ionic charges 120, 121, *174*
– gap 121
Ionicity 21, 118–125
–, critical 123, 125
– of octet compounds 124
–, pressure dependence 125
– scale 119
– – based on XPS valence band spectra 121
– –, dielectric theory 119
– –, Pauling 119
Ionization potential (photoemission threshold) 128, 133, *17, 25, 49*
Ion neutralization spectroscopy 132
Ir, core line shape *229*
Itinerant, electrons 192, 258
– ferromagnet 202

Jellium model *33, 34, 43*
Joint (optical) density of states 41, *86, 238*

KBr valence band spectrum (XPS) 125
–, work function *49*
KCl, angular resolved photoemission 334
–, CIS spectra 318
– valence band spectrum (XPS) 125
–, work function *49*
$K_2Cr_2O_7$ 180
–, valence band spectrum (XPS) 181
Keldysh formalism *109*
Kelvin method *17, 22*
KF valence band spectrum (XPS) 125
–, work function *49*
KI, CIS spectra 321
– valence band spectrum (XPS) 125
–, work function *49*
Kohn anomaly 38
– variational principle *156*
Koopmans' states 226
– theorem *65–67, 174*
Koster-Kronig transitions *79*
Kramers-Kronig analysis 42
Krogmann salt 36
Kr *177*
–, photoionization cross section 68
–, solid UPS spectra (synchrotron radiation) 332
K-TCNQ 282
–, UPS spectrum 281
K, UPS spectra 369
–, work function *38*
K-edge (X-ray emission), Li *215*
– – –, Al, Mg 224

La 230, 240
La, valence band spectrum (XPS) 230
– halides 237, 238
– –, valence band spectra (XPS) 238
–, RAPW calculation 230
LaB_6 246
–, XPS spectrum 245
Langmuir *58*
LaSb, XPS spectrum 239, 241, 251–253
Layer compounds 26, 32, 48, 72, 75, *251, 253–255*
Lead chalcogenides *see* PbS, PbSe, PbTe
LEED *see* Low energy electron diffraction
Li *76, 211–214*
–, work function *39*
LiF, absorption coefficient and penetration depth vs angle of incidence 325
–, core excitons 337
–, cutoff 218
–, reflectivity vs angle of incidence 324
–, valence band spectrum (XPS) 183
–, yield spectrum vs angle of incidence 325
Ligand chemical shift 104
Like-atom bonds 101
Linear alkanes 266
– –, valence band spectra (XPS) 267
Linewidth, phonon contribution 335, *15, 212, 215, 243*
Liquid metals, yield spectra 329
Local density of states 99
Localized orbitals, photoemission *130*
– states 114, 118
Lone pairs 31, 111
Long range order 86, 114
Low energy electron diffraction (LEED) 132, 135, 141–144, 148, 151, *9, 55, 117, 241, 253*
Lu 152

LuB_n, valence band spectrum (XPS) 248
L_{23}-X-ray edge, Al *223, 224*
– –, Mg *223*
– –, Na *222*

Madelung constant 127
– potential *62, 178*
Mahan-Nozières-De Dominicis effect 40, 350, *198, 199*
Many body features in photoemission 177–179, 193, 352–354, *109, 117, 165*
– – perturbation theory *156*
Mean free path, electrons *see* Electron escape depth
Metal non-metal transition: VO_2 188
– semiconductor interface 154
Metals, *d*-band 192–206
–, free electron like 357–369
–, organic 280–287
Methane *see* CH_4
Mg 350
–, core level spectra 358, 359, *190, 218*
–, density of states 368
–, plasmons 358–362
–, valence band spectrum (XPS) 368
–, work function *39*
Mg_2Ge 24
MgO (:Ni) 183
Mg_2Pb 24
Mg_2Si 24
–, energy bands 26
–, valence band spectrum (XPS) 62
–, X-ray emission spectrum 62
Mg_2Sn 24
Mg_2X (X = Si, Ge, Sn, Pb) 61
–, density of valence states 26
Microcrystal model of amorphous phase 86
Microfields *30*
Mixed valence in rare earths 254–257, *172*
Mn *169*
MnF_2 *168, 170*
MnO 180, 183
–, UPS spectra and partial *d*-, *p*-components 182
Mo, angular resolved photoemission spectroscopy *261*
–, work-function *19*
Mobility gap 114
Model densities of states for amorphous semiconductors 94, 96
Modulation spectroscopy 12, 14, 40
Monochromatization, X-rays 227, *15, 53*
Monochromators for synchrotron radiation 311
Mooser and Pearson plot 123
Mortals, ordinary *9*
MoS_2 33, 34, 35, *251, 254*
–, UPS spectrum 73
MoTe 74
$MoTe_2$, UPS spectrum 73
Mott insulator 176, 183
– transition, Ce 235, 237
– –, VO_2 188
Multichannnel detector *51*
Multidetecting systems *244, 245*
Multiplet splitting *14, 166, 174*
– – in rare earths 220, 223, 226, 234, 250, *171–173*
–, intensities at 1.5 keV 218, 219
– – in transition metals 179–183, *167–170*
– structure *143*

Na 350
–, absorption coefficient (EXAFS) 147, 148
–, core level spectra 358, *216*
–, density of states 366, 368
–, yield spectrum, absorption spectrum for 2*p* transitions 330
Na2*p* linewidth vs temperature 337
–, plasmons 358, 360, 361
–, soft X-ray emission spectrum 365
–, valence band spectra 365, 366, 368
–, work function *38, 39*
NaBr valence band spectrum (XPS) 125
NaCl *74, 77, 80*
–, constant final-state spectra 338
–, core excitons 338
–, surface core excitons 341–343
–, valence band spectrum (XPS) 125
–, work function *49*
NaF valence band spectrum (XPS) 125
NaI valence band spectrum (XPS) 125
–, work function *49*
Naphtalene, UPS spectrum 269
–, vapor pressure 263
NbO_2 189
NbS_2 34, 35
$NbSe_2$, UPS spectrum 73
Nb_3Sn 212
Nd 230
–, valence band spectrum (XPS) 230
NdB_6 246
–, valence band spectrum (XPS) 247
NdBi, valence band spectrum (XPS) 244
NdS, valence band spectrum (XPS) 244
NdSb, valence band spectrum (XPS) 244
Negative electron affinity *7, 25*
Ne, solid, UPS spectra (synchrotron radiation) 332

Ni 200, 202–205
–, angular resolved photoemission spectroscopy *261*
–, angular resolved photoemission spectroscopy, band dispersion 204
–, bandwidth 202–205
–, core line *223*
–, correlation energy 177
–, density of valence states 204
–, valence band spectra (XPS) 204
NiAl, absorption spectrum 329
NiO 176, 179, 183–187
–, band-structure calculations 187
–, UPS spectra and partial *d*-, *p*-components 182
–, UPS spectra with synchrotron radiation 186
–, valence bands 187
–, XPS spectra 184, 185
NiS 176
Noble metals 194–200
Nonane, valence band spectrum (XPS) 267
Nondirect transitions 83, 92, 314, *262*
Non-local pseudopotential 52

Occupied and empty states in photoemission 330
One-dimensional singularities 48, 198
Optical absorption 12, 40
Organometallic phenyl compounds 270–272
Orthogonality catastrophe *199*
Orthogonalized plane waves (OPW) 16
Orthorhombic structure 107, 111
Oxidation of Al, synchrotron radiation spectroscopy 343
– of Ge 52

Partial densities of states 12, 67–72, 186, 366–369
Partial yield spectroscopy 79, 80
– – spectrum of GaAs 80
Passive electrons *185*
Patches *18, 20, 21*
Pb 106
–, core levels *228*
–, work function *39*
PbI_2 33
–, valence band spectra 76
PbS 28
–, angular resolved spectra 47
–, critical ionicity 125
–, – points 65, 66
–, valence band spectra 47, 63
PbSe, critical points 65, 66
–, phase transition 126
–, UPS spectra 63
PbTe, band structure 29
–, critical points 65, 66
–, phase transition 126
–, valence band spectra 63, 65
Pd, core lines *232*
–, valence band spectrum (XPS) and theoretical density of states 201
–, work function *19*
PdAg alloys 207
– –, valence band spectra 207
– –, virtual bound state parameters 208
Peierts transition 36
Peltier effect *31*
Penn model 123
Pentane, valence band spectrum (XPS) 267
Phase shift, Coulomb *141*
– shifts *199, 201, 204, 219*
– –, sum rule *199, 219, 226, 227*
Phonon broadening 335–337, *212, 215*
– – in EuO 243
Photoabsorption measurements *135*
Photocathode, solar blind *7*
Photocathodes *6, 7*
Photoconductivity, surface 132
Photoeffect, surface *3*
Photoelectric cross sections *see* Cross section
– effect *3*
– –, surface vectorial *3, 9*
Photoelectron spectroscopy, complementary methods 40
Photoemission, angle resolved 80–85, 199, 204, *4, 9, 237–263*
–, – – from surface 138
–, formal theory 48, *105–131, 252–254*
– from biological materials 278–280
– from organic molecular crystals 262
– of semiconductor surface 130
–, three-step model *84–92, 122–128, 247*
– threshold *see* Ionization potential
–, time resolved 277
Photohole, localization 287, 356
Photoionization cross sections *see* Cross sections
Photoyield near threshold *22–26*
Phthalocyanines 278, 279
– (H_2, Mg, Pb, Cu), UPS spectra 279
Physisorption *57*
Pinning of E_F in Si 134, 137
Plasmon frequency 44
– dispersion 355, 356
Plasmons 45, 89, 351–369, *175, 189–191*
–, Al 358–363, *211*
–, Be 357, 360
– and adsorbates *192*
–, energies 360

Plasmons, GaAs 45
–, Ge 89
–, intrinsic, extrinsic 191, 351, 352, 354, 357, *201, 207*
–, Li *211*
–, Mg 358–363, *190, 217*
–, Na 358–363, *216*
–, Si 89
–, surface 190, 356, 360, 363, 367
–, width 360
Polarization energy *74*
– shift 127, 291
Polk model 87, 95, 98
– – density of states 100
Polytypes of Ge, Si 97, 98
Porphyrin 279
Positron annihilation *34*
Potential model for core level shifts 127, *69, 70*
Propane, valence band spectrum (XPS) 267
Pr, valence band spectrum (XPS) 230
Pseudopotential method 17, *246*
Pt, core line *231*
–, valence band spectrum (XPS) and theoretical density of states 201

Quadratic response *106*
Quantized description of radiation *114*
Quantum efficiency (yield) *6, 27, 130*

Racah method 221
Radial distribution function 86
Random phase approximation (RPA) *119, 156*
Rare earth borides 245–249
– – –, 4*f*-lifetime 249
– – –, interatomic Auger transitions 245
– – –, structure 245
– – –, valence band spectra (XPS) 245, 247–249
– – chalcogenides 238–243
– – fluorides *171*
– – halides 237, 238
– – intermetallics 249
– – ions, divalent 221
– – –, trivalent 221
– – metals 229–237, *174*
– – pnictides 238–243
Rare earths, 3*d* and 4*d* electrons 251–253
– –, photoabsorption cross sections *158, 159*
– – trifluorides 234, 237, 238
Rare gas line source *52*
– – solids 330–333
– – –, valence bands 330–333
Referencing of binding energies 128, *13*
Reflectance, normal incidence 43
Reflection and transmission amplitudes for photoemission spectroscopy *125*
Refractive index 41
Relativistic dehybridization 105
Relaxation *37, 118, 174*
– energy 127, 267, *63, 64, 68, 69, 71, 72, 118, 175–182*
– –, 4*f* electrons 226, 253
– – in anthracene 273
–, extraatomic *63, 177*
– in free molecules *178*
– in metals *180*
–, intraatomic *63, 176*
– of *k*-conservation 92
– processes 337–341
Renormalization energy *70, 71, 75*
Renormalized atom scheme 221, 225, 237
ReO_3 176, 189, 190
–, calculated density of states 190
–, valence band spectrum (XPS) 190
Resolution 227, *52*
Richardson plot *20*
Rigid band model 36, 206
Rings, fivefold 97–99
–, odd-membered 98, 100
–, sixfold 95, 96, 98, 107
R-matrix theory *156*
Rotating anodes 227
Russell-Saunders coupling 221

S, monoclinic 111
–, orthorhombic 111, 112
–, –, valence band spectrum (XPS) 113
–, S_8 rings 112
Sample preparation 228, 229, *57*
Satellites 177
–, charge transfer 177
–, core levels *76, 141, 175*
–, Kotani-Toyozawa 179
–, multielectron peaks 184, 185, *182–189*
–, shake up/off 226, 252, *182–189*
Sb 104
–, amorphous 105
–, – Raman spectrum 105
–, –, valence band spectrum (XPS) 106
Sb_2Se_3 111
Scattering time 53, *90–92*
Schottky Barrier 134, 154–156
– effect *21*
Screening of core holes *204*
see also Relaxation
Se 30, 86
–, amorphous 112
–, – dielectric constant 90
–, – valence band spectrum 110

–, anisotropy in absorption coefficient 326, 327
–, energy bands 30
–, monoclinic 31, 113
–, – valence band spectrum (XPS) 113
–, Se_8 rings 31, 112, 113
Secondary electrons 79, 127, 264
– –, energy distribution *85*
– – in organic compounds 264
Secondary emission processes 319, 320
Self-energy of the electron *117*
– of the electron, imaginary part *118*
Semiconductors 11
–, amorphous 85–118
– surfaces 130–158
Semimetals 11, 104
Shake-off *see* Satellites
Shake-up *see* Satellites
Short range order 87
Si 14
Si II 97, 98
Si 2H-4, density of states 99
Si III 97, 98
Si, amorphous 87–100
–, – calculated density of states 91
–, – dielectric constant 90, 92
–, – films 89
–, – valence band spectra 88, 92
–, – valence band spectra, gap states 116, 117
–, BC-8, density of states 99
–, density of states 88, 99
–, electron affinity 133
–, ionization potential (photoemission threshold) 133, 134, 143, *46*
–, oxidation 343
–, photoionization cross section 68
–, polytypes 97, 98
–, – densities of states 99
–, ST-12, densities of states 99
– surface, band bending 138
– –, chemisorbed hydrogen 151–154
– –, – hydrogen density of states 153
– –, – – photoemission spectra 153
– –, electronic structure 141, 147
– –, geometry 142
– – and In photoemission spectra 155
– –, pinning of E_F 137
– –, relaxation 142
– – states, charge density 144
– – –, density of states 145–147
– – –, electronic theory
– – –, photoemission 136, 138, 140, 146, 147
– – – states, photoemission, angle resolved 139
– – –, unreconstructed 146
– – –, yield spectrum 133, 135, 140
– –, vacancies 153
– valence band spectra (XPS) 56, 88
–, X-ray emission spectrum 46
–, work function 133, 134, 143, *49*
SiC 23
SiH_4 56
Simple metals 349–370, *34*, *38*
SiO, SiO_2 86
Slater integrals F and G 224, 250, *166*
Sm 225, 240, 252, *173*
–, XPS spectrum 231, 251
$SmAl_2$ *173*
Small angle scattering 89
SmB_6, valence band spectrum (XPS) 247, 251
SmS 237, 258
SmSb, XPS spectrum 239, 242, 251
SmTe, XPS spectrum 239, 242, 251
$Sm_{1-x}Y_xS$ 255
Sn, core levels *228*
–, photoabsorption cross section *155*
$(SN)_x$ 280, 285, 287
–, band dispersion 286
–, band structure 285
–, UPS spectrum 286
–, – –, angular resolved 286
SnS 67
SnS_2 33, 75
SnSe 67
$SnSe_2$ 75
–, valence band spectrum (XPS) 76
SnTe 29
–, critical ionicity 125
–, critical points 65, 66
–, valence band spectrum (XPS) 63
Space charge layer 132–134, *14*
Spin-orbit splitting 21, 29, 67
– –, Bi 105
– –, core levels 43
– –, in rare earths 234
– – of 4*d* electrons 251
– – of virtual bound states 208
– –, PbS, PbSe, PbTe 84
– –, reversal 21
Spin polarization: bulk vs surface 203
– – in EuS 217
– – in Ni 202
– polarized photoemission 257, *2*, *9*
– – – in EuO and $Eu_{1-x}Gd_xO$ 258
Sputtering *58*, *59*
Staggered configuration 23
Step edges, Si surface 138, 146
Sticking coefficient *58*
Stoner-Wohlfahrt model 200, 202, 203
Storage rings and synchrotrons available 308

Structure factor 89
Sudden approximation *175*
Sum rule, Lundquist *175, 181*
– –, Manne and Åberg *181*
Superionic conductors 21
Surface chemistry of semiconductors 151
– effects at threshold *26, 27*
– phase transitions 257, *46*
– photoelectric effect *3, 9, 262*
– plasmons *130, 190*
– reconstruction 132, 148, *46*
– relaxation 46–48, 132, 148
– resonance 131, 140, *129*
– sensitivity of photoemission *192*
– states 9, 14, 44, 47, 51, 122, *129*
– –, III-V compounds 148–151
– –, effects in photoemission *128, 129*
– – of GaAs 149
– – – GaAs, density of states 150
– – – GaAs, energy loss 150
– – – GaAs, excitonic shift 150
– – – GaAs, UPS spectrum 150
– – – GaAs, yield spectrum 150
– – of Si 133, 134
– – – Si, density of states 145–147
– – – Si, dispersion 139
– – – Si, infrared spectroscopy 144
– – – Si, UPS spectrum 136, 138–140, 146, 147
– –, synchrotron radiation experiments 341–343
– transition term in photoemission 174, *126*
Surfaces, semiconductor 130–150
SXPS, soft X-ray photoemission spectra 174
Synchrotron radiation 43, 44, 51, 205, 218, 299–344
– –, angular emission pattern 301
– –, available or projected sources 308, 309
– –, compared with other sources 305, 306, *9, 54, 255, 260, 262*
– –, its uses 299
– –, laboratory layout 311
– –, monochromators 311, 313
– –, polarization 302
– –, properties 301–305
– –, spectroscopic techniques 313

TaC, calculated density of states 190
–, valence band spectrum (XPS) 190
Tantalus I, Experimental layout 311
TaS_2 35, 39
–, angular resolved UPS spectrum *253, 254*
–, UPS spectrum 73
$TaSe_2$ 38
–, angular resolved UPS spectrum *254*
TbB_6, valence band spectrum (XPS) 247
Tb, valence band spectrum (XPS) 234
TCNQ, molecular orbital calculation 281
–, UPS spectrum 281
Te 30, 31
–, amorphous 112, 113
–, – valence band spectrum (XPS) 110
–, valence band spectrum 110
Terrace site, Si(111) 138, 146
Tetracene, UPS spectrum 263, 269
–, vapour pressure 263
Tetrahedral coordination, semiconductors 18
Tetraphenyl tin (Ph_4Sn) 271
– –, partial cross sections 271
– –, photoemission spectrum 271
ThB_6 246
–, valence band spectrum (XPS) 245
Theory of photoemission, independent particle model *105–131*
Thermionic emission *4, 19, 108*
Thermionic emitters *7*
Thomas-Fermi model *34, 143*
Three step model 351, *8, 84–89, 190*
Thymine 280
Tight-binding method (LCAO) 17
TiO_2, valence band spectrum (XPS) 185
TiS_2 35
–, valence band spectra 73, 74
$TiSe_2$ 35
–, angular resolved UPS spectrum *255*
TlCl 28
Tm 240, 252
–, valence band spectrum (XPS) 234
TmB_n, valence bands spectrum (XPS) 248
TmSb, valence bands spectrum (XPS) 241
TmSe, valence band spectra 255
Transistor 14
Transitional metal, chlorides 188
– – compounds 176–191
– – dichalcogenides 32, 33, 36, 72–75
– – –, stacking modifications 33, 34
– –, fluorides 188
– –, oxides 183–191
– metals 192–206, *45, 167, 170*
– operator method *67–69*
– potential model *70*
– probability, dipole *78, 138*
Transitions, direct *8, 25, 26*
–, indirect *8, 25, 26*
Transmission probability 174, *125*
Tridecane, valence band spectrum (XPS) 267
TSeF 283
–, UPS spectrum 284
TTF, molecular orbital calculation 282
–, UPS spectrum 281, 284

TTF-TCNQ 280, 281, 287
–, charge transfer 280, 281
–, core level spectra 292, 293
–, valence band spectrum 281
–, valence charges, self-consistent calculation 291

Ultra high vacuum *58*
Unfilled inner shells: rare earths and compounds 217
– – – : transition metal compounds 173
UPS regime 174
Urical 280
US, UPS spectrum 73

Vacancies 114
–, Si(111) surface 153
Vacuum incoming wave components *111*
– – – state *111, 112, 117, 121*
– level *16*
Valence charges, effect of molecular polarization 290, 291
Van Vleck expression for multiplet splitting *166, 169, 171*
Vapor deposition *58, 59*
Vapor pressure, elements 59
Virtual bound state parameters of transition metal alloys 208
VO_2 176, 188
–, valence band spectra (XPS) 183
Voids 89, 114
Volume effects in photoemission *129, 130*
Volume limit of photoemission *122*
Volume photoemission: angular integrated 47
V_3Si 212

W angular resolved UPS spectrum *260, 261*
Wigglers 307
Wigner-Seitz cells (spheres) *32, 33, 35*
– radius *220*
WO_3 189
Work function *3, 16*
– – determination, break point of retarding potential curve *22*
– – –, calorimetric method *31*
– – –, effusion method *31*
– – –, electron beam method *22*
– – –, field emission *29*
– – –, Fowler plot *24*
– – –, isochromat method 27
– – –, Kelvin method *22*
– – –, photoyield near threshold *23*
– – –, thermionic emission *19*
– – –, threshold of EDC *28*
– – –, total photoelectric yield *28*

– –, semiconductors, insulators *46*
– –, temperature dependence *21, 41, 42*
– –, theory *32, 40*
– –, transition metals *44, 45*
– –, volume dependence *41, 42*
Wrong bonds 100, 102
Wurtzite 23

X_α cluster calculations 34, *67*
Xe in Ar, UPS spectra 333
Xe-doped Ar, yield spectra 340
– Ne, UPS spectra 340
Xe-like ions *186*
Xenon, photoionization cross section 68, *144, 145, 152–155, 157*
–, solid, UPS spectra (synchrotron radiation) 332
XPS *10, 12*
–, angular resolved *16, 249–252*
– regime 51, 62, 67, 174
X-ray absorption edge, vibrational broadening *76*
– – spectroscopy *10*
– edge *198*
– – anomaly *see* Mahan-Nozières-De Dominicis effect
– – threshold exponent *198, 199, 201, 204, 223, 224*
– emission spectroscopy 40, 45–47, *10*
X-rays, monochromatized *12*

Yb 225, 240, 252
–, valence band spectrum (XPS) 234, 256
$YbAl_3$, valence band spectrum (XPS) 256
YbTe 243
Yield spectroscopy 80, 150, 263, 322–330
– –, applications 326
– –, oblique incidence 323
YM_ξ anodes (sources) *54*
YS 243
–, valence band spectrum (XPS) 244

Zn_3As_2 24
$ZnGeP_2$ 24
–, charge distributions 61
–, density of valence states 25, 60
–, energy band structure 25
–, valence band spectrum (XPS) 60
ZnO 11, 23
ZnS 23
ZnSe, valence band spectrum (XPS) 122
ZrC, calculated density of states 190
–, valence band spectrum (XPS) 190
ZrS_2 34
ZrS_2, valence band spectra 73, 74

Springer Series in Solid-State Sciences

This series is devoted to single- and multi-author graduate-level monographs and textbooks in the areas of solid-state physics, solid-state chemistry, and solid-state technology. Also covered are semiconductor physics and technology as well as surface physics. In addition, conference proceedings which delineate the directions for significant future research are considered for publication in the series.

Volume 1
C. P. Slichter
Principles of Magnetic Resonance
2nd, revised and expanded edtion 1978.
115 figures. X, 397 pages
ISBN 3-540-08476-2

Volume 2
O. Madelung
Introduction to Solid-State Theory
Translated from the German by
B. C. Taylor
1978. 144 figures. XI, 486 pages
ISBN 3-540-08516-5

Volume 3
Z. G. Pinsker
Dynamical Scattering of X-Rays in Crystals
1978. 124 figures, 12 tables.
XII, 511 pages
ISBN 3-540-08564-5

Volume 4
Inelastic Electron Tunneling Spectroscopy
Proceedings of the International Conference and Symposium on Electron Tunneling, University of Missouri – Columbia, USA, May 25–27, 1977
Editor: T. Wolfram
1978. 126 figures, 7 tables. VIII, 242 pages
ISBN 3-540-08691-9

Volume 5
F. Rosenberger
Fundamentals of Crystal Growth
Macroscopic Equilibrium and Transport Concepts
ISBN 3-540-09023-1
In preparation

Volume 6
R. P. Huebener
Magnetic Flux Structures in Superconductors
1979. 99 figures, 5 tables. Approx.
270 pages
ISBN 3-540-09213-7
In preparation

Volume 7
E. N. Economou
Green's Functions in Quantum Physics
1979. 49 figures, 2 tables. Approx.
270 pages
ISBN 3-540-09154-8

Volume 8
Solitons and Condensed Matter Physics
Proceedings of the Symposium on Nonlinear (Soliton) Structure and Dynamics in Condensed Matter, Oxford, England, June 27–29, 1978
Editors: A. R. Bishop, T. Schneider
1978. 120 figures. XI, 341 pages
ISBN 3-540-09138-6